Lecture Notes in Networks and Systems 1780

The series "Lecture Notes in Networks and Systems" publishes the latest developments in Networks and Systems—quickly, informally and with high quality. Original research reported in proceedings and post-proceedings represents the core of LNNS.

Volumes published in LNNS embrace all aspects and subfields of, as well as new challenges in, Networks and Systems.

The series contains proceedings and edited volumes in systems and networks, spanning the areas of Cyber-Physical Systems, Autonomous Systems, Sensor Networks, Control Systems, Energy Systems, Automotive Systems, Biological Systems, Vehicular Networking and Connected Vehicles, Aerospace Systems, Automation, Manufacturing, Smart Grids, Nonlinear Systems, Power Systems, Robotics, Social Systems, Economic Systems and other. Of particular value to both the contributors and the readership are the short publication timeframe and the world-wide distribution and exposure which enable both a wide and rapid dissemination of research output.

The series covers the theory, applications, and perspectives on the state of the art and future developments relevant to systems and networks, decision making, control, complex processes and related areas, as embedded in the fields of interdisciplinary and applied sciences, engineering, computer science, physics, economics, social, and life sciences, as well as the paradigms and methodologies behind them.

Indexed by SCOPUS, EI Compendex, INSPEC, WTI Frankfurt eG, zbMATH, SCImago.

All books published in the series are submitted for consideration in Web of Science.

For proposals from Asia please contact Aninda Bose (aninda.bose@springer.com).

Jose Luis Ballesteros ·
Andres Emiro Diez Restrepo ·
Julio Normey ·
Hugo Espinosa
Editors

Systems, Smart Technologies, and Innovation for Society

Proceedings of CITIS'2025

Springer

Editors
Jose Luis Ballesteros
Politecnica Salesiana University
Guayaquil, Ecuador

Andres Emiro Diez Restrepo
Pontifical Bolivarian University
Medellín, Colombia

Julio Normey
Federal University of Santa Catarina
Florianópolis, Brazil

Hugo Espinosa
Griffith University
Nathan, QLD, Australia

ISSN 2367-3370 ISSN 2367-3389 (electronic)
Lecture Notes in Networks and Systems
ISBN 978-3-032-14286-3 ISBN 978-3-032-14287-0 (eBook)
https://doi.org/10.1007/978-3-032-14287-0

This Springer imprint is published by the registered company Springer Nature Switzerland AG
The registered company address is: Gewerbestrasse 11, 6330 Cham, Switzerland

To all researchers, educators, and innovators who dedicate their work to advancing knowledge and improving lives.

To the new generations of scientists and professionals who believe that science, technology, and innovation can transform societies for the better.

And to the community that made CITIS 2025 possible—authors, reviewers, organizers, and participants—whose commitment and collaboration give meaning to this collective effort.

Foreword

It is with great pleasure that we present the proceedings of the XI International Conference on Science, Technology, and Innovation for Society (CITIS 2025), held in Guayaquil, Ecuador, from June 4–6, 2025. CITIS has become a recognized platform for fostering dialogue and collaboration among researchers, educators, professionals, and students from around the world.

Over the years, CITIS has consistently highlighted the role of science, technology, and innovation in addressing pressing societal challenges. The XI edition continues this tradition by presenting high-quality research in areas including sustainability and environment, artificial intelligence and digital networks, technologies for health and wellness, and renewable energies and industry. The papers collected in this volume demonstrate the diversity, creativity, and rigor of contemporary research, offering solutions that are both scientifically sound and socially relevant.

We are deeply grateful to the authors for their contributions, to the Program Committee and reviewers for their meticulous evaluation of all submissions, and to the organizing institutions for their support in making this conference possible. We hope that the research presented here will inspire further collaboration and innovation, and that these proceedings will serve as a valuable resource for the scientific community and beyond.

Preface

This book is composed of the papers written in English and accepted for presentation and discussion at the XI International Conference on Science, Technology, and Innovation for Society (CITIS 2025). This conference, supported by the Universidad Politécnica Salesiana (UPS), took place in Guayaquil, Ecuador, during June 4–6, 2025.

In its eleventh edition, CITIS once again brought together researchers, academics, professionals, and students from around the world, offering a unified platform for dialogue and the exchange of ideas. The conference aimed to address both theoretical and practical challenges of major impact in today's society, focusing on sustainability and environment, artificial intelligence and digital networks, technologies for health and wellness, and renewable energies and industry. These themes reflect the urgent need to integrate scientific knowledge and technological innovation into solutions that contribute to human well-being and sustainable development.

The Program Committee of CITIS 2025 was formed by a diverse and multidisciplinary group of experts from different countries, who dedicated their time and expertise to evaluating, through a rigorous double-blind review process, the many papers submitted to the conference. Their careful work ensured the academic quality of this volume and enriched the scientific dialogue of the event.

This year, CITIS 2025 received 283 contributions from 20 countries around the world, a clear demonstration of its growing international scope and relevance. The accepted papers, presented and discussed during the conference, are now published by Springer in this book and will be submitted for indexing by Scopus and Google Scholar, among others. We hope that these proceedings will serve not only as a record of the knowledge shared during CITIS 2025 but also as inspiration for future research and collaborations that continue to link science, technology, and innovation with the needs of society.

Acknowledgments

We would like to express our sincere gratitude to all those who made the XI International Conference on Science, Technology, and Innovation for Society (CITIS 2025) possible.

First and foremost, we thank the authors, whose contributions form the essence of this volume. Their commitment to advancing knowledge and sharing results with the global community has been invaluable. We also extend our appreciation to the reviewers and members of the Program Committee, whose dedication and expertise ensured a rigorous and constructive evaluation process, thus guaranteeing the quality of the accepted papers.

Our heartfelt thanks go to the Universidad Politécnica Salesiana (UPS) for its institutional support and for hosting CITIS 2025 in Guayaquil, Ecuador. We also acknowledge the tireless efforts of the Organizing Committee, whose planning, coordination, and teamwork were crucial to the success of the conference.

Finally, we wish to thank all participants—researchers, professionals, and students—who joined us from around the world. Their presence and active engagement enriched the discussions and strengthened the spirit of collaboration that defines CITIS.

Organization

Honorary Committee

P. Marcelo Farfán	Chancellor, Universidad Politécnica Salesiana, Ecuador
P. Juan Cárdenas Tapia	Rector, Universidad Politécnica Salesiana, Ecuador
Fernando Pesántez Avilés	Executive Vice-Rector, Universidad Politécnica Salesiana, Ecuador
Susana Lam	Vice-Rector, Guayaquil Campus, Universidad Politécnica Salesiana, Ecuador
Esteban Inga Ortega	Vice-Rector for Research, Universidad Politécnica Salesiana, Ecuador
Angela Flores Ortiz	Vice-Rector for Academic Affairs, Universidad Politécnica Salesiana, Ecuador
Vladimir Robles Bykbaev	Vice-Rector for Postgraduate Studies, Universidad Politécnica Salesiana, Ecuador

Executive Committee

Esteban Inga Ortega
José Luis Ballesteros
Xavier Serrano Guerrero

Track Chairs

Joe Llerena Izquierdo
Diego Carrión Galarza
Jaime Naranjo Morán
Yaroslava Robles Bykvaeb

Advisory Committee

Raquel Ayala Carabajo	Universidad Politécnica Salesiana, Ecuador
Silvio Simani	University of Ferrara, Italy

José Cardier	Instituto Venezolano de Investigaciones Científicas, Venezuela
Sandra Céspedes	Concordia University, Canada
Alexis Kalergis	Pontificia Universidad Católica de Chile, Chile
Andreu Moreno Vendrell	IQS School of Engineering, Ramon Llull University, Spain
Marcelo López-Lastra	Pontificia Universidad Católica de Chile, Chile
Gianni Sacchetti	University of Ferrara, Italy
Mónica Huerta	Universidad Politécnica Salesiana, Ecuador
Carlos Luis Barzola Iza	Wageningen University & Research, Netherlands

Contents

Sustainability and Environment

Technologies for Health and Wellness

List of Contributors

Echeverría Guzmán Ángel Universidad Bolivariana del Ecuador, Durán, Ecuador

Diana Aguirre-Ruiz Escuela Superior Politécnica de Chimborazo, Riobamba, Ecuador

Diego Almeida School of Biological Sciences and Engineering, Yachay Tech University, Urcuquí, Hacienda San José, Ecuador

Diego Almeida-Galárraga School of Biological Sciences and Engineering, Universidad Yachay Tech, San Miguel de Urcuquí, Ecuador

Oscar Analuiza Universidad Técnica de Ambato, Ambato, Ecuador

Thalía San Antonio Universidad Técnica de Ambato, Ambato, Ecuador

Carlos Alberto Flórez Arias UIS, Bucaramanga, Colombia

Elmer Arias-Montaño eX-MoVeT, Research Group, Facultad de la Energía, las Industrias y los Recursos Naturales no Renovables, Universidad Nacional de Loja, Loja, Ecuador

Jorge I. Armijos-Rivera Grupo de Investigación en Genética y Biología Molecular, Universidad Nacional de Loja, Loja, Ecuador;
Centro de Biotecnología, Universidad Nacional de Loja, Loja, Ecuador

Carlos Avilés-Cruz Departamento de Electrónica, Universidad Autónoma Metropolitana, México, Ciudad de México, Mexico

Constantino Miguel Nieves Barreto Universidad Nacional del Callao, Bellavista, Peru

Teodiano Bastos-Filho Federal University of Espirito Santo, Vitoria, Brazil

Cesar Benavides-Alvarez Departamento de Electrónica, Universidad Autónoma Metropolitana, México, Ciudad de México, Mexico

Francis-Nicole Benitez-Montalvo Unidad Académica de Informática, Ciencias de La Computación e Innovación Tecnológica, Universidad Católica de Cuenca, Cuenca, Ecuador

Andrés Benito Regional Faculty of Mendoza, National Technology University, Islamabad, Pakistan

Edgar Briones-Bitar Facultad de Ciencias Sociales y Humanísticas, ESPOL Polytechnic University, ESPOL, Guayaquil, Ecuador;
Centro de Investigaciones Rurales (CIR), ESPOL Polytechnic University, ESPOL, Guayaquil, Ecuador

Josué Briones-Bitar Centro de Investigación y Proyectos Aplicados a las Ciencias de la Tierra, ESPOL Polytechnic University, ESPOL, Guayaquil, Ecuador;
Facultad de Ingeniería en Ciencias de La Tierra, ESPOL Polytechnic University, ESPOL, Guayaquil, Ecuador

Pablo-Andrés Buestán-Andrade Unidad Académica de Informática, Ciencias de La Computación e Innovación Tecnológica, Universidad Católica de Cuenca, Cuenca, Ecuador

Maye Catalina Neira Cárdenas Corporación Universitaria Minuto de Dios - UNIMINUTO, Bogotá, Colombia

Myriam Ximena Mancheno Cárdenas Universidad Politécnica Salesiana, Quito, Ecuador

Rosangela Caicedo-Quiroz Universidad Bolivariana del Ecuador, Durán, Ecuador

Edisson Calderón Departamento de Materiales, Escuela Politécnica Nacional, Quito, Ecuador

Esther Carlín-Chávez Universidad Bolivariana del Ecuador, Duran, Ecuador;
Facultad de Posgrados, Universidad Estatal de Milagro, Milagro, Ecuador

Diego Carrión Department of Electrical Engineering, GIREI Research Group, Universidad Politécnica Salesiana, Quito, Ecuador

Paúl Carrión-Mero Centro de Investigación y Proyectos Aplicados a las Ciencias de la Tierra, ESPOL Polytechnic University, ESPOL, Guayaquil, Ecuador;
Facultad de Ingeniería en Ciencias de La Tierra, ESPOL Polytechnic University, ESPOL, Guayaquil, Ecuador

Darwin Castillo Departamento de Química y Ciencias Exactas, Universidad Técnica Particular de Loja, Loja, Ecuador

Jairo Castillo-Calderón eX-MoVeT, Research Group, Facultad de la Energía, las Industrias y los Recursos Naturales no Renovables, Universidad Nacional de Loja, Loja, Ecuador

Pedro Castro-Verdezoto Faculty of Industrial Engineering, University of Guayaquil, Guayaquil, Ecuador

Lia Cedeño-Zambrano IPGCS Group, Faculty of Engineering, Industry, and Architecture, Universidad Laica Eloy Alfaro de Manabí, Manta, Ecuador

Lorenzo Cevallos-Torres Universidad Bolivariana del Ecuador, Duran, Ecuador;
Facultad de Ciencias Matemáticas y Físicas, Universidad de Guayaquil, Guayaquil, Ecuador

Luis Enrique Chávez-Rubio Facultad de Diseño y Arquitectura, Universidad Técnica de Ambato, Ambato, Tungurahua, Ecuador

Wilmer Chaca-Espinoza Faculty of Design and Architecture, Technical University of Ambato, Ambato, Ecuador

Anderson Steve Caranqui Chanalata Carrera de Ingeniería Industrial, Universidad Politécnica Salesiana, Quito, Ecuador

Angel Chicaiza Laboratorio de Diseño Sostenible de Espacios Habitables, Universidad Regional Amazónica Ikiam, Tena, Napo, Ecuador

Juan Chuquin-Vasco Escuela Superior Politécnica de Chimborazo, Riobamba, Ecuador

Silvia Clavijo Faculty of Applied Sciences to Industry, National University of Cuyo, San Rafael, Argentina

Xavier Cornejo Herbario GUAY, Departamento de Botánica, Facultad de Ciencias Naturales, Universidad de Guayaquil, Guayaquil, Ecuador

Luis Vásquez Cortez Faculty of Agricultural Sciences, Technical University of Babahoyo, Babahoyo, Los Ríos, Ecuador

Anthony Crespo School of Mathematical and Computational Sciences, Yachay Tech University, Urcuquí, Ecuador

Bernardo Vega Crespo Facultad de Ciencias Médicas, Universidad de Cuenca, Cuenca, Ecuador

Fabricio Crespo School of Mathematical and Computational Sciences, Yachay Tech University, Urcuquí, Ecuador

Jonathan Cruz School of Biological Sciences and Engineering, Yachay Tech University, Urcuquí, Hacienda San José, Ecuador

Jonathan Cruz-Varela School of Biological Sciences and Engineering, Universidad Yachay Tech, San Miguel de Urcuquí, Ecuador

Alejandra Gómez Cuartas Facultad de Farmacia y Bioquímica, Universidad de Buenos Aires, Consejo Nacional de Investigaciones Científicas y Técnicas, Instituto de Química y Fisicoquímica Biológicas (IQUIFIB), Buenos Aires, Argentina

Cristian Cuji Universidad Politécnica Salesiana, Quito, Ecuador

Marcelo Pereira Da Cunha University of Campinas, Institute of Economics, Campinas, Brazil

Cabrera Daniel Facultad de Ciencia E Ingeniería en Alimentos y Biotecnología, Universidad Técnica de Ambato, Ambato, Ecuador

Raul Davalos School of Biological Sciences and Engineering, Universidad Yachay Tech, Urcuquí, Ecuador;
Biological and Environmental Science and Engineering, King Abdullah University of Science and Technology, Thuwal, Saudi Arabia

Gabriel de Jesus Celis-Escudero Departamento de Electrónica, Universidad Autónoma Metropolitana, México, Ciudad de México, Mexico

Guissella de los Ángeles Rivera Yunga Universidad Politécnica Salesiana, Quito, Ecuador

Diego Díaz Sinche eX-MoVeT, Research Group, Facultad de la Energía, las Industrias y los Recursos Naturales no Renovables, Universidad Nacional de Loja, Loja, Ecuador

Johanna Dueñas Carrera de Ingeniería en Riesgos de Desastres, Universidad Estatal de Bolívar, Guaranda, Ecuador

Luis Duque-Cordova Universidad del Pacífico, Guayaquil, Ecuador

Lisbeth Encalada Centro de Investigaciones Biotecnológicas del Ecuador, ESPOL Polytechnic University, Escuela Superior Politécnica del Litoral, ESPOL, Guayaquil, Ecuador

Kenny Escobar-Segovia Escuela Superior Politécnica del Litoral, Guayaquil, Ecuador

Mónica Maldonado Espinosa Universidad Nacional de Loja, Loja, Ecuador

Pamela Espinosa Cuayal School of Biological Sciences and Engineering, Universidad Yachay Tech, San Miguel de Urcuquí, Ecuador

Bryan Chulde Fernández School of Biological Sciences and Engineering, Yachay Tech University, San Miguel de Urcuquí, Ecuador

Gonzalo Xavier Vizuete Fiallos Facultad de Ciencias de La Ingeniería y Aplicadas, Universidad Técnica de Cotopaxi, Latacunga, Ecuador

Jessica Nataly Castillo Fiallos Facultad de Ciencias de La Ingeniería y Aplicadas, Universidad Técnica de Cotopaxi, Latacunga, Ecuador

M. J. Lovato- Fortun YAinnova-Ecuador. Ideas para el Agro, Quito, Ecuador

Diego Carrillo Faculty of Life Sciences and Technologies, Laica University Eloy Alfaro of Manabí, Manta, Ecuador;
Faculty of Applied Sciences to Industry, National University of Cuyo, San Rafael, Argentina

Cinthya Game-Varas Universidad Espíritu Santo, Samborondón, Ecuador

Marcelo García Department of Electrical Engineering, GIREI Research Group, Universidad Politécnica Salesiana, Quito, Ecuador

Scarlet Annabel Gutierrez Gavilanez Universidad de Guayaquil, Guayaquil, Ecuador

Alfonso González-Briones BISITE Research Group, Edifice I+D+i, University of Salamanca, Salamanca, Spain

Cesar Ángel Durand Gonzales Universidad Nacional del Callao, Bellavista, Peru

Cintya Gordon-Falconí Pontificia Universidad Católica de Ecuador Sede Ibarra, Escuela de Ciencias Agrícolas y Ambientales (ECAA), Ibarra, Ecuador

Andrea Goyes-Balladares Department of Architecture Graphic Expression, Universitat Politécnica de Valencia, Valencia, Spain

María Cristina Grandal-López Universidad Bolivariana del Ecuador, Duran, Ecuador

María Daniela Groppa Facultad de Farmacia y Bioquímica, Universidad de Buenos Aires, Consejo Nacional de Investigaciones Científicas y Técnicas, Instituto de Química y Fisicoquímica Biológicas (IQUIFIB), Buenos Aires, Argentina

Ramiro Israel Vivanco Gualán Departamento de Ciencias de la Computación y Electrónica, Universidad Técnica Particular de Loja, Loja, Ecuador

Lorenzo Guerrini Dipartimento Territorio e Sistemi Agro-Forestali (TESAF), Università degli Studi di Padova, Padua, Italy

David Guevara Universidad Técnica de Ambato, Ambato, Ecuador

Marco Orlando Fuel Herrera Bioquímica y Farmacia, Universidad Central del Ecuador, Quito, Ecuador;
Enfermería, Universidad Politécnica Salesiana, Quito, Ecuador

María Florencia Iannone Facultad de Farmacia y Bioquímica, Universidad de Buenos Aires, Consejo Nacional de Investigaciones Científicas y Técnicas, Instituto de Química y Fisicoquímica Biológicas (IQUIFIB), Buenos Aires, Argentina

Erick Iles School of Biological Sciences and Engineering, Universidad Yachay Tech, San Miguel de Urcuquí, Ecuador

Carolina Játiva School of Biological Sciences and Engineering, Yachay Tech University, Urcuquí, Hacienda San José, Ecuador

Mero Jaritza Facultad de Ciencia E Ingeniería en Alimentos y Biotecnología, Universidad Técnica de Ambato, Ambato, Ecuador

Yuliana Jiménez-Gaona Departamento de Química y Ciencias Exactas, Universidad Técnica Particular de Loja, Loja, Ecuador

Francisco Jurado Departamento de Ingeniería Eléctrica, Universidad de Jaén, Jaén, Spain

Alejandra Lascano Universidad Técnica de Ambato, Ambato, Ecuador

Juan Carlos Lata-García Universidad Politécnica Salesiana, Guayas, Ecuador

Rogelio Santiago León-Japa Universidad Nacional de Loja, Loja, Ecuador

Inés Catalina Echeverría Llumipanta Bioquímica y Farmacia, Universidad Central del Ecuador, Quito, Ecuador

Juan P. López-Goyez Universidad Politécnica Estatal del Carchi, Tulcán, Ecuador; BISITE Research Group, Edifice I+D+i, University of Salamanca, Salamanca, Spain

Gustavo Machado-Merino Universidad Técnica de Ambato, Ambato, Ecuador

Sánchez Manoella Facultad de Ciencia E Ingeniería en Alimentos y Biotecnología, Universidad Técnica de Ambato, Ambato, Ecuador

Luigi Orlando Freire Martinez Facultad de Ciencias de La Ingeniería y Aplicadas, Universidad Técnica de Cotopaxi, Latacunga, Ecuador

Bertha Mazon-Olivo AutoMathTIC Research Group, Facultad de Ingeniería Civil, Universidad Técnica de Machala, Machala, Ecuador

Blanca Lilia Arias Medina Universidad Técnica de Cotopaxi, Latacunga, Ecuador

Andrés Medina-Moncayo Faculty of Design and Architecture, Technical University of Ambato, Ambato, Ecuador

Jorge Mina-Ortega Universidad Politécnica Estatal del Carchi, Tulcán, Ecuador

Víctor Molina-Dueñas Faculty of Design and Architecture, Technical University of Ambato, Ambato, Ecuador

Juan Gabriel Mollocana-Lara Grupo de Investigación Ambiental en el Desarrollo Sustentable GIADES, Carrera de Ingeniería Industrial, Universidad Politécnica Salesiana, Quito, Ecuador

Claudia Moncada School of Mathematical and Computational Sciences, Yachay Tech University, Urcuquí, Ecuador

Javier Montalvo Universidad Politécnica Estatal del Carchi, Tulcán, Ecuador

Arturo Montejo-Ráez Universidad de Jaén, Jaén, Spain

Yuliana I. Mora-Ochoa School of Biological Sciences and Engineering, Universidad Yachay Tech, Urcuquí, Ecuador

Alex Moreta-Masaquiza Departamento de Automatización y Control Industrial, Escuela Politécnica Nacional, Quito, Ecuador

Manuel Eugenio Morocho-Cayamcela School of Mathematical and Computational Sciences, Deep Learning for Autonomous Driving, Robotics, and Computer Vision Research Group (DeepARC Research), Yachay Tech University, Urcuquí, Ecuador; Faculty of Technical Sciences, School of Computer Science, Universidad Internacional del Ecuador, Quito, Ecuador

Manuel Eugenio Morocho-Cayamcela School of Mathematical and Computational Sciences, Yachay Tech University, Urcuquí, Ecuador

Roberto Moya-Jiménez Faculty of Design and Architecture, Research and Development Directorate, Technical University of Ambato, Ambato, Ecuador

Juan Pablo Medina Namicela Universidad Nacional de Loja, Loja, Ecuador

Paulo Navas School of Biological Sciences and Engineering, Yachay Tech University, San Miguel de Urcuquí, Ecuador

Katiuska Jajaira Obando-Proaño Dirección Técnica de Ecosistemas de Emprendimiento e Innovación, Universidad Politécnica Salesiana, Quito, Ecuador

Sara Ochoa Economías Sociales, Universidad Regional Amazónica Ikiam, Tena, Napo, Ecuador

Albeiro Aguilar Olivera Corporación Universitaria Minuto de Dios - UNIMINUTO, Bogotá, Colombia

Ivette Judith Zamora Olvera Universidad de Guayaquil, Guayaquil, Ecuador

Samantha Natasha Marcillo Ontaneda Bioquímica y Farmacia, Universidad Central del Ecuador, Quito, Ecuador

Roberto Ortega Instituto Superior Universitario Central Técnico, Quito, Ecuador

Bryan David Llerena Ortiz Facultad de Ciencias de La Ingeniería y Aplicadas, Universidad Técnica de Cotopaxi, Latacunga, Ecuador

Jenny Ortiz-Zambrano Universidad de Guayaquil, Guayaquil, Ecuador

Aguirre Pamela Facultad de Ciencia E Ingeniería en Alimentos y Biotecnología, Universidad Técnica de Ambato, Ambato, Ecuador

Milton Paredes-Aguirre Facultad de Ciencias Sociales y Humanísticas, ESPOL Polytechnic University, ESPOL, Guayaquil, Ecuador

Aura Paucar-Cabrera Museo de Zoología LOUNAZ, Universidad Nacional de Loja, Loja, Ecuador

Daniela Paz-Barzola Escuela Superior Politécnica del Litoral, Guayaquil, Ecuador

Julie Astrid Solano Pérez Corporación Universitaria Minuto de Dios - UNIMINUTO, Bogotá, Colombia

Nathalia-Michelle Peralta-Vasconez Unidad Académica de Informática, Ciencias de La Computación e Innovación Tecnológica, Universidad Católica de Cuenca, Cuenca, Ecuador

Julio Pérez Naranjo Departamento de Energía Eléctrica, Escuela Politécnica Nacional, Quito, Ecuador

Carlos Borras Pinilla UIS, Bucaramanga, Colombia

Ivis Yania Piovet-Hernández Universidad Bolivariana del Ecuador, Duran, Ecuador

Andrés Palma Ponce Escuela Politécnica Nacional, Quito, Ecuador

Andres Eduardo Ponce-Macas School of Mathematical and Computational Sciences, Deep Learning for Autonomous Driving, Robotics, and Computer Vision Research Group (DeepARC Research), Yachay Tech University, Urcuquí, Ecuador

Marcelo Pozo-Palma Departamento de Automatización y Control Industrial, Escuela Politécnica Nacional, Quito, Ecuador

Edison Quimbiamba Escuela Politécnica Nacional, Quito, Ecuador

Sebastián Quizhpi-Salamea eX-MoVeT, Research Group, Facultad de la Energía, las Industrias y los Recursos Naturales no Renovables, Universidad Nacional de Loja, Loja, Ecuador

Juan Sebastian Rada Rey UIS, Bucaramanga, Colombia

Lenin J. Ramirez-Cando School of Biological Sciences and Engineering, Universidad Yachay Tech, Urcuquí, Ecuador

Dayron Rumbaut Rangel Universidad Bolivariana del Ecuador, Durán, Ecuador

Lindsay Rangel-Anchundia IPGCS Group, Faculty of Engineering, Industry, and Architecture, Universidad Laica Eloy Alfaro de Manabí, Manta, Ecuador

Lorena Recalde Escuela Politécnica Nacional, Quito, Ecuador

Cristian Redroban Escuela Superior Politécnica de Chimborazo, Riobamba, Ecuador

Segundo Reyes Faculty of Life Sciences and Technologies, Laica University Eloy Alfaro of Manabí, Manta, Ecuador

Iván Galo Reyes-Chacón Faculty of Technical Sciences, School of Computer Science, Universidad Internacional del Ecuador, Quito, Ecuador

Kevin Reyna-Guarderas Departamento de Automatización y Control Industrial, Escuela Politécnica Nacional, Quito, Ecuador

Wilmer Rivas-Asanza AutoMathTIC Research Group, Facultad de Ingeniería Civil, Universidad Técnica de Machala, Machala, Ecuador

E. Fabián Rivera Universidad Estatal de Bolívar, Guaranda, Ecuador

Jimmy Ronald Riojas Rivera Universidad César Vallejo, Trujillo, Spain

Luis Rojas-Oviedo Escuela Superior Politécnica de Chimborazo, Riobamba, Ecuador

Da niela Román-Cáceres Universidad Nacional de Loja, Loja, Ecuador;
Grupo de Investigación en Genética y Biología Molecular, Universidad Nacional de Loja, Loja, Ecuador

Daniela Román-Cáceres Grupo de Investigación en Genética y Biología Molecular, Universidad Nacional de Loja, Loja, Ecuador;
Museo de Zoología LOUNAZ, Universidad Nacional de Loja, Loja, Ecuador

María Alejandra Chugá Rosero International SEK University, Quito, Ecuador

Milton Ruiz Department of Electrical Engineering, GIREI Research Group, Universidad Politécnica Salesiana, Quito, Ecuador

Melissa Sánchez Economías Sociales, Universidad Regional Amazónica Ikiam, Tena, Napo, Ecuador

Pablo Andres Salamea-Cordero Biomedical Engineering Research Group, Universidad Politécnica Salesiana, Cuenca, Ecuador

Leticia E. Silvestre Universidad Austral de Chile, Valdivia, Chile

Mirian Solis-Ramón Universidad Católica de Santiago de Guayaquil, Guayaquil, Ecuador

Fernanda Souza Federal University of Espirito Santo, Vitoria, Brazil

Richard Sucuy Facultad de Ingeniería Civil, Universidad Técnica de Machala, Machala, Ecuador

Gabriela Suntaxi Escuela Politécnica Nacional, Quito, Ecuador

Joe Dylan Cáceres Taya Bioquímica y Farmacia, Universidad Central del Ecuador, Quito, Ecuador

Gabriel Tello-Hidalgo Facultad de Ciencias Agrícolas, Escuela de Ingeniería Agronómica, Universidad Central del Ecuador, Quito, Ecuador

Freddy Tinoco Universidad Nacional de Loja, Loja, Ecuador;
Grupo de Investigación en Genética y Biología Molecular, Universidad Nacional de Loja, Loja, Ecuador

Sandino Torres Departamento de Materiales, Escuela Politécnica Nacional, Quito, Ecuador

Eduardo Tusa AutoMathTIC Research Group, Facultad de Ingeniería Civil, Universidad Técnica de Machala, Machala, Ecuador

Enrique Pinchi Ugarte Universidad Nacional del Callao, Bellavista, Peru

Adamary Vásquez-Tituana Grupo de Investigación en Genética y Biología Molecular, Universidad Nacional de Loja, Loja, Ecuador

Anahi Vaca School of Biological Sciences and Engineering, Yachay Tech University, Urcuquí, Hacienda San José, Ecuador

Clara Elena Robayo Valencia Universidad Técnica de Ambato, Ambato, Ecuador

Renato Valle Carrera de Comunicación, Universidad Estatal de Bolívar, Guaranda, Ecuador

Jonathan Cruz Varela School of Biological Sciences and Engineering, Yachay Tech University, San Miguel de Urcuquí, Ecuador

Jonathan Vasco Department of Electrical Engineering, GIREI Research Group, Universidad Politécnica Salesiana, Quito, Ecuador

Sleyther Arturo De La Cruz Vega Universidad Cesar Vallejo, Trujillo, Peru

Santiago Velastegui School of Biological Sciences and Engineering, Yachay Tech University, San Miguel de Urcuquí, Ecuador

Alexandra Veloz Universidad Estatal de Bolívar, Guaranda, Ecuador

Veronique Verhoeven Family Medicine and Population Health, University of Antwerp, Antwerp, Belgium

Segundo F. Vilema-Escudero Universidad ECOTEC, Samborondón, Guayas, Ecuador

Rómulo Villamar-Triviño Universidad del Pacífico, Guayaquil, Ecuador

Ana Cecilia Villa-Parra Biomedical Engineering Research Group, Universidad Politécnica Salesiana, Cuenca, Ecuador

Rafael Viteri Centro de Investigaciones Biotecnológicas del Ecuador, ESPOL Polytechnic University, Escuela Superior Politécnica del Litoral, ESPOL, Guayaquil, Ecuador

Paulina Vizcaíno Faculty of Technical Sciences, School of Computer Science, Universidad Internacional del Ecuador, Quito, Ecuador

Ignacio P. Chamorro -Warnken Agroimpulsa, Talca, Chile

Sebastián Yánez-Segovia Escuela de Ciencias Agrícolas y Ambientales, Pontificia Universidad Católica del Ecuador-Sede Ibarra, Ibarra, Ecuador

Gary Xavier Reyes Zambrano Universidad Bolivariana del Ecuador, Durán, Ecuador

Karen Zambrano IPGCS Group, Faculty of Engineering, Industry, and Architecture, Universidad Laica Eloy Alfaro de Manabí, Manta, Ecuador

Antonio Zavala-Alcívar IPGCS Group, Faculty of Engineering, Industry, and Architecture, Universidad Laica Eloy Alfaro de Manabí, Manta, Ecuador

Arturo Zúñiga-López Departamento de Electrónica, Universidad Autónoma Metropolitana, México, Ciudad de México, Mexico

Diana Belén Peralta Zurita International SEK University, Quito, Ecuador

List of Abbreviations

AI	Artificial Intelligence
CITIS	International Conference on Science, Technology, and Innovation for Society
ICT	Information and Communication Technologies
ML	Machine Learning
IoT	Internet of Things
SDG	Sustainable Development Goals
UPS	Universidad Politécnica Salesiana
R&D	Research and Development
HE	Higher Education
ICTN	Information and Communication Technology Networks
H&W	Health and Wellness
RE	Renewable Energy
ESG	Environmental, Social, and Governance
BCI	Brain-Computer Interface
IoMT	Internet of Medical Things
NLP	Natural Language Processing
ANN	Artificial Neural Network
PV	Photovoltaic
AIoT	Artificial Intelligence of Things
GIS	Geographic Information System

Sustainability and Environment

Critical Analysis of the Shear Stress Formula in the Ecuadorian Bamboo Construction Code: Error Identification and Technical Recommendations

Angel Chicaiza[1](✉), Sara Ochoa[2], Luis Enrique Chávez-Rubio[3], and Melissa Sánchez[2]

[1] Laboratorio de Diseño Sostenible de Espacios Habitables, Universidad Regional Amazónica Ikiam, Tena, Napo 150102, Ecuador
angelchicaiza1@gmail.com
[2] Economías Sociales, Universidad Regional Amazónica Ikiam, Tena, Napo 150102, Ecuador
[3] Facultad de Diseño y Arquitectura, Universidad Técnica de Ambato, Ambato, Tungurahua 180207, Ecuador

Abstract. The present research identifies and reports critical errors in the shear stress calculation formula (Equation 9) provided by the Ecuadorian Construction Standard for Bamboo Structures (NEC-SE-Guadúa) for *Guadua angustifolia* Kunth (GaK) structures. Significant inconsistencies in the equation are demonstrated through rigorous mathematical analysis, comprehensive literature review, and comparison with international standards (ISO 22156 and Colombia's NSR-10). These inconsistencies produce unreliable results that compromise the structural integrity of bamboo buildings by affecting the proper dimensioning of load-bearing elements. This paper presents specific technical recommendations to address these deficiencies. The findings underscore the urgent need for revision and update of the NEC-SE-Guadúa to enhance the reliability of structural designs using *Gak*, an important sustainable construction resource widely utilized in Ecuador and neighboring countries. This work improves safety standards in bamboo construction, supporting the sustainable use of this valuable natural material.

Keywords: *Guadua* · Beams · NEC-SE-Guadúa

1 Introduction

The planet is confronting an unprecedented climate crisis marked by escalating CO_2 emissions [1], unsustainable energy consumption [2], and overwhelming solid waste generation [3]. In this context, the construction industry is decisive in generating greenhouse gases [4]. This environmental impact necessitates an urgent transition toward more sustainable building materials [5, 6] and construction methodologies.

It is important to focus on the engineering and construction sectors to ensure the proper use of resources needed to produce or preserve the built environment, as well as

J. L. Ballesteros et al. (Eds.): CITIS 2025, LNNS 1780, pp. 3–12, 2027.
https://doi.org/10.1007/978-3-032-14287-0_1

to consider the amount of pollution and waste generated by construction [7]. A viable alternative to non-natural building materials such as steel and concrete is bamboo due to its mechanical and processing properties. In places where bamboo is a natural resource, it is common to find buildings that use it as an easily accessible material [8].

Bamboo has emerged as a promising renewable [9–11] and biodegradable [12] alternative that substantially [13] reduces the carbon footprint of architectural projects [5, 10] while delivering considerable socioeconomic benefits to local communities [14]. In Ecuador, GaK is particularly noteworthy for its exceptional physical-mechanical properties [11, 13, 15], impressive growth rate (reaching structural maturity in just 3 to 5 years) [16], and significant contribution to circular economy principles [10, 12, 17]. These characteristics position GaK as an ideal resource for constructing residential buildings, bridges, educational facilities, recreational spaces, commercial establishments, and sustainable furnishings [5, 13, 18–20].

Despite its remarkable advantages, the widespread adoption of GaK in construction faces multiple challenges [5, 21, 22]. These include insufficient knowledge regarding proper treatment protocols, optimal harvesting practices, appropriate storage conditions, efficient transportation methods, accurate structural analysis techniques [5, 21], and effective on-site implementation. A significant barrier is the inherent variability in bamboo material properties, both between species (interspecies) and within the same species (intraspecies), which requires individual characterization of bamboo resources [22]. Additionally, its technical use is relatively recent [6], so standards, codes, regulations, formulas, and calculation methodologies are constantly evolving [5, 6, 21].

The Ecuadorian Construction Standard for Bamboo (NEC-SE-Guadúa) [15] represents a significant advancement in establishing technical guidelines for bamboo construction. However, as with any emerging technical standard, certain inconsistencies [5, 23] may exist in its mathematical formulations that could potentially compromise structural integrity and safety. The present research critically examines the shear stress calculation formula specified in the NEC-SE-Guadúa [15], aiming to identify potential errors, propose technical corrections, and contribute to strengthening the regulatory framework for sustainable bamboo structures in Ecuador and beyond.

2 Methods

This research employed a systematic mixed methods approach to evaluate the reliability and mathematical consistency of the shear stress formula for bamboo beams in the NEC-SE-Guadúa [15], with a particular focus on Gak. The methodology was structured into three sequential phases designed to comprehensively analyze the formula from theoretical, comparative, and analytical perspectives.

2.1 Literature Review

A review of specialized literature on the structural applications of GaK and other bamboo species was conducted using established scientific databases including Scopus, Web of Science, and Google Scholar. Additional sources were gathered from national and international academic repositories, technical reports, and graduate theses. Key regulatory documents examined included the Ecuadorian NEC-SE-Guadúa [15], International Organization for Standardization (ISO) 22156:2021 [24], and Colombia's NSR-10 [25]. The literature review established the current state of knowledge regarding bamboo structural mechanics and identified critical knowledge gaps [5] and potential inconsistencies in existing research and regulations.

2.2 Verification of Units

The mathematical consistency of the formulas was examined by verifying dimensional homogeneity and unit compatibility [26]. Additionally, the theoretical principles of materials science and structural mechanics that underpin this formula were assessed for their applicability to the anisotropic nature of bamboo. This phase included a detailed review of the constitutive equation and their alignment with the mechanical behavior of GaK.

2.3 Comparative Analysis of Standards and Regulations

A systematic cross-comparative analysis was performed to evaluate the NEC-SE-Guadúa [15] against other established standards, primarily Colombia's NSR-10 [25] and the international standard ISO 22156 [24]. This analysis methodically compared calculation methodologies, mechanical parameters, and design assumptions across these standards. The comparative framework specifically examined how different regulatory bodies address the unique mechanical characteristics of bamboo and formulate shear stress calculations for structural elements.

The final phase centered on assessing the reliability of the formula used to compute the shear force in bamboo beams. This assessment incorporated both analytical and numerical methods. On the analytical side, the focus was on verifying the formula's mathematical consistency and theoretical underpinnings. Meanwhile, the numerical component compared outcomes across the Ecuadorian NEC, the Colombian Regulation, and ISO 22156:2021, offering a comprehensive understanding of the formula's performance within different normative frameworks.

3 Results

3.1 Literature Review Findings

The comprehensive review of scientific literature revealed that ISO 22156.2021 represents the most recent standard for calculating *Guadua* bamboo structures. The Colombian Regulation (NSR-10) stands as the pioneering reference that served as the foundation for the development of both Ecuador's NEC-SE-Guadúa and Peru's *Guadua* Standard (NORMA E.100).

A significant finding from the literature analysis was that the NEC-SE-GUADÚA only addresses one type of shear stress, whereas in practice, three distinct types are recognized in bamboo structural mechanics: shear stress parallel to fibers, shear stress perpendicular to fibers, and punching shear. This simplification in the Ecuadorian standard represents a notable limitation that could affect the accuracy of structural calculations.

The variables considered in Eqs. (1), (2) and (3) represent essential geometric and load parameters for determining shear stress parallel to fibers in structural elements made of *Guadua* bamboo. In these equations, f_v is the acting shear stress (MPa); V represents the shear force applied to the analyzed section (N); A corresponds to the cross-sectional area of the round *Guadua* bamboo element (mm^2); D is the average external diameter of the Guadua element (mm); and t refers to the average thickness of the *Guadua* cross-sectional wall (mm). These variables allow evaluating, through different normative approaches (Colombian NSR-10, Ecuadorian NEC-SE-Guadúa, and ISO 22156:2021 [24]), the structural capacity and behavior of *Guadua* bamboo elements subjected to shear loading.

From the Colombian Regulation (NSR-10) [25], Eq. (1) was obtained:

$$f_v = \frac{2 * V}{3 * A} * \left(\frac{3 * D^2 - 4 * D * t + 4 * t^2}{D^2 - 2 * D * t + 2 * t^2} \right) \tag{1}$$

From the Ecuadorian NEC-SE-Guadúa [15], Eq. (2) was obtained:

$$f_v = \frac{2 * V}{3 * A} * \left(\frac{3 * D^2 - 6 * D * t + 4 * t^2}{D^2 + 2 * D * t + 2 * t} \right) \tag{2}$$

By rearranging the shear stress formula from ISO 22156:2021 [24], Eq. (3) was derived:

$$f_v = \frac{8 * V}{3 * \pi * t} * \left(\frac{D^3 - (D - 2 * t)^3}{D^4 - (D - 2 * t)^4} \right) \tag{3}$$

3.2 Verification of Units

In this section, a dimensional inconsistency was identified in the NEC-SE-Guadúa equation due to the term $D^2 + 2 * D * t + 2 * t$. In this term, both D and t are expressed in millimeters; however, it mixes expressions in mm^2 ($D^2 + 2 * D * t$) with a term in mm ($2 * t$). This inconsistency renders the denominator dimensionally non-homogeneous and, consequently, leads to shear stress (f_v) results that do not accurately reflect the actual behavior of the structural element.

In the following section, a more in-depth comparative evaluation of the results from this and other regulatory equations is carried out to determine their reliability in the structural design of Guadua elements.

3.3 Comparative Analysis of Shear Stress Calculations

The numerical analysis demonstrated that for identical values of external diameter (D), wall thickness (t), and shear load (V), the calculated shear stress varies considerably depending on which standard is applied. As shown in Table 1 and Fig. 1, these variations are systematic and significant across the full range of evaluated dimensions.

When comparing the three standards for a bamboo culm with an external diameter of 124 mm, wall thickness of 10 mm, and shear force of 3400 N, the calculated shear stresses were:

ISO 22156:2021: 1.89 MPa

NSR-10 (Colombia): 2.01 MPa

NEC-15 (Ecuador): 1.38 MPa which represents a relative error of 26.7% compared to ISO 22156:2021, and 31.1% compared to NSR-10.

This substantial discrepancy between the NEC-15 and other standards persists throughout the entire analysis range, with the Ecuadorian standard consistently yielding shear stress values lower than those calculated using the Colombian standard and ISO 22156:2021.

The graphical representation (Fig. 1) illustrates the linear relationship between shear force and calculated shear stress across all standards, but with significantly different slopes. The allowable shear stress value is also included for reference, highlighting that the NEC calculation methodology could lead to under-designed structural elements that may fail to meet safety requirements when subjected to actual loading conditions.

These findings suggest a fundamental inconsistency in the NEC-SE-GUADÚA shear stress formula that requires immediate attention to ensure structural safety in bamboo construction projects throughout Ecuador.

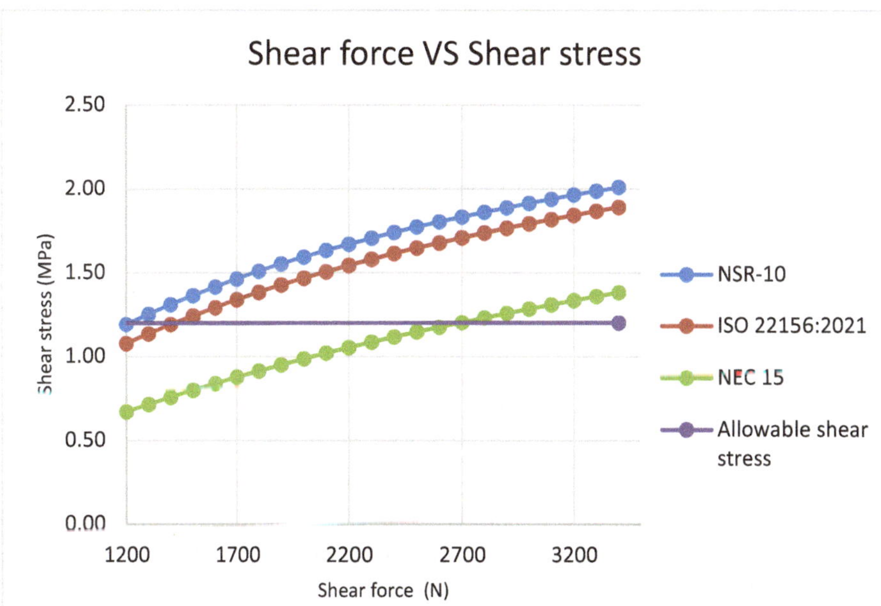

Fig. 1. Shear force VS Shear stress obtained with NSR 10, ISO 22156:2021 and NEC 15, for culms of different diameters

Table 2 and Fig. 2 present the results obtained from the comparative analysis of shear stress calculations for a standardized *Guadua* bamboo culm with an external diameter

Table 1. Shear stress values obtained with NSR 10, ISO 22156:2021 and NEC 15, for culms of different diameters

V	D	t	$d = D - 2t$	Area A	f_v NSR10	f_v ISO 22156:2021	f_v NEC15
N	mm	mm	mm	mm^2	MPa	MPa	MPa
1200	80	10	60	2199	**1.19**	1.08	0.67
1300	82	10	62	2262	1.25	1.13	0.72
1400	84	10	64	2325	1.31	**1.19**	0.76
1500	86	10	66	2388	1.36	1.24	0.80
1600	88	10	68	2450	1.42	1.29	0.84
1700	90	10	70	2513	1.46	1.34	0.88
1800	92	10	72	2576	1.51	1.38	0.91
1900	94	10	74	2639	1.55	1.43	0.95
2000	96	10	76	2702	1.59	1.47	0.99
2100	98	10	78	2765	1.63	1.51	1.02
2200	100	10	80	2827	1.67	1.54	1.05
2300	102	10	82	2890	1.71	1.58	1.09
2400	104	10	84	2953	1.74	1.61	1.12
2500	106	10	86	3016	1.77	1.65	1.15
2600	108	10	88	3079	1.80	1.68	1.18
2700	110	10	90	3142	1.83	1.71	**1.20**
2800	112	10	92	3204	1.86	1.74	1.23
2900	114	10	94	3267	1.89	1.76	1.26
3000	116	10	96	3330	1.91	1.79	1.29
3100	118	10	98	3393	1.94	1.82	1.31
3200	120	10	100	3456	1.96	1.84	1.34
3300	122	10	102	3519	1.99	1.87	1.36
3400	124	10	104	3581	2.01	1.89	1.38

of 100 mm and wall thickness of 10 mm. The analysis was performed across a range of shear forces (1200–3400 N) using three different standards: the Ecuadorian Construction Standard (NEC-15), the Colombian Regulation (NSR-10), and the International Standard ISO 22156:2021.

The results demonstrate significant variations in calculated shear stress values depending on which standard is applied. For a shear force of 3000 N, the calculated shear stresses were:

NSR-10 (Colombia): 2.28 MPa
ISO 22156:2021: 2.10 MPa

NEC-15 (Ecuador): 1.44 MPa which represents a relative error of 31.8% compared to ISO 22156:2021, and 36.9% compared to NSR-10.

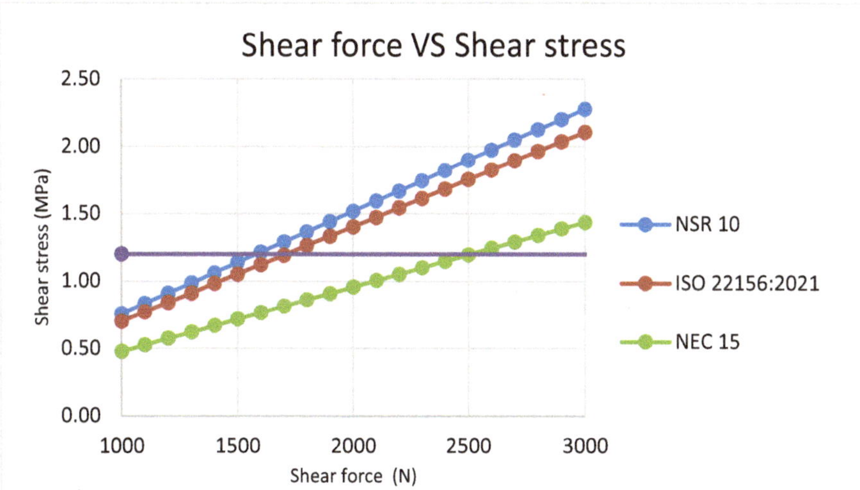

Fig. 2. Shear force VS Shear stress obtained with NSR 10, ISO 22156:2021 and NEC 15, for culms with a diameter of 100 mm and a thickness of 10 mm.

Table 2. Shear stress values obtained with NSR 10, ISO 22156:2021 and NEC 15, for culms with a diameter of 100 mm and a thickness of 10 mm.

V	D	t	$d = D - 2t$	Area A	*fv* NSR 10	*fv* ISO 22156:2021	*fv* NEC 15
N	mm	mm	mm	mm^2	MPa	MPa	MPa
1000	100	10	80	2827	0.76	0.70	0.48
1100	100	10	80	2827	0.84	0.77	0.53
1200	100	10	80	2827	0.91	0.84	0.57
1300	100	10	80	2827	0.99	0.91	0.62
1400	100	10	80	2827	1.06	0.98	0.67
1500	100	10	80	2827	**1.14**	1.05	0.72
1600	100	10	80	2827	1.21	1.12	0.77
1700	100	10	80	2827	1.29	**1.19**	0.81
1800	100	10	80	2827	1.37	1.26	0.86
1900	100	10	80	2827	1.44	1.33	0.91
2000	100	10	80	2827	1.52	1.40	0.96
2100	100	10	80	2827	1.59	1.47	1.01

(*continued*)

Table 2. (*continued*)

V	D	t	$d = D - 2t$	Area A	*fv* NSR 10	*fv* ISO 22156:2021	*fv* NEC 15
2200	100	10	80	2827	1.67	1.54	1.05
2300	100	10	80	2827	1.75	1.61	1.10
2400	100	10	80	2827	1.82	1.68	1.15
2500	100	10	80	2827	1.90	1.75	**1.20**
2600	100	10	80	2827	1.97	1.82	1.24
2700	100	10	80	2827	2.05	1.89	1.29
2800	100	10	80	2827	2.13	1.96	1.34
2900	100	10	80	2827	2.20	2.03	1.39
3000	100	10	80	2827	2.28	2.10	1.44

4 Discussion

From the results obtained, it is evident that the calculated shear stress values in Guadua bamboo culms differ substantially depending on which formula is employed—whether from NEC 15, the Colombian Regulation, or ISO 22156:2021. In fact, discrepancies of up to 36.9% have been observed when comparing the NEC 15 results to those of the other two standards. Such a gap can lead to significant underestimation or overestimation of the actual shear capacity in Guadua structural elements.

Upon reviewing the formulations, it becomes apparent that there is an inconsistency in the units used in the NEC formula's denominator. Specifically, one term (2 t) is expressed in millimeters, while the other terms ($D^2 + 2D * t$) are in mm^2. This unit inconsistency creates challenges when performing calculations and may lead to erroneous interpretations of shear strength values in bamboo structural applications.

5 Conclusions

In conclusion, significant variations were observed in the results obtained using the NEC, ISO 22156:2021, and Colombian regulations. The inconsistencies identified in the units of the shear stress formula in the NEC highlight a critical issue that affects calculation accuracy and structural design reliability. These findings underscore the urgent need for immediate review and comprehensive updating of the Ecuadorian Construction Standard (NEC) for *Guadua* bamboo structures. Such revisions would ensure greater consistency with international standards, improve the accuracy of structural calculations, and ultimately enhance the safety and efficiency of bamboo construction practices in Ecuador.

References

1. Oliva, D., et al.: ITNG 2024: 21st International Conference on Information Technology-New Generations, p. 265. Springer Nature (2024)
2. Chevalier, J.-M.: The new energy crisis. In: The New Energy Crisis: Climate, Economics and Geopolitics, pp. 6–59. Springer (2009)
3. Gichamo, T., Gökçekuş, H.: Interrelation between climate change and solid waste. J. Environ. Pollut. Control. **2**, 104 (2019)
4. Crawford, R.H.: Greenhouse gas emissions of global construction industries. In: IOP Conference Series: Materials Science and Engineering, p. 12047. IOP Publishing (2022)
5. Liu, K., et al.: "Bamboo: a very sustainable construction material"-2021 international online seminar summary report. Sustain. Struct. **2**, 15 (2022)
6. Gatóo, A., Sharma, B., Bock, M., Mulligan, H., Ramage, M.: Sustainable structures: bamboo standards and building codes. Proc. ICE Eng. Sustain. **167**, 189–196 (2014). https://doi.org/10.1680/ensu.14.00009
7. Tiza, T.M., Singh, S.K., Kumar, L., Shettar, M.P., Singh, S.P.: Assessing the potentials of bamboo and sheep wool fiber as sustainable construction materials: a review. Mater. Today Proc. **47**, 4484–4489 (2021)
8. He, Z., Zhou, H., Zhou, X., Du, Y., Shi, L.: Shear failure characteristics and stress-strain relationship of Moso bamboo parallel to the fibers. Ind. Crop. Prod. **222**, 119694 (2024)
9. Ramirez, F., Correal, J., Yamin, L., Atoche, J., Piscal, C.: Dowel-bearing strength behavior of glued laminated Guadua bamboo. J. Mater. Civ. Eng. **24**, 1378–1387 (2012). https://doi.org/10.1061/(ASCE)MT.1943-5533.0000515
10. Archila-Santos, H.F., Ansell, M.P., Walker, P.: Low carbon construction using Guadua bamboo in Colombia. Key Eng. Mater. **517**, 127–134 (2012)
11. Palombini, F.L., Nogueira, F.M.: Bamboo and Sustainable Construction. Springer (2023)
12. Chen, L., Han, S., Li, D., Deng, J., Chen, F., Wang, G.: Vibration performance of bamboo bundle/wood veneer composite floor slabs for joist-type floor coverings. Buildings. **13**, 1265 (2023). https://doi.org/10.3390/buildings13051265
13. Garcia, F.B.: Bamboo Housing Structural Design (2021).
14. Janssen, J.J.A.: Designing and building with bamboo. Int. Netw. Bamboo Rattan Neth. (2000)
15. MIDUVI: NEC Norma Ecuatoriana de la Construcción. Estructuras de Guadúa (GaK) (2016).
16. Moreno, L.A.C., Rojas, W.G.H., Junco, O.J.G.: Caracterización de la Guadua Angustifolia Kunth cultivada en Miraflores (Boyacá) de acuerdo con la NSR-10. Fac. Ing. **21**, 53–71 (2012)
17. van der Lugt, P., van den Dobbelsteen, A.A.J.F., Janssen, J.J.A.: An environmental, economic and practical assessment of bamboo as a building material for supporting structures. Constr. Build. Mater. **20**, 648–656 (2006). https://doi.org/10.1016/j.conbuildmat.2005.02.023
18. Liese, W.: The Anatomy of Bamboo Culms. Brill (1998)
19. Arce, O.A.: Fundamentals of the Design of Bamboo Structures. (1993).
20. Seixas, M.A., Ripper, J.L.M., Ghavami, K.: Prefabricated bamboo structure and textile canvas pavilions. In: Proceedings of IASS Annual Symposia, pp. 1–12. International Association for Shell and Spatial Structures (IASS) (2014)
21. Harries, K., et al.: Knowledge Gaps and Research Needs in Bamboo Construction (2022). https://doi.org/10.5281/zenodo.6569784.
22. Harries, K.A.: Adoption of full-culm bamboo as a mainstream structural material. npj Mater. Sustain. **3**, 4 (2025)
23. Chicaiza, A., Ortiz, M., Chávez, L., Toaza, M.: Desafíos y evolución de las normativas en el diseño de estructuras de bambú: Un estudio comparativo global y la perspectiva ecuatoriana (2023).

24. International Organization for Standardization (ISO): ISO 22156:2021 Bamboo Structures – Bamboo culms – Structural design, (2021).
25. Asociación Colombiana de Ingeniería Sísmica: Reglamento colombiano de construcción sismo resistente NSR-10. Capítulo G.12 - Estructuras de guadua. Bogotá, Colombia (2010).
26. Martín, L.E.: Mediciones e instrumentación: Metrología, modelamiento sensórica. Alpha Editorial (2019).

Optimizing Reuse Cycles for Glass and PET Bottles: A Comparative Life Cycle Assessment Addressing Recycled Content Variability and Consumer Behavior in Ecuador

Lenin J. Ramirez-Cando[1(✉)], Yuliana I. Mora-Ochoa[1], Raul Davalos[1,3], and Lorenzo Guerrini[2]

[1] School of Biological Sciences and Engineering, Universidad Yachay Tech, Urcuquí, Ecuador
lramirez@yachaytech.edu.ec

[2] Dipartimento Territorio e Sistemi Agro-Forestali (TESAF), Università degli Studi di Padova, Viale Dell'Università, Padua, Italy

[3] Biological and Environmental Science and Engineering, King Abdullah University of Science and Technology, Thuwal, Saudi Arabia

Abstract. This study evaluates the Global Warming Potential (GWP) reduction potential of reusing glass and polyethylene terephthalate (PET) beverage bottles, incorporating critical real-world constraints such as recycled content variability, chemical toxicity risks (for PET), and consumer non-return behavior (for glass). Using a cradle-to-grave Life Cycle Assessment (LCA) compliant with ISO 14040/44, we model 1–25 reuse cycles for glass and 1–9 cycles for PET, adjusting for regional energy mixes in Ecuador and end-of-life uncertainties. Results demonstrate that glass reuse achieves 87–95% GWP reduction at 18–25 cycles, outperforming PET (72–86% reduction at 5–9 cycles), but requires high return rates ($>70\%$) and cullet incorporation ($\geq 25\%$). PET's toxicity risks limit reuse to 9 cycles, with recycled resin variability increasing emissions by 36%. Policy interventions—including standardized recycled content mandates, deposit schemes, and non-toxic additives—are essential to balance safety and sustainability. This work provides actionable insights for advancing circular economy transitions in beverage packaging systems.

Keywords: LCA · Glass Bottle; PET bottle · GWP

1 Introduction

Single-use PET (polyethylene terephthalate) and glass bottles significantly contribute to global waste generation and greenhouse gas emissions. While recycling and reuse are promoted as circular economy solutions, their efficacy is constrained by material-specific challenges [1, 2]: PET bottles face chemical leaching risks during repeated use, and glass production is energy-intensive. Existing studies often overlook systemic factors such as variability in recycled content, consumer behavior, and region-specific infrastructure, leading to overestimated environmental benefits [2–5].

J. L. Ballesteros et al. (Eds.): CITIS 2025, LNNS 1780, pp. 13–19, 2027.
https://doi.org/10.1007/978-3-032-14287-0_2

Consumer behavior towards the reuse of glass and PET materials is increasingly influenced by environmental awareness and sustainability trends. Many consumers are now more inclined to choose products packaged in reusable or recyclable materials, driven by a desire to reduce their ecological footprint [6, 7]. Glass, being infinitely recyclable without loss of quality, is often preferred for its durability and perceived purity. PET, while also recyclable, is valued for its lightweight and shatter-resistant properties, making it convenient for on-the-go use. However, the effectiveness of reuse initiatives largely depends on consumer education and the availability of recycling facilities [8–10]. As more brands adopt eco-friendly packaging and governments implement stricter recycling regulations, consumer behavior is expected to continue shifting towards more sustainable practices [11]. On the contrary, consumers in developmental countries tend to give these kinds of containers differences uses that its intend use, reducing reuse and recovery rates along the life cycle of a single bottle.

This study addresses these gaps by quantifying GWP reductions for glass and PET reuse under realistic scenarios, including recycled content variability (5–25% for PET resin and glass cullet), toxicity-driven reuse limits for PET ($\leq$9 cycles), and consumer non-return behavior for glass (30% loss). The findings aim to inform regionally adaptable policies that maximize sustainability outcomes while addressing practical limitations.

2 Methods

The study employs a cradle-to-grave LCA framework, encompassing raw material extraction, production, transportation, washing, and end-of-life (EoL) stages. Functional units include 1 bottle of beverage delivered annually for PET bottles in Ecuador and 1 glass bottle (1250 mL) in Ecuador. Scenarios for PET incorporate 1–9 reuse cycles with recycled resin ratios of 5–25% and a 50% landfill rate post-use due to toxicity concerns. For glass, effective reuse cycles are adjusted from 1–25 to 1–18 to account for a 30% non-return rate. Data were sourced from SimaPro 9.2 using the ReCiPe 2016 Midpoint method particularly, in this analysis GWP was considered as main aim to reduce, this study was regionalized for Ecuador's energy mix. Key inputs include virgin PET resin (0.386 kg CO_2 eq), recycled PET (0.023 kg CO_2 eq), glass production (0.903 kg CO_2 eq), and cullet recycling credits (-0.241 kg CO_2 eq). Sensitivity analyses test recycled content variability, non-return rates, and material degradation. The GWP per- use (Eq. 1) formula integrates production, washing, transport, and end of life EoL impacts:

$$\mathrm{GWPper-use} = \frac{\mathrm{Production} \times (1 - \mathrm{r}) + \mathrm{Recycled} \times \mathrm{r} + \mathrm{N} \times (\mathrm{Washing} + \mathrm{Transport}) + \mathrm{EoL}}{N} \tag{1}$$

where r represents the recycled fraction and N denotes reuse cycles.

Data Analysis, all data were processed in RStudio, and the uncertainties were estimated as de standard deviation of full error contribution to the scenarios analyzed and confidence intervals 95%.

3 Results

The comparative analysis of PET and glass bottles reveals critical insights into their Global Warming Potential (GWP) under revised scenarios that account for recycled content variability, consumer behavior, and end-of-life (EoL) management.

The GWP of PET bottles is heavily influenced by the fraction of recycled resin (r) integrated into production. For instance, increasing recycled content to 25% reduces virgin PET emissions by 25%, lowering the per-bottle footprint from 0.386 kg CO_2 eq to 0.289 kg CO_2 eq. While recycled resin itself contributes minimally to emissions (0.006 kg CO_2 eq at r = 25%), its variability poses a significant challenge. In regions with limited recycling infrastructure—where recycled content may drop to 5%—virgin material dependency escalates emissions by 36%, underscoring the environmental cost of underdeveloped circular systems (see Table 1).

Table 1. PET bottle analysis for GWP.

Parameter	*Original Value*	*Revised Value*
Virgin PET Production	0.386 kg CO_2 eq	0.386 × (1 - r) kg CO_2 eq
Recycled PET Production	0.023 kg CO_2 eq	0.023 × r kg CO_2 eq
Recycling Credit (EoL)	-0.0179kg CO_2 eq	-0.0179 × 0.5 kg CO_2 eq
Landfill Emissions (EoL)	0 kg CO_2 eq	0.053 kg CO_2 eq (methane)

End-of-life management further complicates PET's sustainability profile. Toxicity risks from chemical leaching (e.g., antimony) restrict reuse to a maximum of 9 cycles and reduce post-consumer recyclability. Consequently, only half of retired bottles are recycled, slashing the recycling credit by 50% (−0.009 kg CO_2 eq) and introducing landfill emissions (0.053 kg CO_2 eq) from methane release. This dual burden—constrained recycling and landfill penalties—highlights the urgent need for standardized material safety protocols and infrastructure investments to mitigate PET's lifecycle impacts

Glass production, while energy-intensive (0.903 kg CO_2 eq per bottle), benefits markedly from cullet integration. A 25% cullet fraction (*c*) reduces virgin material demand by 25%, curbing emissions to 0.677 kg CO_2 eq and leveraging recycling credits (−0.060 kg CO_2 eq). However, low cullet usage (5%) diminishes these benefits, amplifying reliance on virgin resources and increasing net emissions by 18%.

Consumer behavior further shapes glass's GWP trajectory. A 30% non-return rate—where bottles are repurposed or discarded—reduces effective reuse cycles from 25 to 18, inflating per-use emissions by 42% (0.043 to 0.061 kg CO_2 eq). Non-returned bottles also incur landfill penalties (0.016 kg CO_2 eq), emphasizing the environmental toll of inconsistent consumer participation in return systems (see Table 2).

While PET's lower initial emissions (0.386 vs. 0.903 kg CO_2 eq for glass) suggest short-term advantages, its toxicity risks and recycled content dependency limit long-term viability. Glass, conversely, achieves superior GWP reductions (>90%) at high reuse cycles but requires robust return systems and cullet mandates to offset its carbon-intensive production.

To optimize sustainability outcomes (see Fig. 1.):

- **PET systems** should mandate ≥ 25% recycled resin and prioritize non-toxic additives to safely extend reuse cycles.

Table 2. Glass bottle analysis for GWP.

Parameter	*Original Value*	*Revised Value*
Virgin Glass Production	0.903 kg CO_2 eq	0.903 × (1 - c) kg CO_2 eq
Recycled Glass (Cullet)	0.241 kg CO_2 eq credit	0.241 × c kg CO_2 eq credit
Effective Reuse Cycles	1–25	1–18 (30% non-return rate)
Landfill Emissions	0.053 kg CO_2 eq	0.053 × 0.3 kg CO_2 eq (non-returned bottles)

- **Glass systems** require ≥ 25% cullet integration and deposit-return schemes to reduce non-return rates below 10%.

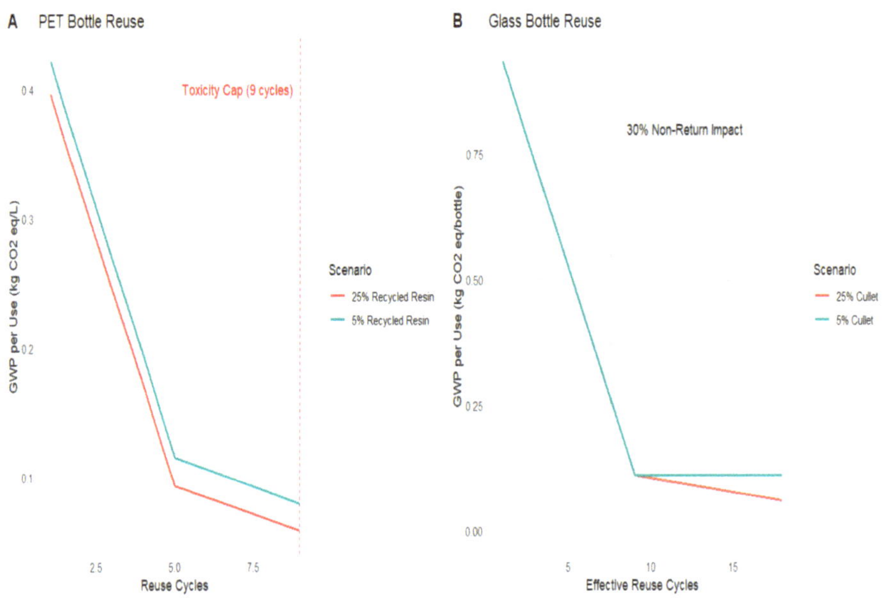

Fig. 1. Scenario analysis for GWP reduction as function of reuse cycles.

Subcategories like *Noise* and *Time* suggest sensitivity to operational fluctuations and long-term trends, such as inefficiencies in equipment or evolving energy grids. Equally critical is the *Non-Return Rate*, a metric reflecting inefficiencies in recycling loops. Higher rates imply increased reliance on virgin materials, amplifying emissions from extraction and production. *Recycled Resin*, by contrast, offers a pathway to lower GWP, though uncertainties (*Noise*) and temporal shifts in recycling infrastructure could temper its benefits.

The *Transport Distance* parameter reinforces the role of logistics, where longer hauls—particularly via fossil fuel-dependent modes—inflate emissions. Meanwhile, the

Material comparison between *Glass* and *PET* encapsulates a classic sustainability trade-off: glass production demands high energy but boasts recyclability, whereas PET, lighter and less energy-intensive to produce, faces challenges in end-of-life management. The interplay of these factors hinges on system boundaries, such as whether recycling rates or landfill leakage are prioritized (Fig. 2).

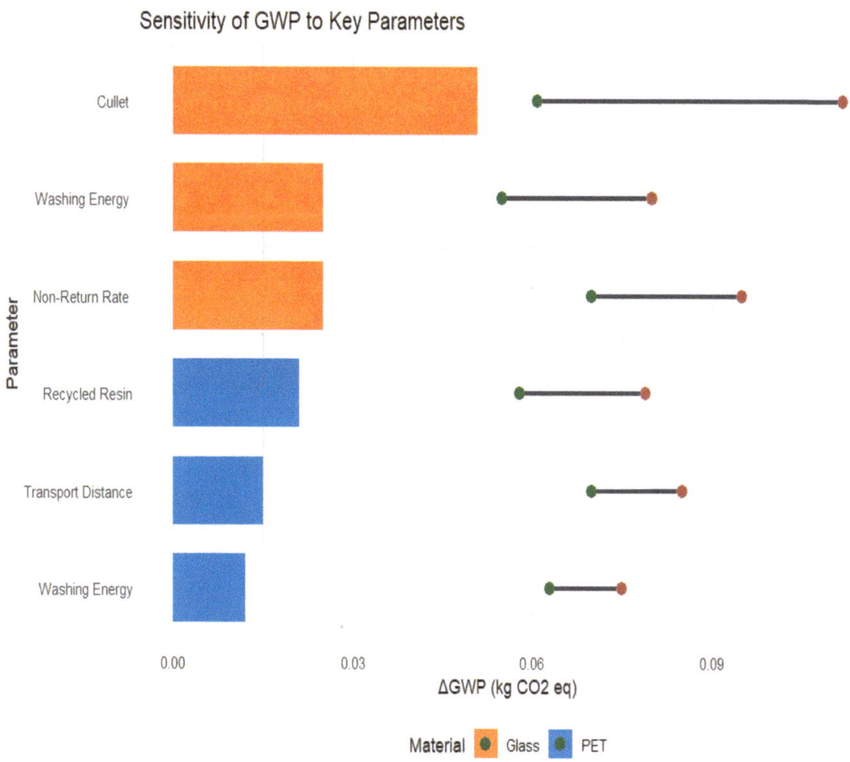

Fig. 2. Uncertainty analysis for Glass and PET Bottles GWP.

4 Discussion

The trade-offs between PET and glass highlight material-specific challenges. PET's lower initial emissions (0.421 vs. 0.903 kg CO_2 eq for glass) are offset by toxicity risks and dependency on recycled resin quality [1, 12–14]. Mandating ≥25% recycled resin stabilizes PET emissions but requires regulatory enforcement, nowadays in Ecuador it is mandatory to use 25% of recycled materials in bottles production [15, 16]. Glass, despite its energy-intensive production, achieves superior long-term GWP reductions when supported by return systems and cullet integration. Policy measures such as deposit schemes for glass (to reduce non-return rates to <10%) and standardized non-toxic additives for PET are critical. Limitations include regional biases (Ecuador's fossil-dependent grid inflates washing/transport impacts) and behavioral assumptions (non-return rates based on National data reported).

To mitigate GWP, prioritizing Washing Energy optimization and scaling Recycled Resin adoption emerges as a logical starting point. Streamlining logistics to reduce Transport Distance—through localized supply chains or cleaner fuels—could further curb emissions. Addressing the duplicated Washing Energy entry is also essential to ensure data integrity. For materials like Glass and PET, lifecycle-wide assessments are needed to balance production emissions with circular economic potential [17–19].

While the framework provides a valuable qualitative roadmap, its utility is constrained without quantitative data. Numerical values would enable ranking parameters by impact magnitude, validating assumptions, and identifying high-leverage interventions. Future work should integrate empirical metrics to transform this conceptual model into an actionable tool for decarbonization.

5 Conclusions

This study demonstrates that glass reuse, despite higher initial emissions, outperforms PET in long-term GWP reduction when supported by robust return systems and cullet mandates. PET's toxicity risks necessitate a 9-cycle cap, emphasizing the need for material innovation. Policymakers must prioritize dual strategies: incentivizing glass circularity through deposit schemes and enforcing PET safety protocols. Future research should integrate social-behavioral models to refine scalability and explore hybrid packaging systems.

References

1. Uthayakumar, A. Life Cycle Assessment of Glass Bottle (2020)
2. Gambino, I., et al.: Pet-bottled water consumption in view of a circular economy: the case study of Salento (South Italy). Sustainability (Switzerland) **12**, 1–13 (2020). https://doi.org/10.3390/su12197988
3. Ferrara, C., De Feo, G., Picone, V.: Lca of glass versus pet mineral water bottles: an italian case study. Recycling **1**, 6 (2021). https://doi.org/10.3390/recycling6030050
4. Solano, C.M., Roldán, R.F., Carvajal, M.F., Gómez, A.J., Mattos, S., Vives, J.I.: Reverse logistic processes for glass container reuse. Environ. Processes **8** (2021). https://doi.org/10.1007/s40710-020-00482-z
5. Joshi, S.V., Drzal, L.T., Mohanty, A.K., Arora, S.: Are natural fiber composites environmentally superior to glass fiber reinforced composites? Compos. Part A Appl. Sci. Manuf. **35**, 371–376 (2004). https://doi.org/10.1016/j.compositesa.2003.09.016
6. Grießhammer, R., Benoît, C., Dreyer, L.C., Flysjö, A.: Feasibility Study: Integration of Social Aspects into LCA. Main 1–14 (2006)
7. Oliveral, A., Cristobal, S., Saizar, C.: Análisis de Ciclo de Vida Ambiental. Económico y Social. Innotec **7**, 20–27 (2016)
8. Chan, H.K.: Green process and product design in practice. Procedia Soc. Behav. Sci. **25**, 398–402 (2011). https://doi.org/10.1016/j.sbspro.2012.02.050
9. Dreyer, L.C., Hauschild, M.Z., Schierbeck, J.: Societal assessment (subject editor : david hunkeler) a framework for social life cycle impact assessment. Int. J. **11**, 88–97 (2006). https://doi.org/10.1065/lca2005.08.223

10. Fonte, M., Cucco, I.: Cooperatives and Alternative food networks in italy. the long road towards a social economy in agriculture. J Rural Stud. **53**, 291–302 (2017). https://doi.org/10.1016/j.jrurstud.2017.01.019
11. Ramírez-Cando, L.J., et al.: Environmental assessment of flax straw production for non-wood pulp mills. Chem. Eng. Trans. **58** (2017). https://doi.org/10.3303/CET1758132
12. Pradhan, A., Shrestha, D.S., McAloon, A., Yee, W., Haas, M., Duffield, J.A.: Others energy life-cycle assessment of soybean biodiesel -revisited. Am. Soc. Agric. Biol. Eng. **54**, 1031–1039 (2011)
13. Finkbeiner, M., Schau, E.M., Lehmann, A., Traverso, M.: Towards Life Cycle Sustainability Assessment. Sustainability **2** (2010). https://doi.org/10.3390/su2103309
14. Lueddeckens, S., Saling, P., Guenther, E.: Temporal issues in life cycle assessment— a systematic review. Int. J. Life Cycle Assess. **25** (2020)
15. MAE Tercera Comunicación Nacional Del Ecuador a La Convención Marco de Las Naciones Unidas Sobre El Cambio Climático (2017)
16. IPCC 2006 IPCC Guidelines for National Greenhouse Gas Inventories. 2006 IPCC Guidelines for National Greenhouse Gas Inventories 3, pp. 1–40 (2006)
17. Cheng, W.: Study of residential energy-saving green technology. Appl. Mech. Mater. **446–447**, 1552–1557 (2013). https://doi.org/10.4028/www.scien-tific.net/AMM.446-447.1552
18. Iyke, B.N.: Climate change, energy security risk, and clean energy investment. Energy Econ. **129** (2024). https://doi.org/10.1016/j.eneco.2023.107225
19. Meneses-Jácome, A., Diaz-Chavez, R., Velásquez-Arredondo, H.I., Cárdenas-Chávez, D.L., Parra, R., Ruiz-Colorado, A.A.: Sustainable energy from agro-industrial wastewaters in Latin-America. Renew. Sustain. Energy Rev. **56** (2016)

Nanoencapsulation of Neem Oil via Ionic Gelation for Application in Food Preservation

Diego Carrillo[1,2](✉), Silvia Clavijo[2], Segundo Reyes[1], and Luis Vásquez Cortez[3]

[1] Faculty of Life Sciences and Technologies, Laica University Eloy Alfaro of Manabí, Manta, Ecuador
diego.carrillo@uleam.edu.ec

[2] Faculty of Applied Sciences to Industry, National University of Cuyo, San Rafael M5600APG, Argentina
scclavijo@fcai.uncu.edu.ar

[3] Faculty of Agricultural Sciences, Technical University of Babahoyo, Babahoyo, Los Ríos, Ecuador
lvazquezc@utb.edu.ec

Abstract. The encapsulation of plant extracts through ionic gelation presents challenges in obtaining stable, high-quality materials. Optimizing key parameters is essential for preserving the biological activity of active compounds. The optimal nanoencapsulation of neem oil (1.7%) and phosphatidylcholine (4%) in sodium alginate was achieved using a homogeneous emulsion with glycerol and a buffer at pH 7.0, which was sonicated in an ultrasonic disruptor. This process resulted in a yield of 90% and an efficiency of 99.6%. To evaluate the optimal conditions, 25 mW, láser was utilized for spectrum acquisition, and Minitab 19 software was employed for response surface analysis and contour plots. Particle size and Z-potential were measured using the Brookhaven 90Plus Nanoparticle Size Analyzer, while morphology was analyzed using scanning electron microscopy (SEM). The results indicated that the capsules had a Z potential of −35.41 mV and a particle size of 101.53 nm. The lyophilized liposomes exhibited diameters ranging from 130 to 150 nm, with an average diameter of 10.68 nm. In conclusion, effective encapsulation was successfully achieved using ionic gelation with sodium alginate and phosphatidylcholine.

Keywords: FTIR · neem · azadirachtin · encapsulation

1 Introduction

Nanoencapsulation of oils is a rapidly expanding area of research due to its wide range of applications in the food, pharmaceutical, and cosmetic industries. Among these, neem oil has garnered particular attention because of its well-documented antimicrobial, antioxidant, insecticide, and biopesticide properties, making it a promising natural alternative to synthetic preservatives and pesticides [1, 2]. The encapsulation of neem oil using ionic gelation has been explored as an effective strategy to enhance its stability and biological

J. L. Ballesteros et al. (Eds.): CITIS 2025, LNNS 1780, pp. 20–29, 2027.
https://doi.org/10.1007/978-3-032-14287-0_3

activity. For instance, Singh et al. demonstrated that this technique improves both the antimicrobial properties and the release profile of the active compounds in neem oil [3]. Ionic gelation relies on electrostatic interactions between charged macromolecules (e.g., polysaccharides) and multivalent ions to form gel-like matrices. This mild and non-toxic method protects sensitive bioactive compounds and enables their controlled release [4, 5]. In Ecuador, numerous studies have focused on the encapsulation of natural oils to improve their functional properties. Pacheco et al. investigated the encapsulation of babassu oil through complex coacervation, reporting that this method preserved the oil's physicochemical integrity and enhanced its surface activity [6]. Although the oil studied was different, the findings supported the broader application of encapsulation methods for improving the functionality of plant-based oils. A recent study by Carillo Freire et al. analyzed the nanoencapsulation of neem oil in a sodium alginate matrix and evaluated the impact of UV-C radiation on its stability and bioavailability. Their results showed that the encapsulated oil exhibited improved resistance to photodegradation and enhanced release behavior, highlighting its potential use in agricultural and food systems [7]. Analytical techniques such as Fourier-transform infrared spectroscopy (FT-IR) have played a critical role in characterizing encapsulated systems, as they allow for the identification of functional groups and interactions based on the absorption of infrared radiation [8, 9]. As encapsulation technologies continue to evolve, ionic gelation remains one of the most promising methods for delivering and protecting natural oils, particularly in the context of reducing chemical inputs in food and agriculture [10, 11].

2 Methods

2.1 FTIR Methodology

FTIR analysis was conducted using a Perkin Elmer Spotlight 200 spectrometer equipped with an infrared light source, a KBr gas cell, and a high-sensitivity detector. Measurements were taken in the wavenumber range of 4000 cm^{-1} to 400 cm^{-1} with a resolution of 4 cm^{-1}. Solid samples for FTIR analysis were prepared using the KBr pellet technique. The pressed sample was then analyzed using the FTIR spectrometer. At least three measurements were performed for each sample, and the resulting spectra was averaged. The FTIR spectra obtained was processed using Thermo Scientific OMNIC 8.0 software. A Savitzky-Golay filter with a 9-point window was applied to smooth the spectral signals. To identify the compounds, present in the samples, the FTIR spectra was compared to a spectral library of known compounds using the KnowItAll® Informatics System. Characteristic peaks in the FTIR spectrum were assigned based on comparison with the available literature.

2.2 Purification of Soy Phosphatidylcholine

To obtain soy phosphatidylcholine, 10 g of crude soybean lecithin were weighed and dissolved in 50 mL of ethyl acetate. Then, 2 mL of distilled water were slowly added under stirring, resulting in the formation of two phases. The upper phase was retained, and the lower phase, which appeared gel-like, was subsequently mixed with 30 mL of acetone, resulting in a clumpy solution. The mixture was then stirred with a glass rod, and the acetone was removed by decantation. This process was repeated by adding 30 mL of fresh acetone. Finally, the precipitate was vacuum filtered and placed in a desiccator for 24 h to remove solvent residues.

2.3 Ionic Gelation Encapsulation

Encapsulation was carried out by mixing phosphatidylcholine, alginate, and neem oil to obtain a homogeneous emulsion. The mixture was prepared with glycerol and phosphate buffer at pH 7.0, then sonicated using an ultrasonic cell disruptor (5 pulses, 750 watts, 100% amplitude for 60 s). Ionic gelation was then performed in an alginate matrix. To determine the optimal microencapsulation conditions, compound concentrations, UV-C light feed flow rate, and particle size were evaluated.

2.4 Liposome Preparation Method

Different concentrations of sodium alginate, sodium phosphatidylcholine, glycerol, neem oil, and 10 mM phosphate buffer (pH 7.0) were weighed. The mixture was lightly stirred and heated in a water bath at 60 °C for 1 h. Finally, a $CaCl_2$ solution (100:1.7 vol. ratio of film-forming solution to 0.02 N $CaCl_2$) was stirred at 500 rpm. The valve was then opened at a frequency of 0.64 Hz and a volumetric flow rate of 24 mL/min, allowing the formed capsules to fall into a lower beaker containing the $CaCl_2$ solution. After sedimentation, the capsules were collected for further analysis. The capsule size was determined using a Q-Scope digital microscope.

2.5 Particle Size

The Brookhaven 90Plus nanoparticle size analyzer was used to measure capsule particle size with a precision of $\pm$1–2% for monodisperse samples and repeatability of $\pm$1–2% for dust-free samples. Laser: 15 mW. solid-state laser. Temperature ranges from 5 °C to 75 °C in 0.1 °C increments. Sample volume: 50 µL to 3 mL; 40 µL flow cell available. Temperature control: 6 °C to 100 °C, $\pm$0.1 °C. Laser: 35 mW., solid-state, standard.

2.6 Peak Analysis in KnowItAll® Informatics System

This software enables the accurate identification and characterization of specific compounds in a sample by using a database of spectra and absorption patterns. It compares the results obtained through FTIR spectroscopy with previously identified compounds. For this study, a sample consisting of dried neem leaves, stems, and seeds was analyzed via FTIR spectroscopy in a wavelength range from 4000 to 400 cm^{-1} with a resolution of 4 cm^{-1}. The results allowed for both the identification and quantification of the main components present in the sample.

2.7 SEM Analysis

A sample analysis was performed using a Scanning Electron Microscope (SEM) under the following conditions: electron acceleration at 15 kV and detection in secondary electron mode. Samples were previously fixed using a 2.5% glutaraldehyde solution and dehydrated in a graded ethanol series. Subsequently, they were coated with a thin layer of gold to prevent charge accumulation. Images were acquired in secondary electron mode (SE) with a resolution of up to 5 nm, allowing detailed visualization of the surface topography.

2.8 Experimental Design and Statistical Analysis

Response surface analysis was conducted using Octave version 9.2.0 to generate the corresponding plot. This approach allowed for the evaluation of the efficiency of different neem oil encapsulation formulations, providing a robust statistical tool to identify significant differences between groups. Furthermore, it facilitated the determination of the formulation that optimizes encapsulation efficiency, contributing to the design and improvement of encapsulation systems for specific applications (Tables 1, 2, 3).

3 Results

3.1 FTIR Assessment in Husk

In FTIR spectroscopy, several types of biomolecules can be found by seeing characteristic peaks in specific regions of the spectrum. For instance, lipids are identified by peaks in the 2800 to 3300 cm^{-1} range, which correspond to the vibrations of hydrocarbon chains and functional groups. Proteins, on the other hand, exhibit peaks in the 1600 to 1700 cm^{-1} range, corresponding to the vibrations of peptide bonds. Carbohydrates are detected by peaks in the 1000 to 1300 cm^{-1} range, which are related to the functional groups of sugars. Steroids can also be identified by peaks in the 1000 to 1300 cm^{-1} range, corresponding to the vibrations of steroid rings and functional groups. Additionally, lignin can be detected by peaks in the 1600 to 1700 cm^{-1} range, associated with aromatic bond vibrations and functional groups, while cellulose is identified by peaks in the 800 to 900 cm^{-1} range, corresponding to the vibrations of cellulose chains (Graphs 1 and 2).

3.2 FTIR Spectroscopic Assessment of Neem Seed

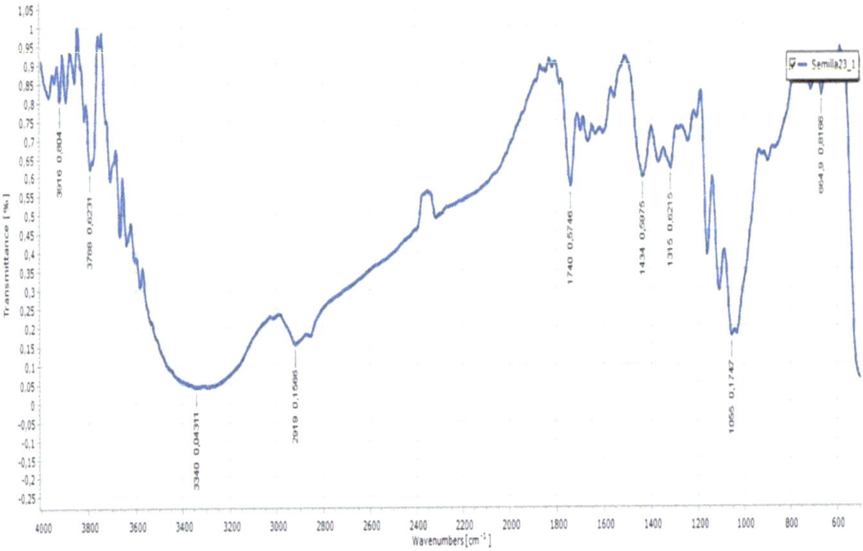

Fig. 1. FTIR spectrum of the neem seed sample analyzed using the KnowItAll® Informatics System. The characteristic absorption bands were identified and compared with reference spectral libraries to assign the principal functional groups associated with the bioactive constituents of Azadirachta indica.

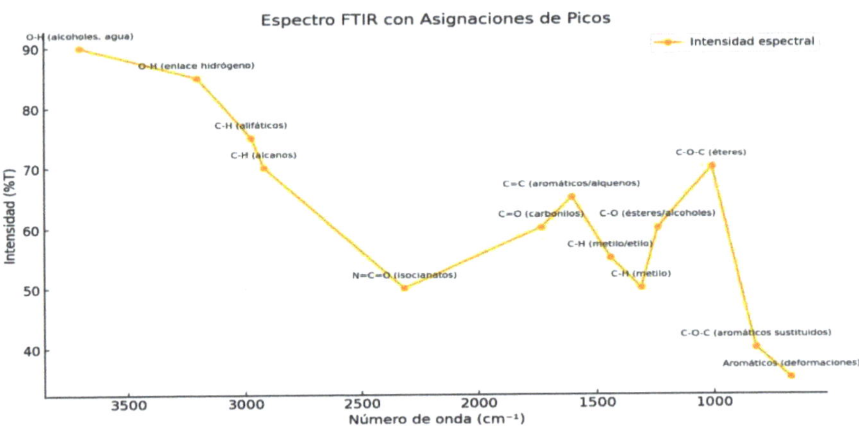

Fig. 2. Characterization of the principal FTIR spectral markers identified in Azadirachta indica samples. The figure highlights the distribution and relative intensity of the absorption bands in different infrared regions, allowing the assignment of functional groups associated with lipids, hydroxyl, carbonyl, carbohydrate, and other bioactive compounds.

3.3 Comparison of Experimental Data with Bibliographic References

Table 1. Comparative values between bibliographic citations and experimental data (maximum values of the evaluated spectra)

Bibliographic Reference	υmax / cm^{-1}	Analysis EPN-UNCUYO
Ref 1 to Ref 12: Geetha	3376, 2976, 2880, 1760, 1708, 1696, 1651, 1596, 1488, 1465, 1379, 1315, 1273, 1068, 1001, 969, 880, 838, 646, 611	RAMAN SPECTRA: Seed - 1315, Husk - 773, Leaves – 1006
Ref. 7: Carrillo et al.	3434, 2935, 1719, 1650, 1622, 1272, 1046 cm^{-1}; 3407, 2956, 1729, 1650, 1383, 1038 cm^{-1}	FT-IR SPECTRA: Seed - 1697.83 (%T = 50.75), 1315 (%T = 32.35), Husk - unspecified, Leaves - 1728.82 (%T = 36.21)

3.4 Vibration Frequencies

The study conducted by Geetha presents a list of vibration frequencies in cm^{-1} for various functional groups in a sample of interest [12]. In contrast, the Notes Journal of Natural Products offers two lists of vibration frequencies in cm^{-1} for different functional groups in a sample [13]. This suggests that the studies vary in the number of vibration frequencies analyzed and the functional groups examined.

3.5 Characteristics of FT-IR Spectra

The paragraph states that Raman spectra are defined by light-induced vibrational frequencies, whereas FT-IR spectra are characterized by the absorption of infrared light by the sample. This information is not referenced in connection with the studies published in the Geetha and Notes Journal of Natural Products, suggesting that the paragraph offers supplementary details regarding the characteristics of the analyzed spectrum.

3.6 Encapsulation Efficiency

Table 2. Experimental Data: Evaluation of Encapsulation Efficiency

W total (g)	Eficiencia (g)	Contenido del principio activo (%)	Extracto encapsulado (%)	Longitud de esferas (μm)
43,00	99,33	11,71	68,06	47,05
43,13	99,62	9,78	47,61	57,48

(continued)

Table 2. (*continued*)

W total (g)	Eficiencia (g)	Contenido del principio activo (%)	Extracto encapsulado (%)	Longitud de esferas (μm)
42,90	99,09	8,00	31,70	52,18
42,95	99,15	10,25	55,00	50,00
43,05	99,40	12,00	60,00	49,50
42,80	98,90	9,50	40,00	45,80
43,20	99,70	10,75	65,00	48,20
43,10	99,50	9,00	38,50	46,50
42,85	99,05	8,75	30,00	55,00
43,25	99,75	11,50	66,50	47,80
43,15	99,60	10,00	50,00	48,90
42,95	99,20	8,50	35,50	53,50

3.7 Analysis of Variance

Table 3. Regression Analysis for Efficiency as a Function of Wtotal and Encapsulated Extract: Effects of Active Compound Content

Source	DF	Adj SS	Adj MS	F Value	P Value
Regression	2	0.854455	0.427228	307.59	0.000
W total	1	0.530193	0.530193	381.73	0.000
Encapsulated extract	1	0.000694	0.000694	0.50	0.498
Error	9	0.012500	0.001389		
Total	11	0.866956			

Model Summary S: 0.0372684 R^2: 98.56% R^2 (adj): 98.24% R^2 (pred): 97.30%
Regression Model:
Efficiency = 13.14 + 2.005 × Wtotal −0.00074 × Encapsulated Extract

Model Fit: Model (1) has a high R^2 value (98.56%) and an adjusted R^2 of 98.24%, indicating the model explains most of the variability in efficiency, and the residuals are small, suggesting a good fit. The regression has an F-value of 307.59 with a p value of 0.000, meaning that at least one of the predictors (in this case, Wtotal) has a significant impact on the dependent variable. The encapsulated extract shows no significant F-value (0.50) and a p-value of 0.498, confirming its lack of relevance in the model.

3.8 SEM Analysis

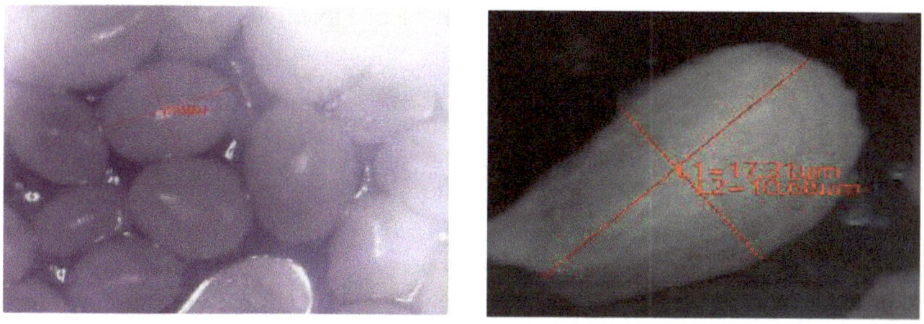

a b

Fig. 3. Morphological characterization of neem oil nanocapsules prepared by ionic gelation. (a) Scanning electron microscopy (SEM) image showing the surface morphology and spherical structure of the nanocapsules. (b) Q-Scope digital microscope image illustrating the particle size distribution and morphology of the encapsulated particles.

Figures 1 and 2. Morphological and particle size analysis of liposomes using scanning electron microscopy (SEM) and Q-Scope digital microscope

A comprehensive analysis of liposome morphology and particle size was conducted using scanning electron microscopy (SEM) and Q-Scope digital microscopy techniques. The results demonstrated a homogeneous distribution of liposomes, with an average size of 100 nm and a predominantly spherical morphology. These findings are essential for accurately characterizing liposomes and understanding their behavior in medical and pharmaceutical applications. Previous studies have indicated that the stability and functionality of liposomes are directly correlated with their size and morphology, which in turn influence their pharmacokinetics and biodistribution. This study facilitated the evaluation of liposomal particle morphology and size through SEM and Q-Scope digital microscopy. The analysis confirmed a uniform particle distribution, with an average diameter of approximately 100 nm and primarily spherical morphology, as illustrated in Figs. 3a and b. These results are significant for comprehending the physicochemical properties of liposomes, particularly concerning their behavior in biomedical and pharmaceutical contexts. However, a more in-depth discussion is warranted regarding the existing literature on the optimal structural characteristics of liposomes for therapeutic applications.

3.9 Encapsulation Efficiency

Treatment 2 demonstrates the highest encapsulation efficiency, featuring an active compound content of 47.6% and an efficiency rate of 99.62%. This indicates the proportion of active compound present in the encapsulated formulation, serving as a crucial indicator of the formulation's quality and effectiveness.

4 Conclusion

FTIR spectroscopic analysis of the "Semilla23_1" sample showed a molecular profile abundant in functional groups commonly found in bioactive compounds from neem oil. The broad absorption bands near 3340 cm^{-1}, along with specific peaks at 2911 and 2854 cm^{-1}, suggest the presence of hydroxyl groups (–OH) and aliphatic $-CH_2-$ stretching vibrations, respectively. Furthermore, a prominent absorption band at 1740 cm^{-1} indicates the stretching vibration of carbonyl groups (C=O), which are typical of esters and ketones. These spectral characteristics align with the presence of azadirachtin, a tetranortriterpenoid limonoid found in neem (*Azadirachta indica*), known for its strong antimicrobial, insecticidal, and antioxidant properties. The identification of azadirachtin highlights its potential use as a natural preservative in food systems, providing a sustainable alternative to synthetic additives, especially in products prone to microbial spoilage like meats, seafood, and minimally processed fresh foods.

The concentration of the active compound, measured as a percentage, varies between 7.5% and 12% based on different experimental data. The highest concentrations are found at points 1, 4, and 12, while the lowest are at points 2, 6, and 11. This variation indicates that the concentration of the active compound fluctuates, possibly due to changes in experimental conditions or formulation techniques. In contrast, efficiency remains stable, hovering around 99% at most points, except for point 12, where it slightly drops to 98.5%. This consistency suggests that the process of incorporating the active compound is generally highly effective, with minor variations likely due to operational factors or natural process variability. Conversely, encapsulation efficiency shows more variation, ranging from 31.7% to 68%. While encapsulation efficiency is generally lower than incorporation efficiency, a positive correlation has been observed between higher concentrations of the active compound and better retention within the encapsulating matrix. This study also demonstrated the successful nanoencapsulation of neem oil using ionic gelation with sodium alginate and soy lecithin, resulting in nanocapsules that exhibited high encapsulation efficiency and stability.

References

1. Mishra, B., Singh, R., Kumar, P.: Neem oil: A natural alternative with antimicrobial and insecticidal potential.
2. Ulrich, A.S.: Biophysical aspects of using liposomes as delivery vehicles. Biosci. Rep. **22**(2), 129–150 (2002).). https://doi.org/10.1023/A:1020170301553
3. Singh, R., Kumar, P., Sharma, B.: Encapsulation of neem oil using ionic gelation: a novel approach for enhanced bioactivity and controlled release. J. Agric. Food Chem. **69**(14), 4021–4032 (2021)
4. Gibbs, B.F., Kermasha, S., Alli, I., Mulligan, C.N.: Encapsulation in the food industry: a review. Int. J. Food Sci. Nutr. **50**(3), 213–224 (1999). https://doi.org/10.1080/096374899101256
5. Rodríguez, A., Fernández, M.J., López, C.A.: Stability and controlled release of encapsulated bioactive compounds using coacervation and ionic gelation techniques. Food Hydrocoll. **101**, 105472 (2020)

6. Pacheco, A., Pérez-Santín, E., Bordenave-Juchereau, S., Arnaudin, I., Gómez-Guillén, M.C., Montero, P.: Encapsulation of babassu oil via complex coacervation. Food Res. Int. **44**, 1044–1051 (2011). https://doi.org/10.1016/j.foodres.2010.03.004
7. Carillo Freire, D., Clavijo, S.C., Reyes Solorzano, S., Vásquez Cortez, L.: Encapsulación de aceite de neem en una matriz de alginato de sodio: evaluación del impacto de la luz ultravioleta tipo C en la estabilidad, biodisponibilidad y fotoprotección. J. Sci. Res. **9**(CININGEC) (2024)
8. Silva, R., Little, C., Ferreira, H., Cavaco-Paulo, A.: Incorporation of peptides in phospholipid aggregates using ultrasound. Ultrason. Sonochem. **15**, 1026–1032 (2008). https://doi.org/10.1016/j.ultsonch.2008.03.010
9. Villamar, C.A.: Acuicultura orgánica-ecológica: aplicación de productos naturales en sustitución de químicos en los procesos de cría de camarones en cautiverio. Revista AquaTIC. **10**, 27–30 (2000)
10. Abad, M.J., Guerra, J.A., Bermejo, P., Irurzun, A., Carrasco, L.: Search for antiviral activity in higher plant extracts. Phytother. Res. **14**, 604–607 (2000). https://doi.org/10.1002/1099-1573(200012)14:8<604::AID-PTR678>3.0.CO;2-L
11. Zhang, Y., Patel, M., Zhou, Y.: Advances in ionic gelation for encapsulation of bioactives.
12. Geetha, K.: Vibrational spectroscopy and functional groups. J. Mol. Struct., 560, 45–52 (2001). Notes. Journal of Natural Products, 62(7), 987–990 (1999).

What is the Cost of Going Green? A Hybrid Model Applied to Climate Change Policies

Pedro Castro-Verdezoto[1], Marcelo Pereira Da Cunha[2], and Andrés Benito[3](✉)

[1] Faculty of Industrial Engineering, University of Guayaquil, Guayaquil, Ecuador
[2] University of Campinas, Institute of Economics, Campinas, Brazil
[3] Regional Faculty of Mendoza, National Technology University, Islamabad, Pakistan

Abstract. The Ecuadorian government has ratified its nationally determined contribution (NDC) to reduce GHG emissions. Since the energy category is the leading GHG emitter, the NDC initiatives prioritize renewable energy sources and energy efficiency. The traditional scheme for environmental assessments uses the bottom-up (BU) and top-down (TD) models. This study uses a hybrid model, called MPC3e, that considers the energy-economic-environmental aspects of the mitigation policies, as well as their positive and negative implications. More specifically, the MPC3e is a hybrid between the LEAP model (a BU model) and the CGE model (a TD model) through a "linking process".

The Ecuadorian NDC assessment was developed by using the MPC3e model by comparing BAU and NDC scenarios. So, its implementation implies GHG reduction linked to GDP growth and price index decrease. In addition to macroeconomic implications due to the reduction of fossil fuel consumption, other industrial sectors would be affected, implying lower tax revenues and exports, without changes in the economy's export structure. The capital supply in the NDC scenario is lower than in the BAU scenario, obtaining the same levels of economic growth. So, the economy requires less effort to generate wealth as a result mitigation policies, making a less polluting transition feasible.

Keywords: Hybridization modeling · Ecuadorian NDC · Energy forecasting · GHG reductions

1 Introduction

The fifth national communication on climate change estimated Ecuadorian GHG emissions of 88,262 GgCO2-eq. The energy sector was the main emitter, followed by land use (47% and 29%, respectively) (MATTE, 2024). As a consequence, the energy sector gained importance by defining mitigation policies. The governmental policy focused on changing the energy mix and diversifying generation (MERNNR, 2023). The energy sector has a major role in the Ecuadorian economy, specifically the production and export of crude oil and production of refined products for domestic consumption. The relationship between economic growth and energy consumption was explored by Ortega-Pacheco et al. (2021).

J. L. Ballesteros et al. (Eds.): CITIS 2025, LNNS 1780, pp. 30–39, 2027.
https://doi.org/10.1007/978-3-032-14287-0_4

Following the Paris agreements, the Ecuadorian government proposed the nationally determined contribution (NDC). Several questions have been raised about the NDC implementation: What are its implications for economic growth? Which sectors would benefit and which would be harmed? Would these policies generate new jobs? Or, a simpler question: Is the country's energy structure appropriate to achieve the NDC objectives? How should the objectives be achieved?.

Previous studies have explored alternatives for GHG reduction and productive transitions (Castro et al., 2018; Villamar et al., 2021), emphasizing the country's largely untapped hydroelectric potential (Carvajal et al., 2019; Hurtado & Castro, 2023; Ramirez et al., 2020). In both cases, the authors used a bottom-up approach. Nevertheless, it is important to assess energy-environment-economy aspects (E3) to provide relevant insights into GHG reduction, especially to evaluate the socioeconomic impacts of environmental policies (Lv et al., 2024).

This study contributes to the debate with a new tool to evaluate climate change policies, through a hybrid of bottom-up (BU) and top-down (TD) models. The MPC3e is a soft-link between the Long Energy Alternative Planning (LEAP) model and a Computable General Equilibrium (CGE) model, both developed for Ecuadorian energy-economic conditions. Policies have positive and negative aspects, and mitigation policies are no exception. The results allowed answering the main question: What is the cost of going green?

2 Ecuadorian NDC Review

The Republic of Ecuador responded positively to the invitation extended by the COP 21 to intensify domestic actions ratified through NDC. The NDC considers two sections: one exclusively for the land use and land-use change (LULUFC) and another for energy from other categories (MAE, 2019). So, we conducted a broad analysis of the NDC on the energy sector. In 2010 (the NDC baseline year), Ecuador's GHG emissions were 56,038 GgCO2-eq, of which the energy sector had a share of 64%. NDC implementation implies a reduction of 9,130 GgCO2-eq by 2025, equivalent to a GHG reduction of 11.87%. To achieve these objectives, Ecuador's NDC establishes lines of action and initiatives aiming at reducing GHG emissions.

Table 1. NDC mitigation initiatives. Source: MAE (2019)

Initiatives	Description
Hydropower development	Resources for electricity generation
Optimization of power generation	Use of natural gas for power generation
Unconventional renewable energy	Wind and solar energy expansion
Efficient cooking program	Induction stoves for cooking
Efficient public transport	Subways lines

The first initiative consists of the construction of eight hydropower plants, which will be gradually incorporated into the national system, increasing installed capacity from 2,4 to 5,2 GW. The Optimization of power generation project consists of power generation from associated natural gas from crude oil production. In the third initiative, unconventional renewable sources (wind and solar) have marginal participation in the energy mix. The wind potential in the short term is estimated at 884 MW and has an effective installed capacity of 21.15 MW. The solar potential is 911 MW, but the country only has 26 MW of installed capacity.

The fourth initiative refers to the Efficiency Cooking Program. Its main goal is to replace the use of LPG with electricity for cooking and water heating. This is achieved by replacing LPG stoves with induction stoves in 1.5 million households. It also expects growing use of electrical equipment for water heating (e.g., electric shower heads) in 750,000 homes. The last initiative considers two local projects: a tramway in the city of Cuenca and a subway line in the nation's capital, Quito. Both projects seek to improve urban mobility and traffic. It is expected that these will transport 400,000 and 120,000 passengers per day in Quito and Cuenca, respectively.

3 Literature Review

The engineering models focus on detailed analysis of technological options and potential for technical changes (A. O. Benito et al., 2024; Pérez-Gelves et al., 2024). In turn, the macroeconomic models analyze the consequences of policies in terms of public finance, economic competitiveness and employment (Gerber Machado et al., 2023). The UNFCCC raises the importance of integrated assessment models (IAM) and energy-environment-economy models (E3) to provide policy-relevant insights into global environmental change, several models have been proposed in this context (UNFCCC, 2025). In Latin America were developed several prospective energy consumption and transition scenarios for GHG reduction using the LEAP model (A. Benito et al., 2023; Perez & Díaz, 2021). About the LEAP model for Ecuador, since the transport sector is the main energy consumer, Chiriboga et al. (2023) and Rivera-González et al. (2019) analyzed scenarios to reduce fossil fuel in this sector.

The CGE models have been widely used for the analysis of the economic impacts due to mitigation policies, in Latin American countries, Benavides et al. (2015) analyzed the economic implications of carbon taxes on the Chilean electricity generation sector and Ribeiro and da Cunha (2022) evaluated the potential of biofuels to reduce Brazilian GHG emissions. There are no specific studies using CGE models for Ecuador regarding energy and environmental assessments. Here, we propose the MPC3e model, which is a hybrid between the LEAP and CGE models. Both are applied to Ecuadorian reality, aiming to assess the impacts of the NDC implementation.

There are two global studies about NDC implications (Fragkos et al., 2018; Siriwardana & Nong, 2021). The former used a climate-change policy version of the GTAP-E model, while the latter employed an integrated assessment model (IMAGE) to estimate the annual reduction cost to achieve NDC targets. The latter also used an economy-wide global CGE model to quantify policy impacts of NDC. Based on the energy prospective approach, Postic et al. (2017) evaluated the impact of NDC in Latin American countries.

4 Methodology

The LEAP_EC model uses a bottom-up approach to determine the Ecuadorian energy consumption of diverse sectors, as well as the competition of all sectors to obtain the different energy resources available (Castro Verdezoto et al., 2019). The CGE_EC model is based on the general equilibrium model to evaluate energy-economic policies for Ecuador (Castro-Verdezoto et al., 2024), so the current version is built on the GTAP and Ecuadorian Central Bank databases.

The energy consumption and GHG emissions are responses to the LEAP_EC variables. Its outputs allow us to calculate the variations in the economic-energy structure, so the variations become the CGE_EC's Input variables using an energy link module. This module uses the economic-energy ratios between the variables of the LEAP_EC and CGE_EC models. The ratios obtained must be consistent with internal and external sales prices, production costs, and import costs. It also accounts for annual variations of energy consumption by sectors quantified in monetary terms, because these are inputs that determine the shock on the CGE_EC's exogenous variables (Fig. 1).

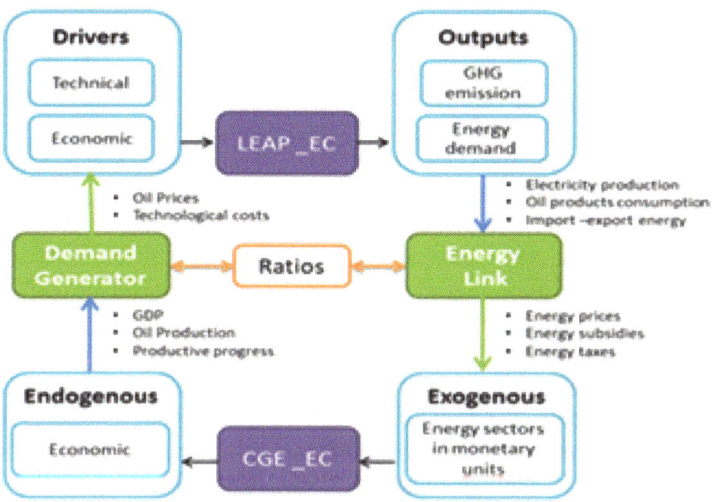

Fig. 1. Schematic view of MPC3e soft-link framework

The shock on the CGE_EC's exogenous variables reflects changes in consumers' and producers' preferences, since society is always reaching a new balance point. After defining these new conditions, the CGE_EC's exogenous variables reflect variations in the main macroeconomic variables. These macroeconomic variations become the drivers in the LEAP_EC model by using a demand generator module. This module uses technological changes and price variations to create new LEAP_EC conditions.

These new conditions will define the new levels of energy consumption and GHG emissions in the LEAP_EC model. The GDP variation is the focus, as it is the main input variable in the LEAP_EC model and the main response variable of the CGE_EC model. The loop between models ends when we obtain the same levels of year-to-year variation in GDP between them. Once the MPC3e model is structured, we use two scenarios to assess the NDC implications: the business-as-usual (BAU) scenario and the NDC scenario. The former considers the energy structure and energy expansion plans from 2010 (NDC base year) to 2025 (NDC horizon year).

5 Results and Discussion

In the BUA scenario, GHG emissions are estimated at 53,150 $GgCO_{2\text{-eq}}$ for 2025, corresponding to a 1.57-fold increase from base year emissions. NDC initiatives' implementation implies a reduction of 7,396 $GgCO_{2\text{-eq}}$ (13.93%) by 2025 (NDC scenario). This result is close to the NDC goals, which propose a GHG reduction of 13.47% in 2025. There is consistency between the results obtained by the MPC3e model and the NDC goals, included in the assertiveness of the NDC initiatives for the energy sector. The MPC3e model foresees that energy initiatives would represent 81.01% of the total GHG reductions achieved by 2025. The energy initiatives' implementation is necessary to determine the goals proposed in the NDC.

5.1 Environmental Effects

The carbon dioxide emissions are preponderant among emissions related to the energy sector. This condition does not change for the two scenarios. For 2025, the CO_2 emissions represent 98.51% and 98.33% of the total GHG in the BAU and NDC scenarios, respectively. The reduction achieved by the NDC initiative is 7,396 $GgCO_{2\text{-eq}}$, 99.65% of which comes from CO_2 reduction, while the remainder is related to methane and nitrous dioxide reductions.

The NDC will have a greater impact on energy transformation processes than energy demand. GHG emission transformation in the BAU scenario would be 14.821 $GgCO_{2\text{-eq}}$. NDC implementation would reduce it to 6,482 $GgCO_{2\text{-eq}}$ by 2025, representing 71% of the GHG reduction planned in the NDC. The final demand would not present a GHG reduction trend until 2020. It would be desirable to include more initiatives regarding final energy consumption in the NDC, with emphasis on the transportation sector. In fact, there would be a slight increase in emissions in 2015, probably due to the expansion of electrification and supply of repressed demand.

Power plants are the main GHG emitters in the energy transformation scenario, representing 92.49% of the total. NDC implementation would imply a GHG reduction of 47% in consumption of power plants by 2025, the equivalent of 6,686 $GgCO_{2\text{-eq}}$ in relation to the BAU scenario. Nevertheless, the OGE&EE project implies an increase of 48% in gas emissions (Table 1), the equivalent of 204 $GgCO_{2\text{-eq}}$, due to the associated gas used for electricity generation. Despite this undesirable condition, there is a positive balance due to reduced fuel oil use to generate electricity.

5.2 Energy Results

The NDC implies a reduction of 2.29 million BOE in final energy demand, a reduction of 2.08% with respect to the BAU scenario in 2025. Failure of its implementation would imply energy consumption of 110 million BOE in 2025, or 1.52 times the energy demand in 2010. The results would be accentuated from 2015 on, as there would be a transition period five years from the implementation. Energy demand reductions are not favorable when compared to GHG reductions, because GHG reductions are not only due to energy consumption, but also to the consumption of less polluting fuels, such as the reduction of LGP consumption in the residential sector, instead using electricity for cooking.

The residential sector would concentrate 92.19% of the energy savings from final consumption (summarized by the PEC program), obtaining energy savings of 2.14 million BOE for 2025. We highlight that the main beneficiaries of this program would be households located in urban areas, with the highest concentration of inhabitants, as well as the greatest electricity coverage, implying the existence of previous electrical infrastructure to facilitate penetration of the proposed technology.

The NDC scenario envisions increased hydropower generation of 13,858 GWh, equivalent to 1.21 times the production in the BAU scenario, and thermal power generation would be reduced by 48.76%. In addition, the NDC scenario prioritizes a 100% increase in renewable energy, but this represents only 0.17% of all energy consumption, so the NDC initiatives need to be more aggressive regarding renewable energy sources. Oil products' consumption is the main input for power generation in the BAU scenario, representing 53.55%.

The NDC scenario implies a 20.12% reduction in power plant inputs, the equivalent to 10.24 million BOE, concentrated in a reduction of oil products, because hydropower generation is a more efficient transformation process than thermal power generation, so its increase in the energy mix implies more efficient power generation. This involves two significant benefits: a stabilization of the input curve for power generation and achieving electricity self-sufficiency with surpluses.

However, this self-sufficiency would be limited to a period of 10 years. Thereafter, electricity importation will have to resume to supply domestic demand, mainly due to the growth of energy consumption in the industrial and transport sectors. It is necessary to define energy efficiency policies in these sectors or alternative technologies to take advantage of the surplus of available electricity. Due to NDC implementation, Ecuador will become self-sufficient in electricity, reducing imports to zero and generating surpluses for export to neighboring countries. In this context, the energy exports would increase by 7.63%. However, this does not mean a change in the export structure, since crude oil will continue to be the main export product, representing 78% (NDC) or 84% (BAU) of total exports. Consequently, there are no changes in the crude oil requirements or its export levels in both scenarios.

5.3 Economic Implications

The NDC implementation projects an increase of 38.5% in GDP in comparison with the base year, equivalent to US$ 27.07 billion in nominal terms, a real increase rate of 2.2% per year. This GDP growth is consistent with the historical average, which is around 3% a year, so the NDC initiatives are not counterproductive regarding economic growth. The BAU scenario implies GDP growth of 37.51% compared to the base year, linked to GHG emissions of 53,150 $GgCO_{2\text{-eq}}$. The NDC scenario implies GDP of US$96.63 billion in 2025, 1.03% higher than in the BAU scenario, and a GHG reduction of 7,396

$GgCO_{2\text{-eq}}$. The BAU scenario implies unemployment rate reduction of 31.56%, while the NDC scenario implies a reduction of 28.66%. Both scenarios have favorable results for this indicator, but the BAU scenario has more favorable results. Both scenarios envision reductions in the price index of 20.16% and 18.65% for the BAU and NDC scenarios, respectively. So, the actions to meet the Ecuadorian NDC in the energy sector have positive impacts on the overall economy.

As seen in Table 2, not all elements of aggregate demand are benefited by GHG reduction policies. Household and government consumption show a slight reduction compared to baseline conditions (BAU scenario), while the GFCF increases 3.54% due to higher investment levels. The main implications are in the external market, where there is a generalized decrease of imports and exports when compared to the BAU scenario. The first case is due to the reduction of domestic consumption of refined oil products, representing 20.91% of the total imports. The NDC scenario reduces exports by US$1,393 million, the equivalent of 5.34%, a reduction concentrated in secondary sectors. Despite this reduction, the exports of oil products present an increase of 30% (US$230 million), mainly due to the surplus of fuel oil for export.

Table 2. Aggregate demand compared by scenario

Expenditures	BAU scenario (US$ Million)	NDC scenario (US$ Million)	Variation (BAU – NDC)
Household consumption	58,854	58,758	−0.16%
GFCF	23,652	24,490	3.54%
Government consumption	12,791	12,724	−0.52%
Exports	26,080	24,686	−5.34%
Imports	29,666	27,956	−5.76%
Stock variation	3,931	3,924	−0.17%

In the NDC scenario, the electricity exports would increase by 268% in comparison with the BAU scenario. In monetary terms, this increase is equal to US$ 121 million. Despite significant increases in electricity, the NDC and BAU scenarios have similar export structure results for 2025. In both cases, 70% of exports are concentrated in the crude oil, farming and food industry sectors. So NDC implementation would not change the economy's export structure.

As a result of lower oil reserves in both scenarios, there is a reduction in capital supply in the crude oil sector. This is more accentuated in the NDC scenario than the BAU scenario. A similar trend exists for the labor factor, with a reduction of 10.81% for the BAU scenario and 11.07% for the NDC scenario. There will be reduced employment in this sector regardless of NDC implementation. It is thus necessary to define public policies to mitigate this situation.

6 Conclusions

The Ecuadorian NDC initiatives in the energy sector are feasible and framed in the country's economic-productive structure. GHG reduction strategies through the use of less polluting energy sources do not imply a reduction in GDP. Instead, there is a slight increase (when compared to trend conditions), so a less polluting transition is feasible. The GHG reduction is concentrated in carbon dioxide emissions, accentuated since 2020 when new hydropower plants started operation.

The energy transformation sector is the main contributor to GHG reduction. The household sector is the main beneficiary of energy savings due to more energy efficient cooking. There are reductions in the transportation sector, but these are marginal when considering the energy consumption levels in this sector. The productive sectors do not have significant energy savings.

Ecuador would benefit from both economic growth and GHG reduction. In fact, the NDC scenario is less intensive than the BAU scenario. In other words, the economy would require less effort to generate wealth as result of mitigation policies. However, it must be recognized that there would be no change in the country's productive and consumption structures.

The macroeconomic impacts are concentrated in the external market. Electricity exports must guarantee the same income levels as crude oil exports. This will be a challenge, considering the physical limitations and low demand from bordering countries. Electricity surpluses must be guaranteed either for export or domestic consumption. Otherwise, NDC implementation will be counterproductive.

The hybrid LEAP/CGE model applied to Ecuador was successful. It was possible to link the technical model with the macroeconomic one and their convergence was achieved by an adequate iterative process, using soft-link modeling based on energy-economic ratio and input-output variable connections.

References

Benavides, C., et al.: The impact of a carbon tax on the chilean electricity generation sector. Energies. **8**(4), 2674–2700 (2015). https://doi.org/10.3390/en8042674

Benito, A., Castro Verdezoto, P.L., Rodríguez, P.: Prospectiva sobre la integración energética de sistemas de generación distribuida. Reciamuc. **7**(2), 375–389 (2023). https://doi.org/10.26820/reciamuc/7.(2).abril.2023.375-389

Benito, A.O., Castro Verdezoto, P.L., Burlot, A., Arena, A.P.: Hybrid power system for distributed energy deploying biogas from municipal solid waste and photovoltaic solar energy in Mendoza, Argentina. E3S Web Conf. **532** (2024). https://doi.org/10.1051/e3sconf/202453201001

Carvajal, P.E., Francis, L., Soria, R., Cronin, J., Anandarajah, G., Mulugetta, Y.: Large hydropower, decarbonisation and climate change uncertainty: modelling power sector pathways for Ecuador. Energ. Strat. Rev. **23**(2018), 86–99 (2019). https://doi.org/10.1016/j.esr.2018.12.008

Castro, P., Castro, M., Cunha, M.: Análisis Comparativo de Indicadores Energéticos de Países Miembros de la Comunidad Andina de Naciones. Revista Técnica "Energía," 14, 236–245 (2018). ISSN 1390–5074

Castro Verdezoto, P.L., Vidoza, J.A., Gallo, W.L.R.: Analysis and projection of energy consumption in Ecuador: energy efficiency policies in the transportation sector. Energy Policy. **134**(2018) (2019). https://doi.org/10.1016/j.enpol.2019.110948

Castro-Verdezoto, P.L., Da Cunha, M.P., García-Gutiérrez, Á., Montaño, W.Q.: Energy policy implications of Ecuador's NDC. J. Infrast. Policy. Dev. **8**(13), 1–19 (2024). https://doi.org/10.24294/jipd7542

Chiriboga, G., Chamba, R., Garcia, A., Heredia-Fonseca, R., Montero-Calderón, C., Carvajal, C.G.: Useful energy is a meaningful approach to building the decarbonization: A case of study of the Ecuadorian transport sector. Transp. Policy. **132**(2022), 76–87 (2023). https://doi.org/10.1016/j.tranpol.2022.12.019

Fragkos, P., et al.: Coupling national and global models to explore policy impacts of NDCs. Energy Policy. **118**, 462–473 (2018). https://doi.org/10.1016/j.enpol.2018.04.002

Gerber Machado, P., de Oliveira Ribeiro, C., Oller do Nascimento, C. A.: Risk analysis in energy projects using Bayesian networks: A systematic review. Energ. Strat. Rev. **47**(2022), 101097 (2023). https://doi.org/10.1016/j.esr.2023.101097

Hurtado, J., Castro, P.: Feasibility study of a small hydroelectric plant on a tributary of the Guayas River - Ecuador. Polo Del Conocimiento. **8**(12), 1226–1238 (2023). https://doi.org/10.23857/pc.v8i12

Lv, F., Wu, Q., Ren, H., Zhou, W., Li, Q.: On the design and analysis of long-term low-carbon roadmaps: A review and evaluation of available energy-economy-environment models. Renew. Sust. Energ. Rev. **189**(PA), 113899 (2024). https://doi.org/10.1016/j.rser.2023.113899

MAE: Primera contribución determinada a nivel nacional (2019).

MATTE: Quinta comunicación Nacional del Ecuador sobre Cambio Climático. In: Ministerio de Ambiente, Agua y Transición Ecológica (2024)

MERNNR: Plan Maestro de Electricidad. In: Ministerio de Energía y Recursos Naturales No Renovables (2023)

Ortega-Pacheco, D.V., Castro-Verdezoto, P.L., Mendoza-Jiménez, M.J., Almeida Benalcázar, E., Castro, M.-P.: Social and economic contribution of the bioeconomic sector in Ecuador. In: Venkatramanan, R., Shah, V., Prasad, S. (eds.) Sustainable Bioeconomy, pp. 35–65 (2021). https://doi.org/10.1007/978-981-15-7321-7_3

Perez, J., Díaz, G.: Energy transition on public transportation during COVID-19 in a large city: a study in Bogota. SSRN (2021). https://doi.org/10.2139/ssrn.4153362

Pérez-Gelves, J.J., Castro-Verdezoto, P.L., Alvarado-Cantos, N.M., César, T.A.: Applying fuzzy logic and neural networks to forecasting in efficiency programs. E3S Web Conf. **532** (2024). https://doi.org/10.1051/e3sconf/202453201006

Postic, S., Selosse, S., Maïzi, N.: Energy contribution to Latin American INDCs: Analyzing sub-regional trends with a TIMES model. Energy Policy. **101**(2016), 170–184 (2017). https://doi.org/10.1016/j.enpol.2016.11.023

Ramirez, A.D., Boero, A., Rivela, B., Melendres, A.M., Espinoza, S., Salas, D.A.: Life cycle methods to analyze the environmental sustainability of electricity generation in Ecuador: is decarbonization the right path? Renew. Sust. Energ. Rev. **134**, 110373 (2020). https://doi.org/10.1016/j.rser.2020.110373

Ribeiro, C.H., da Cunha, M.P.: The economic and environmental impacts of Brazilian National Biofuel Policy. Biofuels Bioprod. Biorefin. **16**(2), 413–434 (2022). https://doi.org/10.1002/bbb.2326

Rivera-González, L., Bolonio, D., Mazadiego, L.F., Valencia-Chapi, R.: Long-term electricity supply and demand forecast (2018-2040): A LEAP model application towards a sustainable power generation system in Ecuador. Sustainability (Switzerland). **11**(19) (2019). https://doi.org/10.3390/su11195316

Siriwardana, M., Nong, D.: Nationally determined contributions (NDCs) to decarbonise the world: A transitional impact evaluation. Energy Econ. **97**(2017), 105184 (2021). https://doi.org/10.1016/j.eneco.2021.105184

UNFCCC: Integrated Assessment Models. UNFCCC (2025). https://doi.org/10.1002/9781118786352.wbieg2062

Villamar, D., et al.: Long-term deep decarbonisation pathways for Ecuador: insights from an integrated assessment model. Energ. Strat. Rev. **35**(2020), 100637 (2021). https://doi.org/10.1016/j.esr.2021.100637

Smart Irrigation Strategy and Monitoring of Edaphological and Meteorological Variables Based on IoT

María Alejandra Chugá Rosero(✉) and Diana Belén Peralta Zurita

International SEK University, Quito, Ecuador
malejandrachuga@gmail.com, diana.peralta@uisek.edu.ec

Abstract. This project aims to design and implement an intelligent irrigation system using IoT to assess water stress and determine irrigation times for urban agriculture.

The prototype incorporates humidity and temperature sensors adapted to the application, providing data for a closed-loop system. A centralizing module collects this information and transmits it wirelessly to a cloud-based interface. Additionally, a water storage tank, monitored by two level sensors, ensures reserve management.

The collected data is visualized in a cloud application developed in ThingSpeak, allowing remote access for users to manually control the irrigation system. The platform also stores data for further crop behavior analysis.

The system's effectiveness was validated by comparing crops using intelligent irrigation versus traditional methods, demonstrating a 45% difference in radish weight and a 28% reduction in water usage, thus achieving greater productivity and efficient resource utilization.

Keywords: Internet of Things · MQTT protocol · cloud application

1 Introduction

In 2022, Ecuador's agricultural sector contributed 7.57% to the national GDP, making it the fourth most significant sector in the economy. Despite this importance, many agricultural practices still rely on traditional methods that lack efficiency and make minimal use of modern technologies [1, 2]. Key barriers to technological adoption include the high costs of systems designed for large-scale farming and a general lack of awareness about their benefits [3].

In contrast, developed countries have embraced the principles of the Fourth Industrial Revolution, integrating advanced technologies such as internet-connected systems for real-time data acquisition, analysis, and remote control of agricultural processes [4]. Ecuador, by comparison, still operates within the technological framework of the Second Industrial Revolution, with a lag of at least two decades. However, globally, the trend continues toward optimizing industrial processes through automation, smart systems, and efficient resource management.

J. L. Ballesteros et al. (Eds.): CITIS 2025, LNNS 1780, pp. 40–50, 2027.
https://doi.org/10.1007/978-3-032-14287-0_5

Industry 4.0 emphasizes real-time information gathering through sensors, centralized cloud databases, and accessible data visualization tools for end users [5]. These systems support informed decision-making that enhances productivity, reduces operational time, and promotes sustainable resource use [6]. Based on these principles, a solution using IoT is proposed for radish (Raphanus sativus) cultivation in Ecuador. This approach integrates dynamic soil data acquisition, intuitive data presentation, and storage, aiming to develop a smart irrigation system that improves production and resource efficiency [7].

2 Materials Y Methods

The project is carried out in four stages. The first involves selecting the crop type and its location. The second focuses on choosing temperature and humidity sensors, actuators to complete the control loop, and a central device to collect field data. The third stage includes selecting a cloud platform for visualizing data, sending control commands, and performing analyses. The fourth stage involves system testing.

2.1 Crop Selection

As a case study, a radish plantation has been chosen, belonging to the group of vegetables, selected for its short harvest time. Table 1 describes the main characteristics of this vegetable [8]. The crop spans approximately 2.25 m^2 and is located in the Pichincha province, Quito canton, in the Tumbaco parish, in the Sierra region of Ecuador. It is situated at an altitude of 2427 m above sea level, where the soil is of the "cangahua" type, consisting of approximately 30% sand, 55% silt, and 15% clay, thus considered a loamy (coarse) soil [9].

Table 1. Radish Characteristics

Crop type	Vegetable	Plant pH	5,5-6,8
Optimal Temperature	18 °C–22 °C	**Soil Moisture**	60% al 65%
Average Harvest Time	30 days	**Irrigation Type**	Drip irrigation
Soil Type	Any soil type		

2.2 Selection of Sensors, Actuators, Centralizing Device, and Data Acquisition Platform.

Main measurement variables are temperature and humidity. These variables allow the system to be activated or deactivated efficiently. For this reason, selecting the appropriate sensor for this application will be the starting point for the system design [10].

Table 2. Characteristics of Temperature and Humidity Sensors.

	SHT200	DHT11	DHT22	SHT10
Interface Type	I2C	Single-bus	Single-bus	Digital
Temperature Measurement Range	−40 °C a + 125 °C	0 a 50 °C	−40 °C a 80 °C	−10 °C a 80 °C
Humidity Measurement Range	0%–100%	0%–100%	0%–100%	0%–100%
Humidity Accuracy	±3%RH	±5%RH	±2%RH	±5%RH
Temperature Accuracy	±3 °C	±2 °C	<±0.5 °C	±0.5 °C

Table 2 details the most suitable temperature and humidity sensors for this type of application, along with their key characteristics [9–11].

The DHT22 sensor is chosen to measure air temperature and humidity based on its temperature measurement range and accuracy, providing reliable data for the process [11]. The SHT10 sensor is selected to measure soil temperature and humidity in the area where the crop will be located, primarily for its robustness characteristics for this type of application. The irrigation system interface will be a solenoid valve. Table 3 shows the main characteristics of the solenoid valve that will be used as an actuator. This component was selected because it meets the water supply needs for the selected.

Table 3. Characteristics of the Solenoid Valve

Power Supply Voltage	12Vdc	Operating Mode	Nornmally closed
Current Consumption	0,6A	**Pressure**	0,02 a 0.8 mPa
Power	8 W	**Input and Output**	½" Hose

The control system operates with an ESP32 microprocessor created by the company Espressif, focused on IoT applications, as it features a chip capable of connecting to Wi-Fi and Bluetooth. Additionally, it is characterized by having input and output ports that convert analog to digital and vice versa, 16 PWM output ports, and communication interfaces such as I2C, UART, and SPI [12].

A microcontroller-based system was designed to monitor and control key environmental variables in Raphanus sativus cultivation. Temperature and soil moisture sensors provide real-time data to the microcontroller, which triggers irrigation based on set thresholds. Data are transmitted via MQTT and stored on the ThingSpeak cloud platform, allowing remote monitoring and analysis through MATLAB scripts. This IoT-based system enhances precision agriculture by optimizing irrigation, conserving water, and improving crop management in both soil and controlled environments [13].

2.3 System Architecture

Figure 1 presents the intelligent irrigation system divided into four subsystems, described in detail below.

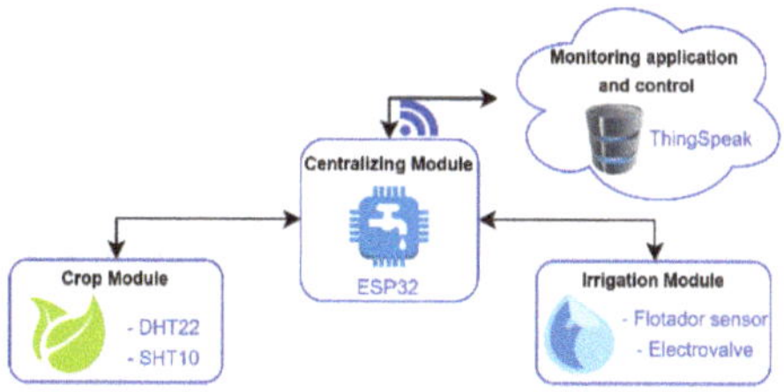

Fig. 1. Intelligent Irrigation System Diagram

Crop Module

The system consists of an ambient and soil temperature and humidity sensor. These sensors will transmit information to the centralizing module via wired communication, without the need for a conditioning circuit.

Irrigation Module

This module monitors the water level in the storage tank and controls the inlet and outlet valves using two level sensors and two solenoid valves connected to a central module. A storage tank with a water reserve was implemented to prevent issues from supply interruptions, using sensors for maximum and minimum levels.

The ZPC1 sensor, a magnetic type with a Reed Switch, provides ON/OFF signals through contact with a foam float [14]. A solenoid valve controls the tank filling, activating at low levels and deactivating at high levels. Irrigation is managed by another solenoid valve linked to a microcontroller, which activates based on soil humidity needs. A drip irrigation system is used over a 2.25 m^2 area with three 150 cm-long planting lines spaced 50 cm apart, using ½" PVC pipes. Drippers with a flow rate of up to 70 l/h are installed along the pipes, and a final solenoid valve regulates water flow to the distribution lines [15].

Centralizing Module

The developed module controls irrigation using temperature and humidity sensors. It operates autonomously by managing irrigation activation, water tank filling, and real-time communication with a cloud platform. Figure 2 shows the system's flow diagram for both operating modes.

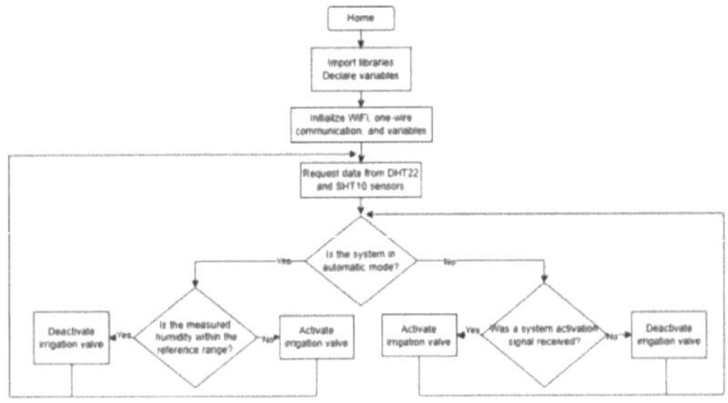

Fig. 2. Flow Diagram on the Main Program of the Irrigation System.

In automatic mode, it checks soil humidity hourly and irrigates when levels fall outside predefined thresholds, taking measurements every second during irrigation. In manual mode, the user remotely controls irrigation via a cloud-based interface using a simple on/off command.

The irrigation control for the crop was designed based on a closed-loop control system. This type of control system is characterized by having sensors that obtain the actual measurement of the parameters to be controlled and compare them with a reference value specific to the case study [16, 17]. Figure 3 shows this control system and the variables interacting within it.

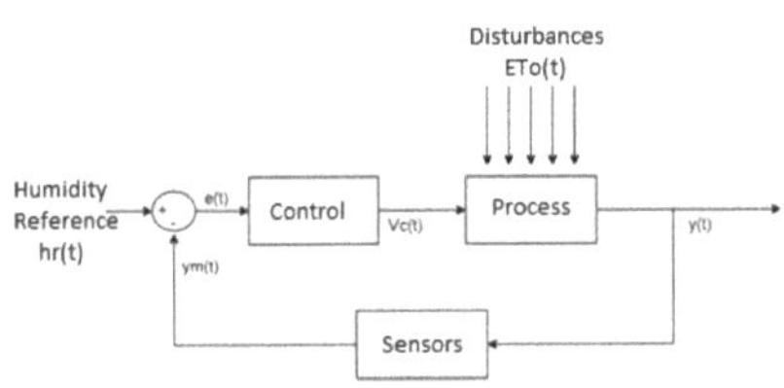

Fig. 3. Control Loop of the Irrigation System

Where: hr.(t) = Reference humidity of the crop; e(t) = Error signal of the system; Vc(t) = Control variable of the system; ETo(t) = System disturbance; y(t) = Output signal of the system; ym(t) = Measured output signal of the system.

Based on the control loop of the irrigation system, the error signal is the result of subtracting the reference humidity from the measured humidity output from the soil humidity sensor SHT10. This is represented by Equation (1).

$$e(t) = hr(t) - ym(t) \quad (1\$\$ (1) \$\$)$$

The disturbance in this system corresponds to evapotranspiration, as these are phenomena that directly affect the percentage of humidity in a given crop. This equation takes the environmental temperature value obtained from the DHT22 sensor as input.

2.4 User Application in the Cloud

The ESP32 microcontroller was configured with key data fields—such as temperature, humidity, irrigation, and tank levels—to enable real-time monitoring and remote control via the IoT platform, ensuring efficient visualization and environmental management. Table 4 details each of the graphs displayed in the monitoring application.

Table 4. Visualization Graphs in ThingSpeak.

ITEM	GRAPHIC	DESCRIPTION
1	Soil Temperature	A scatter plot showing the soil temperature values acquired every 10 min by the SHT10 sensor.
2	Soil Humidity	A needle-type visualizer indicating the soil humidity percentage acquired every 10 min by the SHT10 sensor.
3	Ambient Temperature	A scatter plot showing the ambient temperature values acquired every 10 min by the DHT22 sensor.
4	Ambient Humidity	A needle-type visualizer indicating the ambient humidity percentage acquired every 10 min by the DHT22 sensor.
5	Irrigation Running	A lamp-type visualizer that lights up indicating that the irrigation system is activated, and remains off when irrigation is not active.
6	Tank Filling	A lamp-type visualizer that lights up indicating the tank is in filling mode when the minimum water level is reached, and remains off when the tank is full or above the minimum water level.
7	Manual Activation/Deactivation	A point-type graphic that shows the status of the manual activation/deactivation option for irrigation. • 1 if irrigation is manually activated. • 0 if irrigation is deactivated.

In Fig. 4, the display screen of the application developed on ThingSpeak can be seen. This screen consists of different types of graphs depending on the variable being measured.

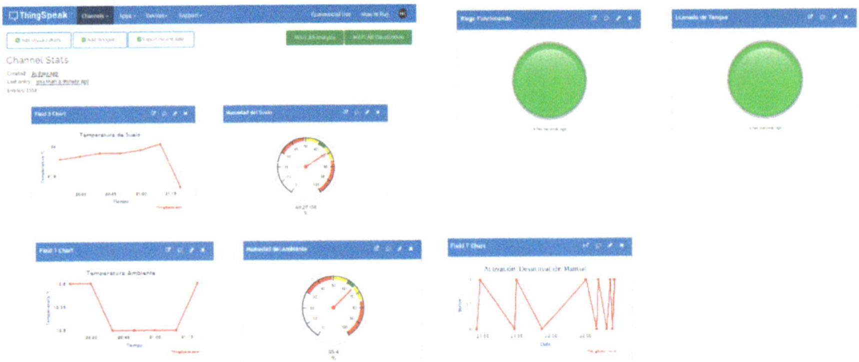

Fig. 4. Main screen of the application developed on ThingSpeak.

The manual activation and deactivation of the irrigation system is performed from the application using a URL generated by ThingSpeak. The field used is Field7, where the value 1 activates the irrigation and 0 deactivates it.

3 Results and Discussion

Figure 5 shows the complete system installed at the location selected for this project.

One of the objectives of the project is to develop a low-cost and highly reliable intelligent irrigation system for urban agriculture. The total installation cost was $189.90, excluding engineering costs.

Fig. 5. Prototype installation.

To evaluate the results, two areas were compared: one with traditional irrigation and the other with the intelligent irrigation system. Below are images showing the evolution of the crop over 5 weeks until harvest (Figs. 6, 7, 8).

Fig. 6. Evolution of the crop after 15 days.

Fig. 7. Evolution of the crop after 28 days.

Fig. 8. Evolution of the crop after 35 days.

Figure 9 shows visual and quantitative comparisons indicating better vegetable growth with the intelligent irrigation system.

Fig. 9. Radish sample for metric comparisons.

Five radishes from both the manual and automated groups were harvested simultaneously, confirming improved development in the IoT-controlled crop (Tables 5, 6).

Table 5. Dimensions and weight data of radish simples without irritation system.

	Radish #1	Radish #2	Radish #3	Radish #4	Radish #5
Weight (grams)	21.4	12.8	8.4	13.4	15.3
Length (cm)	5.4	4.9	3.7	4	4
Width (cm)	3.2	2.4	1.8	2.5	2.2

Table 6. Dimensions and weight data of radish simples with the irrigation system.

	Radish #1	Radish #2	Radish #3	Radish #4	Radish #5
Weight (grams)	32.1	31.6	23.4	22.7	21.6
Length (cm)	7.5	5.3	5.2	4.9	5.5
Width (cm)	3.2	3.7	4.1	3.6	2.6

Using IBM SPSS Statistics software, the mean, standard deviation, and Levene's Test were applied. The results are shown in Table 7.

Table 7. Results obtained in IBM SPSS Statistics.

	Irrigation System	Mean	Standard Deviation	Levene's Test for Equality of Variances
Radish Weight	**Without system**	14.26	4.72	0.436
	With system	26.28	5.13	
Radish Lenght	**Without system**	4.4	7.2	0.692
	With system	5.68	1.0	
Radish Width	**Without system**	2.42	0.5	0.674
	With system	3.44	0.6	

The results indicate differences of about 10 g in weight and 1.2 cm in length and width. Levene's Test, with a value below 0.5, confirms significant differences (Fig. 10), suggesting that the intelligent irrigation system affects radish weight compared to traditional irrigation. This aligns with findings by Zhang et al. (2024), who reported a 34.6% to 94.1% increase in radish yield under stable soil moisture. Although this study uses an IoT-based system, the observed 45% increase in weight supports that optimized irrigation significantly enhances crop growth [20].

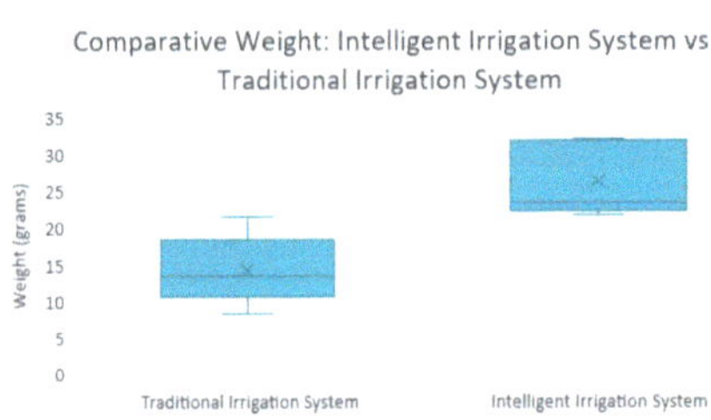

Fig. 10. Comparative weight.

The smart irrigation system reduced water consumption by 28.3% (27.6 L vs. 38.5 L in 35 days), demonstrating the efficiency of IoT in optimizing irrigation. This significant reduction conserves water resources and lowers environmental impact by minimizing runoff and soil degradation. Economically, it reduces costs related to water use and energy consumption for pumping. The system uses real-time monitoring and maintains soil moisture within optimal thresholds (30–50%), enhancing water use efficiency and supporting sustainable, precision urban agriculture.

4 Conclusions

The radish harvest results confirm the effectiveness of the IoT-based intelligent irrigation system, showing a 45.74% increase in weight, 22.54% in length, and 29.66% in width compared to manual irrigation. The system improves productivity and optimizes water and soil use.

Using readily available sensors, actuators, and affordable electronics enables reliable real-time soil data collection. This supports low-cost production analysis and makes the system scalable for urban agriculture and competitive with other solutions in resource efficiency.

The system uses MQTT and HTTPS protocols for secure data exchange and remote access, enabling climate and water use predictions that enhance agricultural productivity and efficiency.

However, the study has some limitations. It was conducted in a specific urban location (Quito, Ecuador), over a relatively short period (35 days), and with a limited sample size, which may restrict the generalizability of the results. External factors such as soil variability, sensor precision, and climate fluctuations could also have influenced the findings.

Future research should explore other crops and environments to assess the system's adaptability. Long-term studies could reveal its impact on soil health and water sustainability. Including more environmental variables like solar radiation, wind, and temperature and improving the user interface would enhance system precision and usability for small-scale urban farmers.

References

1. Zhao, M., Gomez-Rosero, S., Nouraei, H., Zych, C., Capretz, M.A., Sadhu, A.: Toward prediction of energy consumption peaks and timestamping in commercial supermarkets using deep learning. Energies. **17**(7), 1672 (2024)
2. Ipiales, O., Cuichán, M.: Survey of Agricultural Surface and Production (ESPAC). INEC, Ecuador (2024)
3. Handal, A.J., et al.: The study of environmental exposure of mothers and infants impacted by large-scale agriculture (SEMILLA): Description of the aims and methods of a community-based birth cohort study. Children. **11**(9), 1045 (2024)
4. Wang, L., Zhou, J., Wang, P., Pan, C., Li, H.: Green gentrification and urban parks: exploring the complex dynamics of commercial transformation. J. Knowl. Econ., 1–33 (2024)
5. Sang, T.M., Anh, N.Q.: Adopting digital technology for sustainable performance: circular economy practice in the transportation sector. Discov. Sustain. **6**(1), 166 (2025)
6. Garcia-Cedeno, A., et al.: PLATANO: Intelligent technological support platform for azuay province farmers in Ecuador. In: 2019 IEEE International Conference on Engineering Veracruz, ICEV 2019 Artículo 8920501 (2019 IEEE International Conference on Engineering

Veracruz, ICEV 2019). Institute of Electrical and Electronics Engineers Inc. (2019). https://doi.org/10.1109/ICEV.2019.8920501
7. Ahmed, H.A., Wang, Z., Li, Y.: Life cycle sustainability assessment and optimization of seismic retrofit solutions for RC frame structures. Case Stud. Constr. Mater., e04315 (2025)
8. Bustamante-V, D.F., Ramirez, M.E., Peralta-Zurita, D.B., Llanes-Cedeño, E.A., Molina-Osejos, J.V.: Innovation in figure skating through technological transformation: IoT exploration for real-time refereeing. In: International Conference on Computer Science, Electronics and Industrial Engineering (CSEI), pp. 546–563. Springer Nature Switzerland, Cham (2023)
9. Zakaria, Z.: Meeting the sustainable development goals in a tropical climate. J. Chin. Chem. Soc. **67**(12), 2241–2245 (2020)
10. Palacios Orejuela, I., Ushiña, D., Carrera Villacrés, D.: GIS Techniques to Cangahuosos Soil Recovery, for Community Agriculture Purposes in the Ilaló (2020)
11. Moya, V., Quito, A., Pilco, A., Vásconez, J.P., Vargas, C.: Crop detection and maturity classification using a yolov5-based image analysis. Emerg. Sci. J. **8**(2), 496–512 (2024)
12. Mobaraki, B., Komarizadehasl, S., Pascual, F.J.C., Galant, J.A.L.: Open source platforms for monitoring thermal parameters of structures. In: Bridge Maintenance, Safety, Management, Life-Cycle Sustainability and Innovations, pp. 3892–3896. CRC Press (2021)
13. Al-Obaidi, M.A.M., Radhi, M.A.H., Ibrahim, R.S., Sutikno, T.: Technique smart control soil moisture system to watering plant based on IoT with arduino uno. Bull. Electr. Eng. Inform. **9**(5), 2038–2044 (2020)
14. Guachimboza-Davalos, J.I., Llanes-Cedeño, E.A., Rubio-Aguiar, R.J., Peralta-Zurita, D.B., Núñez-Barrionuevo, O.F.: Prediction of monthly electricity consumption by cantons in Ecuador through neural networks: a case study. In: Systems and Information Sciences: Proceedings of ICCIS 2020, pp. 28–37. Springer International Publishing (2021)
15. Sangprasert, N., Inthavisas, K.: A smart marine debris management system for a sustainable coastal city: an IoT-based application. Cleaner Waste Syst., 100262 (2025)
16. Nadakuditi, G., Venakteshwarlu, S., Kumar, S., Pulluri, H., Preeti, P., Goud, B.S.: Thingspeak cloud server based IoT equipped firefighting robot. In: E3S Web of Conferences, vol. 616, p. 03022. EDP Sciences (2025)
17. Hoglund, S.R., Rathke, S.J., Fidel, R.B., Blankinship, J.C.: Contrasting effects of biochar application rate in an alkaline desert cropland soil. J. Arid Environ. **215**, 105011 (2023)
18. Regalado-Pezua, O., Christofle, S.: Future research opportunities on sustainable smart cities: bibliometric analysis and network visualization. Discov. Sustain. **6**(1), 187 (2025)
19. Li, X., et al.: Unraveling long-term water consumption changes and agro-ecological responses to agricultural practices in arid irrigation districts of the upper Yellow River basin (2000–2021). J. Hydrol., 133222 (2025)
20. Quispe-Vilca, J.L., Moreno-Cardenas, E., Sacoto-Cabrera, E.J., Moreno-Cardenas, Y.: Integrating IoT and image processing for crop monitoring: a LoRa-based solution for citrus Pest detection. Electronics. **13**(24), 4863 (2024)
21. Zhang, Y., Wei, Y., Zhang, Q., Liu, M., Shi, X.: Effects of stable and fluctuating soil water on the agronomic and physiological characteristics of cherry radish (Raphanus sativus L.). Front. Plant Sci. **15**, 1325078 (2024). https://doi.org/10.3389/fpls.2024.1325078

Bio-Based Approach for Paraquat Removal: Silver Nanoparticle-Functionalized Cellulose Extracted from Yerba Mate Waste

Cintya Gordon-Falconí[1](✉), Alejandra Gómez Cuartas[2], María Florencia Iannone[2], and María Daniela Groppa[2]

[1] Pontificia Universidad Católica de Ecuador Sede Ibarra, Escuela de Ciencias Agrícolas y Ambientales (ECAA), Av. Jorge Guzmán Rueda y Av, Aurelio Espinosa, Ibarra, Ecuador
cmgordon@pucesi.edu.ec

[2] Facultad de Farmacia y Bioquímica, Universidad de Buenos Aires, Consejo Nacional de Investigaciones Científicas y Técnicas, Instituto de Química y Fisicoquímica Biológicas (IQUIFIB), Junín 956 (1113), Buenos Aires, Argentina
{mfiannone,mgroppa}@ffyb.uba.ar

Abstract. Cellulose obtained from natural sources can serve as versatile and cost-effective adsorbents. The functionalization of cellulose fibers with nanoparticles (NPs) has gained significant attention in recent years because it promotes the production of polymers with unique properties and a wide range of applications. In this work, a sequential method (CNsec) for obtaining a cellulose-silver nanoparticle composite was compared with an *in situ* (CNins) NPs synthesis method to select the most suitable cellulose-AgNP composite for evaluating the removal of the herbicide 1,1′-dimethyl-4,4′-dipyridyl dichloride (Paraquat) from water. The NPs obtained in the cellulose fibers were monodisperse in both procedures (PDI:0.012–0.013). XRD patterns showed no significant differences between the two processes, and the crystallinity index (~60%) revealed no significant changes between the purified cellulose fibers (CeB) and the treated samples, suggesting that the inclusion of NPs did not alter the crystalline and/or amorphous domains of the cellulose fibers. The main difference was that CNins method achieved a more uniform and homogeneous NPs distribution on the fiber surface (observed via SEM) and a higher silver content. Therefore, CNins fibers were chosen for paraquat removal study compared with non-functionalized cellulose fibers. It was demonstrated that longer contact time did not improve the paraquat adsorption capacity (Q_e: 9 mg/g), suggesting that the adsorption equilibrium was reached up to 30 min. Additionally, the initial herbicide concentration also influenced the removal percentage, likely due to active sites saturation on the polymer. Further research is needed to optimize synthesis methods, enhance scalability, and evaluate the long-term environmental impacts of these functionalized composites.

Keywords: Bioadsorbent · Cellulose · AgNPs

J. L. Ballesteros et al. (Eds.): CITIS 2025, LNNS 1780, pp. 51–60, 2027.
https://doi.org/10.1007/978-3-032-14287-0_6

1 Introduction

Cellulose from natural sources, including plant residues, can be versatile and cost-effective adsorbents. Physicochemical modifications of cellulose can modify its mechanical and physicochemical properties. These modifications may change their ability to adsorb pollutants [1, 2]. The functionalization of cellulose fibers with nanoparticles (NPs) has gained significant attention because it promotes the production of polymers with unique properties and a wide range of applications in environmental remediation, biomedicine, and material science [2–4]. Several studies have shown that incorporating nanoparticles—such as silver (AgNPs), gold (AuNPs), titanium dioxide (TiO_2), and iron oxide (Fe_3O_4)—can significantly enhance the performance of cellulose-based materials [5]. For example, the functionalization with AgNPs has been widely investigated due to their antimicrobial and antifungal properties, making cellulose-AgNP composites well-suited for medical applications and water purification [1, 6]. These modifications can improve characteristics such as contaminant adsorption, mechanical strength, thermal stability, and reusability [7].

Agrochemicals are widely found in water systems due to their widespread use with concomitant risks to ecosystems and human health. Paraquat, 1,1′-dimethyl-4,4′-dipyridyl dichloride, a highly toxic non-selective herbicide, is utilized in over 100 countries for crop management and weed control. In Ecuador, the use of this herbicide has raised considerable concerns regarding its high toxicity and persistence in agricultural soils. Its improper handling and application have resulted in soil degradation, water contamination, detrimental effects on biodiversity, and severe health problems [8]. This study compares two methods for obtaining a cellulose-AgNPs composite and evaluates its use to remove paraquat from water.

2 Methods

2.1 Chemicals and Reagents

The herbicide 1,1′-dimethyl-4,4′-dipyridyl dichloride (Paraquat HD) was procured from Syngenta Agro S.A. under the trade name Cerillo. Sodium hypochlorite (reagent grade), and sodium hydroxide (reagent grade) were acquired from Fisher Scientific. Milli-Q water was used in all experiments.

2.2 Extraction Process of Cellulose from Yerba Mate Wastes and Characterization of the Cellulose Obtained

The extraction of cellulose fibers (CeB) from the yerba mate waste and their characterization was conducted based on the methodologies described in [1].

2.3 Obtaining a Cellulosic-Silver Nanoparticles Composite

The incorporation of AgNPs into previously purified and moisture-free cellulose was performed using two different procedures (sequential and *in situ*). For the sequential procedure (hereafter, CNsec), cellulose fibers (0.1 g) were immersed in 10 mL of a solution of previously synthesized nanoparticles [9]. The fibers were kept under magnetic stirring for 60 min and then filtered and washed until the supernatant was found to be free from nanoparticles (verified by UV-Vis spectroscopy). Finally, the fibers were dried in the oven (50 ± 5 °C) until they reached a constant weight. The *in situ* procedure (hereafter, CNins) was performed by immersing 0.02 g of cellulose fibers in a solution of silver nitrate (1 mM $AgNO_3$). Subsequently, the synthesis of AgNPs was performed following the procedure in [9]. The mixture was kept under stirring for 30 min and then subjected to 2 sonication cycles in a sonicator bath (Ultrasonic, Cole-Parmer.Mexico) for 30 min (each cycle). Finally, the samples were filtered, washed, and dried to constant weight (50 ± 5 °C).

The formation and stability of the nanoparticles in the presence of the cellulose fibers were analyzed. Silver reduction was studied by UV-Vis spectroscopy. The samples were diluted in a 1/10 ratio with Mili-Q water and the absorbance was analyzed over a wavelength range of 200 to 700 nm. The hydrodynamic size distribution was analyzed by DLS under the following analysis parameters: temperature 25 °C, viscosity of the medium 0.9190 mPa/s, refractive index of the medium 1.333, refractive index of the silver 0.280.

The degree of crystallinity of the cellulose fibers was analyzed through X-ray diffraction patterns. In addition, the presence of characteristic silver peaks was evaluated, and the crystallinity index (CI) was calculated. The morphology and surface of the cellulose fibers coated with AgNPs were studied through micrographs acquired by scanning electron microscopy (SEM). The chemical composition was analyzed using the energy-dispersive X-ray spectrum detector included in the microscope, and ten spots on the sample were analyzed for 60 s each.

2.4 Capacity of Paraquat (PQ) Removal

The adsorption capacity of cellulose fibers (CeB and CNins) was studied using the herbicide PQ at 10 and 20 mg/L prepared with deionized water type I (electrical conductivity 0.56 μS/cm). The adsorption process was studied by batch tests with 0.4 mg cellulose per mL of PQ solution. Constant orbital agitation at 150 rpm, dark conditions, and room temperature (20 ± 5 °C) were applied throughout the process. PQ concentration changes in the solution were monitored via the reduction in PQ absorbance at $\lambda max = 260$ nm, measured by UV-Vis spectrophotometry. In addition, pH was monitored at the beginning and end of each test. For the design of experiments, the methodologies detailed in [10, 11] were considered. The removal efficiency (%) and equilibrium adsorption capacity (Qe) were determined according to the following equations:

$$\text{Efficiency}\ (\%) = ((\text{Co} - \text{Ce})/\text{Co}) * 100$$

$$\text{Qe} = ((\text{Co} - \text{Ce})/\text{P}) * \text{V}$$

Where: Co and Ce are the initial and equilibrium concentrations of the pollutant compounds (mg/L), respectively. The volume of the solution is V (L), and P is the weight of the adsorbent material used (g).

2.5 Statistics

Statistical evaluations were conducted using one-way ANOVA, with significance defined at $p < 0.05$. Triplicate samples were processed for all determinations. Results were presented as means ± standard deviations (SD).

3 Results

3.1 Obtaining Cellulose-Silver Nanoparticles Polymer

In the CNsec procedure, the incorporation of previously synthesized AgNPs was observed by the change in color of the cellulose from white to gray (Fig. 1A-C). In the CNins process, AgNPs were synthesized in the presence of cellulose [1]. In both cases, an absorption peak at 414 nm (Fig. 1D) was observed.

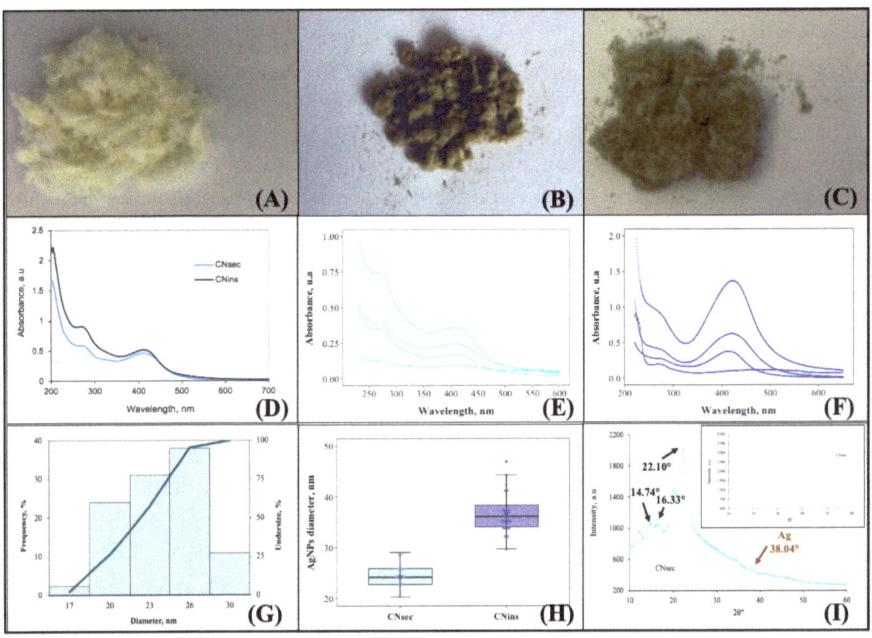

Fig. 1. CeB: previously purified and moisture-free cellulose extracted from the yerba mate waste (A). CNsec: cellulose coated with AgNPs obtained by sequential treatment (B). CNins: cellulose with AgNPs obtained by *in situ* synthesis (C). UV-Vis absorption spectra of the supernatant from CNsec and CNins processes (D). Evaluation of desorption processes of AgNPs on cellulose fibers in CNsec (E) and CNins (F) treatments. Histogram of AgNPs diameter distribution in CNsec according to DLS characterization (G). Comparison of the hydrodynamic diameter of AgNPs according to the treatment applied (H). X-ray diffraction spectra of CNsec. The insert illustrates the CNins treatment, as described in a previous work (I) [1].

The possible desorption processes of the AgNPs were evaluated by UV-Vis spectrophotometry. During the washing process, the absorbance in the supernatant was measured. In both CNsec and CNins, a reduction in the intensity of the absorption peak at 414 nm was observed in the washing solution (Fig. 1E-F) up to 4 washing cycles. The hydrodynamic diameter and distribution profiles of the silver nanoparticles in solution were studied by DLS. In CNsec, the diameter was 24.32 ± 2.7 nm (Fig. 1G) significantly lower than that observed in CNins [1] (Fig. 1H). The polydispersity index (PDI) for CNsec and CNins was 0.012 and 0.013, respectively.

The XRD patterns of the CNsec fibers showed four peaks at 2θ° at: 14.74°, 16.33°, 22.10° and 38.04° (Fig. 1I). The crystallinity index of CNins and CNsec did not differ significantly from each other (Table 1).

Table 1. Effect of AgNPs addition on the crystallinity structure of cellulose derived from yerba mate waste

	Crystallinity index CI, %
CeB	60.43 ± 2.03
CNsec	57.55 ± 2.82
CNins	60.00 ± 6.08

The surface, the morphology of the cellulose fibers, and the presence of nanoparticles were examined using scanning electron microscopy (SEM) micrographs (Fig. 2A). Fig. 2A shows CNsec fibers that incorporated nanoparticles by immersion in a solution of previously synthesized AgNPs. The irregular morphology of the cellulose fibers and the characteristic brightness of the silver atoms deposited on the surface of the cellulose fibers are observed. Chemical mapping suggests that the distribution of the NPs on the fibers is non-uniform, heterogeneous, and aggregates are observed. The elemental composition of the cellulose fibers was performed by energy dispersive spectroscopy (EDS). For CNsec fibers, a spectral signal at approximately 3 keV was observed (Fig. 2B), corresponding to the silver (Ag) region.

3.2 Paraquat Remotion by CNins Fibers

Paraquat removal was studied only with CNins fibers as these fibers showed a more homogeneous distribution of nanoparticles and a higher content of Ag. The adsorption capacity of paraquat on cellulose fibers (CeB and CNins) was evaluated (Fig. 3). The maximum percentage adsorption achieved by CeB was 32.18 ± 0.80% after 60 min of contact, while the adsorption achieved by CNins under the same conditions was 21.34 ± 0.23%. Only 30 min of contact was sufficient to achieve maximal adsorptions (Q_e) of 9.26 mg/g and 7.35 mg/g with CeB and CNins, respectively.

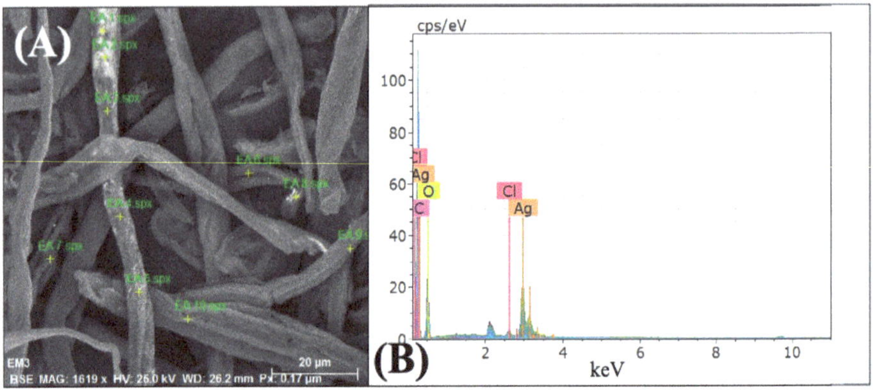

Fig. 2. SEM image of CNsec (A) and EDS spectrum of the corresponding elemental composition of CNsec fibers (B).

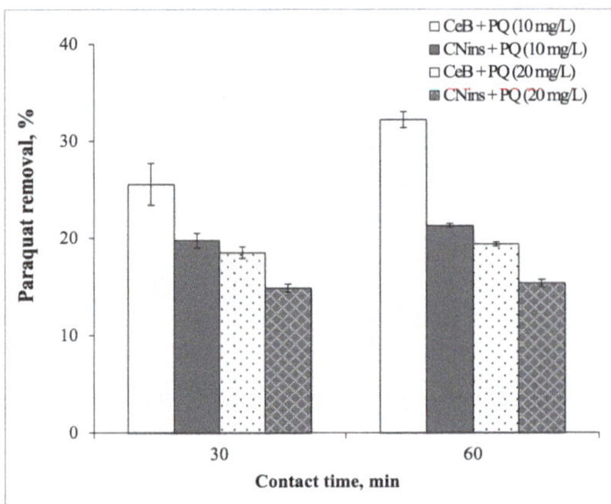

Fig. 3. Effect of the contact time and the initial concentration of paraquat removal efficiency using CeB and CNins as absorbent materials

4 Discussion

In this work, a sequential method for obtaining a cellulose-silver nanoparticle composite was compared with an *in situ* NPs synthesis method. In CNsec a color change similar to that observed in the *in situ* synthesis method was noted [1]. Similar results were obtained by [10], who observed greenish and dark shades on the fibers after the process of nanomaterial incorporation. This darkening of the cellulose fibers, together with the appearance of a peak at 414 nm, characteristic of the surface plasmon resonance of AgNPs [12], confirms the presence of Ag nanoparticles in the fibers.

With both synthesis methods, the release of nanoparticles from the composite is observed up to the fourth washing cycle. Previous research indicates that the release of silver ions occurs through diffusion phenomena, and the process depends on several factors, such as the chemical nature, morphology, and concentration of silver present in the polymer [13, 14]. The amount of silver released in products functionalized with AgNPs depends on the incorporation method applied and can be in ionic form (Ag^+) or as nano-sized particles [15].

In CNsec, the nanoparticles retained the same diameter as that obtained during their NPs synthesis process [9], which means that no modifications occurred during the composite formation process. This diameter was significantly smaller than that of CNins (36.94 ± 4.3 nm) [1], probably because the presence of cellulose fibers during the *in situ* NPs synthesis method could modify the NPs growth process. The *in situ* reduction process takes advantage of the physicochemical characteristics of cellulose. Ag^+ ions have the possibility to enter the active sites of the fiber, thanks to their small ionic radius (~0.1 nm), and during the synthesis process, the polymer can provide support for the slow growth of AgNPs as they would be wrapped in the skeleton of the cellulosic material [16]. The polydispersity index (PDI) established that the nanoparticles obtained in both procedures performed are monodisperse (PDI < 0.05) according to the classification described by [17] and accompanying references.

Both composites showed similar peaks in the XRD analysis [1]. The first three peaks observed are attributed to the (110), (110), and (200) crystalline planes, characteristic of type I cellulose [18]. The peak identified at approximately 38° is characteristic of silver and can be attributed to the crystalline plane (111) that ensures the presence of face-centered cubic (FCC) silver nanocrystals [11] (ICSD, https://icsd.fiz-karlsruhe.de), this peak). Similar results were obtained by Deeksha et al. (2021), who developed fabrics with antibacterial properties based on cellulose and AgNPs [19]. Although the peak intensity was low in both composites, the higher intensity observed in CNins suggests a higher content of nanoparticles in the cellulose fibers. This is explained by the fact that during the *in situ* reduction procedure, the physicochemical characteristics of cellulose are exploited, and Ag^+ ions can enter the interior of the fiber, increasing the presence of nanoparticles after the synthesis process, whereas, in the sequential treatment, the natural adsorption properties predominate [16]. On the other hand, no significant changes in the crystallinity index were observed between the purified cellulose fibers (CeB) and the treatments applied for nanoparticle incorporation. Therefore, it can be assumed that the inclusion of nanoparticles does not alter the crystalline and/or amorphous domains of the cellulose fibers.

SEM micrographs (Fig. 2A) show that some cellulose fibers (CNsec) incorporated nanoparticles by immersion in a solution of previously synthesized AgNPs, but several did not. The irregular morphology of the cellulose fibers and the characteristic brightness of the silver atoms deposited on the surface of the cellulose fibers are observed. Chemical mapping (Fig. 2B) suggests that the distribution of nanoparticles on the fibers is non-uniform, heterogeneous, and aggregates are observed. These results are similar to the data reported by Ali et al. (2018) [20] and opposite to those observed in CNins fibers [1], in which the NPs were uniformly and homogeneously distributed over the surface of the fibers. The presence of Ag was confirmed by EDS analysis. Since AgNPs were more homogeneously distributed in the CNins fibers than in CNsec, the CNins fibers were used for the paraquat herbicide removal study compared to cellulose fibers without NPs.

The initial concentration of herbicide influenced the percentage of herbicide removal, as was also observed with cationic dyes [1], probably due to the saturation of the active sites of the polymer since, at the molecular level, electrostatic interactions likely dominate the adsorption process, supported by hydrogen bonding between paraquat molecule and cellulose hydroxyl groups, along with contributions from van der Waals forces—particularly in less oxidized or unmodified cellulose [21]. In addition, longer contact time does not improve paraquat adsorption capacity, which would indicate that the adsorption equilibrium is reached up to 30 min of contact of the herbicide with the adsorbent (Q_e: 9 mg/g). Although the adsorption capacity of this system exceeded that of others, such as ground plant ash (Q_e: 1.86 mg/g) or magnetically modified 4-sulfonatocalix [n] arenes nanomaterial (Q_e: 0.8 mg/g), it was inferior to that of many others such as carbonated jujube seeds (Q_e: 425 mg/g) or modified rice husk (Q_e: 17 mg/g). A range of advanced adsorption technologies have been explored for paraquat removal, using materials such as silicates, activated carbon, and functionalized biomaterials with Q_{max} between 0.8 and 425 mg/g [22]. Efficient paraquat removal (up to 100 mg/g) has been achieved through the use of TEMPO-Oxidized Cellulose Nanofibers (TOCNs) due to the increased presence of carboxylate groups and the reduced fiber diameter [23].

Functionalization of cellulose fibers can improve the mechanical and physicochemical properties of cellulose. The incorporation of AgNPs into cellulose fibers increased the tensile strength and enhanced the antibacterial properties of the adsorbent [1]. However, the incorporation of NPs in the cellulose fibers decreased the herbicide adsorption (Q_e: 7 mg/g). Similar results were previously observed with the cationic dyes [1], probably because AgNPs could be occupying active adsorption sites in the cellulose fibers.

5 Conclusions

In this work, we compare two methods for obtaining a cellulose-AgNPs composite and its ability to adsorb the herbicide paraquat. Both processes of obtaining the cellulose-AgNPs composite (*in situ*: AgNPs are synthesized in the presence of the cellulose fibers or sequential by contact of cellulose fibers with previously synthesized AgNPs) allowed obtaining cellulose fibers functionalized with AgNPs. However, the *in situ* method achieved a more uniform and homogeneous distribution of the NPs on the fiber surface. Although the presence of NPs improved the mechanical and bactericidal properties of the fibers, the adsorption capacity of the fibers decreased. While cellulose fibers alone achieved approximately 30% of paraquat remotion, CNins achieved 15–20% of removal depending on the initial concentration of herbicide. After 30 min, saturation of the system was already observed, and no further removal was possible at longer contact times.

Cellulose–AgNP composites suggest promising applications in water treatment, especially in agricultural regions affected by paraquat contamination. These materials can be adapted into filtration systems, adsorbent cartridges, or immobilized hydrogels. However, to be viable at scale, further optimization of the synthesis method is necessary to enhance sorption efficiency without compromising other functional properties. In addition, the use of cellulose derived from agricultural waste and the potential for green synthesis of AgNPs offer environmentally responsible routes for large-scale production. However, assessing the long-term environmental impacts of these functionalized composites is also needed.

References

1. Gordon-Falconí, C., Iannone, M.F., Zawoznik, M., Debut, A., Groppa, M.D.: Discarded yerba mate as a source of cellulose fibers with promising applications for drinking water decontamination. Ind. Crop. Prod. **211**(118253), 1–11 (2024)
2. Ramesh, S., Radhakrishnan, P.: Cellulose nanoparticles from agro-industrial waste for the development of active packaging. Appl. Surf. Sci. **484**, 1274–1281 (2019)
3. Salama, A., Abouzeid, R., Owda, M., Cruz-Maya, I., Guarino, V.: Cellulose–silver composites materials: preparation and applications. Biomolecules **1684**(11), 1–29 (2021)
4. Bader, Q., et al.: Cellulose nanomaterials in oil and gas industry and bio-manufacture: current situation and future outlook. Case Stud. Chem. Environ. Eng. **10**(100993), 1–29 (2024)
5. Carazo, E., et al.: Advanced inorganic Nanosystems for skin drug delivery. Chem. Rec. **18**(7–8), 891–899 (2018)
6. Naseem, T., Waseem, M.: A comprehensive review on the role of some important nanocomposites for antimicrobial and wastewater applications. Int. J. Environ. Sci. Technol. **19**(3), 2221–2246 (2021)
7. Fouda-Mbanga, B.G., Onotu, O.P., Tywabi-Ngeva, Z.: Advantages of the reuse of spent adsorbents and potential applications in environmental remediation: a review. Green Anal. Chem. **11**(100156), 1–19 (2024)
8. Huang, Y., Zhan, H., Bhatt, P., Chen, S.: Paraquat degradation from contaminated environments: current achievements and perspectives. Front. Microbiol. **10**(460070), 1–10 (2019)
9. Gordon-Falconí, C., Iannone, M.F., Zawoznik, M.S., Cumbal, L., Debut, A., Groppa, M.D.: Synthesis of silver nanoparticles with remediative potential using discarded yerba mate: an eco-friendly approach. J. Environ. Chem. Eng. **8**(6), 104425 (2020)
10. Ahmed, H., Emam, H., Mashaly, H., Rehan, M.: Nanosilver leverage on reactive dyeing of cellulose fibers: color shading, color fastness and biocidal potentials. Carbohydr. Polym. **186**, 310–320 (2018)
11. Biliuta, G., Coseri, S.: Cellulose: a ubiquitous platform for ecofriendly metal nanoparticles preparation. Coord. Chem. Rev. **383**, 155–173 (2019)
12. Thakur, P., Verma, V.: A review on green synthesis, characterization and anticancer application of metallic nanoparticles. Appl. Biochem. Biotechnol. **193**(7), 2357–2378 (2021)
13. Maneerung, T., Tokura, S., Rujiravanit, R.: Impregnation of silver nanoparticles into bacterial cellulose for antimicrobial wound dressing. Carbohydr. Polym. **72**(1), 43–51 (2008)
14. Pinto, R., Marques, P., Neto, C., Trindade, T., Daina, S., Sadocco, P.: Antibacterial activity of nanocomposites of silver and bacterial or vegetable cellulosic fibers. Acta Biomater **5**(6), 2279–2289 (2009)
15. McGillicuddy, E., et al.: Silver nanoparticles in the environment: sources, detection, and ecotoxicology. Sci. Total Environ. **575**, 231–246 (2017)
16. Xu, Y., Li, S., Yue, X., Lu, W.: Review of silver nanoparticles (AgNPs)-cellulose antibacterial composites. Bioresources **13**(1), 2150–2170 (2018)
17. Mudalige, T., Qu, H., Van Haute, D., Ansar, S.M., Paredes, A., Ingle, T.: Characterization of nanomaterials: tools and challenges. Nanomaterials Food Appl., 313–353 (2019)
18. Park, J.M.: Paper and paperboard packaging. Handbook Food Sci. Technol. Eng. **4**, 178–239 (2012)
19. Deeksha, B., Sadanand, V., Hariram, N., Rajulu, A.: Preparation and properties of cellulose nanocomposite fabrics with in situ generated silver nanoparticles by bioreduction method. J. Bioresour. Bioprod. **6**(1), 75–81 (2021)
20. Ali, A., Mannan, A., Hussain, I., Hussain, I., Zia, M.: Effective removal of metal ions from aqueous solution by silver and zinc nanoparticles functionalized cellulose: isotherm, kinetics and statistical supposition of process. Environ. Nanotechnol. Monit. Manag. **9**, 1–11 (2018)

21. Zhang, Q., Zhu, E., Li, T., Zhang, L., Wang, Z.: High-value utilization of cellulose: intriguing and important effects of hydrogen bonding interactions – a mini-review. Biomacromolecules **25**(10), 6296–6318 (2024)
22. Franco, D., Georgin, J., Lima, E., Silva, L.: Advances made in removing paraquat herbicide by adsorption technology: a review. J. Water Process Eng. **49**, 1–14 (2022)
23. Huang, C., Tu, C., Lee, R., Yang, C., Hung, W., Lin, K.: Study of various diameter and functionality of TEMPO-oxidized cellulose nanofibers on paraquat adsorptions. Polym. Degrad. Stab. **161**, 206–212 (2019)

Structural Improvement of Adobe Units Using Cactus Plant Liquid Mucilage

Sleyther Arturo De La Cruz Vega(✉)

Universidad Cesar Vallejo, Trujillo, Peru
sdelacruzv@ucv.edu.pe

Abstract. The objective of this research was to evaluate the incorporation of cac- tus mucilage as a natural additive for the structural improvement of adobe units. An applied methodology was used, with a quantitative approach and experimental design. The study population consisted of all the adobe units produced in the study, with a sample of 60 units. The mechanical and durability properties of the adobes with different proportions of mucilage (25%, 50%, and 75%) were evaluated, comparing them with a control group without the additive. The results indicated that the addition of 75% mucilage was the most effective, achieving a 53.16% increase in unitary compressive strength and a 28.46% increase in axial compressive strength compared to the control group. A 5.68% reduction in water absorption was also observed, demonstrating an improvement in the material's durability under humid conditions. These improvements were progressive as the mucilage concentration increased. In conclusion, the incorporation of cactus mu- cilage constitutes a sustainable and efficient alternative for optimizing the structural properties of adobe.

Keywords: Adobe · Cactus plant mucilage · Sustainable construction

1 Introduction

Earth has been used as a building material since ancient times, forming part of structures ranging from houses to monuments in many parts of the world [1]. The availability and low cost of this material have contributed to its continued use for centuries. Today, the search for building materials that do not harm the planet is not just a fad, but a real necessity. We need to protect the remaining resources, build without destruction, and especially ensure that people live in better conditions [2].

Currently, almost a third of the world's population lives in earth buildings [3]. In Asia, Africa, and South America, earth construction is still widely used, especially in rural areas. The UN affirms the need to promote appropriate technologies with existing resources, as established by the development goals approved by this organization [4].

In South America, earth construction is still prevalent, both due to cultural tradition and economic necessity. In countries such as Bolivia, Ecuador, Colombia, and Peru, the tradition of adobe construction—a simple material made of earth and water, molded by hand and left to dry in the sun—is still alive [5]. However, it is mostly still made exactly

J. L. Ballesteros et al. (Eds.): CITIS 2025, LNNS 1780, pp. 61–71, 2027.
https://doi.org/10.1007/978-3-032-14287-0_7

the same as centuries ago, without any modern improvements or techniques. Therefore, adobe has virtually no durability or structural safety.

In the particular case of Peru, adobe continues to be the preferred material for self-construction of rural homes, especially in the Andean region. According to [6], more than 50% of homes in the high Andean regions use adobe as the base material. This value is due to its low cost, ease of manufacture, and good thermal and inertial properties. However, its performance in earthquakes or extreme weather conditions remains poor. In this context, the cohesion of the soil component becomes one of the properties often not guaranteed by traditional mixtures [7].

Furthermore, earth constructions carry a social stigma, as they are seen as a symbol of poverty or hardship due to their poor performance and resistance to natural phenomena [8]. This image has hindered both their technical evolution and their integration into contemporary urban development strategies.

To improve the mechanical behavior of adobe, a common practice is to include natural fibers such as straw, which increase the material's tensile strength and cohesion [9]. Recently, other sustainable alternatives have been explored, such as the use of natural additives, including plant mucilage, which aim to increase durability, cohesion, and impermeability [10].

Plant mucilage, an extract from plants such as cacti, nopal, and cactus pad, is very rich in polysaccharides that act as natural binders [11]. The action of plant mucilage in a clay mixture improves adhesion between particles, reduces absorption, and forms a kind of plug to seal the voids between the pieces, giving greater strength to the adobe structure [12]. Plant mucilage's have recently begun to be applied in raw earth, lime, and cement mortars in both new construction and heritage restoration projects [13].

For example, the work by [14] shows that linseed mucilage improves the compressive strength of adobe and makes it more impermeable, even at higher doses than traditional stabilizers such as residual oils. On the other hand, it has been shown that cactus mucilage counteracts surface erosion by creating a denser and more stable matrix [15]. Successful applications have been made in the conservation of the original pigments of archaeological artifacts, which also confirms its compatibility with porous materials [16].

In light of this evidence, this experimental study evaluates the incorporation of liquid cactus mucilage in the production of adobe units. Throughout its development, the structural properties of the material will be evaluated, specifically its unitary compressive strength, the strength of masonry blocks, and water absorption. This proposal seeks to provide a natural, economical, and sustainable solution that improves adobe performance while maintaining and improving its qualities, promoting its safe use in rural and urban environments.

The structural improvement of adobe units with liquid plant mucilage extracted from cactus constitutes a natural and sustainable alternative to address the challenges of strength and durability of earth constructions using adobe in the rural mountains [17]. It is relevant in the face of problems such as climate change, structural weakness, and the lack of decent housing. It seeks practical, low-environment alternatives, promotes the use of local resources, and revalues ancestral knowledge combined with sustainable civil engineering technology.

From a social perspective, it is proposed as an option to improve the quality of life of this vulnerable group living in adobe houses exposed to seismic risks. By improving the resistance of adobe with cactus mucilage, safer and more comfortable homes are built. This promotes safer self-construction and the use of accessible materials that can benefit the autonomy of rural communities, reducing dependence on industry, and fostering sustainability [18].

Theoretically, it is based on the study of the physicochemical properties of plant mucilage and its applications in improving this construction material. It provides empirical evidence based on how mucilage components act as natural binders, improving the cohesion and strength of adobe, but also encompasses contemporary approaches to bio additives applied to materials engineering that contribute to expanding knowledge about appropriate technologies for this type of construction [19].

Methodologically, this experimental research presents a rigorous and reproducible design to evaluate the effect of cactus mucilage on the properties of adobe. Laboratory tests are conducted to assess both the properties of cactus mucilage and adobe, such as compressive strength and water absorption, in accordance with relevant technical standards. This methodological guideline also provides useful data for technicians, researchers, and those interested in the search for sustainable construction alternatives for rural housing [20].

From an environmental perspective, this research promotes the use of natural and biodegradable materials, promoting a reduced use of polluting industrial inputs. Cactus mucilage, an abundant resource in arid and semi-arid environments, is a sustainable alternative compared to other traditional synthetic additives, as it does not typically have a particularly high carbon footprint. The proposal also contributes to the principles of green architecture, adapting to climate mitigation strategies and more environmentally friendly construction practices [21].

2 Methodology

2.1 Type of Research

This research is applied, as it aims to obtain scientific results with practical applications [22], stimulating innovation and linking academic knowledge to the real needs of society [23]. In this sense, the aim is to improve the structural properties of adobe by using liquid cactus mucilage, thus proposing a technical and sustainable solution to real-life construction problems, especially in rural areas where this traditional building material is used.

2.2 Research Design

Is experimental, as in this type of research, researchers manipulate one variable to observe its effect on another, following orderly and established protocols [24]. In this case, the addition of cactus liquid mucilage is manipulated to observe the effects on the physicochemical properties of adobe, allowing for the analysis of its performance compared to traditional adobe manufacturing methods.

2.3 Research Approach

This research is quantitative, as it quantifies the data obtained [25] and is based on the collection and analysis of numerical data to address the research questions posed [26]. Standardized measurements were applied to evaluate both the physicochemical properties of cactus liquid mucilage and its effect on the physicochemical properties of adobe units in terms of compressive strength and water absorption, allowing for replicable results that can be statistically verified.

2.4 Population and Sample

The research population was determined by the total set of adobe units made with different proportions of cactus liquid mucilage. The sample consisted of 60 adobe units, divided into a control group and three experimental groups, each with a different percentage of mucilage added.

2.5 Sampling and Analysis Phase

To obtain the liquid cactus mucilage, we collected cactus leaves near Putina-Cucho. For its preparation, the cacti were cut into 5 cm x 5 cm pieces and left to cool for 48 h. They were then ground for five minutes until a homogeneous substance was obtained.

The adobe units were handcrafted in 40 cm x 20 cm x 10 cm molds. From these, four test groups were formed: one of them without mucilage (100% H2O) and three exper- imental groups with different proportions of mucilage (75% H2O + 25% LCM; 50% H2O + 50% LCM; 25% H2O + 75% LCM), so that each group was made up of three adobe samples, which allowed the comparative analysis of the effect of the natural additive that was being tested.

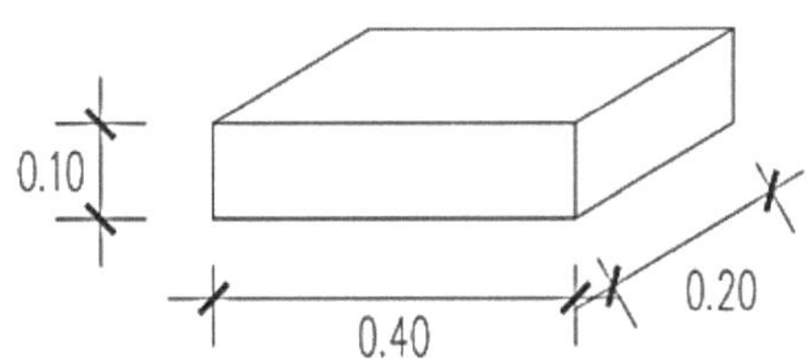

Fig. 1. Adobe unit dimensions.

In turn, tests were carried out in the laboratory to evaluate the properties of both the mucilage and the adobe itself. The mucilage was subjected to analysis using X-ray fluorescence spectrometry, followed by pH, viscosity, and density tests. The adobes were tested for compressive strength and water absorption, all in accordance with the requirements of technical standards such as NTP 399.613 and ASTM C67–14 [27]. Thanks to these tests, it was possible to specify how the material reacts to mechanical stresses and how it interacts with moisture, crucial data for understanding its structural performance.

3 Results

3.1 Soil Used for Making Adobe Bricks

Table 1. Characteristics of the soil used to make adobe bricks

Soil Classification	Soil Sample
SUCS	SM
AASHTO	A-4(1)
% Gravel	2,70
% Sand	57,60
% Passing N.° 200	39,70
Liquid Limit (%)	21,60
Plastic Limit (%)	18,90
Plasticity Index (%)	2,70

The soil used to make adobe bricks is primarily sandy in composition, but with a significant number of fine particles. According to the SUCS system, it is classified as a mixture of sand and silt. However, the AASHTO system considers it suitable for construction due to its properties: its low plasticity and liquid limit facilitate mixing and molding, allowing for good workability without risk of instability. These characteristics make it particularly suitable for the production of earth adobe bricks.

3.2 Cactus Plant Liquid Mucilage

The analysis carried out using X-ray fluorescence spectrometry, a technique that, as explained by [28], is used to identify elements by exciting atoms with X-rays and then detecting their characteristic emissions, gave us the mineral composition of cactus mucilage. The results, which are quantified in the table below, show the chemical components of the mucilage and give us an idea of the role they could play in the properties of adobe.

Analysis of cactus mucilage using X-ray fluorescence spectrometry revealed a water content of 96.43%, demonstrating its primarily aqueous nature. When examining its minerals, potassium emerged as the most abundant component, followed by calcium and magnesium, found primarily in the form of oxides. These compounds play an important role in optimizing essential properties of soil substrates. In addition, trace elements such as phosphorus, sulfur, chlorine, and certain metals were identified, opening intriguing avenues for studying how this additive might function in traditional construction methods, particularly in the production of adobe bricks.

The liquid extracted from the cactus exhibited a thick consistency, suggesting a significant ability to bind and retain water, characteristics beneficial for strengthening the bond in compounds such as clay. Its density, similar to that of water, simplifies its

Table 2. Elemental composition of the liquid plant mucilage of cactus obtained by X-ray fluorescence spectrometry assay [29].

Element	Results (%)	Oxides	Results (%)
Moisture	96.4336	Moisture	96.4336
Potassium (K)	1.781	PotassiumOxide (K_2O)	2.6042
Calcium (Ca)	0.819	Magnesium Oxide (MgO)	0.6014
Magnesium (Mg)	0.5913	CalciumOxide (CaO)	0.311
Chlorine (Cl)	0.2883	Phosphorus Pent- oxide (P_2O_5)	0.0372
Phosphorus (P)	0.0402		
Sulfur (S)	0.0333	Chlorine (Cl)	0.004
Iron (Fe)	0.005	Iron Oxide (Fe_2O_3)	0.0033
Zinc (Zn)	0.0034	SulfurTrioxide (SO_3)	0.0024
Manganese (Mn)	0.0031	Manganese Oxide (MnO)	0.0017
Copper (Cu)	0.0019	Zinc Oxide (ZnO)	0.0008

Table 3. Properties of liquid cactus mucilage.

Property	Results
Viscosity	1087.9 cSt
Density	0.9948 g/cm3
Potential of Hydrogen (pH)	4.61

use and addition to fluid preparations. Additionally, its mild acidity could help maintain the material's health against microorganisms, thus promoting its useful life. In short, these chemical and physical attributes reaffirm its value as a natural component in environmentally friendly building applications.

3.3 Compressive Strength of Adobe Unit

The compressive strength results of the adobe units show that the water-only group obtained an average strength of 7.90 kg/cm^2, and the group with 25% MLC and 75% H2O achieved a higher average strength, reaching 12.10 kg/cm^2, which means an increase of 53.16% compared to the group without mucilage. On the other hand, the mixture with 50% H2O and 50% MLC showed an average strength of 11.60 kg/cm^2, suggesting that the mucilage does help improve the compressive strength of the adobes, but its effectiveness varies depending on the proportion used.

3.4 Axial Compressive Strength of Stacked Adobe Units

Table 4. Compressive strength results of the adobe unit.

ID	Adobe	Proportions		
	100% H2O	75%H2O + 25% LCM	50%H2O + 50 %LCM	25%H2O + 75% LCM
Specimen 1	8.38	10.7	11.5	11.64
Specimen 2	7.64	10.7	11.5	12.66
Specimen 3	7.67	10.6	11.8	11.88
Average	7.9	10.7	11.6	12.10

Table 5. Results of the axial compressive strength of stacked adobe units.

ID	Adobe	Proportions		
	100% H2O	75%H2O + 25% LCM	50%H2O + 50 %LCM	25%H2O + 75% LCM
Specimen 1	2.65	2.79	2.97	3.35
Specimen 2	2.17	2.67	2.95	3.06
Specimen 3	2.57	2.87	2.98	3.07
Average	2.46	2.78	2.97	3.16

The axial compressive strength results of the stacked adobe units show that the 100% H2O group presented an average strength of 2.46 kg/cm^2 and the 25% MLC and 75% H2O group showed the highest average strength, reaching a magnitude of 3.16 kg/cm^2, implying an increase of 28.46% compared to the group that did not contain mucilage. In addition, the other dosages also showed improvements, although to a lesser extent, confirming the positive effect of mucilage in improving the mechanical behavior of adobes when supporting axial loads.

3.5 Water Absorption of Adobe Unit

Water absorption results in adobe units show that the group with 100% H_2O showed an average absorption of 29.9%, while the group with 25% MLC and 75% H_2O presented the lowest value, with 28.2%, indicating a reduction of 5.68%, something positive since lower absorption improves the durability of the material, while the other dosages with mucilage showed a slight increase in absorption, which is not desirable. These results indicate that a greater amount of mucilage can contribute positively to the decrease in water absorption in adobes, thus improving their response to humidity.

Table 6. Water absorption results of adobe units.

ID	Adobe	Proportions		
	100% H2O	75%H2O + 25% LCM	50%H2O + 50 %LCM	25%H2O + 75% LCM
Specimen 1	30.2	30	30.8	29.3
Specimen 2	30.4	31	29.3	28.7
Specimen 3	29.2	30.1	30.9	26.7
Average	29.9	30.4	30.3	28.2

4 Discussion

Regarding the results obtained for the compressive strength of the adobe units, the values obtained in the laboratory tests coincide with those indicated by [30], who increased the dry compressive strength by 61.50%, while in this research, an increase of 53.16% was obtained. For [31], the addition of liquid stabilizers improves the packing of soil particles, as well as reduces porosity and achieves the formation of a more cohesive structure, which allows achieving greater compressive strength. This type of stabilizer, such as plant mucilage, could be a valid alternative for improving the structural behavior of housing in rural and urban contexts, stating that plant-based stabilizers help reduce the erosion of earth units.

Along the same lines, the compressive strength results for the stacked adobe units show an increase compared to the standard group, as reported by [32], who obtained an increase of 185.05% by adding 60% liquid cactus mucilage, while in this study, an increase of 28.46% was obtained. In this regard, [33] considers it important to continue research into natural stabilizing materials such as cactus, due to their low environmental impact.

Finally, regarding the water absorption of the adobe units, it was determined that the application of 75% liquid cactus mucilage reduced absorption to 28.2% (−5.68%). These results are not consistent with those indicated by [2], who report an average water absorption percentage below 17% with only 10% of Opuntia ficus mucilage, this difference being attributable to the creation of a tacit film that occurs due to the contact of the mucilage with the components of the adobe. This is also inconsistent with what was described by [34], who mention that the latex when mixed with the clay would allow to form a film that allows to reduce the micro holes of the adobe, improving its impermeability, which causes a significant decrease in water absorption.

5 Conclusions

The properties of conventional adobe, without the addition of cactus mucilage, showed a unitary compressive strength of 7.90 kg/cm^2, an axial compressive strength of 2.96 kg/cm^2, and a water absorption of 29.90%. These values will serve as the comparative reference base for evaluating the effect of the natural additive.

The addition of liquid cactus mucilage at a ratio of 75% was the most successful, improving the compressive strength of the adobe units by 53.16% compared to the control group, demonstrating its potential as an alternative natural additive for sustainable construction with low environmental impact. Regarding the axial compressive strength of the stacked adobe blocks, the 75% mucilage ratio was found to be ideal, with an increase of 28.46% compared to the control group, while the 25% and 50% ratios achieved smaller increases of 13.01% and 20.73%, respectively, thus demonstrating a gradual increase with the concentration used and controlled in the test.

Regarding water absorption, the 75% mucilage dosage contributed to a 5.68% decrease compared to standard adobe, while the 25% and 50% mixtures showed reductions of 1.67% and 1.34%, respectively, confirming that this mucilage improves the adobe's durability under persistent humidity conditions.

References

1. Viñuales, G.: Tecnología y construcción con tierra. Apuntes: Revista De Estudios Sobre Patrimonio Cultural **20**(2) (2007) https://revistas.javeriana.edu.co/index.php/revApunte-sArq/article/view/8978
2. Guillen, J., García, E., Ortiz, N., Escudero, R., Neftalí, M.: Study of the properties of the Echerhirhu-block made with Opuntia ficus mucilage for use in the construction industry. Case Stud. Constr. Mater. **10**, 1–7 (2019). https://doi.org/10.1016/j.cscm.2019.e00216
3. Rotondaro, R., Herr, C., Correia, M., Guerrero, L., Viñuales, G.: Tierra y cultura. An. Inst. Arte Am. Investig. Estét. 48(1), 9–11 (2018). https://www.scielo.org.ar/sci-elo.php?script=sci_arttext&pid=S2362-20242018000100002
4. Orozco, I.: From Business Ethics to Sustainability. Why should companies care? The Anáhuac Journal 20(1), 76–105 (2020). https://doi.org/10.36105/theanahuac-jour.2020v20n1.03
5. Ureña, M., Llumitasig, S., Siza, A., Coello, D.: Study of compressive strength in artisanal adobe stabilized with cow manure, nopal sap, straw and bull blood. Rev. Téc. Fac. Ing. Univ. Zulia. **42**(3), 112–117 (2019). https://doi.org/10.22209/rt.v42n3a02
6. Bailón, I., Espinosa, I., Acevedo, B.: Soil block stabilized in small format and technology of placement in work, in housing of social interest. Ing. Desarro. **37**(1), 105–127 (2019). https://doi.org/10.14482/inde.37.1.624
7. Quagliarini, E., Lenci, S.: The influence of natural stabilizers and natural fibres on the mechanical properties of ancient Roman adobe bricks. J. Cult. Herit. **11**(3), 309–314 (2010). https://doi.org/10.1016/j.culher.2009.11.012
8. Egenti, C., Khatib, J.: Sustainability of Construction Materials. Universidad de Wolverhampton, Wolverhampton (2016)
9. Costa, C., Cerqueira, Á., Rocha, F., Velosa, A.: The sustainability of adobe construction: past to future. Int. J. Archit. Herit. **13**(5), 639–647 (2019). https://doi.org/10.1080/15583058.2018.1459954
10. Rojas, M., Aquino, E.: Sustainable adobe bricks with construction wastes. Proc. Inst. Civ. Eng.: Waste Resour. Manag. **169**(4), 158–165 (2016)
11. Aranda, Y., Suárez, E.: Cactus stalk waterproof effect in compressed earth blocks. Nova Scientia **6**(11), 311–323 (2013). https://doi.org/10.21640/ns.v6i11.85
12. Barbeta, G., Janer, F., Berthelsen, E.: Estabilización hidrofugante para revocos de tierra con extractos naturales. In: Congresos de arquitectura de tierra en Cuenca de Campos 2014, pp. 277–286 (2015)

13. Velasco, A., Espuna, J., Perez, J., Zuñiga, C., Palacio, A., Suarez, E.: Compressed earth block reinforced with coconut fibers and stabilized with aloe vera and lime. J. Eng. Des. Technol. **19**(3), 1–13 (2020). https://doi.org/10.1108/JEDT-02-2020-0055
14. Fabri, A., Anjo, M., Gomes, R.: Construcción con tierra: adobe con aceites vegetales añadidos. Rev. Nac. Gerenc. Ciud. **7**(53), 12–29 (2019)
15. Kita, Y., Daneels, A., De Vivar, R.: A.: Estudio químico para la identificación del aglu- tinante en muestras arquitectónicas prehispánicas. Mater. Univ. Realid. Locales **13**, 1–12 (2013)
16. Olivero, R., Aguas, Y., Mercado, I., Casas, D., Montes, L.: Utilización de Tuna (Opuntia ficus-indica) como coagulante natural en la clarificación de aguas crudas. Av. Investig. Ing. **11**(1), 70–75 (2014). https://doi.org/10.18041/1794-4953/avances.1.302
17. Velarde, O.: Mejoramiento de la resistencia del adobe estabilizado con mucílago de cactus para efectos de lluvias, Lambayeque, 2021. Universidad Nacional José Faustino Sánchez Carrión (2022). http://hdl.handle.net/20.500.14067/5920
18. Cabalé, E., Rodríguez, G.: Sustainable development in construction activity. Rev. Estud. Desarro. Soc.: Cuba Am. Lat. **5**(2), 40–51 (2017) http://scielo.sld.cu/scielo.php?script=sci_arttext&pid=S2308-01322017000200005
19. Selvakumar, P., Kumarasamy, V.: Investigation on microstructural and compressive strength characteristics of a novel bio-admixture comprising cassava starch-xanthan gum with cement mortar for building application. Matéria (Rio J) **29**, e20230331 (2024)
20. Miranda, J., Narváez, L., Moreno, J.: Initial assessment of the properties of Nopal gum as a possible additive in the conservation of adobe buildings. Intervención (Méx. DF). **13**(25), 161–199 (2022). https://doi.org/10.30763/intervencion.264.v1n25.43.2022
21. González, A.: Programs of environmental education on the climatic change in the formal education and non-formal. Rev. Univ. Soc. 8(3), 99–107 (2016). http://scielo.sld.cu/sci-elo.php?script=sci_arttext&pid=S2218–36202016000300013
22. Quintero, A.: ¿Es necesario hacer ciencia básica? Rev. Científica **17**(5), 431–432 (2007) http://ve.scielo.org/scielo.php?script=sci_arttext&pid=S0798-22592007000500001
23. Mayz, J., Pérez, J.: ¿Para qué hacer investigación científica en las universidades venezolanas? Investig. Postgr. 17(1), 159–171 (2002). http://ve.scielo.org/scielo.php?script=sci_arttext&pid=S1316-00872002000100007
24. Fainete, S.: Los enfoques de investigación en las Ciencias Sociales. Rev. Latinoam. Ogmios **3**(8), 82–95 (2023). https://doi.org/10.53595/rlo.v3.i8.084
25. Huanca, J., Portal, J.: Análisis de contenido cuantitativo sobre gestión del conocimiento en instituciones de educación superior latinoamericanas. Actual. Investig. Educ. **23**(1), 312–343 (2023). https://doi.org/10.15517/aie.v23i1.51513
26. Vega, G., Ávila, J., Vega, A., Camacho, N., Becerril, A., Leo, G.: Paradigmas en la investigación. enfoque cuantitativo y cualitativos. Eur. Sci. J. **10**(15) (2014)
27. ASTM International: Standard Test Methods for Sampling and Testing Brick and Structural Clay Tile [ASTM C67–14] (2014)
28. Torres, D., Marcó, L., Gómez, C., García, Y.: Total X-ray fluorescence (TXRF) as an alternative method for the determination of micronutrients in soils of the Quíbor depression (Venezuela). TecnoLógicas **25**(53), e203 (2022). https://doi.org/10.22430/22565337.2195
29. De la Cruz, S.: Evaluación de las propiedades físico-mecánicas del adobe adicionando mucílago de cactus (Opuntia spp.), Huaura, Lima, 2022. Universidad Nacional del Santa (2023). https://hdl.handle.net/20.500.14278/4374
30. Aranda, Y., Roux, R.: Durability comparison between terracotta and poured earth. Earth Constr. **1**(5), 115–122 (2020)
31. Corrêa, A., Mendes, L., Barbosa, N., Protásio, T., Campos, N., Tonoli, G.: Incorporation of bamboo particles and "synthetic termite saliva" in adobes. Constr. Build. Mater. **98**, 250–256 (2015). https://doi.org/10.1680/jwarm.16.00014

32. Vilcas, C.: Determination of the physical and mechanical properties of compressed earth blocks with the addition of cactus mucilage in the city of Huancayo, 2019. Universidad Continental (2019). https://hdl.handle.net/20.500.12394/8433
33. Mattone, R.: Sisal fibre reinforced soil with cement or cactus pulp in bahareque technique. Cem. Concr. Compos. **27**(5), 611–616 (2005). https://doi.org/10.1016/j.cemcon-comp.2004.09.016
34. Chandra, N., Satish, P., Ram, V.: Performance of natural rubber latex admixed concrete. Int. J. Curr. Eng. Sci. Res. **25**(5), 10–18 (2018). https://doi.org/10.21276/ijcesr

Fatty Acid Composition, Total Phenolic Content, and Antioxidant Activity of *Cayaponia Capitata* Seed Oil

Rafael Viteri[1](✉), Lisbeth Encalada[1], Gabriel Tello-Hidalgo[2], and Xavier Cornejo[3]

[1] Centro de Investigaciones Biotecnológicas del Ecuador, ESPOL Polytechnic University, Escuela Superior Politécnica del Litoral, ESPOL, Guayaquil, Ecuador
raviteri@espol.edu.ec

[2] Facultad de Ciencias Agrícolas, Escuela de Ingeniería Agronómica, Universidad Central del Ecuador, Quito, Ecuador

[3] Herbario GUAY, Departamento de Botánica, Facultad de Ciencias Naturales, Universidad de Guayaquil, Av. Raúl Gómez Lince s.n. y Av. Juan Tanca Marengo (campus Mapasingue), P.O. Box 09-01-10634, Guayaquil, Ecuador

Abstract. This study investigated the chemical composition, fatty acid profile, total phenolic content, and antioxidant capacity of *Cayaponia capitata* seed oil, a poorly studied species of the Cucurbitaceae family. The seeds were subjected to three extraction methods: maceration, ultrasound, and Soxhlet, and evaluated using 1,1-diphenyl-2-picrylhydrazyl (DPPH), 2,2′-azinobis-(3-ethylbenzothiazoline-6-sulfonic acid) diammonium salt (ABTS), and the ferric reducing antioxidant power assay (FRAP) antioxidant assays. The results showed that Soxhlet extraction produced the highest oil yield (62.82%) and the highest phenolic compound content. However, extracts obtained by maceration and ultrasound exhibited better antioxidant capacity in some assays. The lipid profile revealed a high content of unsaturated fatty acids (>81%), with a predominance of oleic, linoleic, and α-linolenic acids, recognized for their nutritional and functional benefits. This is the first comprehensive characterization of *C. capitata* seed oil, highlighting its potential for applications in the nutraceutical, pharmaceutical, and cosmetic industries.

Keywords: *Cayaponia capitata* · Fatty acid composition · Seed oil

1 Introduction

Vegetable oils are essential components of the diet and are used for various commercial purposes. The crops that focus on oil production are palm and soybean. Oil palm is the most significant crop, contributing 40% of the world's total oil production, followed by soybean with 28% [1]. These crops, although highly productive, are associated with challenging environments such as deforestation and biodiversity loss in tropical regions. In this context, the exploration of new alternatives for vegetable oil production is necessary, with species from the Cucurbitaceae family having shown potential for higher oil yields [2].

J. L. Ballesteros et al. (Eds.): CITIS 2025, LNNS 1780, pp. 72–81, 2027.
https://doi.org/10.1007/978-3-032-14287-0_8

Cucurbitaceae is a large family of plants with 120 genera and 965 species. Many of these species have been consumed all around the world and have been used in traditional medicine as a remedy in Chinese and Ayurvedic system [3]. Additionally, the Cucurbitaceae family has an important economical role because of the vegetables and fruits that can be used not only as food, but also in the cosmetic industry. Also, fruits have phenolics compounds, carotenoids are for that reason have excellent antioxidant activity [4]. Whereas seeds belonging to the Cucurbitaceae family are known for their high oil content, are emerging as a promising source of oil production.

The oil seed content of *Cucumis melo* ranges from 28.44 to 49.4% of dry weight. For the *Cucurbita pepo* pumpkin seeds contain between 29.33 to 51.01% dry weight. While the species *Curcurbita moschata*, the oil seed content is 44.42% of the dry weight [2]. These studies show that the seed contains good amounts of oil that can be exploited. *Cayaponia capitata* seeds are distributed in tropical and subtropical regions, such as Amazon rainforest in Ecuador and Peru. The seed is known as cunian peanut by indigenous Shuar communities, and it has utilized as necklace because of the almond shape seeds [5]. However, the extraction of oil from seeds *C. capitata* is an underexploited source, with limited investigation of their fatty acid composition, antioxidant properties and bioactive compounds.

Therefore, the objective of this study was to evaluate the total phenolic content and antioxidant activity of *Cayaponia capitata* seed oil using DPPH, ABTS, and FRAP assays, comparing the effectiveness of three extraction methods: maceration, Soxhlet, and ultrasound. Additionally, this work represents the first report on the fatty acid composition of *C. capitata* seed oil, determined by gas chromatography (GC). The findings aim to support the potential application of this oil in the nutraceutical, pharmaceutical, and food industries.

2 Materials and Methods

2.1 Plant Collection

The *C. capitata* seeds were collected in Morona Santiago Province, Ecuador (Santa María de Tunants, 2°08'S, 78°04'W, 1274 m). The seed *C. capitata* was dried at 60 °C, until we gained constant weight. Dried seeds were grounded in a steel knife mill. Then, the material was stored in a desiccator at room temperature until further use.

2.2 Extraction

The extraction of oil seed was performed by maceration (MC), ultrasound (US) and Soxhlet (SX) technique followed with protocol described by Ghareeb et al., 2018 [6] with slight modifications. 150 mL of hexane over 2 g raw material through ultrasound, maceration and soxhlet extraction with a duration of 30 min, 24 h, and 6 h, respectively. The assays were carried out in triplicate. The solvent was evaporated in rotary evaporator (Heidolph, Laborota 4010, USA) at 50 °C. The seed oil was stored refrigerated at 4 °C until further analysis. Oil yield value was calculated according to the formula (1).

$$\text{Oil yield } (\%) = \frac{m\ (oil\ extracted)}{m(C.\ capitata\ seed)}\ x\ 100\% \tag{1}$$

2.3 Extraction of Compounds Phenolics of the Oil Seed

The extraction of polyphenols was performed treated with methanol and hexane as described by Chanioti et al., [7], with slight modifications. 1 g of oil was dissolved in 1 mL hexane. The extracts were agitated with 2 mL of 80% methanol. The final solution was centrifuged at 4000 rpm for 20 min, then the hexane layer was extracted. The methanolic extracts were blended and evaporated in rotary evaporator (Heidolph, Laborota 4010, USA) to remove solvent. The extract obtained was dissolved with 0.5 mL methanol. The preparation of the phenolics extracts were carried out three times and each extract were analyzed for their properties antioxidants.

2.4 Determination of Total Phenolic Content (TPC)

Total phenolic content was determined by the Folin-Ciocalteu method, as described by Ghareeb et al., 2018 [6]. The results were expressed as a mg gallic acid equivalent (GAE) per gram of oil.

2.5 Antioxidant Assays

DPPH Radical Scavenging Assay
Antioxidant activity was determined by the protocol described by Thaipong et al., 2006 [8], with slight modifications. The absorbance was measured at 517 nm in a microplate reader (CytationTM 5, BioTek). The results were expressed in mg equivalent to Trolox/g of seed oil.

ABTS
ABTS scavenging capacities of seed oils were determined by the method described by Van Den Berg et al., 1999 [9]. Trolox was used as a standard control and the curve calibration was in the range (5–200 μM). The absorbance was measured at 532 nma microplate reader (CytationTM 5, BioTek). The results were expressed in mg equivalent to Trolox/g per 100 g of oil seed.

Frap
Ferric reducing antioxidant power (FRAP) of each sample was determined according to the procedure described by Thaipong et al., 2006 [8] which had been slightly modified. FRAP reagent was generated by mixing 300 Mm acetate buffer (pH 3.6), 10 mM 2,4,6 tripyridyl-s-triazine (TPTZ) in 40 mM HCl, and 20 mM $FeCl_3$ in the proportion of 10:1:1 (v/v). A standard curve was prepared with ferrous sulfate heptahydrate $FeSO_4 \cdot 7H_2O$ in the range 0.05–2 mM. Then, 20 μL of oil samples were mixed with 180 μL of FRAP reagent in 96-well microplate and incubated for 30 min at room temperature in dark. The absorbances at 595 were measured in a microplate reader (CytationTM 5, BioTek). The results were expressed in mmol Fe/100 g of oil seed.

2.6 Determination of Chemical Composition of *C. Capitata* Seed Oil Using Fatty Acid Methyl Ester/ Gas Chromatographic–Mass Spectrometric (FAME/GC-MS) Method

The methods and conditions of FAME preparation were realized according to the protocol described by [10] with slight modifications. 40 μL of seed oil was placed into 15 mL, then 0.7 mL of potassium hydroxide [10 M] solution and 5.3 mL of methanol were added. The mixture was performed at 55 °C for 1.5 h, with mixing 5 s for 20 min. After cooling to room temperature for 5 min, 0.58 mL of sulfuric acid [10 M] solution was added and the reaction was at 55 °C for 1.5 h, with mixing 5 s for 20 min. Then, the sample was cooling to room temperature. Finally, 3 mL of *n*-hexane was added and mixed for 5 min, the tubes were centrifuged for 5 min and the extracts were removed for GC-MS analysis.

2.7 Statistical Analysis

All the assays were performed in triplicate. Data were represented as median ± standard deviation (SD) and statistically analyzed using a one-way ANOVA with Tukey's post hoc test at the significance level of $p < 0.05$, using GraphPad Prism 8.0 (GraphPad Software, Inc., San Diego, California, United States).

3 Results

3.1 Yield Oil Extraction

The oil extraction was carried out through three methods: maceration, ultrasound and Soxhlet. The result showed that the oil yield obtained from Soxhlet extraction method was higher as compared with the other extraction method, with a yield of 62.82%. As regards, maceration extraction obtained a yield of 52.30% while lowest extraction yield was ultrasound extraction with 47.34%.

3.2 Total Phenolics Compounds

Phenolics compounds constitute the main class of secondary metabolites with great importance as potential antioxidant. Their consumption contributes to human health and to prevent oxidation of fat and oils. Therefore, Maceration (MC), ultrasound (US) and Soxhlet (SX) were applied for the extraction of polyphenols compounds which were then determined by Folin Ciocalteu reagent. The total polyphenol content TPC of oil seed ranged from 0.59 ± 0.03–1.10 ± 0.04 mg GAE/100 g oil seed. In which the Soxhlet method showed higher content of total phenol obtained 1.10 ± 0.04 mg GAE/100 g oil seed, followed by ultrasound showed a content of total phenol 0.81 ± 0.04 mg GAE/100 g oil seed, while lowest content of total phenol with 0.59 ± 0.03 mg GAE/ 100 g oil seed were obtained of maceration technique.

3.3 Activity Antioxidant

The antioxidant activity of oil seed was analyzed by DPPH, ABTS and FRAP. The results are shown in Fig. 1. The presence of polyphenols could be responsible for the antioxidant power in seed oil. The DPPH assay was studied by monitoring the reduction of the diphenyl picryl-hydrazyl (DPPH) radical from purple color to yellow, the degree of the discoloration corresponds to interaction with antioxidants compounds [11]. The measured values of scavenging activity in seed oil by maceration, ultrasound and Soxhlet were 1.93 ± 0.05%, 3.45 ± 0.16% and 8.77 ± 0.26%, respectively. According to the results shown in Fig. 1a, the extract higher antioxidant capacity was soxhlet extraction 1.41 ± 0.03 μmol TE/100 g of oil, followed by ultrasound extraction 0.80 ± 0.02 μmol TE/100 g of oil, whereas extraction maceration had the lowest value 0.63 ± 0.01 μmol TE/100 g of oil.

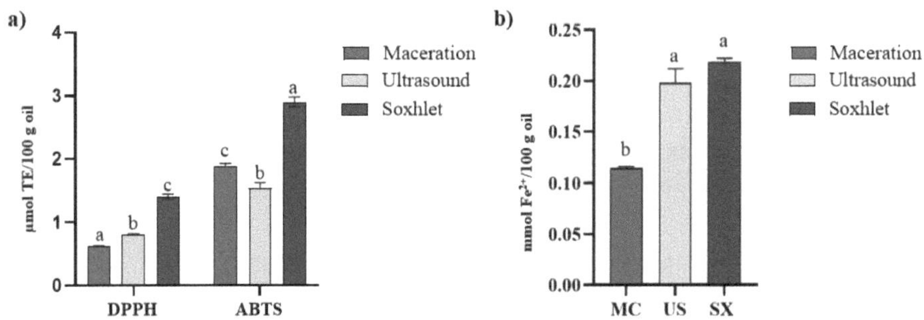

Fig. 1. **a)** DPPH, ABTS free radical scavenging of *C. capitata* seed oil. **b)**. Ferric reducing antioxidant power (FRAP) of *C. capitata* seed oil by Maceration (MC), Ultrasound (US) and Soxhlet (SX) techniques. The letters, a, b, and c, present significant differences among samples ($p < 0.05$, n = 3).

The results of ABTS assay demonstrated that all tested extracts showed activity scavenging against ABTS. The scavenging activity obtained a range from 10.98 ± 0.32% to 19.52 ± 0.52%. In Fig. 1a, the greatest ABTS value was obtained with soxhlet extraction 2.90 ± 0.06 μmol TE/100 g of oil, followed by extraction maceration 1.54 ± 0.07 μmol TE/100 g oil, while the lower value was obtained by ultrasound extraction 1.54 ± 0.07 μmol TE/100 g oil. FRAP assay was determined by measuring the reducing power Fe^{3+} to Fe^{2+}. The antioxidant capacity of different extracts ranged from 0.11 ± 0.001 to 0.21 ± 0.007 mmol $FeSO_4$/100 g of oil (Fig. 1b).

3.4 Fatty Acid Profile

The fatty acid composition of *C. capitata* seed oil was determined using three extraction methods: maceration, ultrasound, and Soxhlet extraction (Table 1). Fatty acids can be classified into saturated and unsaturated fatty acids. Five types of fatty acids were detected in *C. capitata* seed oil: saturated fatty acids such as palmitic acid (C16:0) and stearic acid (C18:0), and unsaturated fatty acids such as oleic acid (C18:1n-9), linoleic acid (C18:2n-6), and α-linolenic acid (C18:3n-3). Oleic acid (18:1n-9) was the majority component in all extractions, representing 36.12 ± 0.60%, 38.03 ± 0.28% and 37.30 ± 0.25% for maceration, ultrasound and Soxhlet, respectively. It was followed by linoleic acid (18:2n-6), with values between 23.84% and 24.12%, and α-linolenic acid (18,3n-3), with proportions ranging from 19.67% to 22.17%. The total content of saturated fatty acids ranged between 17.33% and 18.18%, while unsaturated fatty acids represented the largest fraction, with values above 81% in all methods.

Table 1. Fatty-acid (FA) profile of *C. capitata* seed oil obtained by different extraction methods. ($n = 3$).

Fatty acid	Extraction Method		
	Maceration	Ultrasound	Soxhlet
16:0	10.38 ± 0.09	10.86 ± 0.08	10.40 ± 0.04
18:0	7.46 ± 0.84	7.32 ± 0.07	6.93 ± 0.01
18:1n-9	36.12 ± 0.60	38.03 ± 0.28	37.30 ± 0.25
18:2n-6	23.88 ± 0.33	24.12 ± 0.17	23.84 ± 0.07
18:3n-3	22.17 ± 0.61	19.67 ± 0.19	21.53 ± 0.22
Saturated	17,84 ± 0,69	18,18 ± 0,36	17,33 ± 0,29
Unsaturated	82,17 ± 1,54	81,82 ± 0,64	82,67 ± 0,54

3.5 Correlation Analysis of *C. Capitata* Seed Oil Bioactive Compounds with Activity Antioxidant.

Figure 2 showed that the content phenolics had a high correlation with antioxidant activity of DPPH and FRAP with a correlation coefficient of 0.97 and 0,92 was significant (*p*-value <0.05). Additionally, the stearic acid of seed oil *C. capitata* showed correlation negatively with ABTS, with a correlation coefficient of −0.72, showed a level significant of (*p*-value <0.05). While the components of Linoleic acid were positively associated with palmitic acid ($r = 0.83$; $p < 0.05$)

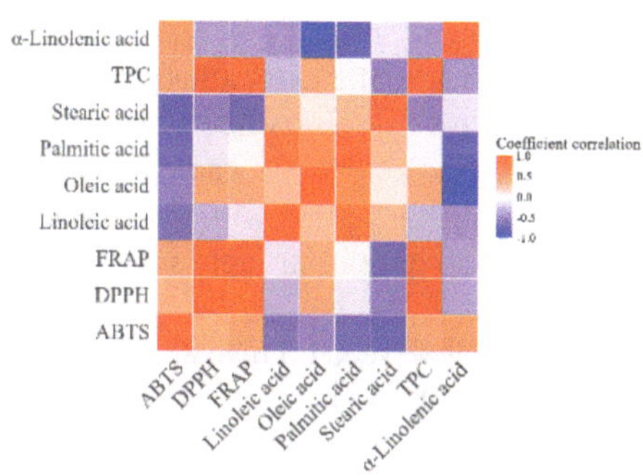

Fig. 2. Analysis correlation of *C. capitata* seed oil bioactive compounds with antioxidant activity ($p < 0.05$).

4 Discussion

The highest yield was obtained using Soxhlet extraction, likely because the sample is repeatedly exposed to fresh portions of solvent, which facilitates the displacement of the transfer equilibrium. Additionally, the extraction temperature can influence the amount of oil yield obtained. As temperature increases, intermolecular interactions within the solvent decrease, resulting in greater molecular motion and improved solubility, which accelerates the extraction process [12]. Moreover, this method effectively disrupts the cellular membrane, increasing its porosity and thereby enhancing oil extraction compared to other methods [13]. On the other hand, the solvent is a critical factor for the extraction, in this work, hexane was used as the extraction solvent, whereas other studies with Soxhlet technique employed petroleum ether and reported oil contents of 41.7% and 41.5% for *Cucumis melo var. inodorus* and *Citrullus lanatus var. lanatus*, respectively [14]. Also, this study found that the content seed oil of *C. capitata* was higher than other species of the family Cucurbitaceae with the same technique and solvent extraction. For example: pumpkin seed (*Cucurbita maxima*) with 42.53%, and the content of oil the seed of *Lagenaria siceraria* with 56.30% [15]. The differences observed in oil yield and composition may be attributed to the pretreatment of the seeds prior to extraction. Previous studies have shown that factors such as drying temperature, seed moisture content, and storage conditions can significantly influence lipid extraction efficiency and the preservation of bioactive compounds [16].

Although this is the first report on *C. capitata*, the scientific information about genre is scarce, however there are some studies about other species of the same family, such as pumpkins (*Cucurbita* spp.). The total phenolic content of pumpkin seed oil has been reported to be 3.76 ± 0.06 mg/100 g oil [17], in contrast to our result which was lower. These results show that the extraction of *C. capitata* seed oil by apolar solvents was inappropriate for obtaining oil with enhanced total phenolics compounds. In previous studies founded that the content of TPC in *Cucurbita maxima* cv. "Béjaoui" seed oil extracted by chloroform/methanol was higher than those with pentane or hexane [18]. Therefore, in this study, the Soxhlet method demonstrated the best extraction, which could be attributed to the effects of temperature and pressure altering the solubility of certain compounds, resulting in a different chemical profile than the maceration method and ultrasound [19].

Our results showed lower antioxidant activity (DPPH and ABTS) compared to previously reported values for *Cucurbita pepo* [20], *C. maxima* [18], and *C. moschata* [21]. This lower activity may be due to the extraction method used, as techniques like mechanical pressing and supercritical CO_2 have shown better preservation of antioxidant compounds [22]. The ABTS assay indicated higher activity in Soxhlet extracts, while the DPPH and FRAP assays showed better results with ultrasound and maceration. These findings confirm that antioxidant activity is strongly influenced by both the extraction method and the type of assay used.

The fatty acid profile of *C. capitata* seed oil revealed a predominance of unsaturated fatty acids, representing over 81% of the total content in all extraction methods evaluated. This finding is relevant, as oils with a high percentage of unsaturated fatty acids are valued for their nutritional properties and health benefits, including cardiovascular risk reduction and anti-inflammatory activity. Among the individual fatty acids, oleic acid (18:1n-9) was the most abundant, exceeding 36% in all techniques. It is considered beneficial for cardiovascular health. Furthermore, its structure with a single cis double bond gives it a lipid profile favorable for human health, helping to reduce LDL ("bad") cholesterol levels and increase HDL ("good") cholesterol, thus contributing to the prevention of cardiovascular disease [23]. Oleic acid is considered to have anti-inflammatory and antioxidant effects [24]. Likewise, the significant presence of polyunsaturated acids, such as linoleic acid (18:2n-6) and α-linolenic acid (18:3n-3), contributes to the oil's functional importance. α-linolenic and linoleic acids are considered essential fatty acids because the human body needs them to stay healthy but cannot synthesize them itself [25]. Linoleic acid, belonging to the omega-6 polyunsaturated fatty acid family, is an essential fatty acid for the human body. Its consumption contributes to the cleansing of blood vessels, helping to lower blood cholesterol levels and prevent the formation of arterial thrombosis [26]. Furthermore, it is recognized for its notable anticancer activity, particularly in the treatment of breast cancer [27]. α-linolenic acid (ALA), an essential fatty acid belonging to the omega-3 family. ALA, found in vegetable oils (chia, flaxseed, canola), some seeds, and nuts, is essential for cardiovascular health, a natural anti-inflammatory, and a precursor to EPA and DHA (other important omega-3 s) [28]. Taken together, these results highlight *C. capitata* oil as a promising source of unsaturated fatty acids, with potential applications in both the nutraceutical industry and the development of high-value cosmetic products. Although this is the first study on oil extracted from *C. capitata* seeds, certain limitations should be acknowledged. The oil extraction was conducted using conventional techniques and limited to organic solvents such as hexane. These constraints suggest the need for future studies that incorporate a broader range of extraction methods, more extensive sampling, and deeper analytical approaches to fully assess the biotechnological and commercial potential of *C. capitata* seed oil

5 Conclusion

Cayaponia capitata seed oil is emerging as a valuable source of functional lipids, characterized by a high content of unsaturated fatty acids and bioactive compounds with antioxidant properties. Soxhlet extraction proved to be more efficient in terms of yield and total phenolic content, although antioxidant activity varied depending on the method and assay used. The predominance of essential fatty acids such as linoleic and α-linolenic acid, along with oleic acid, highlights the oil's nutritional value. These findings contribute to the understanding of underutilized species and support the use of *C. capitata* as a sustainable and promising alternative to produce functional vegetable oils for industrial use.

References

1. Chiriacò, M.V., Galli, N., Santini, M., Rulli, M.C.: Deforestation and greenhouse gas emissions could Arise when replacing palm oil with other vegetable oils. Sci. Total Environ. **914** (2024). https://doi.org/10.1016/j.scitotenv.2023.169486

2. Rezig, L., Gharsallah, K., Mahfoudhi, N. Bioactive Compounds of Cucurbitaceae Seed Oils as Nutraceuticals and Health-Promoting Substances. In Handbook of Research on Advanced Phytochemicals and Plant-Based Drug Discovery; IGI Global, pp. 292–313 (2022). ISBN 9781668451304
3. Rolnik, A., Olas, B. Vegetables from the Cucurbitaceae Family and Their Products: Positive Effect on Human Health. Nutrition 78 (2020)
4. Hussain, A., et al.: Exploration of two Cucurbitaceae fruit (muskmelon and watermelon) seeds for presence of phytochemicals, and antioxidant and antimicrobial activities. Turk. J. Agric. - Food Sci. Technol. **11**, 2493–2498 (2023). https://doi.org/10.24925/turjaf.v11is1.2493-2498.6366
5. Pohle, P., Reinhardt, S.: Indigenous knowledge of plants and their utilization among the Shuar of the lower Tropical Mountain Forest in southern. Lyonia **7**, 133–149 (2004)
6. Ghareeb, M.A., Mohamed, T., Saad, A.M., Refahy, L.A.G., Sobeh, M., Wink, M.: HPLC-DAD-ESI-MS/MS analysis of fruits from Firmiana Simplex (L.) and evaluation of their antioxidant and Antigenotoxic properties. J. Pharm. Pharmacol. **70**, 133–142 (2018). https://doi.org/10.1111/jphp.12843
7. Chanioti, S., Tzia, C.: Optimization of ultrasound-assisted extraction of oil from olive pomace using response surface technology: oil recovery, Unsaponifiable matter, Total phenol content and antioxidant activity. LWT **79**, 178–189 (2017). https://doi.org/10.1016/j.lwt.2017.01.029
8. Thaipong, K., Boonprakob, U., Crosby, K., Cisneros-Zevallos, L., Hawkins Byrne, D.: Comparison of ABTS, DPPH, FRAP, and ORAC assays for estimating antioxidant activity from guava fruit extracts. J. Food Compos. Anal. **19**, 669–675 (2006). https://doi.org/10.1016/j.jfca.2006.01.003
9. Van Den Berg, R., Haenen, G.R.M.M., Van Den Berg, H., Bast, A.: Analytical, nutritional and clinical methods section applicability of an improved Trolox equivalent antioxidant capacity (TEAC) assay for evaluation of antioxidant capacity measurements of mixtures. Food Chem. (1999)
10. Wang, J., Wu, W., Wang, X., Wang, M., Wu, F.: An effective GC method for the determination of the fatty acid composition in silkworm pupae oil using a two-Stepmethylation process. J. Serb. Chem. Soc. **80**, 9–20 (2015). https://doi.org/10.2298/JSC140401073W
11. Marsoul, A., Ijjaali, M., Oumous, I., Bennani, B., Boukir, A.: Determination of polyphenol contents in Papaver Rhoeas L. flowers extracts (Soxhlet, maceration), antioxidant and antibacterial evaluation. In: proceedings of the materials today: proceedings; Elsevier Ltd, Vol. 31, pp. S183–S189 (2020)
12. Nurhadi, B., et al.: Comparison of Curcuma Domestica and Curcuma Xanthorrhiza Oleoresins Extracted Using Maceration, Soxhlet, and Ultrasound-Assisted Extraction (UAE). In Proceedings of the IOP Conference Series: Earth and Environmental Science; Institute of Physics Publishing, Vol. 443 (2020)
13. Singh, A., Kumar, V.: Phyto-chemical and bioactive compounds of pumpkin seed oil as affected by different extraction methods. Food Chem. Adv. **2** (2023). https://doi.org/10.1016/j.focha.2023.100211
14. Jorge, N., Da Silva, A.C., Malacrida, C.R.: Physicochemical characterisation and radical-scavenging activity of Cucurbitaceae seed oils. Nat. Prod. Res. **29**, 2313–2317 (2015). https://doi.org/10.1080/14786419.2015.1007135
15. Olatunji, A.G., Ajerogba, L.I., Olalekan, I.M., Adegoke, A.B., Adepoju, R.A.: Phytochemical screening of seed oils from some Nigerian grown Cucurbitaceae plant seeds. IOSR J. Appl. Chem. (IOSR-JAC) e-ISSN 2021, 14 https://doi.org/10.9790/5736-1401010106.
16. Gutiérrez, L.F., Ratti, C., Belkacemi, K.: Effects of drying method on the extraction yields and quality of oils from Quebec Sea buckthorn (Hippophaë Rhamnoides L.) seeds and pulp. Food Chem. **106**, 896–904 (2008). https://doi.org/10.1016/j.foodchem.2007.06.058

17. Ahmed, I.A.M., AlJuhaimi, F., Karrar, E., Uslu, N., Özcan, M.M.: The effect of roasting on bioactive properties, phenolic compounds and fatty acid profiles of pumpkin (Cucurbita Spp.). Seeds. J. Oleo Sci. **73**, 1457–1465 (2024). https://doi.org/10.5650/jos.ess24153
18. Rezig, L. et al.: Cucurbita Maxima pumpkin seed oil: from the chemical properties to the different extracting techniques. Not Bot Horti Agrobot Cluj Napoca **46**, 663–669 (2018). https://doi.org/10.15835/nbha46211129
19. Uahomo, P.O., Kpaduwa, S., Daniel, C., Ezerioha, C.E.: Characterization and comparative assessment of the essential oil from lime (citrus Aurantifolia) exocarp using maceration and Soxhlet extraction methods. Scholars Int. J. Chem. Mater. Sci. **6**, 126–134 (2023). https://doi.org/10.36348/sijcms.2023.v06i06.002
20. Kulaitiene, J., Černiauskiene, J., Jariene, E., Danilčenko, H., Levickiene, D.: Antioxidant activity and other quality parameters of cold pressing pumpkin seed oil. Not Bot Horti Agrobot Cluj Napoca **46**, 161–166 (2018). https://doi.org/10.15835/nbha46110845
21. Hiranrangsee, L., Kumaree, K.K., Sadiq, M.B., Anal, A.K.: Extraction of anthocyanins from pericarp and lipids from seeds of Mangosteen (Garcinia Mangostana L.) by ultrasound-assisted extraction (UAE) and evaluation of pericarp extract enriched functional ice-cream. J. Food Sci. Technol. **53**, 3806–3813 (2016). https://doi.org/10.1007/s13197-016-2368-8
22. Šamec, D., et al.: The potential of pumpkin seed oil as a functional food—a comprehensive review of chemical composition, health benefits, and safety. Compr. Rev. Food Sci. Food Saf. **21**, 4422–4446 (2022)
23. Pastor, R., Bouzas, C., Tur, J.A.: Beneficial effects of dietary supplementation with olive oil, oleic acid, or Hydroxytyrosol in metabolic syndrome: systematic review and meta-analysis. Free Radic. Biol. Med. **172**, 372–385 (2021). https://doi.org/10.1016/j.freeradbiomed.2021.06.017
24. Lu, Y., et al.: Protective effects of oleic acid and polyphenols in extra virgin olive oil on cardiovascular diseases. Food Sci. Human Wellness **13**, 529–540 (2024). https://doi.org/10.26599/FSHW.2022.9250047
25. Phuong, D.L., Thuy, N.T., Long, P.Q., Kuo, P.-C., Thang, T.D.: Composition of fatty acids, tocopherols, sterols, Total Phenolics, and antioxidant activity of seed oils of Afzelia Xylocarpa and Cassia fistula. Chem. Nat. Compd. **55**, 242–246 (2019). https://doi.org/10.1007/s10600-019-02659-x
26. Chang, J., Kang, X., Yuan, J.: Enhancing emulsification and antioxidant ability of egg albumin by moderately acid hydrolysis: modulating an emulsion-based system for mulberry seed oil. Food Res. Int. **109**, 334–342 (2018). https://doi.org/10.1016/j.foodres.2018.04.055
27. Tao, X.-M., et al.: Enhanced anticancer activity of gemcitabine coupling with conjugated linoleic acid against human breast cancer in vitro and in vivo. Eur. J. Pharm. Biopharm. **82**, 401–409 (2012). https://doi.org/10.1016/j.ejpb.2012.06.007
28. Nunes, E.A., D'Souza, A.C., Steen, J.P., Phillips, S.M.: Lack of evidence for Omega-3 fatty acid supplementation in enhancing lean mass, muscle strength, and physical function in healthy adults and clinical populations: an overview of reviews. Clin. Nutr. ESPEN **67**, 155–165 (2025). https://doi.org/10.1016/j.clnesp.2025.03.012

Mathematical Models to Describe the Growth of *Leucoptera Sinuella* (Reutii) (Lepidoptera: Lyonetiidae) Larvae in a Mediterranean Climate of Central Chile

Sebastián Yánez-Segovia[1](✉), Ignacio P. Chamorro -Warnken[2], Leticia E. Silvestre[3], and M. J. Lovato- Fortun[4]

[1] Escuela de Ciencias Agrícolas y Ambientales, Pontificia Universidad Católica del Ecuador-Sede Ibarra, Ibarra, Ecuador
sebas25ec@yahoo.com
[2] Agroimpulsa, 3460000 Talca, Chile
[3] Universidad Austral dc Chilc, 5091000 Valdivia, Chile
[4] YAinnova-Ecuador. Ideas para el Agro, Quito 170303, Ecuador

Abstract. The growing interest in reducing pesticide use in the agricultural and forestry industries has prompted researchers to study models that more fully describe not only insect biology and behavior, but also how to mathematically model these mechanisms. This study aims to provide a mathematical model to accurately describe the instar and larval growth of *Leucoptera sinuella* (Reutii) (Lepidoptera: Lyonetiidae) and its implications for pest control. The exponential and logistic models, represented by generalizations with third-order polynomial exponent, effectively maintain parsimony in identifying the most appropriate model under the environmental conditions observed during the 2019–2020 season at study sites in central Chile.

Keyword: Differential equations · Cephalic capsule · Exponential model · Logistic model

1 Introduction

Mathematical equations to describe the temporal evolution of a population have aroused the interest of scientists seeking to model population dynamics [1, 2]. Over the years, these models have been widely applied to various cases, including human, animal and cell populations [3, 4]. Especially in the agricultural sector, these tools have been used to predict insect pest populations [5, 6]; as well as to implement control strategies aimed at reducing the use of pesticides on crops [7], thus safeguarding human and environmental health.

Insects are among the most numerous invasive species worldwide, and the number of new pests is increasing exponentially [8], posing a global threat to both the economy and the environment [9, 10]. In North America, more than four hundred and fifty non-native insects have colonised forest and urban trees, with fourteen percent of them

J. L. Ballesteros et al. (Eds.): CITIS 2025, LNNS 1780, pp. 82–93, 2027.
https://doi.org/10.1007/978-3-032-14287-0_9

causing significant damage [11]. An example of recent biological invasion is that which occurred during the 2015 - 2016 season, when the poplar moth, *Leucoptera sinuella* (Reutii) (Lepidoptera: Lyonetiidae), was detected in Chile [12] and Mendoza - Argentina in 2021 [13]. The larvae of this moth cluster by mine on the leaves of Salicaceae, causing losses in the forestry industry [14–16].

A key issue in the use of mathematical models to describe biological processes is the lack of comprenhensive theories for these processes [17], which inevitably leads to simplifications to schematize the problem [18]. In particular, many equations asume that individual growth increases linearly with time [19]. This assumption complicates the application of exponential and/or logistic equations, especially in poikilothermic organisms, which are highly dependent on environmental parameters [18]. Therefore, to generate more realistic models of insect populations in their habitat, it is necessary to consider: *(i)* the possibility of including data provided by field monitoring, *(ii)* models to describe how insects change growth rate depending on metamorphosis and instar at the larval stage, and *(iii)* the description of cephalic capsule growth, as an allometric measure of population age. The main challenge in obtaining information on insect larval growth curves stems in part from technical difficulties. In particular, the existence of different larval instars makes larval growth curves complex [20]; consequently, any instantaneous measurement of growth rate must explicitly consider the exact phase of molt change. The aim of this work is to provide a mathematical model as a decision support system for the control of *L. sinuella* in poplar plants. To develop this model, we will use separable variable differential equations, compare the exponential and logistic models, and finally analyse the daily cephalic capsule growth rate to infer the polynomial that best describes instars as a function of time.

2 Materials and Methods

2.1 Experimental Data

The experimental data used to apply the models were obtained from poplar leaves [*(Populus balsamifera subsp. Trichocarpa x Populus deltoides) x Populus deltoides)].* With the presence of *L. sinuella* larvae in eight-year-old trees arranged at a distance of six by six meters. These trees were located in commercial plantations in the communes of Coinco, Talca and Retiro, between the V and VII regions of Chile in the 2019–2020 season. The plants received the standard cultural practices for the region, including irrigation, thinning, and fertilization.

Ten trees were randomly selected and five leaves with mines were collected from each tree. From these leaves, two larvae were selected from the same mine [15]. For each location and day, one hundred larvae were recorded, ensuring that different trees were assessed during each consecutive daily reading. Measurements began one day after hatching (DAH) and continued daily, recording the growth in micrometers of the cephalic capsule and the diameter between genas, using a Led Motic BA310 binocular microscope and Motic Images Plus 3.0 software, with a Leica 10X-1024X-728 microscope objective. The evaluation period lasted 25 days, when, on average, they started to pupate. We assume that cephalic capsule growth is directly related, as an allometric measure, to body size,

Dyar's rule [21], and this is related to time, so that physiological mathematical models can be generated, providing high accuracy in inferring larval instar and growth.

2.2 Functions and Statistics

Adjustments were made using the NLIN procedure of the STAT module of the SAS statistical program, which employs the Gauss-Newton algorithm with Marquardt correction. The iterative calculation method requires initial values of the coefficients, so approximate estimates were sought. For the final diameter coefficients (Df), one of the largest observed diameter measurements was used, and this value substituted into each of the equations allows them to be reduced, by means of the appropriate transformations, to third-order polynomial models, where, with the usual linear regression methods, the remaining parameters were estimated. These estimates are those that enter the non-linear regression algorithms.

The coefficient of determination (R^2) was used as a comparison statistic and the residuals were checked for compliance with the model assumptions. Using the selected function, individual curves were fitted to each of the cephalic capsule readings as a function of time and variability of each of the parameters.

In addition, the first and second derivatives of the general model were calculated to determine the characteristic points and the parameters of main physiological interest: time to reach the inflection points, corresponding diameter measurements and growth rates. Due to the mathematical complexity of these functions, which makes it impossible to obtain their roots analytically, numerical calculation programs developed in Turbo Pascal language were used.

3 Results

3.1 Exponential Growth Modeling

Our model assumes that the cephalic capsule increases exponentially in size as a function of time. This model can also be interpreted as a description of the growth of the larval stage and its corresponding developmental stages. We present the exponential growth function for the diameter in micrometers *(μm)* of the cephalic capsule, denoted as D, which describes the growth over time *(t)* in days, and assuming constant growth *(k)*, is described by the following general equation:

$$y = a^x \Leftrightarrow D(t) = e^{kt} \tag{1}$$

For this case, it is possible to model by means of differential equations with separable variables, where the derivative function is expressed as follows:

$$\frac{dD}{dt} = kD \tag{2}$$

By rearranging the equation algebraically using the addivitivy property of similar components on both sides, we obtain:

$$dt * \frac{1}{D} * \frac{dD}{dt} = kD * dt * \frac{1}{D} \tag{3}$$

Integrating the both sides of the equation:

$$\int \frac{1}{D} * dD = \int k * dt \tag{4}$$

We apply the resolution of the equation using the properties of logarithms:

$$\ln|D| = kt + C_1 \tag{5}$$

We solve the equation using the properties of exponentials:

$$|D| = e^{kt+C_1} = e^{kt} * e^{C_1} = Ce^{kt} \tag{6}$$

The equation of exponential model that describes the growth of the cephalic capsule as a function of time is described by:

$$D(t) = Ce^{kt} \tag{7}$$

Based on the knowledge generated regarding the physiology and biology of *L. sinuella*; as well as the development of the daily growth rate (Fig. 3), three parameters (*c, d, and f*) are incorporated into the model exponent. These parameters not only condition the linearization coefficient and the inflection points but also describe the variable behavior of the growth rate:

$$D(t) = Ce^{(k+ct+dt^2+ft^3)} \tag{8}$$

Due to the specific of the data, 113.36 μm was considered the actual observational measurement of the cephalic capsule before hatching, corresponding to the value of the intercept at $t = 0$ days. By substituting these values into the Exponential Model (7), we obtain:

$$113.36 = Ce^{k0} \Rightarrow C = 113.36 \tag{9}$$

For linearization processes, our observed data at 11 DAH indicated an average diameter of 222.70 μm. By substituting these values into the exponential model (7) and those obtained in step (9), we have:

$$222.70 = 113.36ek11 \Rightarrow 222.70113.36 = ek11 \Rightarrow ln222.70113.36 = k * 11 \Rightarrow k = 0.0633 \tag{10}$$

Therefore, the exponential model equation for the diameter of the cephalic capsule of *L. sinuella* as a function of time, for the towns of Coinco, Talca and Retiro (central Chile), is described by the following the function:

$$D(t) = 113.36e0.0633t \tag{11}$$

3.2 Logistic Growth Modeling

Our model assumes that the cephalic capsule increases in size logistically as a function of time. We present the logistic growth function for the diameter of the cephalic capsule, denoted as D (measured in micrometers, μm), which describes the growth over time *(t)* in days; a the average final size and b the average initial size; it is described by the following general function:

$$y = \frac{a}{1 + e^{-bt}} \Leftrightarrow D(t) = bD\left(1 - \frac{D}{a}\right) \tag{12}$$

In this case, it is possible to model using differential equations and their antiderivative components, whose derivative function is:

$$\frac{dD}{dt} = bD\left(1 - \frac{D}{a}\right) \Leftrightarrow \frac{1}{D\left(1 - \frac{D}{a}\right)} * \frac{dD}{dt} = b \tag{13}$$

By performing an algebraic arrangement of the expansion in partial fractions and substituting, we obtain:

$$\left(\frac{1}{D} + \frac{\frac{1}{a}}{1 - \frac{D}{a}}\right) * \frac{dD}{dt} = b \tag{14}$$

By applying the property of antiderivatives and factoring, we have:

$$\frac{1}{D} * \frac{dD}{dt} - \frac{-\frac{1}{a}}{1 - \frac{D}{a}} * \frac{dD}{dt} = b \tag{15}$$

We apply the resolution of the equation by the properties of logarithms and assume that $0 < D(t) < b$

$$\ln|D| - \ln\left|1 - \frac{D}{a}\right| + C_1 = bt + C_2 \tag{16}$$

We solve the equation by properties of the difference of logarithms, inverse function and properties of exponents:

$$\ln\left(\frac{D}{1 - \frac{D}{a}}\right) = bt + C \Rightarrow \frac{D}{1 - \frac{D}{a}} = e^{bt+C} \Rightarrow \frac{D}{1 - \frac{D}{a}} = Ce^{bt} \tag{17}$$

Using multiplicative or reciprocal inverses and when $t = 0$ then $D(0) = D_0$, have:

$$\frac{1 - \frac{D}{a}}{D} = C_3 e^{-bt} \Rightarrow D(t) = \frac{1}{C_3 + \frac{1}{a}} \Leftrightarrow D(0) = \frac{1}{C_3 + \frac{1}{a}} = D_0 \Rightarrow C_3 = \frac{1}{D_0} - \frac{1}{a} \tag{18}$$

Equating by common factor (Eq. 18) and substituting (18) in the equation of step (17), where the growth rate is given: $p(t) = b + c.t + d.t^2 + ft^3$:

$$D(t) = \frac{1}{\left(\frac{1}{D_0} - \frac{1}{a}\right) * e^{-(b+c.t+d.t^2+f.t^3)} + \frac{1}{a}} \left(\frac{D_0 a}{D_0 a}\right) \quad (19)$$

The logistic model equation that describes the growth of the head capsule as a function of time is described by:

$$D(t) = \frac{a}{1 + e^{-(b+c.t+d.t^2+f.t^3)}} \quad (20)$$

Hence, with the model of (20) and subsequent validation, the observed referential values of initial diameter are obtained $Di = 127.38$ µm and final diameter $Df = 558.73$; with which it is linearized (Fig. 1):

$$D(t) = \frac{Di * Df}{Di + (Df - Di)e^{-ut}} \Rightarrow D(t) = \frac{127.38 * 558.73}{127.38 + (558.73 - 127.38)e^{-ut}} \quad (21)$$

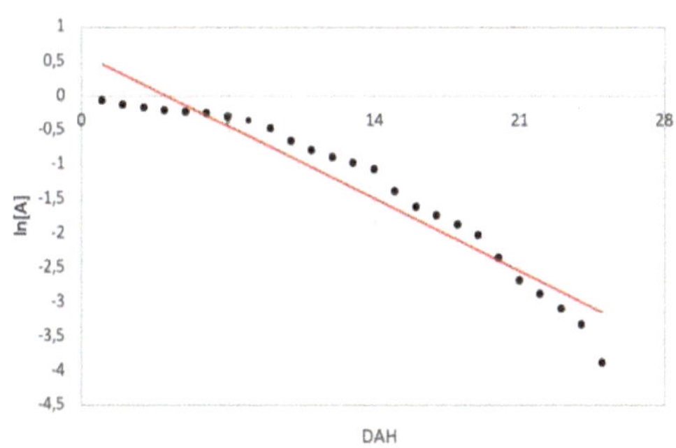

Fig. 1. Linearization of logistic modeling for cephalic capsule growth evaluated for days after hatching (DAH).

The linearization constant of the equation is represented by $-u$; which is present in the equation of step (21), for which we solve by means of a system of linear Eqs. (2 × 2), where at 2 DAH it has a diameter of 140.72 µm and at 19 DAH it has a diameter of 388.03 µm (Table 1):

$$-e^{-ut} = A \Rightarrow A = \left[\frac{Di * Df}{D} - Di\right] * \frac{1}{Df - Di} \quad (22)$$

Days after hatching (DAH).

Since the growth rate depends on the size of the cephalic capsule at each moment, as well as on time. The time dependency is expressed by the derivative of the exponential polynomial:

$$p(t) = c.t + 2d.t^2 + 3f.t^3 = 4.5161t - 0.0109t^2 + 0.00001t^3 \quad (23)$$

Table 1. System of linear Eqs. (2 × 2) reemplace value for 2 and 19 days for cephalic capsule growth.

DAH	Diameter μm	A	ln A	Equations	$\ln A = a + ut$
2	140.72	0.8772074	-0.1310118	1	$-0.1310118 = a + u\,(2)$
19	388.03	0.12990913	-2.0409201	2	$-2.0409201 = a + u\,(19)$
					$u =$ -0.11235

The quadratic function that describes the variable behavior of the growth rate is represented by the parameters c, d and f. The considerations presented here suggest that the generalized exponential model is the most appropriate to describe the growth in size of the cephalic capsules expressed through the diameter measurement. The equation corresponding to the total of all capsules in all generations is:

$$D(t) = \frac{71171.0274}{127.38 + (431.35)e^{-(281.81+4.5161t-0.0109t^2+0.00001t^3)}} \tag{24}$$

3.3 Comparison of Exponential and Logistic Models

The two models, exponential and logistic, fit the experimental data differently (Fig. 2); and have coefficients of determination of 98.21 and 94.10%, respectively (Table 2).

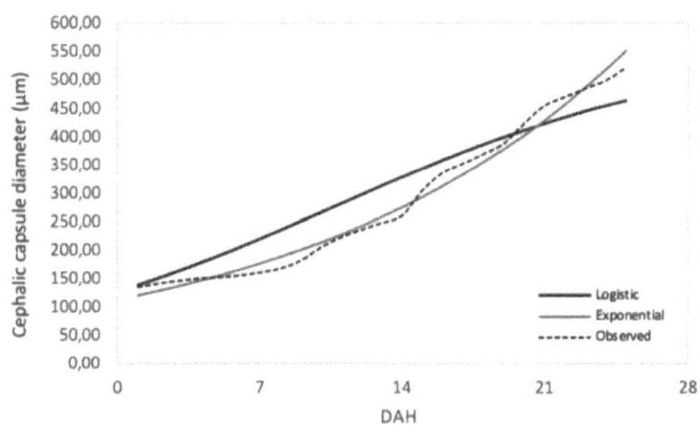

Fig. 2. Evolution of the growth of the cephalic capsule in the exponential and logistic models, and observed values. Day after hatching (DAH).

Since the growth rate is dependent on the size of the cephalic capsule at each instar, as well as on time. The time dependence is expressed by the derivative of the polynomial exponent:

$$p(t) = c.t + 2d.t^2 + 3f.t^3 = 3.67270t - 0.00960t^2 + 0.00001t^3 \tag{25}$$

The equation corresponding to the total of all capsules in all generations is:

$$D(t) = 113.36e^{(-219.04000+3.67270t-0.00960t^2+0.00001t^3)} \tag{26}$$

Table 2. Comparison of models with the estimates obtained from the total generations of the 2019–2020 season.

Models	Larval growth model of *L. sinuella*	R^2	Modification in the exponent polynomial	R^2
Exponential	$D(t) = 113.36e^{0.0633t}$	98.21	$0.00001x^3 - 0.00960x^2 + 3.67270x - 219.04$	99.66
Logistic	$D(t) = \frac{127.38*558.73}{127.38+(558.73-127.38)e^{-0.11235(t)}}$	94.10	$0.00001x^3 - 0.01090x^2 + 4.51610x - 281.81$	99.11

After carrying out mathematical descriptions using differential equations separable variables for the exponential and logistic models concerning the growth of the cephalic capsule of *L. sinuella*, it was determined that the best fit is represented by the exponential model (Table 2, Fig. 2).

The validation of the models, comparing observed *Vs.* Estimated values, with the polynomial exponent, shows similar responses for both models, each achieving a coefficient of determination greater than 99%, which suggests an optimal adjustment of growth (Fig. 3a,b).

3.4 Daily and for Instar Growth Rate

From the analysis of the generalized exponential function, to which the first order derivatives are associated (Fig. 4), the three phases of growth can be observed. The first characterized by a variable rate that reaches its maximum value (23.82 μm per day) around 10 DAH, from there it decreases (12.97 μm per day) when the process is at its midpoint, when the larvae has reached the third larval instar; at the end of this stage there is a constant increase until reaching a maximum growth value throughout the larval metamorphosis process of (43.76 μm per day).

After this, it decreases (16.23 μm daily) at day 17 DAH, to have a growth increase at the end of the fourth instar (35.91 μm daily) and subsequent decrease (13.98 μm daily) and a final increase of I remove stage (23.59 μm daily), before the larvae undergo their final ecdysis for the transition to the pupal stage.

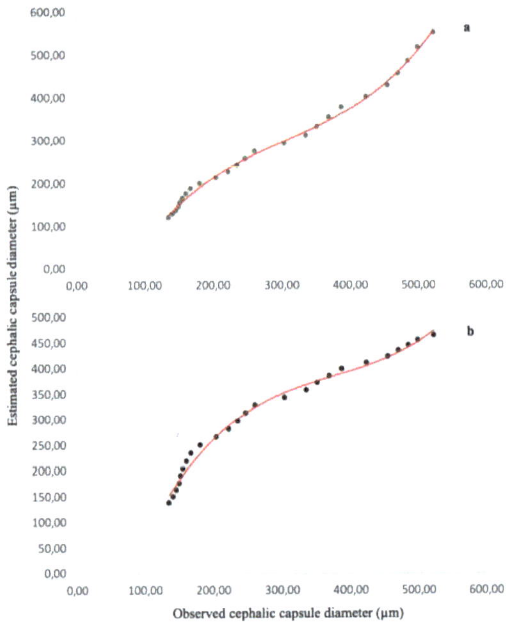

Fig. 3. A: Validation of the exponential model and, **b:** logistic model of the growth of the cephalic capsule as a function of development time.

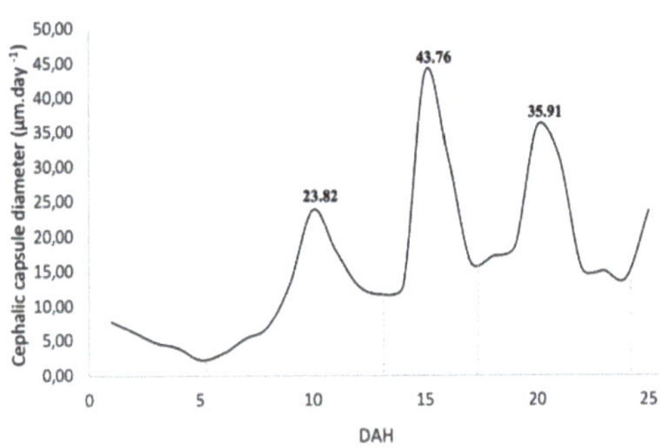

Fig. 4. Dynamic of the growth rate of the cephalic capsule in the exponential model as a function of time and its value at the inflection points. Day after hatching (DAH).

4 Discussion

In this study we present two mathematical models describing the growth of *L. sinuella* larvae by measuring the cephalic capsule, which were optimally fitted, especially when modifying the polynomial exponent (Table 2). Although both equations tested gave satisfactory results, the five-parameter exponential model seems to be the most appropriate due to its higher coefficient of determination as well as fewer problems related to deviation from linearity. The logistic model may be a good alternative, especially in species that have a constant growth rate in the last instar before pupation or show lower β-ecdysone activity in this instar. Although in general there are cases in which the Hessian matrix, formed by second-order derivatives, is singular and therefore it is not possible to obtain asymptotic descriptors of the estimated parameters (standard error and correlations between them), this is less frequent. In the exponential model [6, 22] indicated that, in a life-history context, simple power functions should be sufficient to model individual growth in all Lepidoptera, excepts perhaps the largest species. However, considerable heterogeneity in the estimated values of the allometric exponent discourages the use of a universal relationship for all species [23].

Reynolds et al. [24] noted that for small, fast-growing poikilothermic microorganisms such as insect larvae, it is reasonable to assume that the term describing catabolism may be of limited importance, and that the growth rate is mainly determined by the rate of anabolism, closely associated with the rate of food (energy) intake. Another complication is the extreme sensitivity of larval growth rates to environmental conditions, temperature, and food quality. Therefore, any non-synchronous comparison should be treated with extreme caution, and longitudinal studies on particular individuals cannot be recommended. If it had been verified that the growth rate cancels out at the end of the first instar (Fig. 4), it would indicate that the growth function presents a saddle point at this stage, a point where the first derivative equals zero without presenting a maximum or minimum, corresponding instead to a tangent line with a zero slope that passes through the curve. Clearly, the growth of *L. sinuella* larvae can be simulated with the quadratic equation we have developed, which shows five well-defined inflection points. This allows us to reduce the number of model parameters to four and demonstrates that the larvae have five instars, same instar larvae determinated in studies of the [15] and [16]. The second instar begins with a rebound in speed due to an increase in the rate of consumption and greater activation of the juvenile hormone β-ecdysone, reaching its maximum rate around fifteen DAH. This peak is twice that observed in the first instar (23.82 μm per day) and occurs when the larva is about to finish the third instar, i.e. when the larva has exceeded fifty percent of its definitive size, thereafter growth slows down and increases again for the third instar, until it reaches the size corresponding to twenty percent less than the previous peak. The last instar ends with a definitive size at twenty-five DAH. Studies such as that of [25] proposed theoretical tests to explain body sizes in the framework of demographic optimization, based mainly on juvenile growth curves selected based on model adequacy rather than biological realism.

5 Conclussion

These models have undeniable value as a foundation for conceptial understanding, they are not sufficiently realistic to explain the actual body sizes observed in nature. In our studies, the continuity of experiments under different environmental conditions will allow us to analyses parameters variability and interpret it through climatic factors.

Acknowledgements. This research was made possible thanks to the support of the Centro Tecnológico del Álamo (CTA) of the University of Talca, who provided access to their nursery and the plant material used in the study. Financial support from the Fondo de Innovación para la Competitividad de la Región de O'Higgins (FIC-O'Higgins grant IDI 40008896–0).

References

1. McKendrick, A.: Applications of mathematics to medical problems. In: Proceedings of the Edinburgh Mathematical Society. Edinburgh, EH8 9BT|UK: The Bayes Cenre, 47 Potterrow, pp. 98–130 (1926)
2. Dourado, H., Lercher, M.: An analytical theory of balanced cellular growth. Nat. Commun. **11**(1226), 1–14 (2020). https://doi.org/10.1038/s41467-020-14751-w
3. Guekenheimer, J., Oster, G., Ipaktchi, A.: The dynamics of density dependent population models. J. Math. Biol. **4**, 101–147 (1977)
4. Chou, T., Greenman, C.: A hierarchical kinetic theory of birth, death and fission in age-structured interacting populations. J. Stat. Phys. **164**(1), 49–76 (2016). https://doi.org/10.1007/s10955-016-1524-x
5. Fieldman, R., Curry, G.: A PDE formulation and numerical solution for a boll weevil-cotton crop model. Comput. Math. Appl. **9**(3), 393–402 (1983)
6. Rojano, F., et al.: Application of a modified McKendrick-von foerster equation to predict beetle population dynamics (Xyleborus affinis) under Artificial Medium in Growth Chambers. In: AgEng Conference, pp. 1–8 (2018)
7. Rupnik, R., Kukar, M., Vra, P., Ko, D., Pevec, D., Bosni, Z.: AgroDSS : a decision support system for agriculture and farming. Comput. Electron. Agric. **161**, 260–271 (2019). https://doi.org/10.1016/j.compag.2018.04.001
8. Seebens, H., et al.: Global rise in emerging alien species results from increased accessibility of new source pools. Proc. Natl. Acad. Sci. **115**(10), E2264–E2273 (2018). https://doi.org/10.1073/pnas.1719429115
9. Kenis, M., et al.: Ecological effects of invasive alien insects. Biol. Invasions **11**(1), 21–45 (2009). https://doi.org/10.1007/978-1-4020-9680-8_3
10. Aukema, J., et al.: Economic impacts of non-native forest insects in the continental United States. PLoS ONE **6**(9), 1–7 (2011). https://doi.org/10.1371/journal.pone.0024587
11. Aukema, J., McCullough, D., Von Holle, B., Liebhold, A., Britton, K., Frankel, S.: Historical accumulation of nonindigenous forest pests in the continental United States. Bioscience **60**(11), 886–897 (2010). https://doi.org/10.1525/bio.2010.60.11.5
12. Sandoval, A., Ide, S., Rothmann, S., Zúñiga, E., Bosch, P., Pergallo, M.: Detección de Leucoptera sinuella (Reutti) (Lepidoptera: Lyonetiidae) En Chile, Con La Identificación de Algunos Parasitoides Asociados. Revista Chilena de Entomología **45**(1), 65–77 (2019)
13. San Blas, G., Quiroga, V., Holgado, M.: Detección de La Polilla Del Álamo, Leucoptera sinuella (Lepidoptera: Lyonetiidae), En Argentina. Revista de La Sociedad Entomológica Argentina **81**(1), 4 (2022)
14. Fuentes-Contreras, E., Yánez-Segovia, S., Guajardo, J.: Avances En El Manejo Integrado de La Polilla Del Álamo En Chile. Revista Frutícola **44**(2), 32–38 (2022)
15. Yánez-Segovia, S., Ramírez, C.C., Lindroth, R.L., Fuentes-Contreras, E.: Resistance against Leucoptera sinuella (Lepidoptera: Lyonetiidae) among hybrid clones of Populus spp. in central Chile. J. Econ. Entomol. **116**(5), 1662–1670 (2023)
16. Cañete-Salinas, P., de la Fuente-Sáiz, D., Yánez-Segovia, S., et al.: Use of satellite images to monitor Leucoptera sinuella leaf damage in poplar plantations in central Chile. New For. **55**, 1181–1194 (2024). https://doi.org/10.1007/s11056-024-10029-x
17. Ageno, M.: Le Radici Della Biologia, vol. 62. Feltrinelli, Genova (1986)
18. Rossini, L., Severini, M., Contarini, M., Speranza, S.: A novel modelling approach to describe an insect life cycle vis-à-vis plant protection: description and application in the case study of tuta absoluta. Ecol. Model. **409**, 1–9 (2019). https://doi.org/10.1016/j.ecolmodel.2019.108778

19. Diekmann, O., Lauwerier, H., Aldenberg, T., Metz, A.: Growth, fission and the stable size distribution. J. Math. Biol. **18**, 135–148 (1983)
20. Esperk, T., Tammaru, T.: Does the 'investment principle' model explain moulting strategies in lepidopteran larvae? Physiol. Entomol. **29**(1), 56–66 (2004). https://doi.org/10.1111/j.1365-3032.2004.0365.x
21. Taylor, R.: On 'Dyar'S Rule' and its application to sawfly Larvae. Ann. Entomol. Soc. Am. **24**(3), 451–466 (1931). https://doi.org/10.1093/aesa/24.3.451
22. Day, T., Taylor, P.: Von Bertalanffy's Growth equation should not be used to model age and size at maturity. Am. Nat. **149**(2), 381–393 (1997)
23. Glazier, D.: Beyond the '3/4-Power Law': variation in the intra- and interspecific scaling of metabolic rate in animals. Biol. Rev. Camb. Philos. Soc. **80**(4), 611–662 (2005). https://doi.org/10.1017/S1464793105006834
24. Reynolds, S., Nottingham, S., Stephens, A.: Food and water economy and its relation to growth in fifth-instar larvae of the tobacco hornworm, manduca sexta. J. Insect Physiol. **31**(2), 119–127 (1985). https://doi.org/10.1016/0022-1910(85)90016-2
25. Day, T., Rowe, L.: Developmental thresholds and the evolution of reaction norms for age and size at life-history transitions. Am. Nat. **159**(4), 338–350 (2002). https://doi.org/10.1086/338989

Short-Term Financial Management Tool Applied in Administrative-Financial Management for Sustainability in the Small Shrimp Industry

Milton Paredes-Aguirre[1], Edgar Briones-Bitar[1,2], Josué Briones-Bitar[3,4](✉), and Paúl Carrión-Mero[3,4]

[1] Facultad de Ciencias Sociales y Humanísticas, ESPOL Polytechnic University, ESPOL, Campus Gustavo Galindo, Km. 30.5 Vía Perimetral, Guayaquil 090902, Ecuador

[2] Centro de Investigaciones Rurales (CIR), ESPOL Polytechnic University, ESPOL, Campus Gustavo Galindo, Km. 30.5 Vía Perimetral, Guayaquil 090902, Ecuador

[3] Centro de Investigación y Proyectos Aplicados a las Ciencias de la Tierra, ESPOL Polytechnic University, ESPOL, Campus Gustavo Galindo, Km. 30. 5 Vía Perimetral, Guayaquil 090902, Ecuador

briones@espol.edu.ec

[4] Facultad de Ingeniería en Ciencias de La Tierra, ESPOL Polytechnic University, ESPOL, Campus Gustavo Galindo, Km. 30.5 Vía Perimetral, Guayaquil 090902, Ecuador

Abstract. Global trade is undergoing constant change, and digital transformation is key to the competitiveness of small and medium-sized enterprises (SMEs). In Ecuador, SMEs represent more than 80% of the business fabric, generating jobs, and local economic development. Ecuador is one of the world's leading shrimp exporters, accounting for approximately 20% of non-oil exports. Many SMEs are at an initial level of digitalisation, where administrative financial management is one of the main problems for a company's growth. This research focuses on analysing a proposed Short-Term Financial Management tool (STFM) through its application in a case study for administrative-financial decision-making in SMEs in the shrimp industry, with a view to their sustainability. The method compares financial management models, considering key input factors, to obtain better performance indices. The STFM tool allowed the digitisation of financial statements, optimising 50% of staff time and obtaining a projected annual growth of 2% for the company. Thus, there is great potential for SMEs in the shrimp industry to integrate more efficiently into global value chains by adopting modern technologies (progressive digital transformation). This is within the context of modernising aquaculture production and strengthening the local economy.

Keywords: digital transformation · short term financial management · business development · Sustainable Development Goals · SME

1 Introduction

Global trade bears the imprint of innovation at every stage, from small and medium-sized enterprises to large corporations [1]. Mentioning only online business transactions, overcoming the entire business chain, would occur within the scope of constant digital

J. L. Ballesteros et al. (Eds.): CITIS 2025, LNNS 1780, pp. 94–104, 2027.
https://doi.org/10.1007/978-3-032-14287-0_10

transformation [2]. Today, we are in an era of eminent digital migration of whole documents and ledgers, which have remained on the shelves of the accounting department [3].

Digital transformation represents a new industrial revolution that radically changes business models and performance [4]. The drive towards digital transformation is motivated by the great potential of new technologies to drive innovation and competitive advantage [5, 6]. The Small and Medium Enterprise (SME) sector is increasingly seen as a key driver of employment development, income growth, and job creation [7, 8].

The SME sector plays a vital role in the socioeconomic aspects of a country and is considered the backbone of an economy [9]. It is estimated that SMEs account for more than 90% of all enterprises and 45% of the total employment and contribute significantly to the Gross Domestic Product [10]. Diagnosing the sources of risk in SMEs is the most critical phase of risk management because preventive actions can only be designed to eliminate the identified risk [11]. Financial constraints can be the main issue when considering the problems faced by SMEs [12].

Financial planning is a key tool in economic management that comprises and provides control, improves the decision-making process, and increases the achievement of organisational goals and targets [13]. Planning can be flexible because it can be adapted to different areas and aspects of business and personal life [14]. A company's success depends mainly on its financial management practices, whereas its failure is due to inadequate and ineffective financial management practices [15, 16].

In Latin America, SMEs account for 99% of all non-financial enterprises and generate employment for approximately 70% of the total labour force [17]. Many SMEs have expanded into multiple foreign markets and have derived half or more of their total revenue from abroad [18]. Aquaculture is one of the main drivers of socioeconomic development in rural and coastal communities [19]. Shrimp farming started more than 600 years ago in Indonesia, and a similar practice was developed in Latin America by the Incas of Ecuador 400 years ago in temporarily flooded coastal mangrove lagoons [20]. Despite the importance of shrimp aquaculture in a country's economy, this activity is attributed to adverse environmental impacts, such as increased carbon emissions from mangrove deforestation, water pollution, loss of biodiversity, and salinisation of agricultural land [21, 22].

SMEs in Ecuador play a fundamental role in generating thousands of jobs and boosting the economy [23]. SMEs greatly help a country's economy by giving dynamism to the economy and employment, but the big problem they face is that difficulties in financing diminish growth opportunities. Ecuador's three most important exports are oil, shrimp, bananas, and flowers. Shrimp farming in Ecuador began in 1950 and has become a rapidly growing sector [24]. With a proportionally smaller territory than Brazil and Mexico, Ecuador produces more than 50% of shrimp in the Americas [20].

The study area of this work focuses on the Via a la Costa sector of the city of Guayaquil in the province of Guayas (Fig. 1). This sector is considered to have approximately 20,000 ha, which is distributed in 37.93% of Ecological Reserves (e.g. "Cerro Blanco" Protected Forest and "Manglares del Salado" Reserve) [25], and 25.72% are areas destined for shrimp extraction and farming [26]. The case study was a small company with ten workers and three senior managers (one manager and two assistant managers). The company is engaged in producing shrimp larvae and exporting shrimp pools. The specific name of the institution is kept confidential for security reasons in Ecuador [27] and for the parties involved in the case study.

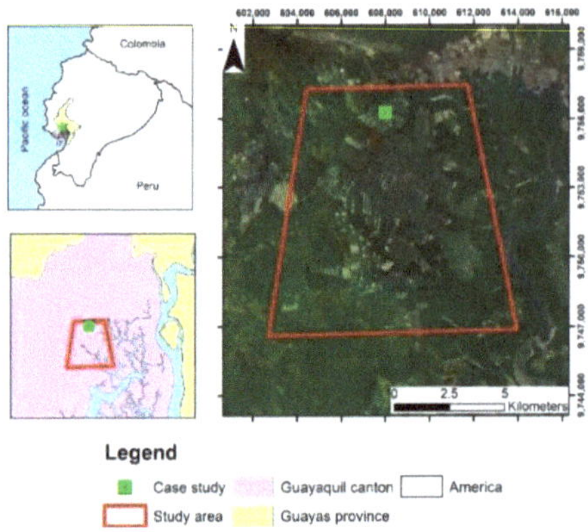

Fig. 1. Geographical location of the study area and the case study research site.

Financial planning allows projects to coordinate and take present actions based on possible scenarios to control and make optimal decisions to achieve the desired financial objectives. Therefore, the following question arises: How does the use of a digital financial management tool designed for the collaboration and collection of administrative-financial data impact strategic decision making for the sustainability of shrimp SMEs?

Therefore, the aim is to propose a digital transformation tool capable of adapting to the needs of SMEs in the shrimp sector in Ecuador through the collection of data from the financial/administrative department of a case study for decision-making in the short- and medium-term by decision-makers that allows for sustainability in shrimp SMEs.

2 Materials and Methods

For the development of this work, the authors used a qualitative-quantitative approach, which determined the variables analysed in the proposed model. With this model, the company was analysed, defining its behaviour to subsequently carry out a simulation and analyse the results obtained within a Short-Term Financial Management tool (STFM). Figure 2 presents a summary of the phases conducted in this study.

2.1 Phase 1: Data Collection for the Diagnosis

Document management is one of the priorities of this system, which focuses on short- and medium-term sustainability and the migration of physical documents (paper use) towards digital storage and communications, allowing savings not only in economic resources but also in office materials.

Phase 1: Data collection	**Phase 2:** Evaluation and administrative and financial assessment	**Phase 3:** Proposal for implementation strategies
Physical documentation: • Balance sheets • Profit and loss statements • Financial reports • Internal communications (e.g., memos or letters)	• Preparation of digital databases and spreadsheets. Short-Term Financial Management tool (STFM): • Estimates of income and expenses for the next five years. • Liquidity, guarantee, financial and debt ratios.	• Information obtained from Phase 1 and 2 • Strategies for digital transformation through STFM implementation • Key actions and tools

Fig. 2. Three-phase methodological structure proposed and followed in this research.

There are different documents in the filing cabinets of the case study company, such as balance sheets, profit and loss statements, treasury reports, and internal company communications (e.g. memos and circulars). The financial documents are separated from the institutional documents to feed the proposed spreadsheet for use by the general manager. For the time being, the task of digitalising and organising the numerical data obtained in the first meeting with the company's physical documents is also defined.

2.2 Phase 2: Financial/administrative Assessment Through Short-Term Financial Management (STFM) Tool

The financial/administrative evaluation was based on a two-year period. An operator assigned by management observed and described the operation of the case study company, facilitating a link between the general manager and the documents provided. The most important were income statements, annual balance sheets, and financial documents provided by an operator in the company.

In the short term, management will be responsible for the design, structuring, detailing, and elaboration of the spreadsheets and the subsequent proposal of the digital collaboration tool. A Short-Term Financial Management (STFM) digital collaboration tool is proposed based on short-term models that generate confidence in the management and needs of the company. It points out just a few key factors, such as the lack of versatility and dynamism that comes with the presentation of physical documentation, and even more so with handwritten documents, to move towards a digital system that translates into a reduction in time spent on productivity and greater communication between the different business spheres in the company in daily operations.

2.3 Phase 3: Proposed Implementation Strategies

Using the information collected in Phases 1 and 2, the authors developed strategies for digital transformation through the implementation of short-term financial management (STFM) tools within SMEs in the shrimp industry in Ecuador. Three strategies were proposed along with their actions and examples of key tools for their application and implementation.

3 Results

3.1 Diagnosis of the Company (Case Study)

The present documentation revealed a lack of interest in continuous improvement, that is, migration from paper to digital recording systems (e.g. spreadsheets), and a lack of environmental care with the abundant use of office supplies. These aspects are essential given the Sustainable Development Goals (SDGs), notably SDG 8 (Decent Work and Economic Growth), SDG 9 (Industry, Innovation, and Infrastructure), SDG 12 (Responsible Consumption and Production), and SDG 13 (Climate Action).

With the documents provided by the financial/administrative office of the case study company, financial information was entered into a spreadsheet (Table 1). The personnel expenses and sales income from the shrimp trade were digitalised, mainly to large exporters, who, looking for the best product, paid the prices mentioned, which were defined by the operational manager at the time of the transfer of the goods. The Profit and Loss Statement section was generated once the historical information was digitalised, which allows for a more specific evaluation of the situation in which the company finds itself.

Table 1. Information on income and expenses related to the company's profit and loss statement.

Description	Year 1 (USD)	Year 2 (USD)	Year 3 (USD)
Net amount	**473,522.41**	**215,544.79**	**389,805.91**
Sales	473,522.41	215,544.79	389,805.91
Procurement	**141,022.54**	**95,817.80**	**118,420.17**
Supplies	141,022.54	95,817.80	118,420.17
Personnel expenses	**304,831.38**	**297,674.96**	**186,508.99**
Salaries	141,706.14	161,526.28	87,812.26
Social benefits	163,125.24	136,148.68	98,696.73
Financial expenses	**445,853.92**	**393,492.76**	**304,929.16**
Profit BAIT	**27,668.49**	**- 177,947.97**	**84,876.75**
15% employees	4,150.27	- 26,692,19	12,731.51
22% income tax	5,174.01	5,174.01	5,174.01
Legal reserve	**1,834.42**	**- 21,353.76**	**10,185.21**
Net profit	**16,509.79**	**- 156,594.21**	**74,691.54**

3.2 Financial/administrative Analysis Using STFM

For the evaluation of the company, we considered two years. For both cases, we took historical records and performed financial ratios to find the critical points to be considered (Table 2).

Table 2. Financial ratios obtained from the analysis of the data provided by the company.

Financial Ratios	Year 1	Year 2	Expected
Liquidity	0.11	0.29	X > 0.50
Guarantee	1.00	1.00	X > 2.00
Cash flow	0.14	0.13	X > 0.50
Debt	0.56	0.45	X < 0.60

Using Table 1 (profit and loss statement), we performed a Monte Carlo simulation that estimated the income and expenses that must be considered to reach a certain level of expected return over time and sustainably (Tables 3 and 4). Sixteen thousand iterations were performed with 95% confidence, a Net Present Value (NPV) of $181,070.35, and an Internal Rate of Return (IRR) of 43.46%, meaning that the proposed conditions were correctly estimated to achieve 2% annual growth.

Table 3. Data entered into the Monte Carlo simulation.

Parameter	Input value
Sales volume (shrimp units)	375,000
Unit price (USD)	1.30
Unit cost (USD)	1.06
Fixed cost (USD)	32,598
Residual value (USD)	1,834
Investment (USD)	5,350
Discount rate	15%
Annual sales growth	2%

Table 4. Estimating revenues and expenditures to achieve sustainable yields.

Description	Year 1	Year 2	Year 3	Year 4	Year 5
Income	487,500.00	497,250.00	507,195.00	517,339.00	527,686.00
Expenditure	430,098.00	438,048.00	446,157.00	454,428.00	462,865.00
Fixed cost	32,598.00	32,598.00	32,598.00	32,598.00	32,598.00
Variable cost	397,500.00	405,450.00	413,559.00	421,830.00	430,267.00

3.3 Strategies to Implement the STFM Tool

Considering the context and reality of the Ecuadorian shrimp industry (e.g. export-dependent and vulnerable to international prices), and with digitalisation still in process,

Table 5 proposes three strategies to implement short-term financial management tools focused on optimising administrative and financial management.

Table 5. Strategies for digital transformation through the implementation of STFM tools.

Strategies	Actions	Key tool
Digitalisation of Cash Flow and Daily Liquidity Control	1. Digital recording of daily income and expenditures (sales, food payments, payroll, and maintenance) 2. Weekly/monthly displays of balance and liquidity projections 3. Automated alerts for accounts payable or cash shortfalls	a. A simple application (mobile or web) b. Customised system (Excel/Google Sheets)
Active Accounts Receivable and Payable Management with Financial Indicators	1. Prioritisation of customers with the best payment history 2. Scheduling payments to suppliers based on actual cash flow 3. Internal refinancing agreements with strategic alliances	a. Digital module or dynamic template to control accounts
Export Cost and Price Simulation Platform	1. Input of key variables by batch/product/export 2. Simulation of profit margins by destination market 3. Definition of minimum operating prices and profitability thresholds	a. Financial dashboard to simulate cost and margin scenarios

4 Discussion of Results

The case study, as well as the SMEs of the shrimp industry on the way to the coast, presents a standard form in family businesses, where the importance of sound administrative and financial planning is lost, making the decision-making process empirical. By incorporating and applying digital collaboration tools (STFM), SMEs are given added value, such as walking towards the digital era and improving security for the new generation of family businesses. Salazar, E. et al. [28] indicated that digitalisation has become a key factor in enhancing the competitiveness and sustainability of SMEs in Ecuador. Naranjo, F. and Barcia, I. [29] mentioned that the economic and social impacts on SMEs generate more sources of employment.

The financial liquidity ratio indicates the capacity of a company to meet its obligations. The case study was below the parameters considered acceptable in years 1 and

2 (with a slight improvement). This is of concern because the liquidity ratio is related to a firm's success or failure, according to Li [30], who used logistic regression for this demonstration. A case study of companies in Sri Lanka proved the direct relationship between these two parameters [31].

The indebtedness ratio indicates the level of debt a company can face without falling into the hands of its creditors. At this point, the company is below the upper acceptable limit, with an improvement in Year 2. Mitton [32] indicates that debt availability plays a role in increasing growth in emerging markets. A similar hypothesis was suggested by Smyth and Hsing [33], who show that economic growth, including the debt ratio, is integrated and maintains a stable relationship in the long run.

By digitising documentation (using STFM) in the case study (e.g. balance sheets, income statements, treasury reports), it was possible to estimate the income and expenses of the company, where it was possible to obtain an expected level of return sustainably (2% annual growth and IRR of 43.46%). Once the short- and medium-term goals have been stabilised and met, it is time to consider the company s long-term growth. Export cost simulations were analysed using the third proposed strategy (Table 5). Ozcalici, M. and Bumin, M. [34] recommend Multi-Criteria Decision-Making (MCDM) to improve the accuracy of stock market forecasts. In addition, Stefvy and Deva Djohan [35] add that financial ratio analysis helps hedge short-term liabilities and cash flows.

5 Conclusions

The digital collaboration tool (STFM) serves in short- and medium-term decision making and is suitable for small shrimp companies. This is a key step towards the digital transformation of recording their business activities efficiently and effectively, leaving aside the classic documentation archive, which is a decisive factor in time, paper-environment savings, and the organisation of data susceptible to other applications. In this case study, a significant reduction of 50% was estimated over time and with the continuous growth of the company (approximately 2% per year). The STFM digital financial management tool allowed for many interactions (16,000, with 95% confidence), obtaining secure short-term conditions for the company (NPV of $181,070.35, IRR of 43.46%).

Three strategies are proposed to continue with digital transformation in the medium and long term: 1) digitalisation of cash flow and daily liquidity control, 2) Active Management of Accounts Receivable and Payable with Financial Indicators, and 3) the development of a simulation platform for export costs and prices.

This research can be replicated in other SME sectors such as bananas, cocoa, and agriculture. The extension and opening of possibilities for customised financial implementation with the STFM tool guarantees digital communication of information that facilitates decision making. This makes it possible to migrate towards the progressive digital transformation of your financial administrative activities in compliance with Sustainable Development Goals 9 (Innovation), 11 (Sustainable Communities), and 17 (Partnerships) of the 2030 Agenda.

Acknowledgments. This work was supported by the research project, " Registro de sitios de interés geológicos del Ecuador para estrategias de desarrollo sostenible", with Code CIPAT-004–2024 from ESPOL Polytechnic University.

References

1. Le Bas, C., Poussing, N.: Are complex innovators more persistent than single innovators? An empirical analysis of innovation persistence drivers. Int. J. Innov. Manag. **18**(01), 1450008 (2014). https://doi.org/10.1142/S136391961450008X
2. Iglesias, O., Markovic, S., Bagherzadeh, M., Singh, J.J.: Co-creation: a key link between corporate social responsibility, customer trust, and customer loyalty. J. Bus. Ethics **163**(1), 151–166 (2020). https://doi.org/10.1007/s10551-018-4015-y
3. Markovic, S., Koporcic, N., Arslanagic-Kalajdzic, M., Kadic-Maglajlic, S., Bagherzadeh, M., Islam, N.: Business-to-business open innovation: COVID-19 lessons for small and medium-sized enterprises from emerging markets. Technol. Forecast. Soc. Change **170**, 120883 (2021). https://doi.org/10.1016/j.techfore.2021.120883
4. Holmström, J., Holweg, M., Lawson, B., Pil, F.K., Wagner, S.M.: The digitalization of operations and supply chain management: theoretical and methodological implications. J. Oper. Manag. **65**(8), 728–734 (2019). https://doi.org/10.1002/joom.1073
5. Solberg, E., Traavik, L.E.M., Wong, S.I.: Digital Mindsets: recognizing and leveraging individual beliefs for digital transformation. Calif. Manage. Rev. **62**(4), 105–124 (2020). https://doi.org/10.1177/0008125620931839
6. Sabando-Vera, D., Montalván-Burbano, N., Yonfá-Medranda, M., Parrales-Guerrero, K.: World research and intellectual structure in digital transformation on SMEs in Covid-19 times. In: Digital and Sustainable Transformations in a Post-COVID World, pp. 347–376. Springer International Publishing, Cham (2023). https://doi.org/10.1007/978-3-031-16677-8_14
7. Zayed, N.M., Mohamed, I.S., Islam, K.M.A., Perevozova, I., Nitsenko, V., Morozova, O.: Factors influencing the financial situation and management of small and medium enterprises. J. Risk Financ. Manag. **15**(12), 554 (2022). https://doi.org/10.3390/jrfm15120554
8. Sabando-Vera, D., Yonfa-Medranda, M., Montalván-Burbano, N., Albors-Garrigos, J., Parrales-Guerrero, K.: Worldwide research open innovation in SMEs. J. Open Innov. Technol. Mark. Complex. **8**(1), 20 (2022). https://doi.org/10.3390/joitmc8010020
9. Heinicke, A.: Performance measurement systems in small and medium-sized enterprises and family firms: a systematic literature review. J. Manag. Control. **28**(4), 457–502 (2018). https://doi.org/10.1007/s00187-017-0254-9
10. Tharmini, T., Lakshan, A.M.I.: Impact of financial management practices on performance of small and medium enterprises – legitimacy theory perspectives. Kelaniya J. Manag. **10**(1), 43–64 (2021). https://doi.org/10.4038/kjm.v10i1.7666
11. Dvorsky, J., Belas, J., Gavurova, B., Brabenec, T.: Business risk management in the context of small and medium-sized enterprises. Econ. Res. Istraživanja **34**(1), 1690–1708 (2021). https://doi.org/10.1080/1331677X.2020.1844588
12. Gorzeń-Mitka, I: .Interpretive structural modeling approach to analyze the interaction among key factors of risk management process in SMEs: Polish experience. Eur. J. Sustain. Dev. 8(1) (2019). https://doi.org/10.14207/ejsd.2019.v8n1p339
13. Kapitsinis, N.: The impact of economic crisis on firm relocation: Greek SME movement to Bulgaria and its effects on business performance. GeoJournal **84**(2), 321–343 (2019). https://doi.org/10.1007/s10708-018-9863-6
14. Zada, M., Yukun, C., Zada, S.: Effect of financial management practices on the development of small-to-medium size forest enterprises: insight from Pakistan. GeoJournal **86**(3), 1073–1088 (2021). https://doi.org/10.1007/s10708-019-10111-4
15. Khan, N.U., Li, S., Safdar, M.N., Khan, Z.U.: The role of entrepreneurial strategy, network ties, human and financial capital in New Venture performance. J. Risk Financ. Manag. **12**(1), 41 (2019). https://doi.org/10.3390/jrfm12010041

16. Yonfá-Medranda, M., Sabando-Vera, D.,Pérez-Gálvez, J., Jara-Alba, C.: Consumer ethnocentrism: bibliometric analysis and literature review through cognitive structure and mapping of research based on web of science (WoS) and scopus. Cogent Bus. Manag. **11**(1) (2024). https://doi.org/10.1080/23311975.2024.2377320
17. Navarrete-Chávez, M., Santillán-Coello, F., Vallejo -Villacís, M.: Ecuador Pymes: accessibility to financial services and generation of financial profitability through new markets. Polo del Conoc **7**, 1232–1245 (2022). https://doi.org/10.23857/pc.v7i6.4134
18. Knight, G., Moen, Ø., Madsen, T.K.: Antecedents to differentiation strategy in the exporting SME. Int. Bus. Rev. **29**(6), 101740 (2020). https://doi.org/10.1016/j.ibusrev.2020.101740
19. Macusi, E.D., Estor, D.E.P., Borazon, E.Q., Clapano, M.B., Santos, M.D.: Environmental and socioeconomic impacts of shrimp farming in the Philippines: a critical analysis using PRISMA. Sustainability **14**(5), 2977 (2022). https://doi.org/10.3390/su14052977
20. Monsalve, E.R., Quiroga, E.: Farmed shrimp aquaculture in coastal wetlands of Latin America — a review of environmental issues. Mar. Pollut. Bull. **183**, 113956 (2022). https://doi.org/10.1016/j.marpolbul.2022.113956
21. Ahmed, N., Thompson, S.: The blue dimensions of aquaculture: a global synthesis. Sci. Total. Environ. **652**, 851–861 (2019). https://doi.org/10.1016/j.scitotenv.2018.10.163
22. de Lima Vieira, J., dos Santos Nunes, L., Rodrigues de Menezes, F.G., Vieira de Mendonça, K., Viana de Sousa, O.: An integrated approach to analyzing the effect of biofloc and probiotic technologies on sustainability and food safety in shrimp farming systems. J. Clean. Prod. **318**, 128618 (2021). https://doi.org/10.1016/j.jclepro.2021.128618
23. Araque Jaramillo, W., Hidalgo Flor, R., Rivera Vásquez, J.: Small and Medium-Sized Enterprises in Ecuador: Evolution and challenges. J. Evol. Stud. Bus. **7**(1), 121–165 (2022). https://doi.org/10.1344/jesb2022.1.j101
24. Viera-Romero, A.M., Diemont, S.A.W., Selfa, T.L., Luzadis, V.A.: The sustainability of shrimp aquaculture: an emergy-based case study in the Gulf of Guayaquil thirty years later. Renew. Sustain. Energy Rev. **194**, 114326 (2024). https://doi.org/10.1016/j.rser.2024.114326
25. MAE-Ecuador, Sistema Nacional de Áreas Protegidas del Ecuador. Accessed: Oct. 10, 2024. [http://areasprotegidas.ambiente.gob.ec/areas-protegidas/reserva-de-producción-faunística-manglares-el-salado
26. Chunga de la Torre, F.E., Rada Alphrecht, R.E, Sandoya Lara, R.A: .Desarrollo urbano estratégico para Vía a la Costa, Guayaquil., UCSG. Accessed 10 Oct 2024. http://repositorio.ucsg.edu.ec/handle/3317/20/browse?type=author&sort_by=1&order=ASC&rpp=65&etal=-1&value=Chunga+de+la+Torre%2C+Félix+Eduardo&starts_with=G
27. Montalvo, D.: Ecuador registra los niveles más altos de crimen, inseguridad y delincuencia del continente, Corporación Participación Ciudadana. Accessed 10 Oct 2024. https://www.participacionciudadana.org/web/wp-content/uploads/2024/02/A1-Ecuador-registra-los-niveles-mas-altos-de-crimen.pdf
28. Salazar, E., Cajas, M., Angulo, D., Rueda, F.: Digitalization of SMEs in Ecuador: Challenges, Opportunities, and Success Stories.. Rev. Científica Multidiscip. GENERANDO **6**(1), 513–524, (2025). https://revista.generando.org/revista/index.php/RCMG/article/view/420/421
29. Naranjo Armijo, F.G, Barcia Zambrano, I.A.: Efecto económico de la innovación en las PYMES del Ecuador," J. Econ. Soc. Sci. Res. 1(1), 61–73 (2021). https://doi.org/10.55813/gaea/jessr/v1/n1/21
30. Li, K.: Liquidity ratios and corporate failures. Account. Financ. **64**(1), 1111–1134 (2024). https://doi.org/10.1111/acfi.13174
31. Madushanka, K.H.I., Jathurika, M.: The impact of liquidity ratios on profitability. Int. Res. J. Adv. Eng. Sci.. **3**(4), 157–161 (2018). https://www.researchgate.net/profile/Jathurika-Gowthaman/publication/331035336_The_Impact_of_Liquidity_Ratios_on_Profitability_With_special_reference_to_Listed_Manufacturing_Companies_in_Sri_Lanka/links/5c6259d545851582c3e18187/The-Impact-of-Liquidity-Rat

32. Mitton, T.: Why have debt ratios increased for firms in emerging markets? Eur. Financ. Manag. **14**(1), 127–151 (2008). https://doi.org/10.1111/j.1468-036X.2007.00430.x
33. Smyth, D.J., Hsing, Y.: In search of an optimal debt ratio for economic growth. Contemp. Econ. Policy **13**(4), 51–59 (1995). https://doi.org/10.1111/j.1465-7287.1995.tb00731.x
34. Ozcalici, M., Bumin, M.: Parameter selection for long short-term memory networks with multi-criteria decision-making tools: an application for G7 countries stock market forecasting. Neural Comput. Appl. **36**(36), 22731–22771 (2024). https://doi.org/10.1007/s00521-024-10433-6
35. Stefvy, D., Deva, D.: The analysis of liquidity ratio as tool to measure financial position in PT. Iplug Indonesia at Medan. Int. J. Heal. Econ. Soc. Sci. **5**(3), 322–327 (2023). https://doi.org/10.56338/ijhess.v5i3.3934

NaturaTECH: Biomimetic Methodology for the Evolutionary Design of Nature-Inspired Products

Roberto Moya-Jiménez[1](✉), Andrea Goyes-Balladares[2], Víctor Molina-Dueñas[3], Andrés Medina-Moncayo[3], and Wilmer Chaca-Espinoza[3]

[1] Faculty of Design and Architecture, Research and Development Directorate, Technical University of Ambato, Ambato 180207, Ecuador
rc.moya@uta.edu.ec

[2] Department of Architecture Graphic Expression, Universitat Politécnica de Valencia, 46022 Valencia, Spain
agoybal@upv.edu.es.com

[3] Faculty of Design and Architecture, Technical University of Ambato, Ambato 180207, Ecuador
{vh.molina,as.medina,wg.chaca}@uta.edu.ec

Abstract. This article introduces NaturaTECH, a specialized biomimetic methodology for product design in the industrial field, developed as part of a research project executed at the Technical University of Ambato, Ecuador. The proposal responds to the need to systematize an approach that allows the integration of functional, structural and adaptive principles of nature in a structured way in creative, technical and productive processes. NaturaTECH is structured in six phases: exploration, functional analysis, abstraction of principles, ideation, prototyping and validation. Its construction is based on an in-depth theoretical review, interviews with experts and case analysis, complemented by experimental phases of application and validation. Theoretical foundations, applications in industrial sectors, methodological strategies, preliminary results and a critical discussion on the implications of this approach are presented. The study provides concrete tools for sustainable design and proposes an innovative framework that dialogues with natural intelligence, aimed at creating efficient, responsible and culturally relevant products.

Keywords: Biomimetics · Industrial Design · Methodology · Sustainability · Bionics · Applied Innovation

1 Introduction

Today's industrial design faces a triple demand: to innovate, to reduce environmental impact and to respond to new ways of living, consuming and producing. Faced with a linear development model that has proven to be environmentally unsustainable, alternative approaches emerge that seek to transform the paradigms of design from their

J. L. Ballesteros et al. (Eds.): CITIS 2025, LNNS 1780, pp. 105–115, 2027.
https://doi.org/10.1007/978-3-032-14287-0_11

epistemological foundations. Biomimicry appears to be one of the most promising, not only because of its technical potential, but also because of its ability to reformulate the relationship between technology and nature.

Taking inspiration from nature has been a constant in the history of design; however, biomimicry goes beyond the symbolic or aesthetic. It is about understanding how nature works, how it solves complex problems with energy efficiency, local materials, and regeneration logics [1] This view becomes especially valuable in a context where efficiency, resilience and sustainability are no longer optional, but fundamental requirements.

Despite its growing relevance, the implementation of biomimicry in industrial design continues to face significant challenges [2]. Existing methodologies are often scattered, piecemeal or focused on formal transfer, rather than functioning. There are epistemological, disciplinary, and technical barriers that hinder its full adoption. That is why there is a need for a specific, step-by-step methodology to guide the designer in the practical application of biomimetic thinking.

This article presents the development, fundamentals and scope of NaturaTECH, a structured methodology that systematizes the transfer of natural principles to product design. This proposal has been built within the framework of a 24-month research project, funded by the Directorate of Research and Development (DIDE) of the Technical University of Ambato. Its main objective is to generate a replicable and scientifically grounded model, which allows addressing design challenges from a logic inspired by ecosystems, the adaptive efficiency of organisms and the intelligent patterns of the natural world.

2 Theoretical Review and State of the Art

2.1 From Natural Inspiration to Biomimetic Thinking

Since the origins of humanity, the observation of nature has been key in the evolution of design and technology. The wings of birds, the shape of shells or the structure of honeycombs have served as models for building tools, houses or vehicles. However, this link has historically been intuitive, empirical, or symbolic. It is in recent decades that a more robust theoretical body has emerged: biomimicry, defined by [3] as the science that studies and emulates the models, systems and elements of nature with the aim of solving human problems.

Biomimicry differs from mere formal inspiration by its emphasis on the functionality and efficiency of natural systems. It is not a question of copying the shape of a leaf, but of understanding how it regulates the exchange of water and gases; not to imitate the appearance of a bone, but to study its internal structure to optimize resistance with the minimum of material [4].

2.2 Evolution of the Biomimetic Approach in Industrial Design

Over the past 20 years, multiple initiatives have emerged that seek to integrate biomimicry into industrial design, from experimental approaches at universities to developments applied at companies such as Mercedes-Benz (aerodynamic shape of boxfish) or Nike (flake-inspired fabrics). In parallel, institutions such as the Biomimicry Institute and the Biomimicry 3.8 network have promoted guides, search platforms and training programs that have expanded the field.

However, as [5] warn, many of these experiences are one-off and lack a consolidated methodological basis. The absence of a common structure has limited the systematization of knowledge, making it difficult to evaluate and reproduce results. Studies such as those by Alshammari & [6, 7] show the need to establish conceptual and methodological frameworks that guide bioinspired design from inception to validation.

2.3 Existing Methodologies: Comparative Review

There are several methodologies that use analogy or biotransfer in the design process. Among them are:

- Biomimetic Design Spiral [8]: structured in six circular stages (identify, interpret, abstract, emulate, evaluate, adjust), it proposes an iterative approach that connects human challenges with natural principles.
- Top-down Design Process [9, 10]: It starts from human needs to identify key functions and then looks for natural solutions that meet them.
- BREED [11]: Methodology for structural applications that incorporates deep biological analysis phases prior to technical synthesis.

These methodologies share a structure based on phases of identification, abstraction and application, but they have common weaknesses: little clarity in the criteria for selecting organisms, poor integration with computer-aided design tools, low applicability in Latin American contexts and limited empirical evidence.

2.4 NaturaTECH's Differential Contributions

Faced with this panorama, NaturaTECH emerges as a situated and applied proposal. Its structure incorporates learning from previous methodologies but adapts them to the needs of industrial design in Latin American environments. Among his main contributions are:

- The use of integrated functional and morphological criteria to select natural referents.
- The explicit inclusion of CAD tools and rapid prototyping from the ideation phase.
- The development of a detailed manual validated by functional prototypes in three industrial sectors.
- The generation of scientific products and intellectual property, ensuring academic impact and applicability.

This approach not only strengthens the scientific dimension of biomimetic thinking but translates it into an operational route that can be replicated, taught, and transferred to companies, educational institutions, and design communities (Table 1).

Table 1. Comparison of biomimetic design methodologies

Methodology	Main Phases	Strengths	Limitations
Biomimetic Design Spiral (B3.8)	Identify, Interpret, Abstract, Emulate, Evaluate, Adjust	Iterative approach, adaptable to different contexts	Limited integration with digital tools and manufacturing technologies
Top-Down Design (López & Berges)	Define Function → Find Analogy → Apply Structure	Enables functional translation from specific challenges	Dependence on the designer's criteria; Low phase standardization
BREED (Fattepur et al.)	Biological Selection → Evaluation → Synthesis → Transfer	Focus on structural engineering; Good technical documentation	High complexity in early stages; requires specialized biological counseling
NaturaTECH	Exploration → Analysis → Abstraction → Ideation → Prototype → Validation	Integration with CAD/CAM, direct applicability in industrial design, validation with real prototypes	Still in the process of longitudinal validation; requires interdisciplinary work

3 General Project Methodology and Applied Phases of NaturaTECH

3.1 Methodological Approach of the Research Project

The project was developed under a qualitative, exploratory, descriptive and applied methodological approach, focused on building practical and grounded knowledge in the field of bio-inspired design. The overall research strategy combined desk review, comparative analysis of existing methodologies, semi-structured interviews with experts, case studies, and prototype experimentation.

The phases of the project were organized as follows:

Bibliographic review: More than 100 scientific sources on biomimicry, bionics, sustainable design and functional transfer strategies from nature to design were identified and systematized.

Selection and analysis of case studies: More than 20 international and regional experiences of products inspired by biological systems (mobility, architecture, packaging, textiles) were studied.

Design of the NaturaTECH methodology: From the previous analysis, a sequential process of six phases was built, with criteria, tools and examples for each stage.

Application of the methodology: The NaturaTECH process was applied in three real cases: ergonomic furniture, biodegradable packaging and technical textile with selfrepairing microstructure.

Validation: Co-evaluation and focus group sessions were held with designers, students, teachers and users to assess the usefulness, replicability and perception of innovation of the methodology.

4 Description of the NaturaTECH Process

The NaturaTECH methodology is made up of six structured and sequential phases, each with specific activities and tools to ensure the correct transfer of natural knowledge to industrial design:

Phase 1: Exploring Natural Phenomena

It begins with the definition of the design challenge and the search for biological references with functional solutions to equivalent problems. Databases such as AskNature, BioSearch and applied biology articles are used here. Example: For an ergonomic seat, protective shells, exoskeletons and support systems in molluscs or insects are studied.

Phase 2: Functional and structural analysis

It consists of breaking down the behavior of the biological system: how it distributes loads, how it adapts to the environment, what materials make it up, and how it selforganizes. Morphogenetic analysis, basic 3D modeling, and functional biomimetic files are applied.

Phase 3: Abstraction of principles

Here the key principles that make the natural system efficient are formulated: economy of materials, optimized geometry, localized flexibility, etc. Irrelevant elements are eliminated, and biological logic is translated into design language. Tools: attribute map, transfer matrix, functional analogies.

Phase 4: Design Solutions Ideation

From the abstracted principles, ideas are generated for products that respond to the original challenge from a biomimetic logic. Creative divergence and the use of techniques such as brainstorming, evolutionary sketching, and multi-criteria selection are encouraged.

Phase 5: Prototyping and experimentation

The selected concept is modeled and manufactured using rapid prototyping technologies (3D printing, laser cutting, structural simulation). This phase allows the technical feasibility and behavior of the design to be validated under controlled conditions.

Phase 6: Validation and feedback

The product undergoes functional testing, user evaluation, comparative validation against traditional solutions, and impact analysis. Feedback allows you to iterate the design or reformulate previous phases.

4.1 Tools Developed within NaturaTECH

To facilitate the application of this methodology, complementary tools were designed to guide the designer in each phase:

Natural Systems Functional Analysis Sheets

- Biological-technical translation matrix
- NaturaTECH Canvas for Process Registration

- Multi-Criteria Ideation and Evaluation Template
- Functional validation and sustainability checklists

These tools have been compiled in a methodological manual in the process of review and publication, with an open academic license for use by universities, design centers and companies.

5 Application of NaturaTECH in Industrial Sectors

The validation of the NaturaTECH methodology was carried out through its implementation in three key fields of industrial design: ergonomic furniture, sustainable packaging and technical textiles. These sectors were selected for their socio-economic relevance, their environmental impact and their potential to integrate bio-inspired solutions with high added value.

5.1 Case 1: Ergonomic Furniture Inspired by Bone Structures and Exoskeletons

Design problem: Developing a chair that adapts to the morphology of the human body, minimizes pressure points, and uses less material without sacrificing structural strength.

Exploration phase: natural structures such as vertebrae, ribs, turtle shells and insect exoskeletons were studied. These structures have double curvature, internal lightening cavities and adaptive load behavior.

Analysis and abstraction: principles such as geometric load distribution, localized stiffness, modularity of parts and alveolar geometries were extracted.

Resulting design: a monostructural chair with honeycomb-inspired honeycomb backrest, flex zones at lumbar contact points, and legs with geometry derived from palm leaf spokes. It was manufactured using 3D printing in reinforced PLA and evaluated by 10 users.

Results: A 22% reduction in the use of material compared to previous designs, better ergonomic adaptation and high perception of comfort and lightness were evidenced.

5.2 Case 2: Biodegradable Packaging with Natural Protection Strategies

Design problem: Creating secondary packaging for fragile products that is biodegradable, structurally efficient, and with functional aesthetics.

Exploration phase: structures such as nut shells (walnut, coconut), seed capsules (moringa, maple), and encapsulated spores were analyzed. All have protection, cushioning, ventilation and ease of opening functions.

Abstraction: Three key principles were identified: double layer with air chamber, compression geometry, adhesive-free closure.

Resulting design: Modular pressed starch packaging, with fins inspired by flying seed wings and mushroom-like vent slots. Made with 3D printed molds and biodegradable materials.

Results: it passed drop and humidity tests and achieved a compostability rate of 85% in 30 days. Its potential in low-impact logistics chains was validated.

5.3 Case 3: Technical Textile with Self-Healing and Thermal Adaptation Capacity

Design problem: to develop a textile surface for technical clothing with thermal microadaptation and surface repair against abrasion.

Exploration phase: reptile skins, butterfly wings, plant cuticles and archer fish structures were studied. These surfaces exhibit mechanisms of scaling, microcapillarity, and protein recomposition.

Functional analysis: principles such as flexible overlays, micro-vented channels, and tensioned fibers that respond to changes in temperature or pressure were extracted.

Resulting design: textile composed of thermoactive two-component fibers, with a flaky surface structure printed in flexible TPU, designed to reconform against damage by mild abrasion.

Results: After wear tests, 34% of the surface texture was self-repaired in 24 h.

A thermal differential of up to 2.8 °C was also measured in climatic chambers.

From these three pilot cases, four key dimensions were assessed (Table 2):

Table 2. Overall results of the NaturaTECH application

Dimension	Leading indicator	Average result
Sustainability	Material Reduction / Biodegradability	Up to 30%
Functionality	Adaptability/Responsiveness to Environment	Loud
Perceived innovation	Expert and user evaluation	8.6 / 10
Methodological replicability	Autonomous use of NaturaTECH phases	80% understanding

These data show that NaturaTECH is not only applicable, but also effective in generating products with high functional, ecological and formal value, without requiring complex technologies or depending exclusively on the aesthetic criteria of the designer.

6 Critical Discussion and Contributions to the Field of Industrial Design

6.1 Methodological Contribution

One of NaturaTECH's main contributions to industrial design is to have systematized a replicable methodological process to apply biomimetic principles in product development. Unlike previous approaches focused exclusively on form or loosely structured analogue processes, NaturaTECH introduces a functional and evolutionary logic that allows the designer to:

- Understand the biological phenomenon in its structural, functional and ecological complexity.

- Abstract key principles from a scientific perspective and not just aesthetics.
- Integrate digital design and prototyping tools from early stages.
- Validate solutions in real contexts, with criteria of efficiency, ergonomics and sustainability.

This methodology is coherently articulated with the trends in user-centred design, ecodesign and behavioural engineering. Its implementation offers a technical and conceptual alternative to rethink design from a more integrative epistemology, where natural knowledge becomes an active resource.

6.2 Relevance in the Latin American Context

In Latin America, where industrial processes often face budgetary, technical and training constraints, bio-inspired design can seem like a luxury. However, NaturaTECH demonstrates that, with proper systematization, it is possible to incorporate natural efficiency logics using medium-access technologies, such as 3D printing, free modeling software, and materials of local biological origin [12].

In addition, by proposing a methodological approach validated in a Latin American context, NaturaTECH breaks with the dependence on models developed exclusively in North America or Europe, and opens the way to a situated biomimicry, adapted to the productive, educational and ecological realities of the region.

7 Pedagogical and Academic Implications

The structure of NaturaTECH has been tested in academic settings at undergraduate and graduate level, generating a better understanding of the relationship between biology and design. The use of analogies, functional sheets, transfer matrices and graphic diagrams has facilitated interdisciplinary work between designers, engineers and biologists [13].

Its implementation as a training module has proven to be effective in improving students' critical thinking, technical innovation, and environmental commitment. In this sense, the methodology can be replicated in other universities as part of sustainable design curricula, applied bionics or bioinspired innovation.

7.1 Study Limitations and Challenges

While NaturaTECH has proven to be effective in its three pilot cases, certain limitations are recognized:

- Complexity in the abstraction phase: requires basic training in biological functional principles, which can be an initial barrier for designers without scientific training.
- Demanding interdisciplinarity: Its optimal application requires close collaboration between designers, engineers, biologists and manufacturers.
- Limited validation: to date, the prototypes developed have been tested on pilot scales and in semi-controlled environments; Its implementation in real large-scale industrial processes is lacking.

- Need for sectoral adaptation: although transferable principles have been identified, some sectors (such as automotive or medical) require greater technical and regulatory requirements that will need to be integrated into future versions of the methodology (Table 3).

Table 3. Figure-summary: NaturaTECH's contributions and limitations

Dimension	Contributions	Limitations
Methodological	Replicable process, clear phases, visual tools	Complexity in biological abstraction
Technique	Integration with CAD/CAM, rapid prototyping, alternative materials	Requires technical knowledge for manufacturing and functional analysis
Academic	Effective didactic transfer, articulation between disciplines	Previous training in functional biology may be limited in design schools
Productive	Applicability in sectors such as furniture, packaging and textiles	Lack of validation in large-scale industrial processes
Environmental/Social	Promotes sustainability, efficiency, and culturally relevant design	Few specific regulations for bio-inspired products

8 Conclusions and Future Projections

8.1 General Conclusions

The development of the NaturaTECH methodology constitutes a significant advance in the formalization of biomimetic thinking as a design strategy applied to the Latin American context. Through an orderly, interdisciplinary and technically feasible process, NaturaTECH allows functional principles of nature to be transferred to real products with criteria of innovation, sustainability and efficiency.

The three pilot cases (furniture, packaging, technical textiles) have demonstrated the technical and conceptual feasibility of the methodology. In addition, its usefulness as a pedagogical, training and professional tool, with a high degree of replicability, has been verified.

In a time of ecological crisis, NaturaTECH offers industrial design an alternative path that not only improves the performance of products, but also proposes a new ethical, ecological and technical framework based on the evolutionary logic of life.

8.2 Projections and Continuity of the Project

The next phases of this project include:

- Expand validation with more functional prototypes in sectors such as construction, urban mobility and medical technology.

- Consolidate the NaturaTECH methodological manual with application guides by sector.
- Create an interactive digital platform that accompanies the methodological process and allows each phase to be documented with examples.
- Initiate processes of intellectual protection of the designs generated and promote their linkage with companies in the productive sector.
- Incorporate NaturaTECH as an elective subject in undergraduate and graduate programs in industrial design, architecture, and engineering (Table 4).

Table 4. Opportunities for impact

Field	Expected impact
Industrial design	Improvement in product quality, adaptability and sustainability
Higher education	Strengthening interdisciplinary training and systems thinking
Business Sector	Technology transfer and differentiation in competitive markets
Environment	Reducing environmental impact through more efficient and durable products
Society	Promoting a design culture based on respect for natural systems

Acknowledgments. Special thanks to the Dirección de Investigación y Desarrollo of the Universidad Técnica de Ambato, which supported this work, as well as to the Facultad de Diseño y Arquitectura, the public and private institutions, people related, researchers, and entire university team that collaborated on and monitored this work.

This work was financed by the Dirección de Investigación y Desarrollo of the Universidad Técnica de Ambato, grant number UTA-CONIN-2024-0198-R.

References

1. Goyes-Balladares, A., Moya-Jiménez, R., Molina-Dueñas, V., Chaca-Espinoza, W., MagalRoyo, T.: What inspires biomimicry in construction? Patterns, trends, and applications. Biomimetics. **10**(5), 259 (2025). https://doi.org/10.3390/biomimetics10050259
2. Moya-Jiménez, R., Goyes-Balladares, A., Rivera-Valenzuela, M., Magal-Royo, T.: User-Centered Evaluation of the ARTH-Aid ExoGlove: Perspectives of Patients and Therapists in Rheumatoid Arthritis Rehabilitation. *Designs (Basel).* **9**(2) (2025). https://doi.org/10.3390/designs9020042
3. Okeke, F.O., Okekeogbu, C.: Biomimicry and Sustainable Architecture: A Review of Existing Literature (2017). [Online]. Available: www.evolo.us
4. Hallstedt, S.: A foundation for sustainable product development a foundation for sustainable
5. Gebeshuber, I.C., Gruber, P., Drack, M.: A gaze into the crystal ball: biomimetics in the year 2059. Proc. Inst. Mech. Eng. C J. Mech. Eng. Sci. **223**(12), 2899–2918 (2009). https://doi.org/10.1243/09544062JMES1563
6. Zari, M.P.: Regenerative Urban Design and Ecosystem Biomimicry (2018). [Online]. Available: www.routledge.com/Routledge-Research-in-Sustainable-Urbanism/book-series/

7. Margariti, E.V.: Biomimetic Approaches to Design a Dissertation Submitted for the Degree of (2020)
8. López Forniés, I., Berges Muro, L.: A top-down biomimetic design process for product concept generation. International Journal of Design and Nature and Ecodynamics. **7**(1), 27–48 (2012). https://doi.org/10.2495/DNE-V7-N1-27-48
9. de Sá, A.A.M., Viana, D.M.: Design and biomimicry: a review of interconnections and creative potentials. Biomimetics. **8**(1) (2023). https://doi.org/10.3390/biomimetics8010061
10. Alshammari, M., Simpson, A.: Towards a Principled Approach for Engineering Privacy by Design
11. Stevens, L., Kopnina, H., Mulder, K., De Vries, M.: Biomimicry design thinking education: a base-line exercise in preconceptions of biological analogies. Int. J. Technol. Des. Educ. **31**(4), 797–814 (2021). https://doi.org/10.1007/s10798-020-09574-1
12. Perricone, V., Langella, C., Santulli, C.: Sustainable biomimetics: a discussion on differences in scale, complexity, and organization between the natural and artificial world. In: Environmental Footprints and Eco-Design of Products and Processes, pp. 171–193. Springer (2022). https://doi.org/10.1007/978-981-19-1812-4_7
13. Vasyaycheva, V.A.: Industrial enterprise innovative design management technology map. Vestnik Samara Univ.. Econ. Manag. **13**(3), 71–78 (2022). https://doi.org/10.18287/2542-0461-2022-13-3-71-78

Establishment of Pure Lines of *Hemileia Vastatrix* Berk. And Br, Causal Agent of Coffee Rust, through Successive Cycles under *in Vitro* Conditions

Adamary Vásquez-Tituana[2], Freddy Tinoco[1,2], Daniela Román-Cáceres[1,2], and Jorge I. Armijos-Rivera[1,2,3](✉)

[1] Universidad Nacional de Loja, Loja, Ecuador
jorge.armijos@unl.edu.ec
[2] Grupo de Investigación en Genética y Biología Molecular, Universidad Nacional de Loja, Loja, Ecuador
[3] Centro de Biotecnología, Universidad Nacional de Loja, Loja, Ecuador

Abstract. This study focused on the development of pure strains of the fungus *Hemileia vastatrix*, which causes coffee rust, using a cyclc *in vitro* culture system on leaf discs of *Coffea arabica var.* Typica. Controlled humidity, temperature, and light conditions were established to ensure leaf disc viability over an extended period. Microtubes containing various concentrations of spores obtained from infected tissue were prepared for inoculation, selecting those with the highest density of mycelium and hyphae. The adapted protocol allowed the development of the fungus to be followed from initial infection to the advanced stages of the pathogenic cycle. Over time, morphological changes and key stages of fungal development were observed, facilitating the creation of a stable system for reproduction and analysis in the laboratory. The methodology used represents a significant advance for future research on the biology of the pathogen, its interaction with the host, and the evaluation of control strategies under reproducible conditions.

Keywords: *Hemileia vastatrix* · *Coffea arabica* · continuous cycles

1 Introduction

The coffee sector is a fundamental pillar of the agricultural economy in more than 75 tropical countries around the world, where its production is highly vigorous and offers significant social and ecological implications [1, 2] More than 9 million metric tons are produced annually, making it the second most traded commodity on the planet [1, 3]. Therefore, it constitutes one of the main sources of income for approximately 300,000 small farmers and is integrated into large agroforestry systems that maintain hundreds of ecosystem services and a high biodiversity of flora and fauna [1].

However, coffee production is exposed to significant threats, and these negative impacts are expected to increase over the years as a result of climate change [4–6].

J. L. Ballesteros et al. (Eds.): CITIS 2025, LNNS 1780, pp. 116–125, 2027.
https://doi.org/10.1007/978-3-032-14287-0_12

Therefore, a decline in coffee quality of approximately 80% is projected, which will increase production costs and reduce farmers' incomes, in addition to posing a risk to the survival of certain coffee varieties [6, 7]. This also increases outbreaks of diseases caused by plant pathogens, and alters the interaction between host and pathogen, which increases the manifestation of emerging pathogenic strains, putting food safety and environmental conditions at risk [6, 8].

One of the most relevant parasites, responsible for multiple epidemics that have invaded the largest coffee-producing areas in the world, is rust, caused by the fungus *Hemileia vastatrix* Berk and Broome [9, 10]. This infectious fungal agent is a biotrophic that was first observed in 1861 in wild coffee plants in East Africa. Eight years later, it would be identified and classified as the causal pathogen of the phytosanitary problem that causes millions of dollars in annual losses [10–12]. The outbreaks that have occurred in America, Asia and Africa over the years have deteriorated the yield and quality of coffee, especially limiting the production chain of the varieties *Coffea arabica* L., 1753 (Arabica coffee) and *Coffea canephora* Pierre ex A. Froehner, 1987 (Robusta coffee), the most cultivated species in all marketing countries [12–14]. Specifically, in Latin America, worrying epidemics caused by rust were reported in 2008 and 2013 in Colombia and Ecuador, respectively [15, 16].

H. vastatrix affects coffee leaves, causing defoliation, and even the branches of the plants may die [16]. The first visible symptoms are orange chlorotic spots. This hemicyclic fungus produces urediniospores, teliospores and basidiospores; however, its life cycle is urediniosporic, being this almost the only source of inoculum [12, 16]. This strategy confers a high infection efficiency to the fungus, but has also limited its study with experimental tools in the laboratory due to its strictly parasitic nature, which makes its growth and cultivation impossible in conventional or artificial media. These conditions have limited progress in the genetic characterization of the parasite, the study of the interaction between the pathogen and its host, the evaluation of its behavior under different environmental parameters, and the search for precise and effective treatments for its control in the context of food safety and good production [17–19].

Faced with these challenges, alternative methodologies have been proposed for the study of *H. vastatrix* under controlled conditions. Therefore, the creation of a functional in vitro system based on continuous inoculation cycles in leaf discs, which allows for reinfection, is proposed. However, this technique must be optimized under the necessary physical and environmental conditions, such as temperature, relative humidity, and photoperiod, which directly affect the stability of fungal structures [20].

This technique would allow the development of pure strains of the pathogen and represent a significant contribution to basic science, as well as paving the way for applied studies on genetic resistance or testing the efficacy of specific molecules for its treatment. Furthermore, the development of more specific management alternatives, less dependent on synthetic fungicides, seeks to reduce the adverse effects of traditional control practices on the environment [21, 22].

2 Materials and Methods

2.1 Sampling Area and Obtaining Plant Material

Plant material was collected from the "El Naranjo" farm, located in the Quilanga canton, Loja province, Ecuador (4°17′53" S, 79°23′55" W). This area is located at 1923 masl and has an average temperature of approximately 19 °C. For sampling, eight leaves from coffee plants of typical varieties were selected, with different levels of severity of *H. vastatrix* infection according to the Sinavef-Lanref scale (Table 1). A total of 1.5 mg of fungal spores were collected with a sterile scalpel and placed in pre-sterilized vials containing 2 ml of distilled water. Healthy coffee leaves were also collected for infection assays. These were sterilized by placing them in a 0.5% sodium hypochlorite solution. They were subsequently washed with 70% alcohol and finally cleaned with distilled water (each process with the respective solutions was carried out for 1 min). After the leaves had dried, cuts were made on the underside using a 2.5 cm circumference metal mold. They were then placed with forceps in Petri dishes (two per box) lined with filter paper moistened with 17 ml of distilled water to keep the plant tissue discs intact.

Table 1. Levels and percentages of severity in the leaf area according to Sinavef-Lanref.

Severity levels	Leaf damage caused by coffee rust	
Low	Slow phase: Translucent (pale) spot	1–5%
Medium	Rapid phase: Reduced number of orange spores	6–29%
High	Terminal phase: Large quantities of spores ready to disperse.	>30%

2.2 Preadaptation of Uninfected Leaf Discs

To accommodate both the uninoculated leaf discs and those containing *H. vastatrix*, controlled conditions were required for the trials, so two incubators were built. These simple incubators were manufactured with the measurements. Their structure contained three LED lamps to simulate natural light, 6 m of transparent plastic, and an electronic timer. Finally, the asepsis of the work area must be continuously ensured to avoid cross-contamination. The Petri dishes containing the uninoculated coffee leaf discs (a total of 72 coffee discs distributed in 36 boxes) were placed in the incubators, previously calibrated to a specific temperature and relative humidity.

2.3 Inoculation of *H. Vastatrix* and Obtaining Pure Lines

From the suspension with the sampled spores, 1 ml was taken to count them, for this purpose a Neubauer. Once the count was completed, inoculation of *H. vastatrix* spores continued, taking 10 microtubes of 1 ml each. From this suspension, it was diluted to obtain a concentration of $10 \times 10\_^{6}$ spores, a quantity that was finally inoculated into the plant tissue discs. After the first 12 h after inoculation, the state of the spores was observed under an optical microscope with 4x, 10x and 40x lenses. Once the spores were adapted, the new pure lines of the fungus causing the rust were extracted, for which continuous infection cycles were carried out. Likewise, to determine the degree of severity of the disease in the leaves, the scales described in Tables 1. The entire process is summarized in Fig. 1.

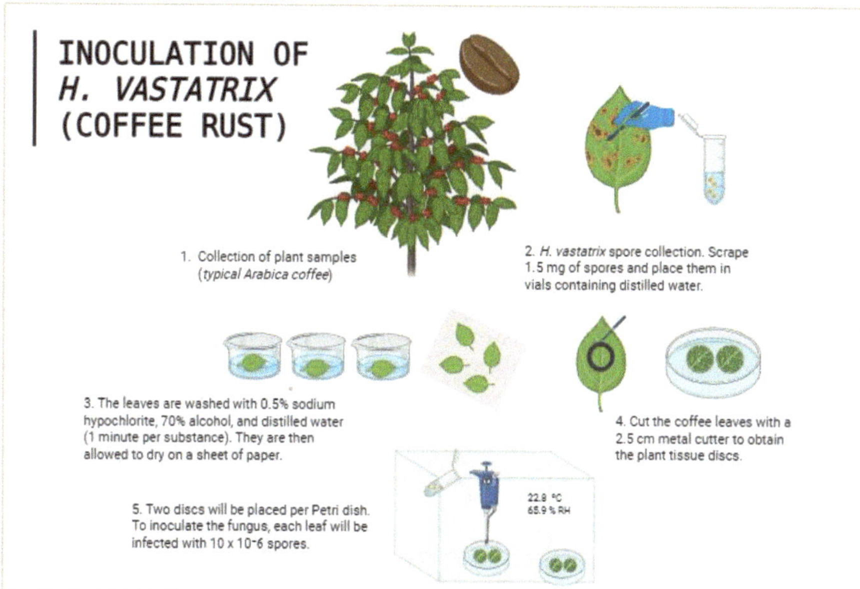

Fig. 1. *H. vastatrix* inoculation process described in 5 graphic steps. Made in Biorender.

3 Results

The spore concentration obtained from *H. vastatrix* scrapings was calculated based on the ratio of spore mass to water volume. Three microtubes were identified as the most productive, with concentrations of 2800, 2900, and 4000 u/ml. The selection criterion was based on the highest concentrations of mycelium and hyphae to initiate the inoculation process in the available plant tissue discs from coffee plants (Fig. 2).

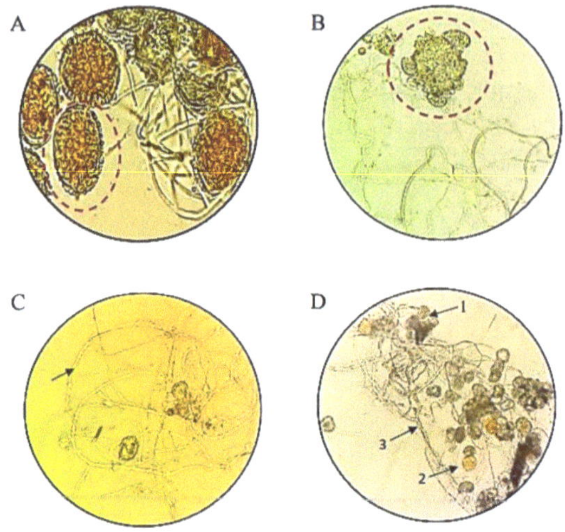

Fig. 2. Microtube contents observed under a microscope. (A) Microtube with spores; (B) with mycelium; (C) with hyphae; (D) a mixture of mycelium, spores, and hyphae, where the following can be identified: (1) mycelium, (2) hyphae, and (3) spores. The precise identification of these structures allowed us to select samples with the highest potential for fungal viability for the inoculation process.

The adaptation of the coffee leaf discs was achieved by establishing an initial temperature of 18.9 °C, obtained by calibrating three LED lamps with daylight intensity and using a 1 mm thick transparent mica sheet. This configuration allowed an optimal temperature of 22.8 °C to be reached. It was observed that increasing the distance between the LED lamps and the Petri dishes decreased the temperature, generating a latent period in the discs. Conversely, reducing this distance increased the temperature, causing dehydration and drying of the coffee leaf discs. Once optimal thermal conditions were established, complementary parameters such as relative humidity and water volume were defined, which contributed to creating a balanced environment to maintain disc viability over time. Under these conditions, adequate disc adaptation was observed, crucial for the reproduction of the phytopathogen *H. vastatrix*. It was observed that coffee discs can maintain their viability and functionality for up to 180 days, a period sufficient to allow the full development of the infection, the severity of which manifests itself after approximately 60 days (Fig. 3).

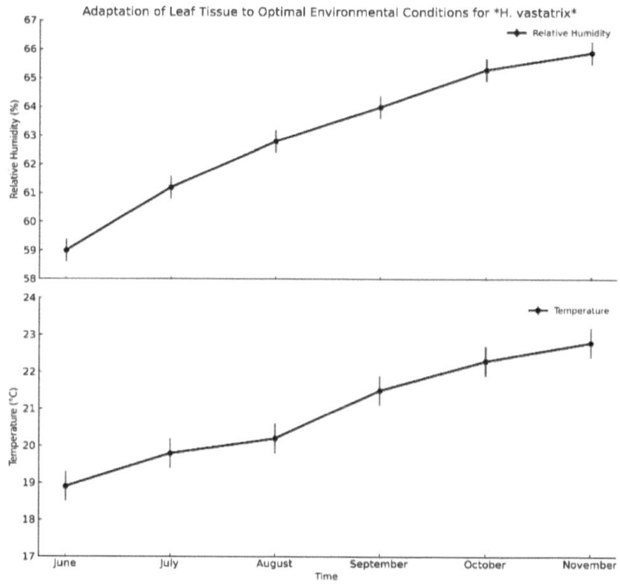

Fig. 3. (A) Relative humidity data were observed inside the incubator over a period of approximately five months. An optimum relative humidity of 65.9% was reached, ideal for the adaptation of *Coffea arabica* var. Tipica leaf tissue discs. (B) Temperature data inside the incubator showed a gradual variation over the same period, reaching an optimum temperature of 22.8 °C for disc adaptation. The values correspond to the monthly mean temperature (n = 4) and the error bars indicate ± the standard error.

Following established criteria, inoculation was carried out in November 2022. Initially, the infection developed slowly, but then moved to an intermediate phase where the infectious process accelerated significantly. This was attributed to the control achieved over temperature and relative humidity. In the final phase of infection, the severity exceeded 28%. After observing a high spore production, it was decided to collect the inoculum for a second round of infection in the cyclic crop. This collection was carried out between February and March, reaching a severity percentage of 15.52%, based on a total damaged leaf area of 1117.5 units, distributed in 72 evaluated discs. According to the Sinavef scale, these values indicate that we are in an accelerated infection phase. The new inoculum was collected after visual inspection, which showed no abnormal coloration or signs of contamination by other organisms. The plant tissue was then scraped to isolate pure lines of *H. vastatrix*. Microscopic analysis of the spore mass revealed that there was no contamination, and the spores observed had morphology matching that of the first inoculum, which is considered a good indicator of viability. Furthermore, the collected spores showed morphological structures typical of *H. vastatrix*.

The inoculum was applied on the leaf discs following the guidelines described in the methodology. Germination took place in 2 mL microtubes, in dark conditions and at room temperature for a period of 72 h. Following the same procedure as in the first inoculation, 45 μL of inoculum were applied on a total of 72 discs. All discs used in this round of cyclic culture infection showed signs of successful establishment of *H. vastatrix* in leaf tissue. Infection began in November, showing slow development during the first 25 days after inoculation (dai). In December, the colonies resulting from the second inoculum showed a severity between 1% and 5%. Between December and February (55 dai), a general increase in severity was observed, corresponding to the acceleration phase of the pathogen. Then, in April, the colonies entered the final phase of infection, according to the same scale. This result made it possible to establish a cyclic culture of *H. vastatrix* under controlled laboratory conditions, with no signs of colonization by opportunistic organisms or contaminants (Fig. 4).

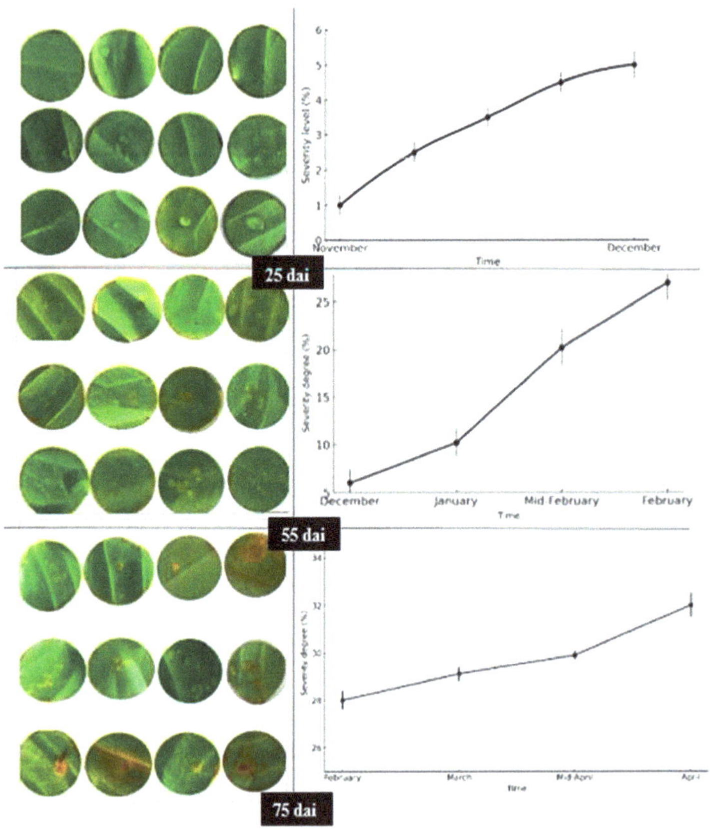

Fig. 4. Percentage severity of *H. vastatrix*. Infection began in november, approximately 25 days after inoculation (dai). From December to February (55 dai), colonies showed, on average, progress toward the acceleration phase. In March and April (75 dai), the final stage of disease development was reached.

4 Discussion

In this study, a cyclical culture of *H. vastatrix*, the agent responsible for coffee rust, was established using a methodology based on repeated inoculations in leaf discs of *Coffea arabica* var. Typica, all under controlled abiotic conditions. The initial collection of viable spores from infected tissue scrapings allowed for the selection of microtubes with high concentrations of spores and fungal structures (2900–4000 u/mL), which were used to initiate the first infection. This selection, based on mycelium and hyphal density, was crucial for the establishment of healthy colonies [23]. The leaf discs were placed in incubators specifically designed to maintain an average temperature of 22.8 °C and a relative humidity of 65.9%, conditions that favored germination and uredospore development. This environmental combination is consistent with previously reported results, which indicate that temperatures between 22 °C and 24 °C, along with constant humidity and no excess liquid water, are ideal for the fungal life cycle. Disc viability was maintained for up to 180 days, confirming and expanding on an impressive finding, as the only study with leaf tissue indicated that viability could last up to 40 days [14, 24, 25].

The infectious process began with the appearance of chlorotic spots approximately 28 days after inoculation (dai). Over the next 55 dai (until February), all colonies advanced to the acceleration phase, showing a progressive increase in the percentage of severity. This severity increased steadily: 28% in December, 29.1% in January, 29.9% in mid-February, and a final value of 32% in April. This progression pattern reflects the typical dynamics of the epidemic cycle described by Sinavef, which coincides with what was observed in Fig. 4, where the gradual increase until reaching the final infection phase is observed.

Throughout the process, the formation of spores and reproductive structures was observed without the presence of contaminating organisms, as confirmed by microscopic analysis [26–28]. The new pure lines obtained after the second inoculation displayed morphological characteristics similar to those of the first inoculum, highlighting their viability and purity. This continuity in infection and colonization reinforces the effectiveness of cyclic culture as an experimental model [29, 30].

The use of leaf discs instead of whole leaves has proven beneficial, allowing for greater repeatability in small spaces and a more accurate estimate of the infected area. Furthermore, the protocol developed for disc handling and acclimatization has helped maintain plant tissue in optimal condition, preventing dehydration thanks to the use of distilled water, mica plastic, and light control with LED lamps. Taken together, these results support the viability of the cyclical cropping model under controlled conditions, not only for assessing the severity dynamics of *H. vastatrix*, but also as a reproducible tool for future resistance testing or control trials. It has been confirmed that, under favorable conditions, the fungal infection process can reach its final phase in approximately 75 days, reaching a severity greater than 30%, laying a solid foundation for pathophysiological studies and fungicidal trials [31–33].

References

1. Liu, X., Tan, Y., Dong, J., Wu, J., Wang, X., Sun, Z.: Assessing habitat selection parameters of Arabica coffee using BWM and BCM methods based on GIS. Sci. Rep. **15**(1), 8 (2025). https://doi.org/10.1038/s41598-024-84073-0
2. Maruthai, S., Selvanarayanan, R., Thanarajan, T., Rajendran, S.: Hybrid vision GNNs based early detection and protection against Pest diseases in coffee plants. Sci. Rep. **15**(1), 11778 (2025). https://doi.org/10.1038/s41598-025-96523-4
3. Salojärvi, J., et al.: Genome and population genomics of allopolyploid Coffea Arabica reveal the diversification history of modern coffee cultivars. Nat. Genet. **56**(4), 721–731 (2024). https://doi.org/10.1038/s41588-024-01695-w
4. Alba-Alejandre, I., Alba-Tercedor, J., Vega, F.E.: Anatomical study of the coffee berry borer (Hypothenemus Hampei) using micro-computed tomography. Sci. Rep. **9**(1), 17150 (2019). https://doi.org/10.1038/s41598-019-53537-z
5. Barford, E.: Crop pests advancing with global warming. Nature (2013). https://doi.org/10.1038/nature.2013.13644
6. Quiroga, S., Suárez, C., Solís, J.D., Martínez-Juárez, P., Fernández Manjarrés, J.: Is coffee transition to cocoa a doubtful adaptation strategy for smallholders in Mesoamerica? Humanit. Soc. Sci. Commun. **12**(1), 1–12 (2025). https://doi.org/10.1057/s41599-024-04261-1
7. Venancio, L.P., et al.: Impact of drought associated with high temperatures on Coffea Canephora plantations: a case study in Espírito Santo state. Brazil. Sci. Rep. **10**(1), 19719 (2020). https://doi.org/10.1038/s41598-020-76713-y
8. Singh, B.K., et al.: Climate change impacts on plant pathogens, food security and paths forward. Nat. Rev. Microbiol. **21**(10), 640–656 (2023). https://doi.org/10.1038/s41579-023-00900-7
9. Gómez-de la Cruz, I., Martínez-Bolaños, M., Chávez-Ramírez, B., Estrada-de los Santos, P.: Biocontrol of Hemileia Vastatrix, the causal agent of coffee leaf rust, by Paenibacillus Sp. NMA1017. Plant Dis. **108**(10), 3163–3169 (2024). https://doi.org/10.1094/PDIS-01-24-0015-RE
10. Guerra-Guimarães, L., et al.: The Chloro-Plast protein HCF164 is predicted to be associated with Coffea SH9 resistance factor against Hemileia Vastatrix. Sci. Rep. **13**(1), 16019 (2023). https://doi.org/10.1038/s41598-023-41950-4
11. Ramírez-Camejo, L.A., et al.: Global analysis of Hemileia Vastatrix populations shows clonal reproduction for the coffee leaf rust pathogen throughout Most of its range. Phytopathology. **112**(3), 643–652 (2022). https://doi.org/10.1094/PHYTO-06-21-0255-R
12. Talhinhas, P., et al.: The coffee leaf rust pathogen Hemileia Vastatrix: one and a half centuries around the tropics. Mol. Plant Pathol. **18**(8), 1039–1051 (2017). https://doi.org/10.1111/mpp.12512
13. de Oliveira Aparecido, L.E., et al.: Future climate suitability of Hemileia Vastatrix in Arabica coffee under CMIP6 scenarios. J. Sci. Food Agric. **105**(1), 394–421 (2025). https://doi.org/10.1002/jsfa.13838
14. Alhudaib, K., Ismail, A.M.: First occurrence of coffee leaf rust caused by Hemileia Vastatrix on coffee in Saudi Arabia. Microbiol. Res. **15**(1), 164–173 (2024). https://doi.org/10.3390/microbiolres15010011
15. Quispe-Apaza, C., Mansilla-Samaniego, R., Espejo-Joya, R., Bernacchia, G., Yabar-Larios, M., López-Bonilla, C.: Spatial and temporal genetic diversity and population structure of Hemileia Vastatrix from Peruvian coffee plantations. Plant Pathol. J. **37**(3), 280–290 (2021). https://doi.org/10.5423/PPJ.OA.10.2020.0192
16. Boudrot, A., et al.: Effects on the dispersal of airborne Hemileia Vastatrix Uredospores. Phytopathology. **106**(6), 572–580 (2016). https://doi.org/10.1094/PHYTO-02-15-0058-R

17. Bettgenhaeuser, J., Gilbert, B., Ayliffe, M., Moscou, M.J.: Nonhost resistance to rust pathogens – a continuation of continua. Front. Plant Sci., 5 (2014). https://doi.org/10.3389/fpls.2014.00664
18. Porto, B.N., et al.: Genome sequencing and transcript analysis of Hemileia Vastatrix reveal expression dynamics of candidate effectors dependent on host compatibility. PLoS One. **14**(4), e0215598 (2019). https://doi.org/10.1371/journal.pone.0215598
19. Morales-Aranibar, L., et al.: Production of new biopesticides from Cymbopogon Citratus for the control of coffee rust (Hemileia Vastatrix) under laboratory and field conditions. Plants. **12**(5), 1166 (2023). https://doi.org/10.3390/plants12051166
20. Koutouleas, A., Jørgen Lyngs Jørgensen, H., Jensen, B., Lillesø, J.-P.B., Junge, A., Ræbild, A.: On the hunt for the alternate host of Hemileia Vastatrix. Ecol. Evol. **9**(23), 13619–13631 (2019). https://doi.org/10.1002/ece3.5755
21. Koutouleas, A., et al.: Shaded-coffee: a nature-based strategy for coffee production under climate change? Rev. Front. Sustain. Food Syst., 6 (2022). https://doi.org/10.3389/fsufs.2022.877476
22. Santiago-Santiago, M., et al.: Essential oils and antagonistic microorganisms as eco-friendly alternatives for coffee leaf rust control. Plants. **12**(20), 3519 (2023). https://doi.org/10.3390/plants12203519
23. Berihun, G., Alemu, K.: Status of coffee leaf rust (Hemileia Vastatrix) and its Management in Ethiopia: a review. Arch. Phytopathol. Plant Protect (2022)
24. Gomes, D.G., et al.: Nanoencapsulation boosts the copper-induced defense responses of a susceptible Coffea Arabica cultivar against Hemileia Vastatrix. Antibiotics. **12**(2), 249 (2023). https://doi.org/10.3390/antibiotics12020249
25. Julca-Otiniano, A., et al.: New races of Hemileia Vastatrix detected in Peruvian coffee fields. Agronomy. **14**(8), 1811 (2024). https://doi.org/10.3390/agronomy14081811
26. Orozco-Arias, S., Candamil, M.S., Jaimes, P.A., Cristancho, M., Tabares-Soto, R., Guyot, R.: Composition and diversity of LTR retrotransposons in the coffee leaf rust genome (Hemileia Vastatrix). Agronomy. **12**(7), 1665 (2022). https://doi.org/10.3390/agronomy12071665
27. Aristizábal, L.F.: Achievements and challenges in controlling coffee leaf rust (Hemileia Vastatrix) in Hawaii. Agrochemicals. **3**(2), 147–163 (2024). https://doi.org/10.3390/agrochemicals3020011
28. Silva, M., et al.: An overview of the mechanisms involved in coffee-Hemileia Vastatrix interactions: plant and pathogen perspectives. Agronomy. **12**(2), 326 (2022). https://doi.org/10.3390/agronomy12020326
29. López-Velázquez, J.C., García-Morales, S., López-Sánchez, G.P., Montero-Cortés, M.I., Uc Várguez, A., Qui-Zapata, J.A.: High-density chitosan induces a biochemical and molecular response in Coffea Arabica during infection with Hemileia Vastatrix. Int. J. Mol. Sci. **24**(22), 16165 (2023). https://doi.org/10.3390/ijms242216165
30. Salazar-Navarro, A., Ruiz-Valdiviezo, V., Joya-Dávila, J., Gonzalez-Mendoza, D.: Coffee leaf rust (Hemileia Vastatrix) disease in coffee plants and perspectives by the disease control. Phyton. **93**(5), 923–949 (2024) https://doi.org/10.32604/phyton.2024.049612
31. Pelayo–Sánchez, G., et al.: Urediniospore Size Variation and Morphological Novelties of Hemileia Vastatrix on Coffea Spp. in Mexico. Research Square (2024). https://doi.org/10.21203/rs.3.rs-4336373/v1.
32. Urbina, H., Aime, M.C., First, U.S.: Continental record of coffee leaf rust (Hemileia Vastatrix) on coffee (Coffea Arabica) in Southwest Florida, U.S.a. Plant Dis. **108**(1), 228 (2024). https://doi.org/10.1094/PDIS-09-23-1869-PDN
33. Góngora, C.E., Silva, M., do C.: Sustainable strategies for the control of crop diseases and pests to reduce pesticides. Agronomy. **14**(9), 2158 (2024). https://doi.org/10.3390/agronomy14092158

Metagenomics for the Study of Microbial Communities in the Madrigal del Podocarpus Reserve, Loja, Ecuador

Daniela Román-Cáceres[1,3], Adamary Vásquez-Tituana[1], Jorge I. Armijos-Rivera[1,2], and Aura Paucar-Cabrera[3(✉)]

[1] Grupo de Investigación en Genética y Biología Molecular, Universidad Nacional de Loja, Loja, Ecuador
[2] Centro de Biotecnología, Universidad Nacional de Loja, Loja, Ecuador
[3] Museo de Zoología LOUNAZ, Universidad Nacional de Loja, Loja, Ecuador
aurapaucar@gmail.com

Abstract. This study shows clear structural differentiation in the bacterial communities present in the soil and gut of dung beetles, depending on the type of vegetation cover and the level of disturbance at the Madrigal del Podocarpus Reserve, in Loja, Ecuador. The relative abundance of genera such as *Paenibacillus* and *Pantoea* in the guts of beetles collected from post-fire environments suggests a selection of stress-tolerant microbial taxa, consistent with reports from fire-impacted soils in other regions. Furthermore, the greater microbial diversity present in natural forest soils supports the idea that undisturbed habitats harbor more complex and functionally redundant microbial networks. Furthermore, the results underscore the role of dung beetles as microbial vectors, facilitating the dispersal of beneficial consortia that could accelerate microbial succession and soil recovery. Together, these findings reinforce the importance of macrofauna-microbiota interactions in the functioning of soil ecosystems. In the context of fire-affected montane tropical landscapes, integrating the gut microbiota of native dung beetles into ecological restoration strategies could be a bioinspired alternative to improve soil resilience.

Keywords: bacteria · dung beetles · Scarabaeinae · soil

1 Introduction

One of the most critical environmental problems, soil degradation, directly affects the ability of ecosystems to maintain their proper functionality and resilience [1]. According to the FAO, 75% of the soils in Latin America and the Caribbean show degradation problems due to various factors such as intensive agriculture, forest fires, deforestation, overgrazing, land-use changes, and other human activities [2]. In tropical areas, such as the southern Andes of Ecuador and especially around protected areas like the Madrigal del Podocarpus Reserve, located southeast of the city of Loja, the concern is even more alarming, as this type of degradation overlaps with areas of high biodiversity and endemism. In 2016, forest fires in this area affected approximately 60 ha of vegetation

J. L. Ballesteros et al. (Eds.): CITIS 2025, LNNS 1780, pp. 126–136, 2027.
https://doi.org/10.1007/978-3-032-14287-0_13

and caused the conversion of the vegetation cover. Ecological soil restoration involves restoring its physical and chemical structure, but also reactivating its complex biological network, where soil microorganisms play a fundamental role [3, 4].

Soil contains one of the most diverse microbial communities in the biosphere, performing functions such as nutrient recycling, aggregate formation, organic matter decomposition, and greenhouse gas regulation [5–7]. However, anthropogenic disturbances tend to reduce the diversity of these microbial communities, impairing their functional capacity and their ability to regenerate natural processes. Therefore, it is crucial to study restoration strategies that use soil biology and its ecological interactions as a framework [8–11]. One of the emerging lines of research in this field is the study of the microbiota that accompanies soil fauna associated with coprophagous insects, such as dung scarabs [12–14].

Within the large network of soil organisms, dung beetles (Coleoptera: Scarabaeidae: Scarabaeinae) are copronecrophagous insects widely distributed in ecosystems, where they play fundamental roles in the fragmentation of organic matter, soil aeration, and the dispersal of seeds and microorganisms [15]. Their burrowing behavior, as well as their mobility within the soil profile [15], makes them a potential vector of microbial redistribution. Furthermore, their use as model organisms in ecological and developmental studies has deepened our understanding of their distribution, functions and phylogenetic relationships [16], making them suitable to candidates for bioremediation and for use as bioindicators. Several studies show that the guts of dung beetles contain a rich diversity of bacteria capable of degrading complex organic compounds such as cellulose, hemicellulose, and lignin [17]. Therefore, these gut bacteria, in addition to performing digestive functions for the host, are capable of spreading through the soil during the construction of tunnels and nesting chambers, acting as potential carriers of microbial restoration. These bacteria contribute to the preservation of soil structure by promoting plant growth. They also facilitate nutrient absorption and even combat the presence of other organisms harmful to soil function [18].

Despite this, very little is known about the relationship between the bacterial communities that inhabit the guts of these dung beetles and those that live in the soil, especially in ecosystems as unique as those of the Madrigal del Podocarpus Reserve, which are at different stages of ecological succession: natural forest, grassland, and post- wildfire forest, which poses challenges to pioneer species. Once this relationship is understood, a solid ecological basis will be available for developing strategies to restore degraded soils by promoting functional excretory insect species.

This work proposes the comparative characterization of the bacterial communities of the soil and digestive tract of dung beetles from the Madrigal del Podocarpus Reserve at each successional stage, using culture-dependent and genomic methods (16S analysis), so that the bacterial genera are identified. This approach not only contributes to the ecological understanding of insect-microbiota-soil interactions, but also opens the door to new possibilities for applying bioengineering concepts to soil restoration programs based on the use of native biological agents as facilitators of functional landscape recovery.

2 Materials and Methods

2.1 Collection of Dung Beetles and Soil Samples

Samples were collected in February 2022 in the Madrigal del Podocarpus Reserve, located in the San Simón micro-basin, approximately 5 km from Loja, Ecuador [19, 20]. Collections were conducted under the Ecuadorian Environmental Ministry permits number and MAAE-ARSFC-2021-1785 and MAATE-ARSFC-2023-3082. The reserve covers an area of approximately 306 ha. Plots measuring 10 x 10 m were established, which were further subdivided into quadrants measuring 1 x 1 m. Three sub-quadrants were randomly collected from each plot to sample three contrasting landscape types, obtaining a total of 9 observation points in the study area.

The first landscape is a natural forest, composed of high plant diversity and moist, brown soils with abundant organic matter, including branches, leaves, and roots. The second landscape was the grassland, characterized by a predominance of grasses and monocots, the presence of few scattered trees, and low livestock activity, with black soils and rock fragments. The third landscape was the post-wildfire landscape, which featured sparse vegetation dominated by scrub, with black soils, few branches, and roots. Three pitfall traps baited with human feces were placed in each sampling plot and left active for 24 h to capture dung beetles [19, 20]. The traps were made from one-liter plastic containers, 14 cm deep, covered with a funnel to prevent escape. The traps were placed on the ground, containing small amounts of local soil, and the bait was hung in 28 g plastic cups. Contact between the bait and the collection container was avoided to prevent contamination with bacteria once collected. The traps were also protected from rain and debris with disposable plates as lids. At the time of collection, the traps contained live Scarabaeinae beetles in perforated plastic containers to allow air entry and were properly labeled. These data were entered into the Epicollect5 application. The specimens were transported alive to the Invertebrate Division of the LOUNAZ Zoology Museum of the Universidad Nacional de Loja, where identification and random selection of individuals were performed. Individuals were selected based on species frequency, prioritizing those detected in all traps, then those that appeared in two, and finally, those that appeared in only one. As a result, two species were chosen per landscape, resulting in a total of 20 specimens selected from the entire study area. Each specimen was placed in 50 ml centrifuge tubes, previously sterilized, capped, and labeled, and transported to the Biotechnology Center of the Universidad Nacional de Loja for subsequent microbiological analysis.

Soil samples were collected at the same observation points where the beetle sampling was conducted. The procedure followed the protocol described by Schweizer Lassaga and consisted of extracting subsamples of the topsoil (approximately 20 cm deep) and backfilling the hole with a shovel [21]. During extraction, pie-shaped cuts were made to remove contaminants (stones, roots, organic debris, residue, etc.). The subsamples obtained at each observation point (3 per plot) were homogenized in a sampling bag, thus forming a representative composite sample, totaling approximately 200 g per plot. The samples were placed and stored in airtight Ziploc bags, sealed, and labeled. Three representative samples were obtained from the landscapes associated with the natural or preserved forest, the grassland, and the area affected by wildfires. The material was transported and stored under controlled conditions at the Biotechnology Laboratory of the Universidad Nacional de Loja, where it was processed. At each sampling site, relative moisture and temperature were recorded.

2.2 Bacterial Sampling from the Digestive Tract of Scarabs and from Soil Samples

The beetles obtained were euthanized by placing them at −20 °C in sterile 50 ml tubes for 24 h. They were subsequently disinfected by successive immersion in sterile distilled water, 70% alcohol, and then in distilled water again, controlling the immersion times. Finally, they were placed in Petri dishes for dissection, separating the intestine from the rectum. Once eviscerated, the intestines were placed in microtubes (2 ml of sterile distilled water), where they were macerated with a sterile metal rod to break the adhesions of remaining microorganisms on the tissues. Serial dilutions of 1:10 were made with the macerated intestinal sample up to a 10–7 dilution using microtubes with sterile peptone water. Specimens were deposited in the Invertebrate Division at the LOUNAZ Museum.

After collecting the soil samples, they were spread out on a flat surface and allowed to dry for approximately three days. Bacterial isolation was then initiated using serial dilutions. For each sample, 10 g of soil was sieved and weighed, then mixed with 90 ml of distilled water in a flask. The mixture was shaken at 340 rpm for 20 min to obtain a stock sample.

2.3 Metagenomic Analysis

Previously extracted and preserved stock samples were sent to an external bio sequencing laboratory (Biosequence, Quito, Ecuador) for molecular analysis using metagenomics. Bacterial identification was conducted through 16S rRNA gene analysis, a highly effective approach for determining the species within mixed microbial samples. Twenty-three stock samples, including dung beetle and soil, were submitted. The process included the extraction of bacterial DNA, which was achieved by decomposing microbial cells using physicochemical methods. The released DNA was isolated using restriction endonucleases; these fragments were ligated into molecular vectors to facilitate their expression, even from bacteria that are not easily cultured in laboratory conditions. Metagenomic libraries were constructed from the sequenced DNA, allowing genetic information to be stored and analyzed. Finally, the sequences were compared against public databases (NCBI) to accurately identify bacteria at species level.

3 Results

The heatmap represented in Fig. 1 shows the 20 most abundant genera in the different vegetation cover areas. This was done to more clearly demonstrate the differentiation between samples, given the large number of readings obtained. Relative composition analysis has revealed unique patterns in how the most abundant bacterial genera are distributed, depending on different combinations of sample type (beetle gut, soil) and plant cover. The genera *Paenibacillus* and *Pantoea* were notably prominent in beetle samples taken from wildfire-affected areas, suggesting a strong connection with post-fire conditions for these insects. On the other hand, genera such as *Pseudomonas*, *Clostridium*, and *Stenotrophomonas* showed moderate abundances, distributed across diverse cover types and samples, which could indicate that they are generalist. Similarly, grassland soil samples showed low representation of dominant genera, indicating a lower predominance of certain taxa or a wider distribution of abundances among less common genera (outside the top 20). In summary, these patterns indicate that certain sample types, especially beetles collected in wildfire-affected areas, harbor bacterial communities that are clearly differentiated from others. This hierarchy, indicated in the figure, reveals a clear ecological structure that depends on the environment and the host.

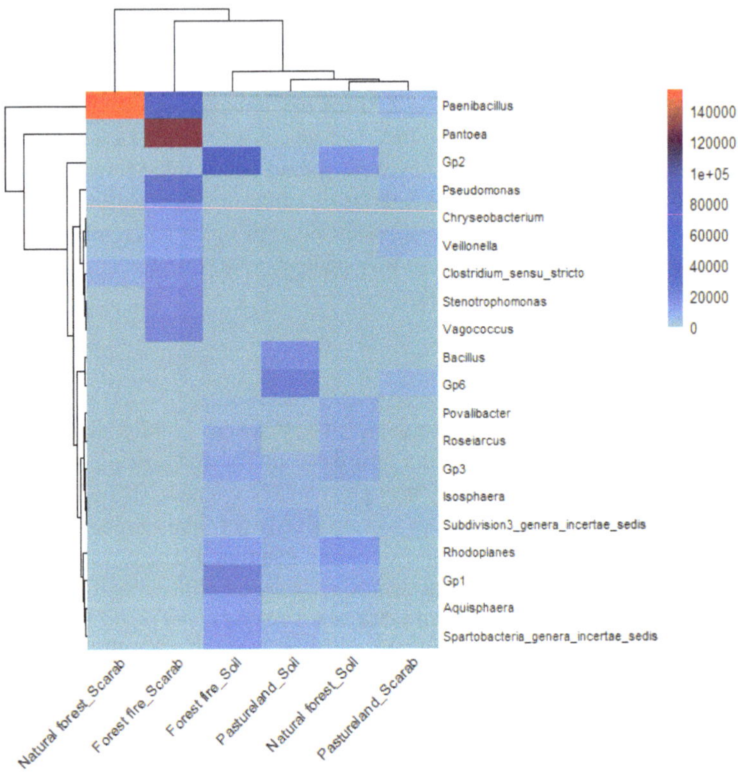

Fig. 1 Heatmap of the abundance of the 20 most common bacterial genera in different combinations of vegetation cover (natural forest, wildfires, and grassland) and sample type (soil and beetle gut). Rows show bacterial genera, and columns represent combinations. Red represents "very high abundance," dark blue "medium abundance," and light blue "low or no abundance." Branches show similar clusters between genera and similar abundance profiles, as well as combinations that share bacterial communities. (Color figure online)

The stacked bar chart in Fig. 2 shows the composition of the 50 most abundant bacterial genera, indicating significant variation between dung beetle species and vegetation cover types. In fire-affected areas, the dung beetles *Onthophagus* 11, *Onthophagus* 13, and *Onthophagus* 2 stood out for having high presence of bacterial genera such as *Paenibacillus, Pantoea*, and *Pseudomonas*. This suggests the possible existence of microbial specialization, associated with the conditions present in these degraded areas. On the other hand, a greater abundance of bacterial genera was found in natural forests, more evenly distributed among dung beetle individuals such as *Uroxys* 9, *Onthophagus* 1, and *Uroxys* 20, with prominent genera such as *Paenibacillus* and *Pantoea*. In the case of grassland dung beetles such as *Onthophagus* 15, *Onthophagus* 19, and *Dichotomius* 15, intermediate profiles are observed, with high but more dispersed abundances among genera such as *Veillonella*, *Streptococcus* and *Pseudomonas*. These differences indicate that both beetle species and cover type play an important role in the structure of the gut microbiota.

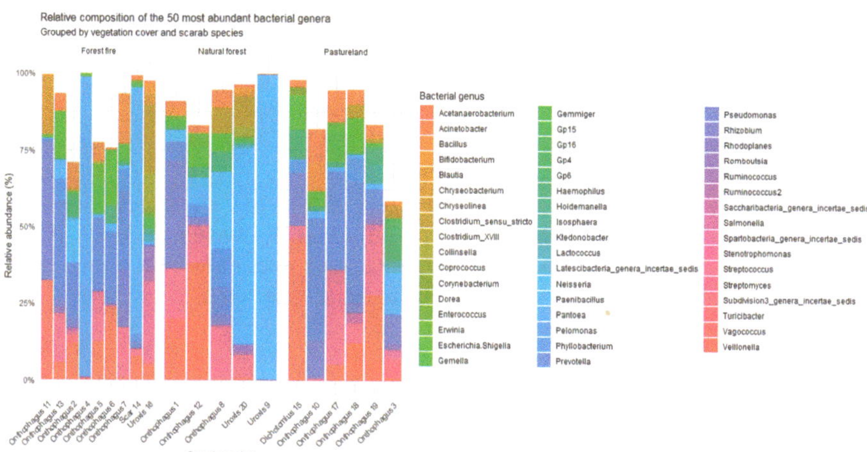

Fig. 2 Relative composition of the 50 most abundant bacterial genera in beetle guts, grouped by species and plant cover. The figure represents relative abundance (%), each bar represents a beetle species and corresponds to a predicted gut microbiota profile, normalized by species. Beetle species names are shown on the x-axis, and the legend on the side color-codes show the bacterial genera.

The heatmap in Fig. 3 shows that most of the top 50 bacterial species had low or even zero abundance in beetle samples collected from the grassland. However, in the wildfire zone and natural forest, high abundance levels were detected for a small group of bacterial species, in particular, with remarkably high abundances, exceeding 75,000 readings, for *Pantoea vagans* (EF688012) and *Paenibacillus pectinilyticus* (EU391157). The prevalence of certain bacterial species likely adapted to environmental stress conditions appears to be higher.

Analysis of the absolute abundance of the 50 most common bacterial species in soil samples (Fig. 4) reveals a clear differentiation in microbial composition according to the type of vegetation cover. The heat map shows a greater accumulation of abundance in the wildfire zone, evidenced by a high representation of species such as *Aquisphaera giovannonii*, Gp2 *Acidobacteria*, and Gp2 *Eubacterium*. In contrast, in natural forest

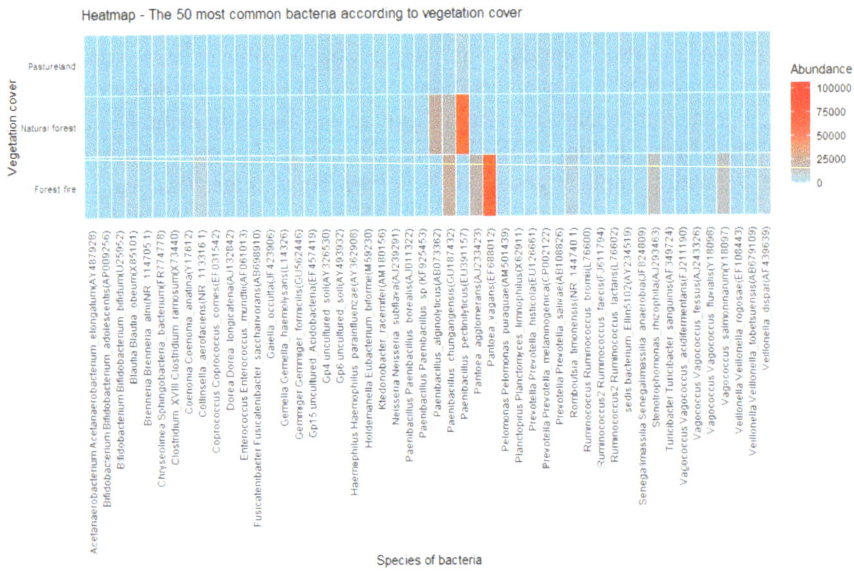

Fig. 3 Heatmap of the 50 most abundant bacterial species in beetle guts, according to plant cover. Identified bacterial species are shown on the X-axis and ordered by overall frequency. The color scale indicates the total number of sequences assigned to each species according to plant cover; deep red tones represent the highest abundance. (Color figure online)

and grassland soils, most species have more homogeneous distributions and lower abundances, which could indicate a low dominance of certain taxa. However, species such as *Rhodoplanes roseus* show relatively high abundance in the natural forest zone.

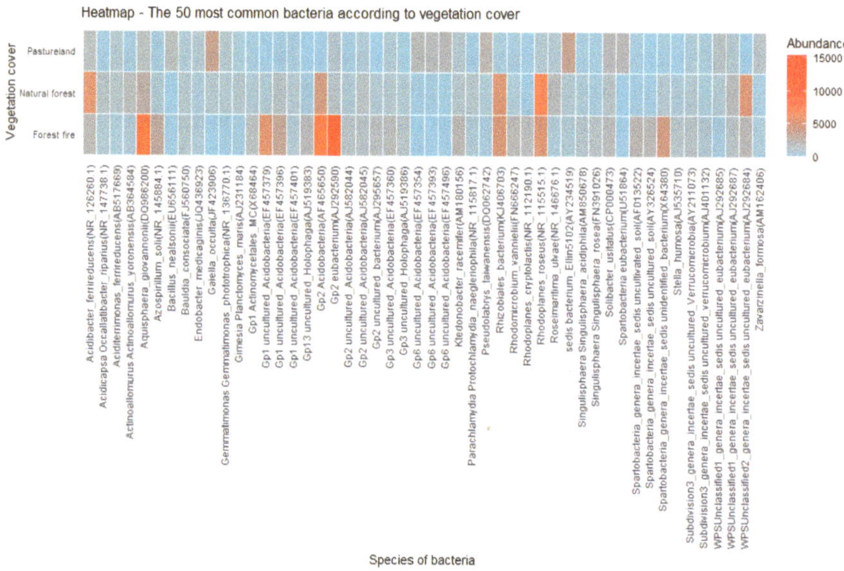

Fig. 4 Heatmap showing the distribution of the 50 most abundant bacterial species in soil samples according to vegetation cover. The color scale represents the intensity of abundance, from blue, representing "low abundance," to red, representing "high abundance". (Color figure online)

4 Discussion

This study reveals a distinct structural differentiation of soil and dung beetle-associated bacterial communities across vegetation types and disturbance gradients within the Madrigal del Podocarpus Reserve. These patterns are consistent with the findings who demonstrated that environmental factors particularly soil pH and disturbance history are key drivers of bacterial community composition at broad ecological scales [22]. The observed dominance of genera such as *Paenibacillus* and *Pantoea* in beetle gut samples collected from post-fire environments suggests selective enrichment of stress-tolerant microbial taxa. This aligns with the work who reported a post disturbance prevalence of *Firmicutes* and *Gammaproteobacteria* in Mediterranean fire-impacted soils, likely linked to their metabolic plasticity and resilience under oxidative stress conditions [23].

Conversely, the greater microbial diversity observed in natural forest soils supports the hypothesis that undisturbed habitats foster more complex and resilient microbial networks. This observation resonates with the findings, who showed a positive correlation between ecosystem integrity and microbial multifunctionality, emphasizing the role of habitat stability in promoting microbial richness and functional redundancy [24].

Furthermore, the role of dung beetles as microbial vectors is reinforced by our results, highlighting their contribution not only to organic matter decomposition and bioturbation but also to the dispersion of beneficial microbial consortia. This supports previous evidence, who demonstrated that coprophagous beetles facilitate the spatial redistribution of soil bacteria, potentially accelerating microbial succession and soil recovery processes [25, 26].

Taken together, our findings contribute to the growing body of literature suggesting that macrofauna–microbiota interactions are central to belowground ecosystem functioning. As proposed incorporating soil biodiversity—both microbial and faunal—into restoration frameworks is critical for enhancing soil resilience. In the context of tropical montane ecosystems affected by wildfires, integrating native dung beetle microbiomes into bioinspired restoration strategies may offer an innovative approach to support natural recovery pathways.

5 Acknowledgements

We are grateful to Marek Castel and the Reserva Madrigal del Podocarpus for providing us access and support that made this research possible. We thank Silvia Falqueto (Universidade Federal de Mato Grosso, Brazil) for her valuable initial guidance on culturing bacteria from scarab beetles. We extend our gratitude to Fernando Vaz de Mello (Universidade Federal de Mato Grosso, Brazil) for his support validating the sampling design. This study was supported by the Research Department of the Universidad Nacional de Loja, through project 07-DI-FARN-2021, awarded to Aura Paucar-Cabrera (PI).

References

1. Schwarz, E., et al.: Spatial control of microbial pesticide degradation in soil: a model-based scenario analysis. Environ. Sci. Technol. **56**(20), 14427–14438 (2022). https://doi.org/10.1021/acs.est.2c03397
2. FAO advierte: tres cuartas partes de los suelos en América Latina y el Caribe están en riesgo. FAO-RLC. https://www.fao.org/americas/news/news-detail/suelos-en-riesgo/es. Accessed 30 Apr 2025
3. Ma, J., Li, C., Hui, L., Wang, J., Fan, Y.: Soil properties under different ecological restoration modes for the quarry in Yanshan Mountains of Hebei Province, China. PeerJ. **10**, e14359 (2022). https://doi.org/10.7717/peerj.14359
4. Su, R., Wang, Y., Huang, S., Chen, R., Wang, J.: Application for ecological restoration of contaminated soil: phytoremediation. Int. J. Environ. Res. Public Health. **19**(20), 13124 (2022). https://doi.org/10.3390/ijerph192013124
5. Lv, C., Wang, C., Cai, A., Zhou, Z.: Global magnitude of rhizosphere effects on soil microbial communities and carbon cycling in natural terrestrial ecosystems. Sci. Total Environ. **856**, 158961 (2023). https://doi.org/10.1016/j.scitotenv.2022.158961
6. Wilpiszeski, R.L., et al.: Soil aggregate microbial communities: towards understanding microbiome interactions at biologically relevant scales. Appl. Environ. Microbiol. **85**(14), e00324–e00319 (2019). https://doi.org/10.1128/AEM.00324-19
7. Philippot, L., Chenu, C., Kappler, A., Rillig, M.C., Fierer, N.: The interplay between microbial communities and soil properties. Nat. Rev. Microbiol. **22**(4), 226–239 (2024). https://doi.org/10.1038/s41579-023-00980-5
8. Zhu, M., et al.: Discrepancy strategies of sediment abundant and rare microbial communities in response to floating microplastic disturbances: study using a microcosmic experiment. Sci. Total Environ. **835**, 155346 (2022). https://doi.org/10.1016/j.scitotenv.2022.155346
9. Eliette, A.-S., Elodie, B., Arnaud, M., Tiffany, R., Aymé, S., Pascal, P.: Idiosyncratic invasion trajectories of human bacterial pathogens facing temperature disturbances in soil microbial communities. Sci. Rep. **14**(1), 12375 (2024). https://doi.org/10.1038/s41598-024-63284-5

10. Philippot, L., Griffiths, B.S., Langenheder, S.: Microbial community resilience across ecosystems and multiple disturbances. Microbiol. Mol. Biol. Rev. **85**(2) (2021). https://doi.org/10.1128/mmbr.00026-20
11. Doering, M., et al.: Microbial communities in floodplain ecosystems in relation to altered flow regimes and experimental flooding. Sci. Total Environ. **788**, 147497 (2021). https://doi.org/10.1016/j.scitotenv.2021.147497
12. Wang, S., Miltner, A., Muskus, A.M., Nowak, K.M.: Microbial activity and Metamitron degrading microbial communities differ between soil and water-sediment systems. J. Hazard. Mater. **408**, 124293 (2021). https://doi.org/10.1016/j.jhazmat.2020.124293
13. Dunn, L., et al.: Soil microbial communities in the face of changing farming practices: a case study in an agricultural landscape in France. PLoS One. **16**(6), e0252216 (2021). https://doi.org/10.1371/journal.pone.0252216
14. Chuong, J.N.: Transplanting soil microbial communities improves climate tolerance in trees. Nat. Biotechnol. **41**(7), 908–908 (2023). https://doi.org/10.1038/s41587-023-01874-w
15. Ferreira, K.R., Puker, A., Correa, C.M.A.: The attraction of Amazonian dung beetles (Coleoptera: Scarabaeidae: Scarabaeinae) to the feces of omnivorous mammals is dependent on their diet: implications for ecological monitoring. Environ. Entomol. **49**(6), 1383–1392 (2020). https://doi.org/10.1093/ee/nvaa106
16. Tarasov, S., Dimitrov, D.: Multigene phylogenetic analysis redefines dung beetles relationships and classification (Coleoptera: Scarabaeidae: Scarabaeinae). BMC Evol. Biol. **16**(1), 257 (2016). https://doi.org/10.1186/s12862-016-0822-x
17. Shukla, S.P., Sanders, J.G., Byrne, M.J., Pierce, N.E.: Gut microbiota of dung beetles correspond to dietary specializations of adults and larvae. Mol. Ecol. **25**(24), 6092–6106 (2016). https://doi.org/10.1111/mec.13901
18. Krell, F.-T., Moon, A.R.: Dung Beetles. Curr. Biol. **29**(12), R554–R555 (2019). https://doi.org/10.1016/j.cub.2019.05.027
19. Jiménez Barba, J.: Caracterización por medios dependientes de cultivo de comunidades bacterianas de escarabajos estercoleros (Coleoptera: Scarabaeinae) y de suelos conservados y alterados por incendios de la Reserva El Madrigal del Podocarpus. Tesis de pregrado, Universidad Nacional de Loja, Loja, Ecuador, Loja (2023)
20. Villamagua Carrión, L.: Microbiota intestinal de escarabajos peloteros (Scarabaeinae) de un área afectada por un incendio forestal en la Reserva Madrigal del Podocarpus. Tesis de pregrado, Universidad Nacional de Loja, Loja, Ecuador, Loja (2024)
21. Schweizer Lassaga, S.: Muestreo y análisis de suelos para diagnóstico de fertilidad. pp. 18. INTA/MAG, San José, Costa Rica (2010). https://www.mag.go.cr/bibliotecavirtual/P33-9965.pdf
22. Bonomo, M.G., et al.: Metagenomic monitoring of soil bacterial community after the construction of a crude oil flowline. Environ. Monit. Assess. **194**(2) (2022). https://doi.org/10.1007/S10661-021-09637-3
23. Naveed, M., et al.: Simultaneous loss of soil biodiversity and functions along a copper contamination gradient: when soil goes to sleep. Soil Sci. Soc. Am. J. **78**(4) (2014). https://doi.org/10.2136/sssaj2014.02.0052
24. Osburn, E.D., Yang, G., Rillig, M.C., Strickland, M.S.: Evaluating the role of bacterial diversity in supporting soil ecosystem functions under anthropogenic stress. ISME Commun. **3**(1), 1–10 (2023). https://doi.org/10.1038/s43705-023-00273-1

25. Sansupa, C., et al.: Soil bacterial communities and their associated functions for forest restoration on a limestone mine in northern Thailand. PLoS One. **16**(4 April) (2021). https://doi.org/10.1371/journal.pone.0248806
26. Suárez-Moo, P., Cruz-Rosales, M., Ibarra-Laclette, E., Desgarennes, D., Huerta, C., Lamelas, A.: Diversity and composition of the gut microbiota in the developmental stages of the dung beetle Copris incertus say (Coleoptera, Scarabaeidae). Front. Microbiol. **11**, 564362 (2020). https://doi.org/10.3389/FMICB.2020.01698/BIBTEX

Artificial Intelligence and Digital Networks

Machine Learning for Fault Diagnosis Using the MAP Sensor Data from an Otto Engine

Juan Pablo Medina Namicela[1(✉)], Rogelio Santiago León-Japa[1], and Francisco Jurado[2]

[1] Universidad Nacional de Loja, Loja, Ecuador
{juan.p.medina.n,rogelio.leon}@unl.edu.ec
[2] Departamento de Ingeniería Eléctrica, Universidad de Jaén, Jaén, Spain
fjurado@ujaen.es

Abstract. In this study, Machine Learning (ML) classification methods were applied and trained using voltage data obtained from a Manifold Air Pressure (MAP) sensor. The experimental unit was an Otto engine bench YESA 3133. The objective was to classify four different engine operating conditions, corresponding to faults intentionally induced in the intake and ignition systems. Data were collected with a NI-USB 6009 data acquisition board at a sampling frequency of 10 kHz. To train the ML models, the most relevant features were extracted through ANOVA, correlation analysis and a Random Forest-based feature selection. The algorithms evaluated included Artificial Neural Networks (ANN), Support Vector Machines (SVM), and baseline ML models available in MATLAB's *Classification Learner* toolbox. The models were trained with both the raw and filtered MAP sensor data. As a result, ANN achieved an accuracy of up to 96.05% with short training times, SVM reached 95%, and K-Nearest Neighbors (KNN) obtained the highest accuracy (92.3%) among the baseline models. Additionally, filtering the sensor data improved the classification performance across most ML algorithms.

Keywords: ANN · SVM · Machine Learning · Fault diagnosis · Otto engine

1 Introduction

Among emerging technologies, Machine Learning (ML) stands out as a powerful tool that can be applied to a wide range of problems with relatively simple coding implementations. In the automotive industry, ML has been employed for fault diagnosis, using algorithms such as Artificial Neural Networks (ANN) [1], Support Vector Machines (SVM) [2], and other models [3] to detect failures in Otto engines.

In the context of Ecuador, current vehicle maintenance processes still face several challenges. These include the lack of updated diagnostic tools, the subjectivity in fault detection, prolonged service times, inefficiencies in fault localization, and even limitations in the On-Board Diagnostics (OBD) system, particularly in detecting failures related to the air intake and ignition systems [1, 4, 5].

Vehicle maintenance is crucial for ensuring safety and reducing pollutant emissions. However, it also requires time and resources to attend and service each unit. According

J. L. Ballesteros et al. (Eds.): CITIS 2025, LNNS 1780, pp. 139–148, 2027.
https://doi.org/10.1007/978-3-032-14287-0_14

to the Asociación de Empresas Automotrices del Ecuador (AEADE), the country's light vehicle fleet increased form 1.816 million in 2022 to 2.624 million by May 2024 [6]. This growth implies that maintenance services must scale accordingly. Yet, given the existing deficiencies in maintenance procedures, the time required per service has increased, potentially making the demand for vehicle maintenance unsustainable in the near future.

1.1 Previous Works

To address the aforementioned challenges, Machine Learning (ML) has emerged as a promising solution. Recent research efforts have focused on developing faster and more effective methods for fault detection by comparing different algorithms, models, and techniques to identify the most suitable [2].

For instance, in the study presented in [3], various ML algorithms such as XGBoost, SVM, Random Forest, Logistic Regression, and a hybrid model combining these techniques were evaluated. The results indicated that XGBoost and SVM outperformed Random Forest and Logistic Regression in terms of effectiveness. Similarly, the research conducted by [2] demonstrated that linear SVM and Radial Basis Function (RBF) algorithms were among the fastest and most efficient. These findings highlight the relevance of SVM as a robust model for classification tasks.

In the context of fault diagnosis, [7] reported that data from the MAP sensor is particularly useful for detecting spark plug failures. Using this data in MATLAB to train an Artificial Neural Network (ANN)-based model, [8] developed an active fault-tolerant control system for managing the air-fuel mixture in Otto engines. Consequently, ANN models show potential not only for diagnostic tasks using MAP sensor data, but also for the development of more complex systems. Additionally, ANN algorithms may involve a lower computational cost compared to SVMs in certain applications [2].

1.2 Research Aim

The main objective of this study is to compare ANN, SVMs, and baseline ML classification algorithms for fault diagnosis in an Otto engine using data from a Manifold Air Pressure (MAP) sensor. The research aims to contribute to the implementation of ML techniques in automotive maintenance by identifying the most suitable algorithm for diagnosing faults such as an obstructed air filter and a faulty spark plug, based on MAP sensor voltage data.

The study consists of an introduction that outlines the current context of Machine Learning and the growth of the vehicle fleet in Ecuador; a methodology section describing the data acquisition process from the MAP sensor and the algorithm training procedures; a results section presenting tables with the precision, recall and F1-score values obtained for each algorithm; a discussion comparing the results with previous studies; and, finally, conclusions based on the findings.

2 Methods

2.1 Study Unit

The experimental unit for this study was a "YESA 3133" engine bench corresponding to a 2002 Hyundai Sonata EF 2.0 L DOHC Otto engine. Prior to data acquisition, the engine was verified to be in optimal condition by inspecting the following components: air filter, fuel pressure, injection timing, engine fluids, spark plugs, cylinder compression, and diagnostic trouble codes (DTCs).

2.2 Experimental Design

A 2^2 factorial experimental design was employed, consisting of two factors with two levels each. These factors and their levels are presented in Table 1.

Table 1. Factors and levels of the experimental design.

Factor	Levels	Low (0)	High (1)
Spark plug	2	Electrodes closed	Electrodes spacing 1 mm
Air filter	2	Air flow 50%	Air flow 100%

The experimental design resulted in four distinct treatments, each representing a different engine operating condition, as described in Table 2.

Table 2. Engine conditions obtained for one replica of the experimental design.

Treatment	Name	Description
1	Engine well	Cylinder 1 spark plug with standard 1 mm gap and a normal air filter.
2	Spark plug & Filter fault	Cylinder 1 spark plug with abnormal 0 mm gap and air filter obstructed by 50%.
3	Air filter fault	Cylinder 1 spark plug with standard 1 mm gap and air filter obstructed by 50%
4	Spark plug fault	Cylinder 1 spark plug with abnormal 0 mm gap and unobstructed air filter.

To improve the accuracy of ML models, [9] mention that more training data can lead to a greater accuracy. Accordingly, 10 replicates per treatment were conducted considering the proposed by [10], yielding a total of 40 tests for data collection. The order of data acquisition followed the sequence specified by the experimental design.

Environmental conditions during data acquisition were considered as uncontrolled factors and were recorded using the weather platform *weather.com*. The values registered during testing were: ambient temperature of 296.15 K, wind speed of 3.611 m/s, relative humidity of 41%, and atmospheric pressure of 100.910 kPa.

2.3 Data Acquisition

Voltage signals from MAP, CMP (Camshaft Position), and CKP (Crankshaft Position) sensors were collected using an NI-USB 6009 data acquisition device and LabVIEW software.

MAP sensor data was collected based on the experimental design. For each test, the engine was run for 3 min at idle speed until it reached its normal operating temperature. Once stabilized, data were recorded for 5 s at a sampling frequency of 10 kHz. A general overview of the process is summarized in Fig. 1.

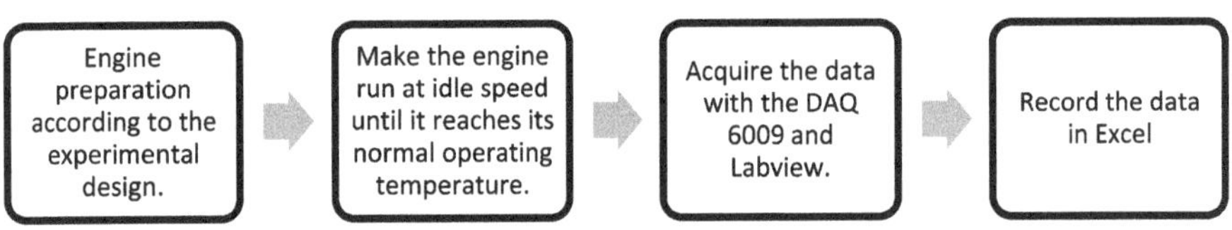

Fig. 1. Data acquisition process of the MAP sensor voltage.

2.4 Processing of Acquired Data

To ensure data quality, the first and last seconds of each recording were discarded, following the observations reported in [1] which identified these periods as potential noise. Since the data were not synchronized with the crankshaft angle, the CMP and CKP signals were used to align the MAP sensor data with the beginning and end of each engine cycle. Specifically, synchronization was set to the top dead center (TDC) of cylinder 1 ignition. This allowed segmentation of the MAP sensor signal into intervals of 720° of crankshaft rotation. This segmentation process was applied to all treatments. A representative 720° segment of the MAP signal is shown in Fig. 2. The total number of segments obtained was: 133 for Treatment 1, 136 for Treatment 2, 136 for Treatment 3, and 137 for Treatment 4.

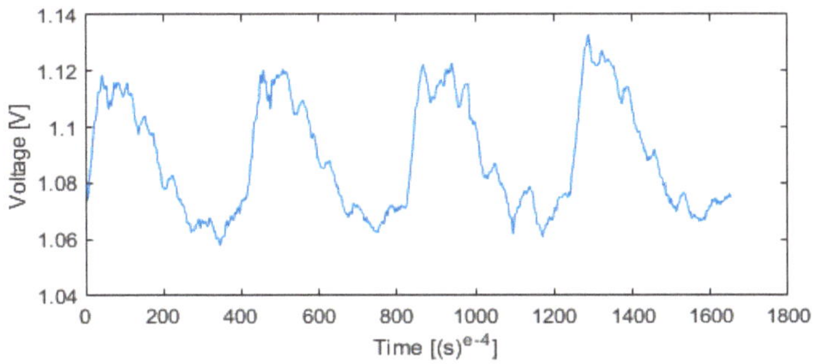

Fig. 2. 720° segment from a MAP sensor signal.

2.5 Feature Extraction from the MAP Sensor Signals

To train the ML algorithms, several features were extracted from each 720° segment of the MAP sensor signal using MATLAB, the following features are marked with a numeric value to understand Table 3: Root Mean Square (RMS) (1), Standard Derivation STD (2), mean (3), variance (4), median (5), mode (6), maximum peak (7), minimum peak (8), peak-to-peak difference (9), energy (10), skewness (11), kurtosis (12), crest factor (13), coefficient of variation (14), 10% trimmed mean (15), Shannon entropy (16) and area (17).

Afterward, multiple statistical analyses and Pareto charts were used to rank the features from most to least significant. Table 3 presents the top 10 features based on their importance in various analyses.

Table 3. Most important features extracted from the MAP sensor signal.

Statistic methods	Extracted features (Most important to the left)									
ANOVA	5	15	3	1	7	8	6	10	2	4
Correlation	7	5	15	1	3	8	6	2	10	4
RF Curvature	16	3	15	1	8	5	17	10	14	12
RF Standard cart	16	10	8	3	17	15	5	6	1	9
RF Interaction Test	16	1	8	5	10	17	12	2	14	7

The final selection of features for training the ML models included only those considered important in four or more evaluations. To prevent overfitting, in cases where two features showed a correlation of 1 or − 1, only one feature was considered. The features ultimately selected for training were: RMS, median, minimum peak, and signal energy.

2.6 Application of Classification Machine Learning Models

A feedforward backpropagation neural network (ANN) was implemented in MATLAB configured with three hidden layers containing 80, 80, 81 neurons respectively.

The SVM algorithm was initially trained using MATLAB's Classification Learner app and then modified to perform 100 iterations, optimizing the "BoxConstraint" (C parameter) and "Kernel Scale" to enhance accuracy.

Baseline ML classification models available in the MATLAB Classification Learner toolbox were also applied. After identifying the best-performing models, the optimizable version of each was trained to improve accuracy.

2.7 Training the Machine Learning Algorithms with the Filtered MAP Data

Additionally, a comparison was made between models trained on raw signal data and those trained on filtered signal data. A fifth-order Butterworth filter was applied to the MAP signal, then, the same segmentation and feature extraction processes were applied, finally the following features were extracted for training on the ML algorithms used in the section above: RMS, median, mode, maximum peak, and signal energy.

2.8 Comparison of the ML Algorithms

To evaluate and compare algorithm performance, the metrics of precision, recall, and F1-score were calculated using the confusion matrices of each model with MATLAB code, following the methodology in [9].

3 Results

After the application of each ML model, a confusion matrix was obtained. Figure 3 shows the confusion matrix for the first ANN model implemented, summarizing the precision and recall per class.

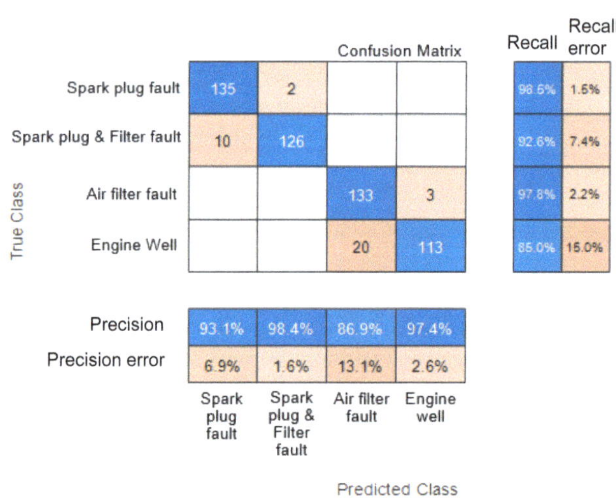

Fig. 3. ANN algorithm confusion matrix.

The configurations of the ANN model that achieved the highest accuracy are summarized in Table 4. The first configuration represents the ANN baseline model, from which further configurations were developed.

Table 4. Results of different modifications of the ANN algorithm.

N°	Learning function	Hidden layers neurons	Precision (%)	Recall (%)	F1-score (%)	Training time (s)
1	trainscg	80 80 81	93.97	93.4	93.5	14
2	trainscg	160 80 80	96.05	95.65	95.67	17
3	trainscg	40 40 41	95.68	95.32	95.34	9
4	traincgf	80 80 81	95.68	95.13	95.14	33
5	trainscg	160 80 40	95.35	94.76	94.78	15
6	traincgp	80 80 81	95.13	94.39	94.4	26

For the SVM algorithm, the top-performing configurations after 100 iterations are shown in Table 5, highlighting the optimal "BoxConstraint" and "Kernel Scale" values.

Table 5. Results for different SVM algorithm configurations.

N°	SVM	C Parameter	Kernel Scale	Code	Precision (%)	Recall (%)	F1-Score (%)	Training time (s)
1	Optimizable	998.2029	1	onevsone	95	95	95	137.01
2	Cubic	11.536	0.78764	onevsall	94.8339	94.73	94.73	227.187
3	Gaussian	964.77	1.5816	onevsall	94.6494	94.55	94.55	70.398
4	Quadratic	0.6671	1	onevsone	94.6494	94.54	94.54	180
5	Cubic	132	1.8002	onevsone	93.9114	93.82	93.82	94.5384

The ML models trained using MATLAB's Classification Learner toolbox got lower accuracy compared to the ANN and SVM algorithms. However, the top-performing models based on precision, recall, and F1-score were: Discriminant Analysis (91.6, 91.3 and 91.3%), Ensemble Classifiers (92.1, 92.1 and 92.1%), K-Nearest Neighbors (KNN) (92.3, 92.3 and 92.3%), and finally Decision Trees (91.2, 91.2 and 91.2%).

3.1 Results of Algorithms Trained with Filtered Data

Table 6 provides a summary of the performance of the top-performing algorithms, this time trained on filtered MAP sensor data.

Table 6. ML algorithms results using filtered data.

ML algorithm	N°	Precision (%)	Recall (%)	F1-Score (%)	Error (%)	Training time (s)
ANN	1	95.53	94.95	94.96	4.47	13
ANN	2	95.46	94.95	94.96	4.54	28
ANN	3	96.48	96.26	96.27	3.52	8
ANN	4	95.78	95.32	95.33	4.22	41
ANN	5	96.56	96.45	96.46	3.44	28
ANN	6	95.21	94.77	94.78	4.79	44
SVM	1	95.5	95.4	95.4	4.5	603.57
SVM	2	94.8339	94.74	94.74	5.1661	245.41
SVM	3	94.4	94.25	94.26	5.6	102.40
SVM	4	94.8339	94.80	94.81	5.1661	445.42
SVM	5	95.19	95.18	95.18	4.81	205.61

The results of precision, recall and F1-score for the filtered-data models from the Classification Learner toolbox were: Discriminant Analysis (89.9, 89.9 and 89.9%), Ensemble Classifiers (92.3, 92.3 and 92.3%), KNN (93.7, 93.7 and 93.7%), and finally Deci-

sion Trees (90.3, 90.2 and 90.2%).

4 Discussion

Based on the results, the ANN was the Machine Learning algorithm that achieved the highest accuracy, aligning with the findings of [11], who also demonstrated that ANN models outperform other ML algorithms in terms of accuracy.

As noted by [12], incorporating more features when training ML models tends to enhance accuracy. This was confirmed in the present study, where the use of filtered MAP sensor data increased the number of relevant features, leading to improved accuracy across most ML models.

A discrepancy was observed when comparing error percentages with other studies using ANNs models trained on MAP sensor data. For example, in the study conducted by [1], their ANN-based algorithm achieved an accuracy close to 100%, despite using similar training process. This difference can be attributed to the type of faults analyzed; while their study focused on different engine issues, this research examined faults such as a clogged air filter and a faulty spark plug, which may be more challenging to detect using MAP sensor data alone. Nevertheless, the precision values obtained in this study are comparable to those found by [13], indicating that both audio signals and voltage signals from the MAP sensor can yield similar precision when an ANN is used.

The SVM algorithms implemented in this study achieved performance comparable to that of the ANN models. However, compared to the study by [13], where SVMs trained on engine audio data reached 100% accuracy, the SVM trained on MAP sensor data in this research reached a maximum of 95.5%. This suggests that audio signals may provide slightly more discriminative features for certain fault classifications.

The models provided by MATLAB's Classification Learner app generally yielded lower accuracy compared to ANN and SVM. However, considering training time, ANN significantly outperformed both SVM and the other algorithms, indicating a lower computational cost. This supports previous findings by [11, 14], which identified ANN as both accurate and efficient. Furthermore, the similar values obtained for precision, recall, and F1-score across models suggest that overfitting did not occur [15, 16].

Although the ML algorithms were trained exclusively using data from the experimental unit, this study successfully demonstrated the feasibility of using MAP sensor voltage data to classify engine faults such as air intake obstruction and faulty spark plugs.

5 Conclusions

This study demonstrated that voltage signals from a MAP sensor can be effectively used to train ML classification algorithms. The process of feature extraction was essential, with RMS, median, minimum peak, and signal energy identified as the most significant features for classification tasks.

The most accurate models were ANN and SVM. However, ANN was considered the best-performing algorithm overall, achieving an accuracy of up to 96.05%, compared to a maximum of 95.0% for SVM. Additionally, ANN exhibited shorter training times, indicating lower computational costs.

The use of filtered MAP sensor data led to the extraction of more relevant features and, consequently, improved performance metrics for most ML models. Specifically, the ANN model's accuracy improved to 96.56% when trained on filtered data, while the best SVM configuration reached a maximum of 95.5%

The findings also suggest that MAP sensor voltage signals, despite being a single-source input, contain sufficient dynamic information to support fault classification in Otto engines. This opens the possibility for low-cost diagnostic systems using existing vehicle sensors combined with ML algorithms, particularly in environments with limited access to more advanced diagnostic equipment

References

1. Contreras, W., Maldonado, J., León, R.: Aplicación de una red neuronal feed-forward backpropagation para el diagnóstico de fallas mecánicas en motores de encendido provocado. Ingenius. Rev. Cienc. Tecnología. **2019**, 32–40 (2019) https://doi.org/10.17163/INGS.N21.2019.03
2. Galindo, E.A., Perdomo, J.A., Figueroa-García, J.C.: Comparative study among multiclass support vector machines, artificial neural networks and self-organized neuro-fuzzy inference system for classification problems. Información Tecnológica. **31**, 273–286 (2020). https://doi.org/10.4067/S0718-07642020000100273
3. García, F.A., Escobar, J.L., Gallegos, C.M., Hernández, E.S.: El enfoque de aprendizaje conjunto en la detección de fallas en cajas de engranajes. Rev. Univ. Sociedad. **15**, 325–333 (2023) https://rus.ucf.edu.cu/index.php/rus/article/view/3751

4. Castillo, J., Díaz, D., Carrión, R., Rivera, D.: Influence of failures not detected by the on- board diagnostic system on the performance and pollutant emissions of a spark ignition engine. Adv. Intell. Syst. Comput. **1326**, 389–403 (2021). https://doi.org/10.1007/978-3-030-68080-0_29
5. Bimboza, J., Cárdenas, L., Mora, C., Mancheno, M.: Calidad del servicio y satisfacción del cliente. el caso del mantenimiento vehicular liviano. Religación: Rev. Cienc. Sociales Humanidades., ISSN-e 2477-9083. **8**(35), 15 8 (2023). https://doi.org/10.46652/rgn.v8i35.1019
6. AEADE: Sector automotor en cifras (2024), https://www.aeade.net/wpcon- tent/uploads/2024/07/6.-Sector-en-Cifras-Resumen-Mayo.pdf
7. Yixin, E.S., Saeid, A.W., Tjong, J.: Fault detection and diagnosis of engine spark plugs using deep learning techniques. SAE Int. J. Engines. **15**, 515–526 (2022). https://doi.org/10.4271/03-15-04-0027
8. Shahbaz, M.H., Amin, A.A.: Design of active fault tolerant control system for air fuel ratio control of internal combustion engines using artificial neural networks. IEEE Access. **9**, 46022–46032 (2021). https://doi.org/10.1109/ACCESS.2021.3068164
9. Li, X., Wang, N., Lyu, Y., Duan, Y., Zhao, J.: Data-driven fault early warning model of automobile engines based on soft classification. Electronics. **12**(3), 511 (2023). https://doi.org/10.3390/electronics12030511
10. Gutiérrez, H., de la Vara, R.: Análisis y diseño de experimentos, 2ª edn. McGraw- Hill Interamericana, México (2008)
11. Tellez, J., et al.: Ai-based prediction of capital structure: performance comparison of ann svm and lr models. Comput. Intell. Neurosci. (2022). https://doi.org/10.1155/2022/8334927
12. Kurani, A., Doshi, P., Vakharia, A., Shah, M.: A comprehensive comparative study of artificial neural network (ANN) and support vector machines (SVM) on stock forecasting. Ann. Data Sci. **10**, 183–208 (2023). https://doi.org/10.1007/S40745-021-00344-X/METRICS
13. Delgado, E.H.: Desarrollo de un algoritmo de diagnóstico para la detección de fallas mecánicas en motores de encendido provocado basados en la transformada wavelet (2018). http://dspace.ups.edu.ec/handle/123456789/15300
14. Gallegos, A.E., Torres, M., Torres, A., de León, E.P.: Contrastación de algoritmos de aprendizaje automático para la clasificación de señales eeg machine Learning algorithms testing for eeg signal classification. Res. Comput. Sci. **149** (2020) https://n9.cl/pu36m
15. Microsoft: Managing machine learning pitfalls. Microsoft Learn (2024). [Online]. Disponible en: https://learn.microsoft.com/es-es/azure/machine-learning/concept-manage-ml-pit-.falls
16. Grandini, M., Bagli, E., Visani, G.: Metrics for multi-class classification: an overview. arXiv preprint arXiv:2008.05756v1 (2020). [Online]. Available in: https://doi.org/10.48550/arXiv.2008.05756

ColpoNet: A Nested Deep Learning Framework for Cervical Lesion Segmentation and Classification Using U-Net and Detectron2

Ramiro Israel Vivanco Gualán[1(✉)], Yuliana Jiménez-Gaona[2], Darwin Castillo[2], Bernardo Vega Crespo[3], and Veronique Verhoeven[4]

[1] Departamento de Ciencias de la Computación y Electrónica, Universidad Técnica Particular de Loja, Loja, Ecuador
rivivanco@utpl.edu.ec

[2] Departamento de Química y Ciencias Exactas, Universidad Técnica Particular de Loja, Loja, Ecuador
{ydjimenez,dpcastillo}@utpl.edu.ec

[3] Facultad de Ciencias Médicas, Universidad de Cuenca, Cuenca, Ecuador
bernardo.vegac@ucuenca.edu.ec

[4] Family Medicine and Population Health, University of Antwerp, Antwerp, Belgium
veronique.verhoeven@uantwerpen.be

Abstract. Cervical cancer is a significant cause of mortality among women, especially in low-resource settings, due to the limited access to timely diagnosis. In this context, this project developed ColpoNet. This computer-aided diagnostic system integrates deep learning and clinical scoring to enhance cervical intraepithelial neoplasia (CIN) detection and classification from colposcopy images. ColpoNet combines a U-Net-based and a Detectron2 instance classification model guided by Swede Score clinical features. The system was trained and validated on both public and private colposcopy datasets. The results achieve strong performance, with an Intersection over Union (IoU) of 0.77 for segmentation and a mean Average Precision (mAP) of 0.74 for lesion classification. These results validate the effectiveness and utility of the integrated framework, ColpoNet, for standardized cervical screening and lesion analysis, reducing diagnostic variability and improving access to high-quality cervical screening, especially in resource-limited environments.

Keywords: Cervical Cancer · Colposcopy · U-Net · Detectron2 · Lesion Segmentation · Classification · Swede Score

1 Introduction

Cervical cancer is the fourth most common type of cancer among women worldwide and a leading cause of mortality in low- and middle-income countries [1]. The disease develops slowly, progressing through precancerous stages collectively known as cervical intraepithelial neoplasia (CIN), which includes three histological grades: CIN1 (mild

J. L. Ballesteros et al. (Eds.): CITIS 2025, LNNS 1780, pp. 149–158, 2027.
https://doi.org/10.1007/978-3-032-14287-0_15

dysplasia), CIN2 (moderate dysplasia), and CIN3 (severe dysplasia or carcinoma in situ) [2, 3]. If not detected and treated early, CIN may evolve into invasive cervical cancer, especially in regions with limited access to medical care.

In many rural areas and remote communities, the lack of trained specialists and difficulty accessing timely diagnostic services significantly reduce the chances of early intervention and successful treatment. Traditional screening methods, such as cytology (Pap smear) and colposcopy examination, are widely used to detect CIN lesions. Although effective, inter-observer variability, subjective interpretation, and the shortage of skilled healthcare professionals often affect these methods. Reported diagnostic sensitivity of colposcopy ranges between 81.4% and 95.7%, while specificity varies from 34.2% to 69%, highlighting the variability in its diagnostic performance [4–6].

Artificial intelligence (AI) has emerged as a promising solution for enhancing medical image analysis in recent years. Convolutional Neural Networks (CNNs), particularly U-Net for image segmentation and Detectron2 for instance-based classification, have achieved remarkable results in lesion detection, segmentation, and classification tasks. Several studies have demonstrated that deep learning models outperform traditional diagnostic methods and experienced clinicians in specific scenarios [7–11].

Despite these advancements, most AI-based approaches do not incorporate standardized clinical frameworks, such as the Swede Score, a scoring system endorsed by the International Federation for Cervical Pathology and Colposcopy (IFCPC) in 2011, which evaluates five key visual criteria: aceto-whiteness, lesion margins, vascular patterns, iodine staining, and lesion size [12].

To address these challenges, this work proposes ColpoNet, an AI-based diagnostic framework designed to automate the segmentation and classification of cervical lesions from colposcopy images. ColpoNet combines a U-Net-based architecture for pixel-wise segmentation of lesions with Detectron2 for instance-level classification, guided by Swede Score criteria, and contributes to developing more accessible and standardized cervical cancer screening systems, particularly in resource-limited environments.

2 Related Work

Several works demonstrated deep learning's effectiveness in colposcopic imaging.

Li et al. [9] developed a semantic segmentation model based on DeepLabv3+ to detect high-grade squamous intraepithelial lesions (HSIL) from colposcopy images and achieved an accuracy of 93.29% and mean Intersection over Union (mIoU) of 80.27%. Kalbhor et al. [13] introduced the CerviCell-detector, an object detection system using YOLOv5, Faster R-CNN, and Detectron2 frameworks on the CRIC dataset. Their binary classification task achieved a mean Average Precision (mAP) of 83%.

Chandran et al. [10] developed the Colposcopy Ensemble Network (CYENET), which outperformed traditional VGG19 models, reaching 92.3% in classification accuracy, demonstrating the advantages of ensemble approaches in improving performance on colposcopy datasets.

Jiménez Gaona et al. [14] created a radiomics-based CAD system for colposcopy image classification using an Unet-SVM pipeline, showing sensitivity and specificity of (70%). Xu et al. [15] proposed an optimized Faster R-CNN model initialized with

COCO-pre-trained weights for cervical cytology image detection with an mAP of 61.6% and an AR of 87.7%. Yuan et al. [6] developed deep learning system classification for colposcopy analysis, achieving high accuracy (AUC = 0.93) and outperforming experts in HSIL detection on standard images. It detected 84.67% of high-grade lesions, outperforming expert colposcopists in some tasks.

3 Methodology

The methodological framework proposed in this study is structured into several stages, as illustrated in Fig. 1.

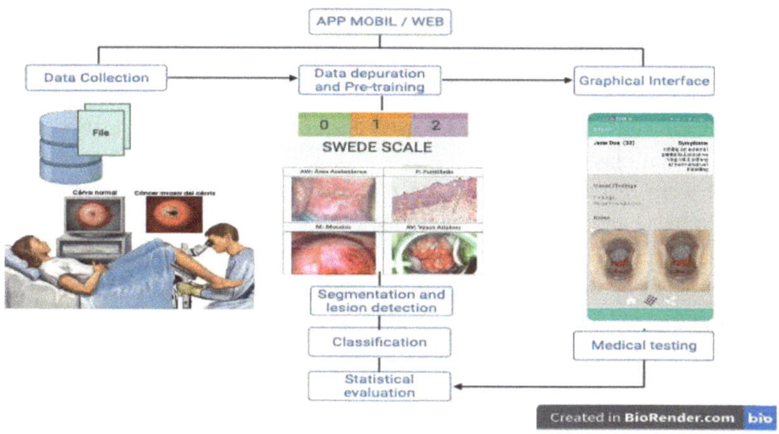

Fig. 1. Overview of the proposed methodological pipeline. The process begins with data collection and depuration, followed by U-Net segmentation and lesion classification using Detectron2. Lesions are identified according to Swede Score criteria, and model performance is evaluated using IoU, Dice, accuracy, and precision.

3.1 Data Collection

The dataset employed in this study originates from a previous multidisciplinary project and comprises both public and private colposcopy images. It includes curated samples from the Intel & MobileODT Cervical Cancer Screening dataset and the CAMIE dataset, the latter of which was collected in collaboration with healthcare centers in Cuenca, Ecuador. Medical specialists clinically validated all images. (See Fig. 1) (Table 1).

During the data preparation phase, all images were curated to eliminate redundancies and irrelevant visual elements. Cervical region masks were manually generated using ImageJ, ensuring that only the area of interest was included while excluding the colposcope and vaginal surroundings. Additionally, lesion annotations were performed with makesense.ai based on the Swede Score criteria (AW, P, M, AV), producing polygonal segmentations stored in JSON format, compatible with VGG and COCO standards. As a result, three key elements were obtained: (1) a curated image set from the

Table 1. Summary of image datasets used in this study.

Dataset Type	Database Name	Number of Images
Public	Intel & MobileODT Cervical Cancer Screening	461
Private	CAMIE (Collected in Cuenca, Ecuador)	20
Total		**481**

Intel and CAMIE datasets, (2) binary masks outlining the cervix region, and (3) JSON files with lesion annotations. These structured data assets enabled the transition to the preprocessing phase.

3.2 Data Preprocessing

Two independent processing pipelines were implemented to support the Cervical Data Repository (CDR) development: one focused on cervix segmentation using U-Net, and the other on lesion classification using Detectron2. Each image and its corresponding binary mask were resized to 256 × 256 pixels for segmentation, with RGB images normalized to [0, 1] and masks preserving their binary structure (0 for background, 255 for region of interest).

For classification, class distribution was evaluated based on valid annotations, with Class 3 (mosaic) and Class 1 (acetowhite areas) being the most prevalent (162 and 156 instances, respectively). In contrast, Classes 2 (punctuation) and 4 (atypical vessels) had moderate presence, and Class 5 appeared in only seven cases. To address this imbalance, data balancing techniques were applied.

3.3 ColpoNet Architecture

ColpoNet is a two-stage deep learning architecture designed for the automated analysis of cervical lesions from colposcopy images (see Fig. 2). The first stage applies a U-Net-based segmentation model that isolates the cervix region by removing irrelevant anatomical structures using manually generated binary masks. The second stage performs lesion classification using an instance segmentation model implemented in Detectron2, based on the Mask R-CNN architecture with a ResNet-50 backbone. The lesions are categorized according to the Swede Score system: acetowhite areas (AW), punctuation (P), mosaic (M), and atypical vessels (AV).

Both models were trained using specific optimization settings. The U-Net model was developed with TensorFlow/Keras and trained on 256 × 256 RGB images using a combined loss function of Binary Crossentropy and Dice Loss, optimized with Adam for 150 epochs. The Detectron2 model was configured with SGD and a WarmupMultiStepLR scheduler, trained over 3000 iterations with polygon annotations in COCO format. In both cases, performance was evaluated using metrics such as IoU, DICE, AP, and mAP. The key hyperparameters used for training are summarized in Table 2.

All experiments were conducted on a TITAN workstation equipped with a 14th-generation Intel Core i7 processor, 64 GB of RAM, and an NVIDIA RTX 4060 GPU with 8 GB of VRAM.

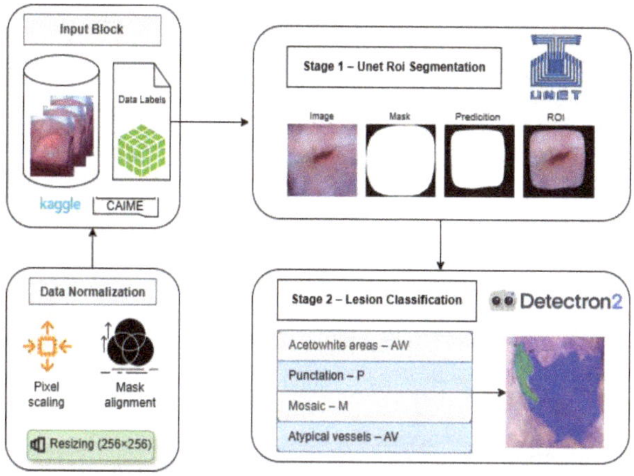

Fig. 2. General architecture of the ColpoNet framework.

Table 2. Detailed hyperparameters used for U-Net and Detectron2 training.

Model	Architecture	Input Format	Optimizer	LR	Batch Size	Epochs / Iter.
U-Net	U-Net	256 × 256 RGB	Adam	1e-4	16	150 epochs
Detectron2	Mask R-CNN (Detectron2, ResNet-50)	COCO-style polygons (Swede)	SGD	0.00025	2	3000 iterations

3.4 Statistical Evaluation

The Kappa value was calculated to determine the level of agreement between the two readers (a total of 420 images were analyzed by two readers with 25 and 4 years of experience, respectively) and the radiomic tool (ColpoNet) during the image classification accuracy percentage. The two readers' choices were classified according to five possibilities and were interpreted as follows: <0.2 slight; 0.21–0.40 fair; 0.41–0.60 moderate; 0.61–0.80 high; 0.81–1.0 almost perfect [17].

4 Results and Discussion

4.1 U-Net Model Segmentation Results

After the training, the final segmentation model was obtained using the curated dataset and corresponding masks. A steady and consistent improvement can be ob-served in IoU and DICE, reaching values close to 0.63 and 0.77, respectively. Like-wise, the loss curves show a continuous decline until stabilization, indicating a good generalization capacity of the model based on the validation data. See Fig. 3.

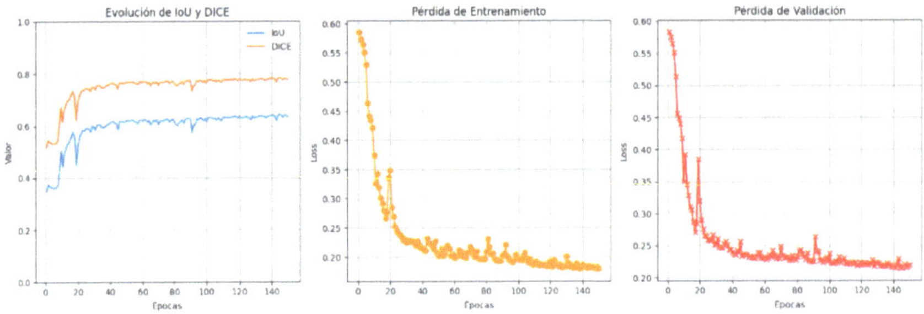

Fig. 3. (Left) Evolution of IoU and DICE metrics during training. (Center) Model total_loss on the training set. (Right) Model total_loss on the validation set.

Performance evaluation was carried out on the test set using the checkpoint from the trained model. Representative samples were selected to assess the model's predictions qualitatively. Figure 4 shows one example, displaying the original RGB image, its ground truth mask, the predicted mask, and the final segmentation overlaid on the original image. This sample achieved an IoU score of 0.77 and a DICE score of 0.87, demonstrating high accuracy in delineating the region of interest (ROI). Visually, the prediction closely aligns with the true mask, with smooth and well-defined boundaries.

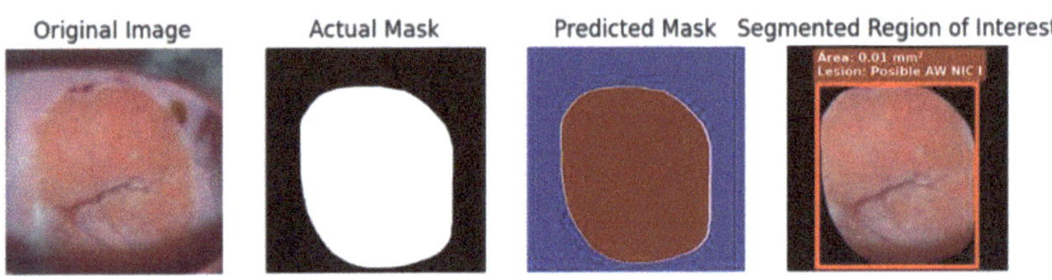

Fig. 4. Representative example from the test set. From left to right: original image, ground truth mask, model prediction, and overlaid segmentation.

4.2 Detectron2 Model, Classification Results

The instance segmentation model based on Detectron2 was evaluated after completing the training process, using the validation set as a reference. Both qualitative visualizations of predictions and quantitative class-wise precision metrics were analyzed to assess their performance, considering different IoU thresholds (See Fig. 5).

Regarding evaluation metrics, Fig. 6 displays a heatmap illustrating the average precision (AP) obtained per class across different IoU thresholds (0.50, 0.75, and 0.90), as well as the overall mean average precision (mAP). The results show the highest performance for classes P (Punctuation) and AW (Acetowhite Areas), with maximum values of 0.89 and 0.78, respectively, at IoU = 0.50. Meanwhile, classes M (Mosaic) and AV (Atypical Vessels) also exhibited acceptable results, although with a slight drop as the IoU threshold increased. The overall mAP values ranged from 0.60 to 0.74 depending on the class.

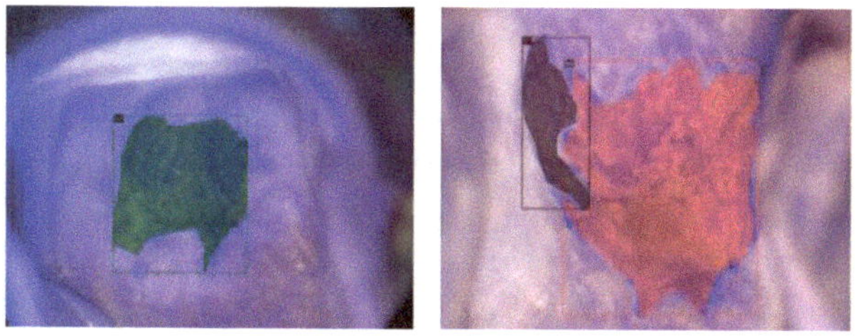

Fig. 5. Prediction examples from the Detectron2 model on colposcopy validation images. These visualizations show the accurate detection and segmentation of lesion regions on the cervix, where the model successfully identifies multiple classes within the same image.

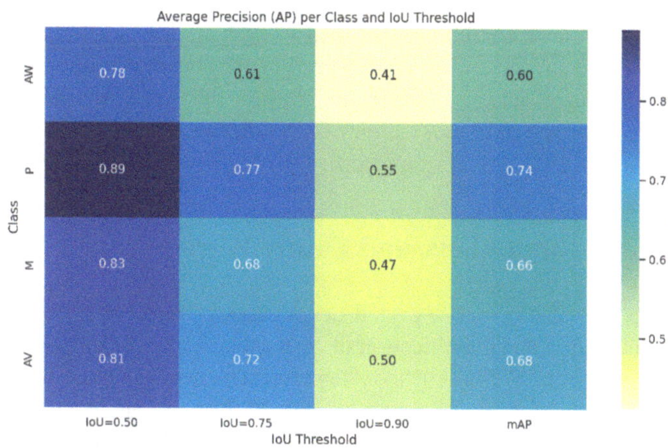

Fig. 6. Heatmap of average precision (AP) per class (AW, P, M, AV) across different IoU thresholds. The mAP value is included as a global reference of model performance.

4.3 Validation of Expert Clinicians

The following codes correspond to the medical evaluations and the applied model: M1A refers to the first physician, M2A to the second physician, and T3A represents the model analysis. The comparison between ColpoTool and human experts revealed varying levels of agreement. Substantial agreement was found between M1A and T3A ($\kappa = 0.729$), while Class M2A showed only slight agreement with the others ($\kappa \approx 0.15$–0.17), suggesting inconsistency in ratings. Overall inter-rater reliability, measured by Fleiss' Kappa ($\kappa = 0.35$) and Krippendorff's Alpha ($\alpha = 0.35$), indicated fair but limited agreement. These findings highlight the need for enhanced rater training. Calibration efforts are strongly recommended.

5 Discussion

Concerning the performance of ColpoNet related to cervical lesion analysis, a comparative evaluation was conducted using several studies. The segmentation module of ColpoNet achieved a Dice score of 87.0% and a mean Intersection over Union (mIoU) of 77.0%, which outperformed previous methods. For instance, Li et al. [9] developed a DeepLabv3 + -based segmentation model that achieved 93.29% accuracy and a mIoU of 80.27%, comparable to ColpoNet's segmentation performance. However, it focused on HSIL lesions only and did not incorporate lesion classification under the Swede Score criteria. In contrast, Jiménez Gaona et al. [14] implemented a U-Net + SVM pipeline with 70% sensitivity and specificity without instance-based segmentation or multiclass lesion detection. In comparison, Yuan et al. [6] reported a significantly lower Dice score of 61.6%.

In terms of classification performance, ColpoNet provides a robust capability to identify and differentiate between lesion types. The classification results demonstrated a mean Average Precision (mAP) of 0.74, with a maximum Average Precision (AP) of 0.89 specifically for the Punctuation lesion class. In this sense, Yuan et al. [6] reported an Area Under the Curve (AUC) of 0.93, and Chandran et al. [10] and Li et al. [9] reported high classification accuracy rates exceeding 92% using VGG19; their methods lacked the integration of lesion segmentation and localization. Jiménez-Gaona et al. [14] reported an accuracy of classification of 58%, further emphasizing the improved diagnostic utility of the ColpoNet framework.

6 Conclusions, Limitations, and Future Work

This work introduced ColpoNet, a nested deep-learning framework designed for the automated segmentation and classification of cervical lesions from colposcopy images. The system integrates a U-Net-based model to perform binary segmentation of the cervix and a Detectron2-based model for lesion classification according to the Swede Score clinical system. Experimental results demonstrated that the U-Net model achieved strong segmentation performance, with an average Intersection over Union (IoU) of 0.77 and a Dice Score of 0.87, indicating high agreement between predicted and ground truth masks. In the classification stage, Detectron2 achieved a mean average precision (mAP) of 0.74, with a maximum average precision (AP) of 0.89 for the Punctuation class.

One of the main limitations identified in this work concerns the volume of the datasets used. Given that the datasets employed contain a limited number of images, Transfer Learning techniques were applied, particularly for classification tasks using Detectron2. Another limitation relates to implementing balancing techniques during training to mitigate dataset imbalance, especially between two lesion categories due to dominant acetowhite and mosaic lesions. In contrast, punctuation, atypical vessels, and other categories appear in significantly fewer samples.

As part of immediate future work, a mobile application will be developed to deploy ColpoNet in real-world clinical settings. The application will incorporate lightweight or online-accessible versions of the segmentation and classification models, using real-time image analysis in field conditions.

Finally, future work will also include the expansion of datasets by collecting new annotated images from various healthcare centers, increasing the diversity of colposcopic lesion types and clinical conditions. This will allow for broader generalization and improved performance across populations.

Funding. Funding was provided by VLIR-OUS as a part of the project entitled "Community engagement and novel technologies: a winning team for the integral management of cervical cancer in a rural context" COENCA; and project "Desarrollo de una interface gráfica movil para clasificación y detección de lesiones en imágenes de mama y cervix"-PROY_INV_QU_2022_3576 from Universidad Técnica Particular de Loja.

Data Availability Statement: All codes that were developed are available as open source at the GitHub site: https://github.com/israv87/ColpoNet.git

Conflicts of Interest: The authors declare no conflict of interest.

References

1. Bruni, L., Albero, G., Serrano, B., et al. (2019). Human Papillomavirus and Related Diseases in the World..
2. Walboomers, J.M.M., et al.: Human papillomavirus is a necessary cause of invasive cervical cancer worldwide. J. Pathol. **189**, 12–19 (1999). https://doi.org/10.1002/(SICI)1096-9896(199909)189:1<12::AID-PATH431>3.0.CO;2-F
3. Arbyn, M., et al.: Accuracy of human papillomavirus testing on self-collected versus clinician-collected samples: a meta-analysis. Lancet Oncol. **15**(2), 172–183 (2014). https://doi.org/10.1016/S1470-2045(13)70570-9
4. Brown, B.H., Tidy, J.A.: A review of research methodology and impact on quality assurance outcomes. **240**, 182–186 (2019). https://doi.org/10.1016/j.ejogrb.2019.07.014
5. Underwood, M., et al.: Accuracy of colposcopy-directed punch biopsies: a systematic review and meta-analysis. BJOG Int. J. Obstet. Gynaecol. **119**(11), 1293–1301 (2012). https://doi.org/10.1111/j.1471-0528.2012.03486.x
6. Yuan, C., Yao, Y., Cheng, B., et al.: The application of deep learning-based diagnostic system to cervical squamous intraepithelial lesions recognition in colposcopy images. Sci. Rep. **10**, 11639 (2020). https://doi.org/10.1038/s41598-020-68252-3
7. Litjens, G., et al.: A survey on deep learning in medical image analysis. Med. Image Anal. **42**, 60–88 (2017). https://doi.org/10.1016/j.media.2017.07.005
8. Sarhangi, A.S., Beigifard, D., Farmani, E., Bolhasani, H.: Deep learning techniques for cervical cancer diagnosis based on pathology and colposcopy images. Inform. Med. Unlocked. **47**, 101503 (2024). https://doi.org/10.1016/j.imu.2024.101503
9. Li, Z., Zeng, C.M., Dong, Y.G., et al.: A segmentation model to detect cervical lesions based on machine learning of colposcopic images. Heliyon. **9**, e21043 (2023). https://doi.org/10.1016/j.heliyon.2023.e21043
10. Chandran, V., Sumithra, M.G., Karthick, A., et al.: Diagnosis of cervical cancer based on ensemble deep learning network using colposcopy images. Biomed. Res. Int. **2021**, 5584004 (2021). https://doi.org/10.1155/2021/5584004
11. Castillo, D., Rodríguez-Álvarez, M.J., Samaniego, R., Lakshminarayanan, V.: Models to identify small brain white matter hyperintensity lesions. Appl. Sci. **15**, 2830 (2025). https://doi.org/10.3390/app15052830
12. Basu P, Sankaranarayanan R. Atlas de colposcopia – Principios y Práctica: IARC CancerBase No. 13 [Internet]. Lyon, Francia: Agencia Internacional para la Investigación sobre el Cáncer (2017). Disponible en: https://screening.iarc.fr/atlascolpo.php. consultado el 23 May 2025
13. Kalbhor, M., Shinde, S., Wajire, P., Jude, H.: CerviCell-detector: an object detection approach for identifying the cancerous cells in pap smear images of cervical cancer. Heliyon. **9**, e22324 (2023). https://doi.org/10.1016/j.heliyon.2023.e22324

14. Jiménez Gaona, Y., Castillo Malla, D., Vega Crespo, B., et al.: Radiomics diagnostic tool based on deep learning for colposcopy image classification. Diagnostics. **12**(7), 1694 (2022). https://doi.org/10.3390/diagnostics12071694
15. Xu, C., Li, M., Li, G., et al.: Cervical cell/clumps detection in cytology images using transfer learning. Diagnostics. **12**(10), 2477 (2022). https://doi.org/10.3390/diagnostics12102477
16. Ben-Or, Y., Kan, W., Sherman, V., et al.: Intel & MobileODT Cervical Cancer Screening Dataset [Dataset]. Kaggle (2017)
17. Jiménez-Gaona, Y., Álvarez, M.J.R., Castillo-Malla, D., et al.: BraNet: a mobil application for breast image classification based on deep learning algorithms. Med. Biol. Eng. Comput. **62**, 2737–2756 (2024). https://doi.org/10.1007/s11517-024-03084-1

Application of Classical NLP Techniques for Toxic Comment Analysis: A Lightweight Alternative to Deep Learning Models

Alexandra Veloz[1], Renato Valle[2], Johanna Dueñas[3], and E. Fabián Rivera[1(✉)]

[1] Universidad Estatal de Bolívar, Guaranda 020102, Ecuador
{eveloz,edgar.rivera}@ueb.edu.ec
[2] Carrera de Comunicación, Universidad Estatal de Bolívar, Guaranda 020102, Ecuador
dvalle@ueb.edu.ec
[3] Carrera de Ingeniería en Riesgos de Desastres, Universidad Estatal de Bolívar, Guaranda 020102, Ecuador
johanna.duenas@ueb.edu.ec

Abstract. The present article focuses on the classification of toxic comments using natural language processing and various machine learning algorithms. A thorough analysis is conducted on a dataset of textual comments labeled as toxic or non-toxic, covering several categories such as insults, threats, obscene language, and identity-based hate. Natural language processing techniques were employed to vectorize the text and train the following models: Logistic Regression, K-Nearest Neighbors (k-NN), Bernoulli Naive Bayes, Multinomial Naive Bayes, Support Vector Machines (SVM), and Random Forests. The results show that the SVM and Logistic Regression models achieve the highest F1-scores, with values of 0.746 and 0.745 respectively, demonstrating a balance between precision and recall. In contrast, the k-NN, model, despite having a good precision of 0.838, exhibits a low recall of 0.288, which reduces its effectiveness in this context. Additional metrics such as ROC and Precision-Recall curves reinforce the findings and allow for evaluating the inherent trade-offs between the different approaches. The analysis suggests that model selection should consider both the most costly type of error and the computational constraints of the deployment environment.

Keywords: Toxic Comments · Machine Learning · Natural Language Processing · Text Vectorization · Artificial Intelligence

1 Introduction

The global expansion of information and communication technologies has profoundly transformed the way people communicate, primarily through the internet and social media platforms [1]. Millions of social interactions now occur via digital platforms, from social networks to opinion forums, facilitating the exchange of ideas but also creating space for new forms of symbolic violence among users. In this context, the presence

J. L. Ballesteros et al. (Eds.): CITIS 2025, LNNS 1780, pp. 159–168, 2027.
https://doi.org/10.1007/978-3-032-14287-0_16

of toxic messages comments that include offensive, aggressive, or even discriminatory language represents a growing challenge that affects digital coexistence, causing negative impacts on users' mental and emotional health and on the quality of public discourse [2]. A toxic comment is one that attacks, insults, denigrates or threatens a person or group, either explicitly or implicitly, and that can generate a hostile environment in digital spaces. According to the NewsCom-TOX study, any message that, regardless of the author's intention, causes verbal or emotional harm, especially if it targets characteristics such as race, gender, religion, sexual orientation or nationality, is considered toxic [3].

The use of toxic messages is an increasingly relevant issue due to the continuous rise in internet access and the active participation of both young people and adults on social media platforms [4]. Despite the democratizing potential of social media, the lack of effective mechanisms to identify and moderate harmful content contributes to the normalization of hostile discourse and the silencing of vulnerable voices. Thus, artificial intelligence algorithms such as and Machine Learning (ML) and Deep Learning (DL) not only offer solutions in fields such as health [5], education [6], and transportation but also provide tools to help identify harmful or negative content in user interactions on social media and other digital platforms. ML algorithms aim to accurately and timely detect toxic messages, especially in contexts where language and culture influence how emotions, criticism, or disagreements are expressed in user interactions [7, 8].

In response to the previously described reality, this study proposes the development and evaluation of ML models for the classification of toxic messages, using natural language processing (NLP) techniques to normalize the text. Various ML models are trained—including Logistic Regression, Naive Bayes, SVM, k-NN, and Random Forest with the goal of evaluating their performance using metrics such as Accuracy and F1-Score, and identifying the most effective classifiers for each category of toxic comment [9]. Additionally, this work contributes tools to efficiently detect harmful content, taking into account both linguistic context and the ethical dimension of the problem. The study presents a systematic and comparative evaluation of classic machine learning algorithms applied to the problem of toxic comment classification in natural language, using a representative set of performance metrics. The research is grounded in the understanding that language has a profound impact on people's lives, and its responsible use is essential to building fairer, safer, and more inclusive digital spaces [10].

This paper is divided into five sections: Sect. 1 presents the Introduction; Sect. 2 explores Related Works through a literature review; Sect. 3 details the Methodological process of the models; Sect. 4 presents the Results; and finally, Sect. 5 highlights the Conclusions and Future Work drawn from this research work.

2 Literature Review

The globalization of digital environments has enabled the massive expression of opinions and thoughts on social networks, forums, video platforms, and other socially accessible digital media. In this context, the freedom of communicative environments has also contributed to the increased circulation of toxic comments, characterized by offensive, discriminatory, and even violent content with alarming social implications. This section presents a brief review of studies that have addressed this issue using various methodological approaches and ML tools capable of detecting and mitigating the impact of toxic comments on social networks and digital platforms.

Regarding toxic comment classification, the study presented in [11] stands out for applying a deep learning approach using a set of algorithms including CNN, LSTM, and their combination, integrating embeddings such as GloVe and fastText to classify comments into six distinct categories. The research reports an outstanding accuracy of 98.46%. However, it is limited to predefined deep architectures and does not evaluate their behavior with imbalanced data. Similarly, studies [12, 13] delve into the use of neural network models for toxicity detection. In [12], LSTM and CNN are applied to tokenized textual data, while [13] contextualizes the issue from a sociotechnological perspective, addressing how the misuse of platforms has amplified toxic behaviors. Both studies agree that the combination of NLP techniques and neural networks enables the identification of harmful patterns. However, they present limitations due to the lack of class balancing strategies or interpretable models that allow decision auditing.

In contrast to these deep learning approaches, other studies have opted for classical and comparative methods. For example, in [14, 15], the impact of data imbalance is analyzed using techniques such as SMOTE and classifiers like Random Forest, Logistic Regression, and a combined model called RVVC. Study [14] demonstrates performance improvements by balancing the data, while [15] reports that T5 combined with Random Forest achieves an F1-Score of 0.91. Nevertheless, both studies lack contextual perspectives on the application environment and do not explore the real impact of toxicity on specific communities. On the other hand, [16] focuses on the issue from an educational standpoint, recognizing that virtual learning environments have been severely affected by toxic comments directed at both students and teachers. Therefore, the study uses a hybrid dataset and models such as Naive Bayes and Random Forest, evaluating the classification using metrics such as ROC and Hamming scores, which is considered a commendable approach.

An updated and state-of-the-art model-focused approach is addressed in [17], which compares pre-trained models such as BERT, RoBERTa, and XLM, concluding that BERT and RoBERTa outperform XLM. Additionally, the study shows that simple downstream classification architectures can be more effective than complex ones. Similarly, [18] centers its research on YouTube comments, highlighting the platform's unique linguistic challenges and achieving strong results with BERT. These studies demonstrate significant advancements; however, they are limited by the lack of mechanisms to handle data imbalance and are not applied to broader communication contexts. Along the same comparative line, [19] proposes a tool based on seven machine learning algorithms to identify toxic comments. Its practical approach allows determining which model is more efficient in terms of accuracy, representing a valuable methodological contribution.

Despite these advances, the reviewed studies share common limitations: insufficient attention to data contextualization, limited incorporation of balancing and generalization techniques, and a weak focus on critical applications such as the educational domain. Furthermore, few works explore hybrid approaches that combine advanced language models with balancing methods to enhance detection in scenarios with limited or highly imbalanced data.

3 Methodology

This Section details the complete data processing workflow, including visualization, text preprocessing, and evaluation of machine learning models aimed at classifying toxic comments and assessing each model's performance. It is divided into several stages, each with a specific purpose within an NLP pipeline.

3.1 Data Loading and Exploratory Analysis

First, the necessary libraries for text processing, visualization, and machine learning are imported. Then, a CSV file is uploaded via Google Colab containing user comments labeled with different categories of toxicity such as toxic, severe toxic, obscene, threat, insult, and identity hate. Information is obtained regarding the dataset's structure, and class proportions for each category are identified. The analysis reveals a significant imbalance across categories, which later motivates the inclusion of a data balancing stage.

3.2 Data Visualization and Preprocessing

Using the matplotlib and seaborn libraries, the script visualizes the distribution of the various types of toxic comments by generating a bar chart, as shown in Fig. 1. This chart represents the total number of records for each toxicity type and serves as a foundational tool for the exploratory analysis, providing guidance for balancing the classes before modeling. It is also observed that the majority of toxic comments fall under the general toxic category, followed by obscene, while threat has a significantly lower representation.

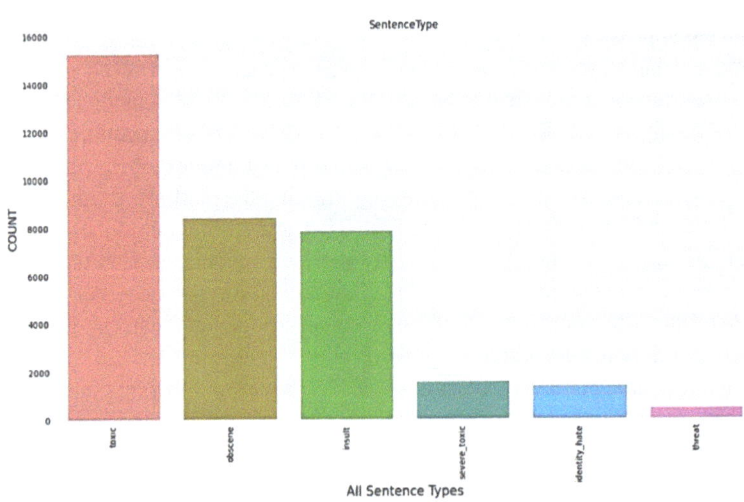

Fig. 1. Distribution of Toxic Comment Types

It is necessary to clean the comments using regular expressions to remove special characters, numbers, punctuation, line breaks, and non-ASCII characters. The text is also converted to lowercase to maintain consistency. This cleaning process is essential for preparing the data for NLP algorithms, as it reduces noise and improves the quality of features to be extracted later.

3.3 Subset Creation and Word Cloud Generation

Subsets of the original dataset are created by separating the comments based on their toxicity type—insult, threat, identity_hate, etc. For each subset, a word cloud is generated using the wordcloud library to visualize the most frequent terms within the toxic comments of each category, as illustrated in Fig. 2. This graphical representation helps to better understand the characteristic language of each type of toxicity and to identify potential lexical patterns useful for classification. It allows for a qualitative visualization of the language used in each form of toxic speech.

Fig. 2. Word Clouds for the different labels

3.4 Data Balancing

Since the categories are imbalanced, it is essential to apply undersampling techniques to balance the number of toxic and non-toxic examples in each subset, given that there are many more non-toxic comments than toxic ones. In this study, for the *toxic* category,

5,000 toxic and 5,000 non-toxic comments are selected. This process is repeated for all categories, ensuring that during training, the classification models are not biased by disproportionate class representation.

3.5 Machine Learning Model Configuration

Multiple classification models are defined for analysis. The machine learning models include Logistic Regression, k-NN, Naive Bayes, SVM, and Random Forest. Text vectorization methods such as CountVectorizer or TfidfVectorizer are specified to convert the preprocessed text into numerical vectors, eliminating stopwords and considering various n-grams. To optimize the hyperparameters of the classical classification models, a grid search technique is applied. Each model is integrated into a pipeline alongside a Tf-idf vectorizer, enabling the transformation of text into numerical representations before training.

For instance, in Logistic Regression, hyperparameters such as C with values [0.1, 1, 10] and solvers liblinear and lbfgs are explored using 3-fold cross-validation and F1-score as the selection criterion. For the SVM model, the C parameter and the kernel type (linear and RBF) are optimized, allowing performance evaluation with different margin configurations and feature space transformations. In the Random Forest model, an exhaustive search is conducted considering three key hyperparameters: the number of trees, the maximum depth of each tree, and the learning rate. The combinations evaluated include between 50 and 100 trees, depths between 3 and 7, and learning rates between 0.01 and 0.3—resulting in a total of 18 possible combinations. All models are trained in parallel to speed up the process and are evaluated based on their performance in predicting the binary toxic label from the training dataset.

4 Results

This Section focuses on the results obtained from the various models applied. Table 1 presents the evaluation metrics for the six classification models used to detect toxic comments, based on four key indicators: Accuracy, Precision, Recall, and F1-Score.

The analysis of the metrics shows that SVM is the most balanced model, achieving the highest F1-Score of 0.7456 and the best Recall of 0.7226—ideal for detecting most toxic comments. Logistic Regression and Random Forest also perform strongly, with high precision and accuracy, making them suitable for minimizing false positives. Additionally, MultinomialNB provides good balance and efficiency, while BernoulliNB, although very precise, suffers from low recall, limiting its effectiveness in capturing toxic comments. Lastly, k-NN, exhibits the worst performance, with very low recall and an insufficient F1-Score, making it unsuitable for this task.

The results from the ROC curve presented in Fig. 3 show that Random Forest and Logistic Regression perform best in classifying toxic comments, with AUC scores of 0.96 and 0.95, respectively. These results indicate that both models possess excellent discriminative capability between toxic and non-toxic comments. The performance of the SVM model is also notable, with an AUC of 0.94, positioning it as a solid alternative.

Table 1. Results of the different models.

N	Model	Accuracy	Precision	Recall	F1-Score
1	Logistic Regression	0.955590	0.846866	0.665240	0.745145
2	k-NN	0.925071	0.837586	0.288099	0.428731
3	BernoulliNB	0.944331	0.879970	0.497432	0.635580
4	MultinomialN	0.950033	0.826835	0.617295	0.706863
5	SVM	0.951872	0.770073	0.722603	0.745583
6	Random Forest	0.954901	0.860458	0.631165	0.728188

The BernoulliNB model achieves an AUC of 0.92, and MultinomialNB obtains 0.90—both offering competitive performance considering their simplicity. In contrast, the k-NN model yields an AUC of only 0.78, reflecting limited ability to distinguish between classes, possibly due to the high dimensionality of the textual data (Fig. 3).

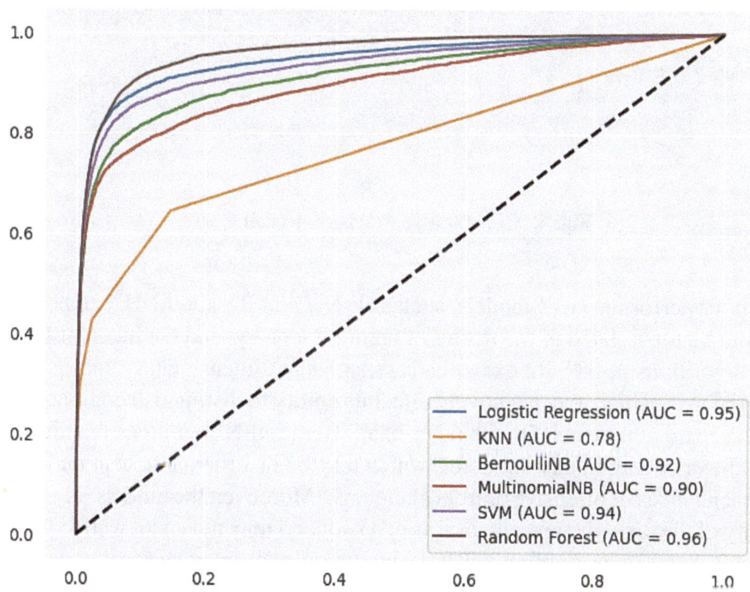

Fig. 3. Comparative ROC curve

The Precision-Recall curve shown in Fig. 4, is especially useful in scenarios with unbalanced classes such as toxic comments, it is again observed that Random Forest presents an AUC of 0.83 and the Logistic Regression model an AUC 0.8; they lead the performance showing a good balance between precision and recall. The SVM model maintains a good performance with an AUC of 0.78, although slightly lower. The Naive Bayes-based models, BernoulliNB with an AUC of 0.75 and MultinomialNB with an

AUC of 0.73, show moderate accuracy, being useful in resource-constrained applications. In contrast, the k-NN model again shows weaknesses, reaching an AUC of only 0.56, indicating that it has significant difficulties in maintaining a good balance between precision and recall in the classification of toxic messages.

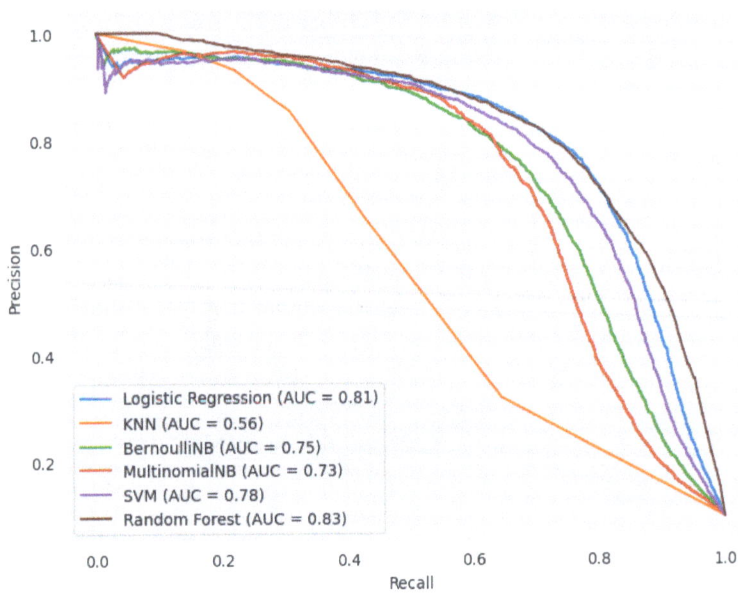

Fig. 4. Comparative Precision-Recall Curve

The poor performance of models, such as k-NN and BernoulliNB, can be attributed to multiple factors related to the nature of natural language and the inherent limitations of these algorithms. k-NN, for example, is sensitive to dimensionality and sparse representation of text, which may negatively affect its ability to distinguish complex semantic patterns in toxic comments. On the other hand, BernoulliNB assumes feature independence and works with binary variables, which may be insufficient to capture lexical and contextual nuances of offensive or ironic language. Moreover, the models rely heavily on the quality of the textual representation as or Tf-idf and may not adapt well to ambiguous expressions, sarcasm or implicit insults, which require more contextual understanding.

5 Conclusions

The results obtained demonstrate that it is possible to classify toxic comments with high effectiveness using classical machine learning models combined with text vectorization techniques. Among the six models evaluated, the SVM model and logistic regression stand out for their balance between accuracy, recall and F1-score, which makes them viable alternatives for automatic content moderation systems. Although all models showed acceptable overall performance, the detailed analysis revealed that those with the highest accuracy did not always achieve the best recall values, a fundamental aspect in applications where it is crucial to identify all cases of toxicity [20].

This work allowed us to clearly visualize the behavior of the classifiers using ROC and Precision-Recall curves, which facilitated the qualitative comparison of performance. It is emphasized that models such as SVM and Random Forest can be effectively implemented on resource-limited platforms without significantly sacrificing classification quality. The exploration of additional metrics beyond accuracy allowed us to detect algorithm-specific strengths and limitations, which reinforces the importance of comprehensive evaluation in sensitive tasks such as toxic language detection.

As future work, we propose to explore deep learning models such as LSTM and BERT, together with advanced natural language processing techniques, to improve text representation and toxic comment detection performance. In addition, we propose to analyze the fairness of the models against different demographic groups, apply class balancing techniques and validate the models in real and multilingual contexts, in order to increase their robustness and applicability in diverse environments.

References

1. Deng, H., Xu, Y., Deng, Y., Lin, J.: Transforming knowledge management in the construction industry through information and communications technology: a 15-year review. Autom. Constr. **142**, 104530 (2022). https://doi.org/10.1016/j.autcon.2022.104530
2. Rawat, A., Kumar, S., Samant, S.S.: Hate speech detection in social media: techniques, recent trends, and future challenges. Wiley Interdisc. Rev.: Comput. Stat. **16**(2), e1648 (2024). https://doi.org/10.1002/wics.1648
3. Taulé, M., Nofre, M., Bargiela, V., Bonet, X.: NewsCom-TOX: a corpus of comments on news articles annotated for toxicity in Spanish. Lang. Resour. Eval. **58**, 1115 (2024). https://doi.org/10.1007/s10579-023-09711-x
4. G. Beknazar-Yuzbashev, R. Jiménez Durán, J. McCrosky, and M. Stalinski: Toxic Content and User Engagement on Social Media: Evidence from a Field Experiment, SSRN Electron. J. (2022). https://doi.org/10.2139/ssrn.4307346.
5. E. D. Mañay, D. Martínez, M. D. Chiliquinga, V. D. Criollo, E. F. Rivera, R. M. Toasa: Low-cost deep learning-based architecture for detecting cardiac arrhythmias in ECG signals, In: Iberian Conference on Information Systems and Technologies, CISTI, (2022). https://doi.org/10.23919/CISTI54924.2022.9820243.
6. E. F. Rivera and V. H. Andaluz: Autonomous Control of an Electric Vehicle by Computer Vision Applied to the Teaching–Learning Process. In: Smart Innovation, Systems and Technologies (2022). https://doi.org/10.1007/978-981-16-5063-5_34.
7. Mullah, N.S., Zainon, W.M.N.W.: Advances in machine learning algorithms for hate speech detection in social media: a review. IEEE Access. **9**, 88364 (2021). https://doi.org/10.1109/ACCESS.2021.3089515
8. M. S. Munir, K. Parveen, U. Farooq, K. Shaalan, A. Z. Abualkishik, A. S. Mohammed: Use of different machine learning algorithms for hate speech detection. In: International Conference on Cyber Resilience, ICCR 2022. (2022). https://doi.org/10.1109/ICCR56254.2022.9995800.
9. Bonetti, A., Martínez-Sober, M., Torres, J.C., Vega, J.M., Pellerin, S., Vila-Francés, J.: Comparison between machine learning and deep learning approaches for the detection of toxic comments on social networks. Appl. Sci. (Switz.). **13**(10), 6038 (2023). https://doi.org/10.3390/app13106038

10. Singh, I., Goyal, G., Chandel, A.: AlexNet architecture based convolutional neural network for toxic comments classification. J. King Saud Univ. - Comput. Inf. Sci. **34**(9), 7547 (2022). https://doi.org/10.1016/j.jksuci.2022.06.007
11. R. Ahuja, A. Banga, S. C. Sharma: Detecting abusive comments using ensemble deep learning algorithms. In: Malware Analysis Using Artificial Intelligence and Deep Learning (2020). https://doi.org/10.1007/978-3-030-62582-5_20.
12. Naidu, B.R., et al.: Toxic comment classification using deep learning. Int. J. Recent Innov. Trends Comput. Commun. **11**(7), 93 (2023). https://doi.org/10.17762/ijritcc.v11i7.7834
13. K. Govinda K. Cengiz: Toxic Comment Classifier. In: Hybridization of Blockchain and Cloud Computing: Overcoming Security Issues in IoT (2023). https://doi.org/10.1201/9781003336624-3
14. Rupapara, V., Rustam, F., Shahzad, H.F., Mehmood, A., Ashraf, I., Choi, G.S.: Impact of SMOTE on imbalanced text features for toxic comments classification using RVVC model. IEEE Access. **9**, 78621 (2021). https://doi.org/10.1109/ACCESS.2021.3083638
15. A. A. Kumar, P. B. Pati, K. Deepa, and S. T. Sangeetha: Toxic comment classification using S-BERT vectorization and random forest algorithm. In: Proceedings of IEEE InC4 2023–2023 IEEE International Conference on Contemporary Computing and Communications (2023). https://doi.org/10.1109/InC457730.2023.10263218
16. M. Vichare, S. Thorat, S. Uberoi, S. Khedekar, S. Jaikar: Toxic comment analysis for online learning. In: ACCESS 2021 - Proceedings of 2021 2nd International Conference on Advances in Computing, Communication, Embedded and Secure Systems (2021). https://doi.org/10.1109/ACCESS51619.2021.9563344.
17. Z. Zhao, Z. Zhang, F. Hopfgartner: A comparative study of using pre-trained language models for toxic comment classification. In: The Web Conference 2021 - Companion of the World Wide Web Conference, WWW 2021 (2021). https://doi.org/10.1145/3442442.3452313
18. S. N. Muhammad, K. H. Hasna: Toxic comment classifier on social media platform," i-manager's J. Comput. Sci., vol. 10, no. 4, 17 2023, doi:https://doi.org/10.26634/jcom.10.4.19307.
19. D. Vodnala, J. Shravya, K. Vishnupriya, V. N. S. Rohit: Machine Learning-Based Tool to Classify Online Toxic Comments (2023). https://doi.org/10.1007/978-981-19-7041-2_10
20. Maslej-Krešňáková, V., Sarnovský, M., Butka, P., Machová, K.: Comparison of deep learning models and various text pre-processing techniques for the toxic comments classification. Appl. Sci. **10**(23), 8631 (2020). https://doi.org/10.3390/app10238631

Pneumonia Detection Using Object Detection with MobileNet and TFLite: Implementation in Android Studio

Pamela Espinosa Cuayal(✉), Erick Iles, Jonathan Cruz-Varela, and Diego Almeida-Galárraga

School of Biological Sciences and Engineering, Universidad Yachay Tech, San Miguel de Urcuquí 100119, Ecuador
{pamela.espinosa,erick.iles,jcruz,dalmeida}@yachaytech.edu.ec

Abstract. Pneumonia is still one of the main causes of illness in areas where medical resources are limited, as is the case in certain regions of Ecuador. Faced with this reality, a portable tool was developed that can be run directly on Android phones, with the aim of facilitating the detection of this condition from chest X-rays. The model was developed using a combination of MobileNet, FPN and RetinaNet, data preparation, model training, export and on-device testing. It achieved a detection precision of 89.2% with mAP@0.75 and 83.3% recall, all in a lightweight 12 MB package. While effective in detecting consolidated cases, its performance in smaller lung lesions remains an area for future improvement. This approach offers a practical and scalable tool to improve early diagnosis in resource-limited healthcare settings.

Keywords: Pneumonia · Object Detection · MobileNet · Android

1 Introduction

Pneumonia is an acute lung infection caused by bacteria, viruses, or fungi, primarily affecting children, the elderly, and immunocompromised individuals especially in developing countries such as Ecuador [1,2]. It is characterized by inflammation of the alveoli and fluid accumulation in the lungs, leading to symptoms such as fever, cough, and difficulty breathing [5]. According to the 2021 Global Burden of Disease reports, pneumonia was responsible for 2.1 million deaths, including more than 500,000 children under five years of age, making it one of the leading causes of child mortality worldwide [5]. Although many of these deaths are preventable, lack of timely diagnosis remains a significant barrier, particularly in regions with under-resourced healthcare systems [5,19].

Chest X-rays (CXR) are the standard tool for diagnosing pneumonia; however, interpretation often depends on the expertise of specialists and may be subject to observer variability [2,11]. In this context, advances in computer vision, particularly through convolutional neural networks (CNNs), have enabled

J. L. Ballesteros et al. (Eds.): CITIS 2025, LNNS 1780, pp. 169–178, 2027.
https://doi.org/10.1007/978-3-032-14287-0_17

the development of automated detection tools. Object detection models such as YOLO, EfficientDet, and RetinaNet have demonstrated high precision in locating lung pathologies, although their models are quite cumbersome compared to lighter models such as the one proposed in this work. This limits their development on mobile devices, which are essential for healthcare in rural and high-risk areas in 86 cantons of Ecuador, including parts of Guayas, the central highlands, and Morona Santiago [1,18].

This work proposes the development of a mobile application based on MediaPipe's MobileNet Multi-AVG model, ideal for environments with limited resources using a database downloaded from Kaggle and manually labeled in Roboflow. This lightweight architecture allows for fast and accurate detection directly on smartphones. This work not only assists the diagnosis specialist, but also drives the development of affordable and sustainable telemedicine solutions for limited-resource communities.

2 Methods

This study corresponds to applied experimental research with a quantitative approach, aimed at the development, training, and validation of an automatic pneumonia detection system using chest X-rays (CXR) through deep learning techniques. The proposed system was optimized for deployment on Android mobile devices, and its methodological development was structured into four main stages: dataset selection and preparation, model training, exportation, and implementation and testing in a mobile environment. The complete workflow is illustrated in Fig. 1.

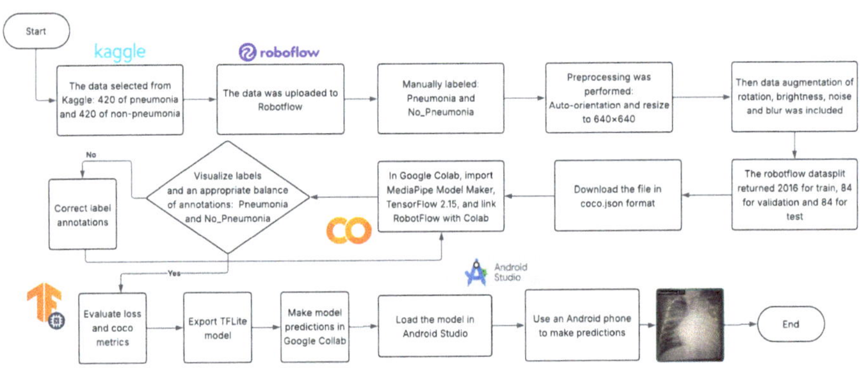

Fig. 1. Process to detect an images as Pneumonia or No Pneumonia in Android Studio and Google colab

2.1 Dataset

"Pneumonia - X-ray - Dataset" was used and downloaded from Kaggle [6]. 840 balanced CXR images were selected, including 420 from patients diagnosed with pneumonia and 420 from patients without pneumonia.

These images were uploaded to the Roboflow platform, where the affected regions were manually annotated using bounding boxes in COCO format. Automated preprocessing was then applied, including resizing to 640×640 pixels and orientation correction.

2.2 Augmentation Techniques

To improve the model's generalization ability and simulate real-world variations in clinical conditions, data augmentation techniques were applied, including:

- $\pm 10°$ random rotation,
- $\pm 15\%$ brightness adjustment,
- Addition of Gaussian noise (0.1% of pixels),
- 2.5 pixel blur.

These transformations were implemented in Roboflow before exporting the annotated dataset. These techniques are shown in Fig. 2.

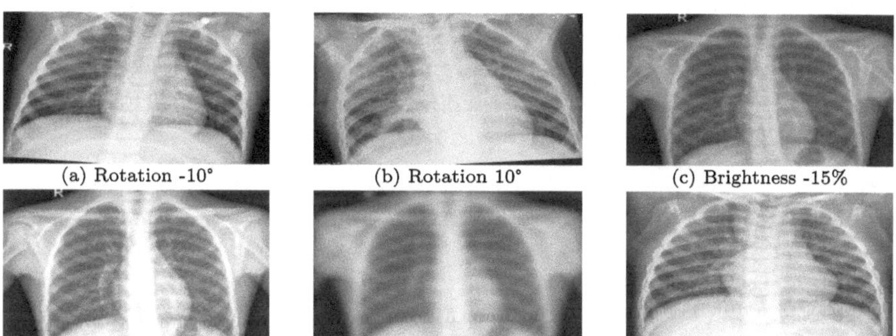

(a) Rotation -10° (b) Rotation 10° (c) Brightness -15%
(d) Brightness 15% (e) Blur 2.5px (f) Noise 0.1%

Fig. 2. Examples of data augmentation

Once the preprocessing and data augmentation were completed, Roboflow automatically generated the dataset partition: 2016 images for training, 84 for validation, and 84 for testing.

2.3 Training the Detection Model

The model was trained in Google Colab, using an NVIDIA T4 GPU and 12.67 GB of RAM. The following tools were used:

- MediaPipe Model Maker with MobileNet Multi-AVG architecture (MobileNet + FPN + RetinaNetHead) [8].
- TensorFlow 2.15.0 as the main deep learning framework.
- Roboflow to integrate the preprocessed dataset and augmentation in Colab.
- OpenCV for image processing.

Training was carried out over 15 epochs, using a batch size of 8, Adam optimizer, and a learning rate of 0.03. After training, the model was evaluated using COCO metrics to verify its results.

After the model was validated, it was converted to TFLite format and then transferred to Android Studio software to create the mobile environment. The final file size was 12 MB, making it ideal for various Android mobile devices.

2.4 Mobile Deployment and Real-Time Testing

The mobile application was developed in Android Studio (Koala Feature Drop version 2024.1.2), integrating:

- MediaPipe Tasks to run the TFLite model,
- CameraX for real-time image capture from the device's camera,
- Model inference directly on the phone.

Our mobile application detects the presence of pneumonia on a previously uploaded chest X-ray or captured with a cell phone camera. It superimposes a bounding box on the image and displays the probability that the image shows pneumonia or not. This approach allows medical specialists to make quick and reliable predictions directly on Android devices, without relying on cloud services or an internet connection.

3 Results

The model demonstrated high reliability in detecting and localizing pneumonia on chest radiographs. At an overlap criterion of 50%, it achieved an average precision of 0.987, implying that 98.7% of the detections were correct. At a more demanding threshold of 75%, it maintained a good performance with an precision of 0.892. In addition, the sensitivity was 83.3%, indicating that it correctly identified most of the actual cases of pneumonia. The Table 1 summarizes the most relevant metrics of the model with respect to the validation data.

The overall average of 0.775 of the AP(0.5:0.95) indicates that, overall, the model is maintaining a 77.5% precision in terms of pneumonia detection according to various levels of stringency. Performance is very good with wider margins, and remains pretty good with extremely strict evaluative criteria, fairly well circumscribing those that are harder to circumscribe. A summary of this can be seen in Fig. 3.

Table 1. Performance Metrics for Pneumonia Detection

Metric	Value	Interpretation
AP 0.50	0.987	Near-perfect precision for detections with ≥50% overlap
AP 0.75	0.892	High precision even with strict overlap criteria ≥75%
AP 0.50:0.95	0.775	Good average performance across multiple IoU thresholds
Recall (AR)	0.833	Detects 83.3% of all actual pneumonia cases
Validation Loss	0.299	Low error

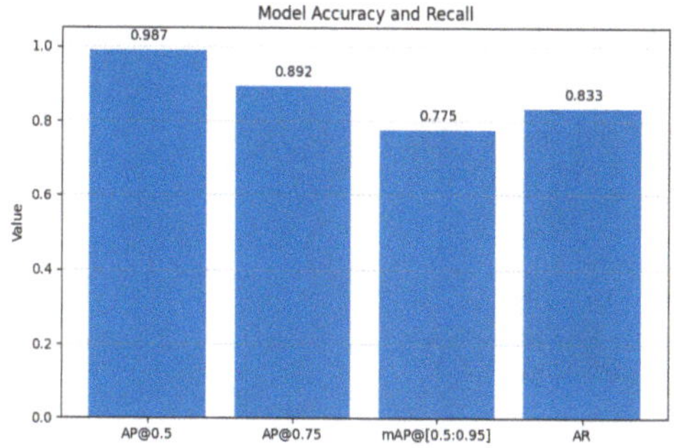

Fig. 3. Performance Metric of precision and recall

3.1 Training Losses

Training and validation showed stable and consistent model performance. Regarding loss values, the total loss decreased to 0.2993, confirming that the model learned effectively without overfitting. The classification loss of 0.1738 and the localization loss of 0.0013, which was almost imperceptible, were particularly low, indicating that the model not only distinguishes well between patients with pneumonia and non-pneumonia, but also locates the affected areas (bounding boxes) with great precision, data that can be identified in Fig. 4(a). These metrics were corroborated by the learning curves, which showed rapid and stable convergence over the 15 training epochs.

The model is accurately differentiating between pneumonia and non-pneumonia, as indicated by the final classification loss of 0.1738. Furthermore, the localization loss validates that the expected bounding boxes nearly precisely matched the associated pneumonia-infected regions in the pictures, which is crucial for generating a precise diagnosis. A summary of this could be found in Table 2.

The parallel convergence of the training and validation curves suggests that the model did not memorize the training data but rather learned generalizable

Table 2. Training Losses

Loss Type	Final Value	Interpretation
Total Loss	0.299	Good overall fit
Classification Loss	0.174	Excellent class distinction
Localization Loss	0.0013	Very precise bounding boxes

patterns to detect pneumonia, avoiding overfitting, which is important due to the reliability of both the diagnosis (classification) and the exact delineation of the lesions (localization) using the bounding boxes. The low final loss values support the quality of the model predictions, as can be seen in Fig. 4(b).

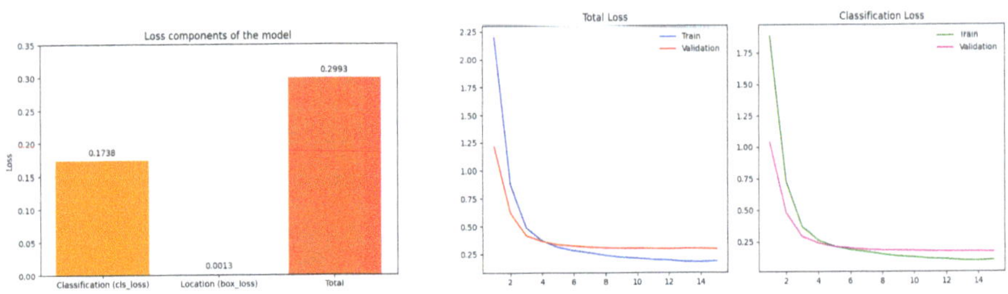

(a) (b)

Fig. 4. (a) Losses during validation (b) Training curves

Analyzing performance by anomaly size, the model showed quite considerable performance in detecting large bounding boxes, APl = 0.775, which are the most obvious anomalies relevant to pneumonia. However, small or medium-sized anomalies APs = −1.0, APm = −1.0 were not evaluated, as the data set focused primarily on clearly visible patterns.

This approach is best suited for diagnosing established cases rather than for early detection. The model implementation on mobile devices maintained this high level of performance. With an optimized size of 12 MB, the system is practical for clinical use in resource-limited settings. Testing showed a good correlation between model predictions and actual annotations.

Evaluation was performed in Google Colab and Android Studio to assess any changes in model sensitivity. Figure 5 shows the successful identification of pneumonia-non_pneumonia and its precision. Two cases with Pneumonia were tested, with values of 77% and 91% in Google Colab and Android Studio, while for No_pneumonia, 84% and 82% were obtained for both Google Colab and Android Studio.

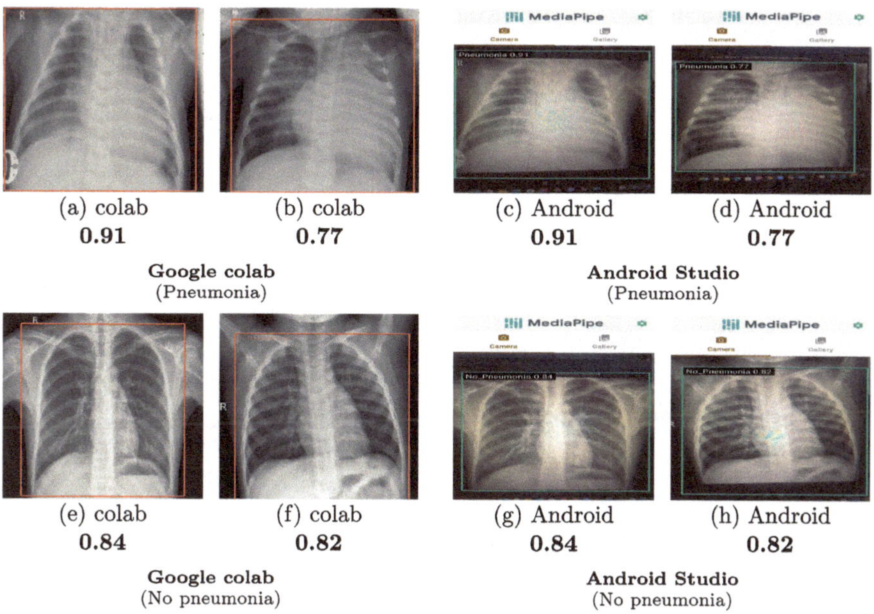

Fig. 5. Comparison of pneumonia detection models. Top: Positive cases. Bottom: Negative cases.

4 Discussion

The results obtained demonstrate a significant improvement over other published studies on the detection and localization of pneumonia in CXR. The MobileNet Multi-AVG architecture used achieved an mAP@0.50 value of 0.987, surpassing the performance of models such as EfficientDet2 [9], whose maximum precision barely reached 0.65. This difference is attributed to the efficient combination of the 3 architectures working together (MobileNetV2, FPN and RetinaNet) in our model, as well as to the quality of the selected data, manual labeling in Roboflow and its subsequent data augmentation that includes random rotation, Gaussian noise and brightness adjustments, which improves the model's ability to adapt to real-world variations in clinical images.

This is further reinforced by an extremely low localization loss (0.0013), indicating a strong ability to identify pathological regions. As [15] points out, accurate localization is essential for assessing the extent of lung disease and guiding clinical decisions. The model's effectiveness is also related to its lightweight design (12 MB) and efficient data augmentation strategy, making it ideal for use in mobile and resource-limited clinical settings.

While models like YOLO V3, V4, and V6 were reported to have slightly higher precision (mAP between 93 and 98%) [12,16], they are difficult to use on mobile devices due to their memory size above 100 MB [20]. On the other hand, our model was developed to work effectively on Android, which makes

Table 3. Comparative table of results of similar works.

Paper	Model	Weight	Results
[16,20]	YOLO V3 YOLO V4 YOLO V6-L YOLO V6-M	123 MB	Precision: 0.84, 0.93, 0.95, 0.95 Recall: 0.91, 0.94, 0.99, 0.99 Average IoU: 67.56, 82.30, 85.19, 88.20 MAP: 93.86, 96.56, 97.86, 98.86
[12]	YOLO	–	Weighted average: 0.71 Macro average: 0.68
[14]	EfficientDet2	7.2 MB	Precisión: 0.45: 0.65 Recall: 0.50: 0.65 mAP@0.5: 0.65 mAP@0.5–0.95: 0.40
[17]	RetinaNet + ResNet-50 (ensemble)	–	mAP: 32.7%.
Ours	MobileNet Multi-AVG: (mobilenet+FPN +RetinaNet)	12 MB	mAP@0.5: 0.987, mAP@0.75: 0.892, mAP@0.5–0.95: 0.775 Recall: 0.833.

it suitable for use in telemedicine applications. This portability enables rapid diagnoses in rural or low-resource regions where access to specialized equipment is limited. As [13] points out, Artificial intelligence in healthcare has the potential to greatly increase access to medical diagnosis. On the other hand, research such as that by [17] obtained an mAP of only 32.7% despite having a large amount of data, highlighting the need to properly balance model architecture, data quality, and preprocessing techniques. In our case, the model achieved a sensitivity of 83.3%, outperforming EfficientDet2 (0.65) and placing itself on par with YOLO variants. Furthermore, it showed remarkable stability across different thresholds, reinforcing its reliability for clinical use as Table 3 shows.

However, the developed model showed positive results, areas for improvement are identified, such as the detection of small and medium-sized lung lesions. This study focused on pneumonia patients rather than early detection, which involves subtle features in small areas and higher quality images with detailed bounding boxes [7]. Adding labels for small and medium-sized lesions could improve performance, since the bounding box size has a greater impact on precision. Therefore, future work could include using size-aware data augmentation techniques to model smaller lesions [10], collecting early-stage pneumonia datasets [12], and developing labeling protocols for better localization [7]. This will contribute to improving the model's effectiveness and reliability.

4.1 Future Work: Towards Clinical Validation

Clinical validation, including preliminary or simulated settings, should be the primary focus of future research. This includes studies that compare model predictions to validated medical diagnoses using anonymized CXR datasets [9]. Radiologists can help evaluate usability and interpretability when deployments take place in controlled telemedicine situations [4]. Collaboration with specialists will allow the model to perform better across a wide range of different images and populations [3]. These steps are critical for minimizing the gap between clinical utility and model performance, hence increasing the reliability and significance of AI in medical diagnosis.

5 Conclusions

This study created a model optimized for Android devices based on MobileNet, FPN and RetinaNet for the identification of pneumonia in thoracic radiographs, achieving a mAP@0.75 of 89.2%, a mAP0.50:0.95 of 77.5% and a recall of 83.3%, with a size of 12 MB, challenging models such as those of YOLO, EfficientDet2 and RetinaNet + ResNet-50. Our application could be suitable as a support for pneumonia diagnosis and its use in rural areas with limited access to radiological specialists, due to its efficiency and offline operation. However, it showed limitations in the detection of small and medium-sized lung lesions, suggesting future improvements that should focus on the use of higher resolution images, better labels and magnification techniques. It will also be essential to move towards clinical validation, comparing the results of the model with real medical diagnoses in simulated environments and collaborating with radiologists to assess its applicability and improve its diagnostic utility in low-resource settings.

References

1. Acosta-España, J.D., Altamirano-Jara, J.B., Arnao-Noboa, A., Rodriguez-Morales, A.J.: Pneumonia, diarrhoea, tuberculosis, malaria, and HIV among children and adolescents in ecuador: a cross-sectional national study, 2015-2021 (2023). https://doi.org/10.21203/rs.3.rs-3001291/v1
2. Arizmendi, C., Pinto, J., Arboleda, A., González, H.: Diagnosis of patients with viral, bacterial, and non-pneumonia based on chest x-ray images using convolutional neural networks. arXiv preprint arXiv:2503.02906 (2025). https://doi.org/10.48550/arXiv.2503.02906
3. Becker, J., et al.: Artificial intelligence-based detection of pneumonia in chest radiographs. Diagnostics **12**(6) (2022). https://doi.org/10.3390/diagnostics12061465. https://www.mdpi.com/2075-4418/12/6/1465
4. de Camargo, T.F., et al.: Clinical validation of an artificial intelligence algorithm for classifying tuberculosis and pulmonary findings in chest radiographs. Front. Artif. Intell. **8**, 1512910 (2025). https://doi.org/10.3389/frai.2025.1512910
5. Cilloniz, C., Dela Cruz, C.S., Dy-Agra, G., Pagcatipunan Jr, R.S., Group, P.S.: World pneumonia day 2024: fighting pneumonia and antimicrobial resistance (2024). https://doi.org/10.1164/rccm.202408-1540ED

6. Gonzales, J.L.: Neumonía x-rays dataset (2023). https://www.kaggle.com/datasets/gonzajl/neumona-x-rays-dataset
7. Guesmi, T.: Detecting pneumonia with a deep learning model and random data augmentation techniques. Int. J. Adv. Comput. Sci. Appl. **14** (2023). https://doi.org/10.14569/IJACSA.2023.01405122
8. Hernández González, N., et al.: Análisis de rendimiento de modelos basados en aprendizaje por transferencia con tensorflow y tensorrt (2023)
9. Hofmeister, J., et al.: Validating the accuracy of deep learning for the diagnosis of pneumonia on chest x-ray against a robust multimodal reference diagnosis: a post hoc analysis of two prospective studies. Eur. Radiol. Exper. **8**(1), 20 (2024). https://doi.org/10.1186/s41747-023-00416-y
10. Islam, S., Ferraioli, G., Pascazio, V., Vitale, S., Amin, M.: Performance analysis of YOLOv3, YOLOv4 and mobilenet SSD for real time object detection **5**, 37–49 (2024)
11. Jain, R., et al.: Detection of COVID-19, pneumonia, and tuberculosis using convolutional neural networks and ensemble of deep learning architectures. ASEAN J. Sci. Technol. Rep. **28**(1), e253662–e253662 (2025). https://doi.org/10.55164/ajstr.v28i1.253662
12. Kailasam, R., Balasubramanian, S.: Deep learning for pneumonia detection: a combined CNN and YOLO approach. Hum.-Centric Intell. Syst., 1–19 (2025). https://doi.org/10.1007/s44230-025-00091-9
13. Kerketta, A., Balasundaram, S.: Leveraging AI tools to bridge the healthcare gap in rural areas in India. medRxiv pp. 2024–07 (2024). https://doi.org/10.1101/2024.07.30.24311228
14. Kimeu, J.M., Kisangiri, M., Mbelwa, H., Leo, J.: Deep learning-based mobile application for the enhancement of pneumonia medical imaging analysis: a case-study of west-meru hospital. Inform. Med. Unlocked **50**, 101582 (2024). https://doi.org/10.1016/j.imu.2024.101582
15. Litjens, G., et al.: A survey on deep learning in medical image analysis. Med. Image Anal. **42**, 60–88 (2017). https://doi.org/10.1016/j.media.2017.07.005
16. Munna, M.S., Chowdhury, R., Siddiqee, A.M.: A comparative study of YOLO models for pneumonia detection. Int. J. Multidiscip. Res. **6**, 1–10 (2024)
17. Rajaraman, S., Guo, P., Xue, Z., Antani, S.K.: A deep modality-specific ensemble for improving pneumonia detection in chest x-rays. Diagnostics **12**(6), 1442 (2022). https://doi.org/10.3390/diagnostics12061442
18. Ríos Quituizaca, P., Calderón, L., Piedra, S., Guerrero, J., Narváez, A.: Propuesta de análisis territorial para enfrentar la pandemia por SARS-COV-2 basado en el perfil de neumonía e influenza en ecuador 2016-2018. Aten. prim.(Barc., Ed. impr.), p. 102021 (2021). https://doi.org/10.1016/j.aprim.2021.102021
19. Romanyuk, L., Volch, I., Kravets, N., Pyatkovskyy, T., Zahrychuk, O.: Retrospective analysis of gender-age structure and comorbid pathology of patients with viral and bacterial pneumonia associated with COVID-19. Int. J. Med. Med. Res. **7**(2), 15–19 (2021). https://doi.org/10.11603/ijmmr.2413-6077.2021.2.12675
20. Wang, C., et al.: Using UAVRs and deep learning to conduct resource surveys of threatened Tibetan medicinal plants in the Qinghai-Tibet plateau. Glob. Ecol. Conserv. **51**, e02884 (2024). https://doi.org/10.1016/j.gecco.2024.e02884

Generative Artificial Intelligence-Based Efficient Airfoil Design Employing a *3-GAN* Model

Gabriel de Jesus Celis-Escudero(✉), Carlos Avilés-Cruz, Arturo Zúñiga-López, and Cesar Benavides-Alvarez

Departamento de Electrónica, Universidad Autónoma Metropolitana, México, Av. San Pablo No. 420 Col. Nueva el Rosario, Ciudad de México, Mexico
{al2232801411,caviles,azl,cesarbenavides}@azc.uam.mx

Abstract. Traditionally, the design of airfoils has relied on empirical approaches and computational numerical analysis. These methods frequently demand substantial computational resources and time, ranging from several days to weeks, contingent on the complexity of the mesh and the expertise of the analyst. Recent advancements in deep learning, particularly in Generative Adversarial Networks (GANs), have rendered the generation of synthetic data, including images and text, a feasible undertaking. In this study, we propose a novel application of GANs for the efficient generation of airfoil geometries, significantly reducing the time required for their design. The proposed model incorporates a three-branch generator, comprising coarse, medium, and fine branches, each with varying numbers of hidden layers, where the fine branch contains the most and the coarse branch the fewest. The discriminator is designed to function independently in order to support this architecture. The training dataset encompasses NACA four- and five-digit airfoils, with each airfoil being represented by 500 coordinate points that delineate the camber line. The generation of new data is achieved through the utilization of a 100-dimensional input vector, which is derived from a standard normal distribution. The generated airfoils are random, yet they contain characteristics of the real airfoils. This initiative signifies the inaugural phase in the development of customized airfoils, which will serve as the foundation for subsequent research endeavors.

Keywords: Deep learning · Airfoils · GAN · NACA

1 Introduction

Traditionally, the design of airfoils has relied on the use of Computational Fluid Dynamics (CFD) to assess lift and drag forces, thereby facilitating the determination of the relevant aerodynamic coefficients. The development of Generative Adversarial Networks (GANs) has created new opportunities for synthetic data generation. GANs have been successfully applied in a wide range of fields. For

J. L. Ballesteros et al. (Eds.): CITIS 2025, LNNS 1780, pp. 179–188, 2027.
https://doi.org/10.1007/978-3-032-14287-0_18

instance, [8] developed a GAN for generating artificial images [15], applied GANs to facial reconstruction [7], focused on synthetic image generation for data augmentation [13], produced artwork in the style of Monet [10], predicted patient stays in emergency rooms [12], facilitated architectural layout design by placing doors. And [17], predicted vegetative drought. Building on this versatility, the present work introduces a three-branch GAN (3-GAN) architecture for the generation of airfoils. This architecture aims to significantly reduce the time required to identify or design suitable airfoils for specific engineering applications.

This research is focused on three key areas where neural networks have been applied to airfoil studies. The first area involves the generation of airfoils. For example, [6,11] proposed generative models based on aerodynamic and pressure coefficient distributions, respectively. Similarly, [18] developed a neural network to design airfoils and fixed-wing geometries from aerodynamic coefficients. While [16], focused on modifying existing airfoils to enhance performance using pressure distribution data. The second area centers on flow field prediction, with studies such as [14,19] employing neural networks to predict flow behavior from images of airfoil geometry and associated pressure and velocity fields. The third area focuses on the prediction of aerodynamic coefficients. It is noteworthy that [5] developed a model to predict lift and drag coefficients, C_l and C_d respectively, from geometric data. As stated in [9], a network was constructed to forecast pressure coefficient distributions based on flow parameters.

This study aims to contribute to this growing field by presenting a GAN-based framework for the efficient generation of airfoils, with potential applications in sectors such as wind energy and aerospace.

The rest of the paper is organized as follows: Sect. 2 presents the methodology. The results and conclusions are presented in Sects. 3 and 4, respectively.

2 Methodology

A 3-GAN was used to generate airfoils based on NACA airfoils. The database used in this study is derived from sources [1,2], corresponding to NACA four- and five-digit airfoil airfoils. For the NACA five-digit series, the designation follows the format NACA LPQXX, where L denotes the design lift coefficient, P indicates the position of maximum camber, and Q specifies the camber type (normal or reflexed). The final two digits (XX) represent the maximum thickness as a percentage of the chord. As illustrated in Fig. 1, the constructed dataset includes examples of both NACA4 and NACA5 airfoils. Each airfoil is represented by two vectors containing 500 coordinate points: one for the upper camber (x_u), shown in blue, and one for the lower camber (x_l), shown in orange.

The methodology is based on the training of a GAN, as illustrated in Fig. 2. The architecture comprises a discriminator, illustrated in the upper brown section of the figure, and a generator, depicted in the lower green section. The following text is intended to provide a comprehensive overview of the subject matter.

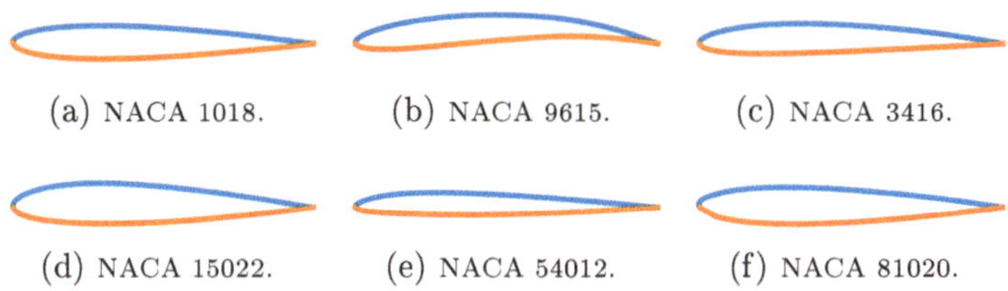

(a) NACA 1018. (b) NACA 9615. (c) NACA 3416.

(d) NACA 15022. (e) NACA 54012. (f) NACA 81020.

Fig. 1. NACA4 and NACA5 examples.

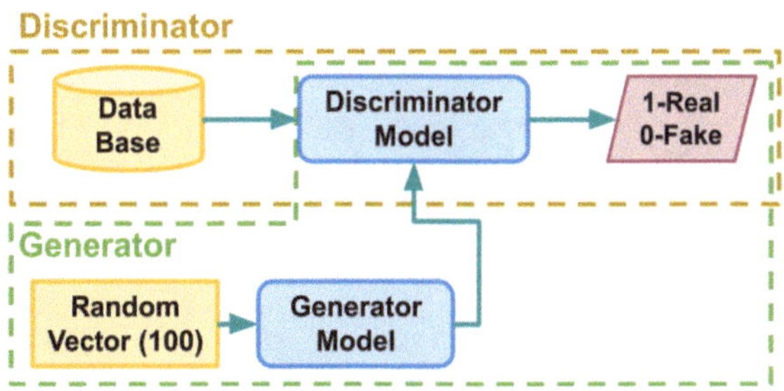

Fig. 2. GAN model.

The notion of implementing a generator with a three-branch architecture is inspired by the work of [4], which draws on the Human Vision System (HVS) paradigm. According to the HVS, the human visual perception process is comprised of three distinct stages. A coarse representation is provided, offering a general overview of the scene that can be readily perceived at a glance. Medium, where the focus is directed toward the individual elements within the scene. In such cases, meticulous examination and a focus on particular objects are imperative. The 3G-AN architecture proposed by Avilés [3] successfully adopted this hierarchical approach for applications such as face and scene generation.

The discriminator is trained using two vectors representing the x-coordinates of the airfoil obtained from the previously described NACA-based database. These are treated as authentic samples during the training process.

The training of the generator involves the sampling of input vectors with a length of 100 from a standard normal distribution. The output of the generator is interpreted as fake airfoils, intended to replicate the structure of the real data. The architectural specifics of both the generator and the discriminator are outlined in the following sections.

2.1 Discriminator Model

The discriminator model is composed of two branches, with an airfoil camber serving as the input. Each branch is comprised of four attribute extraction stages, incorporating a convolution layer and an activation function. Subsequently, a concatenation of the outputs from both branches is performed, followed by an attribute extraction stage that is analogous to the preceding ones. The sequence continues with a flattened structure, followed by three layers that are fully connected. The model culminates with the presence of a full-connected layer comprising a neuron. The model elucidated in this paragraph is illustrated in Fig. 3.

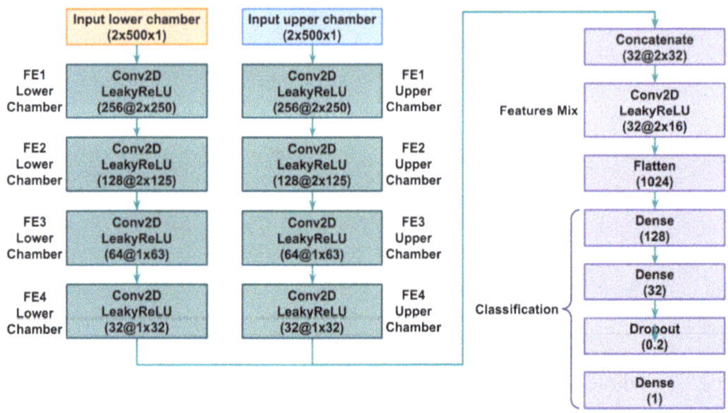

Fig. 3. Discriminator model.

2.2 Generator Model

The generator model comprises three input branches, each with a single input: a random vector of size 100. Each branch incorporates a fully connected layer, a normalization layer, an activation function, a reshape layer, and a Gaussian-noise function. Subsequent to this initial phase, the Fine, Medium, and Coarse branches undergo a sequence of one, two, or four feature-generation stages, respectively. The feature generation stages incorporate a convolution layer, normalization, and an activation function. The weighting values assigned to each branch are 0.5, 0.3, or 0.2, respectively. Subsequent to the application of weights, the values are aggregated, and two terminal branches are generated, each comprising two feature-generation stages. It is important to note that each branch of the tree concludes with an activation function. As illustrated in Fig. 4, the generator model and its associated input and output branches are depicted.

2.3 Generative Adversarial Network Model

The full GAN model integrates the generator and discriminator architectures, where the generator outputs serve as inputs to the discriminator. During the training process, the discriminator parameters are frozen, meaning they are not updated. Consequently, the number of trainable and non-trainable generator parameters is 2,319,746 and 44,044, respectively. The discriminator is comprised of 2,762,913 trainable parameters. The configuration of the GAN model is illustrated in Fig. 5.

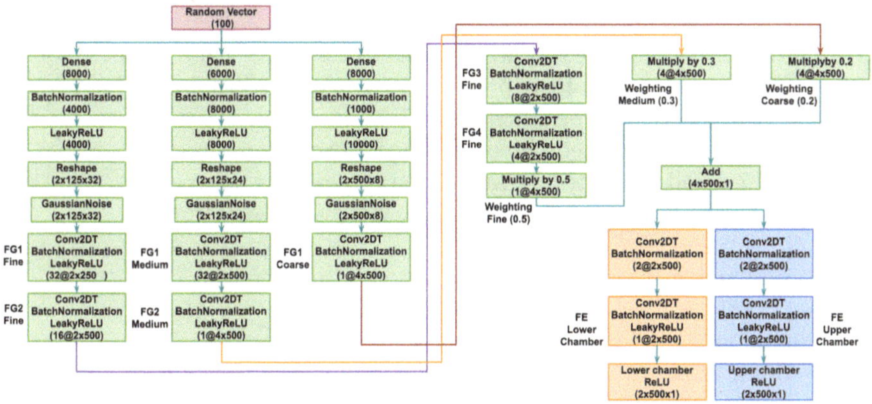

Fig. 4. Generator model.

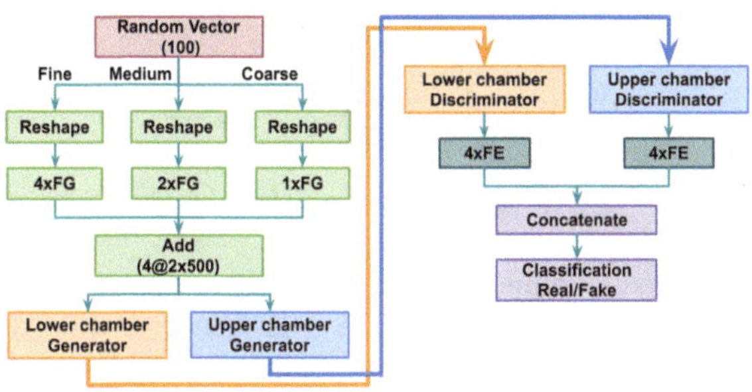

Fig. 5. Proposed Architecture of the Generative Adversarial Network Model.

2.4 *GAN* Training Model

The following hyperparameters were utilized during the training of the model: the Adam optimizer was employed in conjunction with a *CosineDecayRestarts* learning rate schedule. The initial learning rate was set to 0.00002, with a decay rate scaling factor of 2.5 and a learning rate multiplier of 1.5 applied at each restart. The binary cross-entropy loss function was utilized for the training process, and binary accuracy (*BinaryAccuracy*) was designated as the performance metric. For the purpose of data feeding, training was conducted using batches of 64 samples from the dataset that were randomly selected. The cost function of the GAN model, expressed as $\min_{3G};\max_{D};V(D,3G)$, is defined in Eq. (1).

$$\begin{aligned} V(D,3G) &= \mathbb{E}_{X\sim p_{data(x)}}[logD(x)] + \mathbb{E}_{Z\sim p_z(z)}[log(1-D(3G(z)))] \\ &= \mathbb{E}_{X\sim p_{data(x)}}[logD(x)] + \mathbb{E}_{Z\sim p_z(z)}[log(1- \\ &\quad D(\{\alpha\cdot G^C(z)+\beta\cdot G^M(z)+\gamma\cdot G^F(z)\}))], \end{aligned} \tag{1}$$

where:
$D(x)$ = Discriminating network stage for real airfoil.

$\mathbb{E}_x$ is the expected value over true airfoil.
$3G(z)$ is the airfoil generated by whole generator stage, given a noise vector z.
$D(3G(z))$ is the estimator of the probability that the artificial airfoil is real.
$\mathbb{E}_z$ is the expected value over all false airfoil $G(z)$.
$G^C(z)$ is the airfoil generated by coarse stage, given a noise vector z.
$G^M(z)$ is the airfoil generated by medium stage, given a noise vector z.
$G^F(z)$ is the airfoil generated by fine stage, given a noise vector z.
$G^C(z), G^M(z)$, and $G^F(z)$ are $\in \mathbb{R}^{W \times H \times}$.
$W \times H$ are number of data to top and bottom airfoil dimension .
α, β, and γ are weighted constants $\in [0, 1]$, and $\alpha + \beta + \gamma = 1$.
Z is a random noise vector following a standard normal distribution with $mean = 0$ and $standard\ deviation = 1$.

The gradient optimization of the Eq. 1 that takes over a minibatch of size N_B can be expressed as follows (see Eq. 2).

$$\nabla_{\Theta}(D, 3G)\left[\frac{1}{N_B}\sum_{X}[log\ D(x)] + \frac{1}{N_B}\sum_{Z}[log(1- D(\{\alpha \cdot G^C(z) + \beta \cdot G^M(z) + \gamma \cdot G^F(z)\}))]\right]. \quad (2)$$

To minimize the error defined in Eq. 1, and by applying gradient estimation as described in Eq. 2, the optimal condition is achieved when the probability distributions of the real and generated data match, i.e., $p(X) = p(Z)$. Under this condition, the set of generated (fake) airfoils Z becomes indistinguishable from the original airfoils X, effectively deceiving the discriminator into classifying the synthetic samples as real.

Throughout the training process, the parameters Θ are considered trainable and are updated iteratively in both the *Discriminator* and *Generator* stages.

3 Results

The proposed GAN models were implemented and trained in Python using the following libraries: TensorFlow, Keras, NumPy, Matplotlib, and Scikit-learn. The training was carried out on a Dell Alienware Aurora R13 Gaming Desktop, which is equipped with 64 GB of RAM, an Intel Core i9-12900K processor, and a 2 TB NVMe solid-state drive.

As depicted in Figs. 6(a), and 6(a), the loss and accuracy of the discriminator are exhibited for both real and generated airfoils. In both cases, the trend in accuracy is 1 and the loss is 0. The metrics reveal instability in epochs where the generator produces results that challenge the discriminator's capacity to accurately classify airfoils as real or generated. This phenomenon is illustrated in Fig. 7.

As illustrated in Fig. 8, the initially generated airfoil airfoils exhibit noticeable noise along both the upper and lower curves. To address this issue, a low-pass

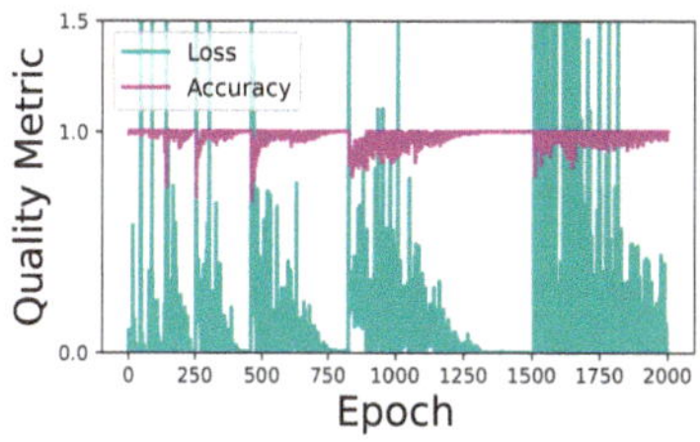

(a) Real airfoils.

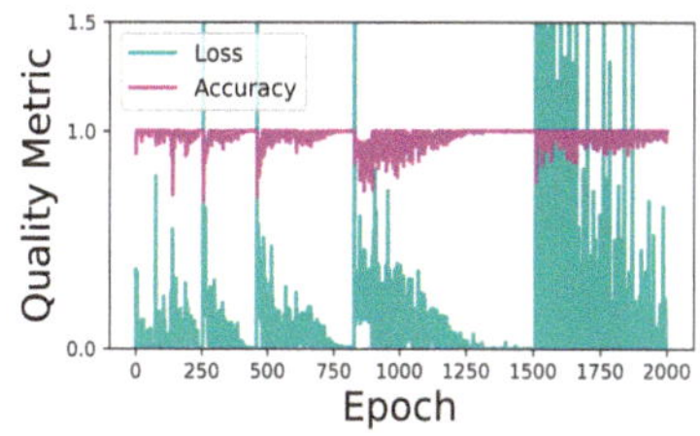

(b) Generated airfoils.

Fig. 6. Discriminator quality metrics.

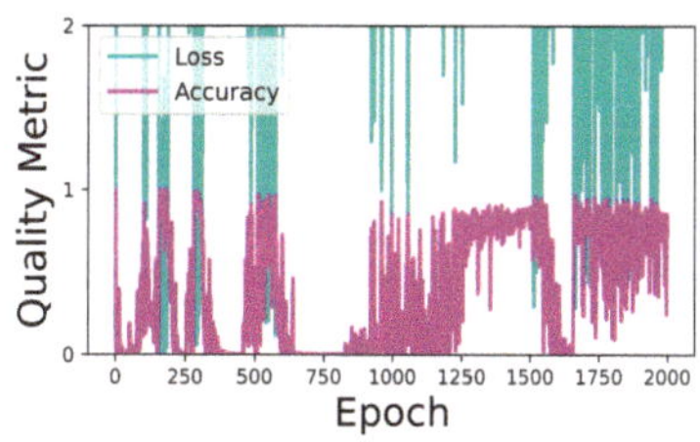

Fig. 7. Generator quality metrics.

filter was applied with the following configuration: a sampling frequency of 500 samples per second, a filter order of 5, and a cutoff frequency of 5 Hz. Furthermore, a rotation and vertical translation were performed to center the airfoils and align the leading and trailing edges along the horizontal axis. Fig. 8(a) displays the unprocessed, noisy airfoils, while Fig. 8(b) presents the same airfoils after filtering and geometric alignment.

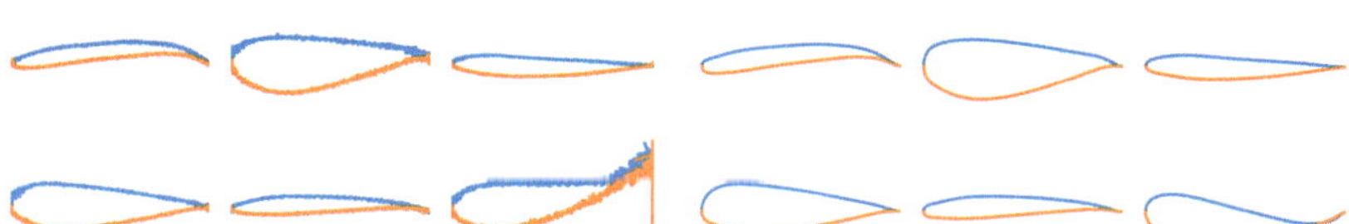

(a) Non-processing. (b) Processing.

Fig. 8. Post-processing of generated airfoils.

Subsequent to the post-processing of the generated airfoils, the calculation of the geometric characteristics was performed, including the area, perimeter, maximum thickness, and average chord. This is to ascertain the location of the generated and real airfoils. As illustrated in Fig. 9(a), the airfoils are represented with three of the four geometric features. To identify the closest airfoil to each generated airfoil, the *KNN* algorithm was implemented. The airfoil pairs are illustrated in Fig. 9(b).

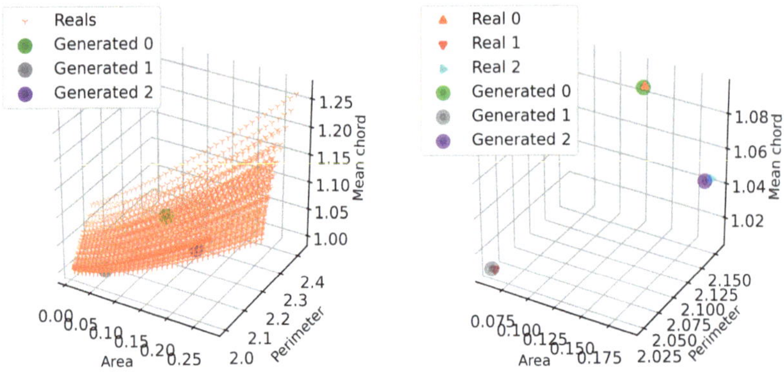

(a) All database. (b) Nearest neighbor.

Fig. 9. Airfoils real and generated.

As illustrated in Fig. 10, the real and generated airfoil pairs are displayed. The sequence of generation was as follows: Generated 0 with NACA 8820, Generated 1 with NACA 1109, and Generated 2 with NACA 2328.

(a) NACA 8820. (b) NACA 1109. (c) NACA 2328.

(d) Generated 0. (e) Generated 1. (f) Generated 2.

Fig. 10. Airfoils real and generated.

4 Conclusions and Further Work

An artificial intelligence model based on the GAN architecture was successfully developed for the generation of new airfoil geometries. A novel three-branch generation structure, comprising coarse, medium, and fine layers with assigned weights, was effectively implemented within the generator. Contrary to conventional methodologies that conceptualize airfoils as images, this study models the airfoil as a structured dataset comprising 500 coordinate points for both the upper and lower surfaces.

The three-branch GAN model generates airfoils that are within the space of real airfoils, after post-processing the point cloud provided by the generator.

Given the preliminary nature of the results, the subsequent step entails conducting a CFD analysis. This will enable the observation of the behavior of the airfoils through their aerodynamic coefficients. Subsequently, a comparative analysis will be conducted with their real counterparts.

Subsequent endeavors will center on refining the model through the exploration of alternative evaluation metrics and loss functions, with the objective of enhancing the quality and precision of the generated airfoils. The hyperparameters of the network will also be optimized to enhance the model's performance. Additionally, the objective is to modify the random vector to specific data from real airfoils, thereby generating a result based on the specific design and usage conditions.

References

1. AirfoilTools.com: Airfoil tools (2024). http://airfoiltools.com/airfoil/naca4digit
2. AirfoilTools.com: Airfoil tools (2024). http://airfoiltools.com/airfoil/naca5digit
3. Avilés-Cruz, C., Celis-Escudero, G.J.: 3G-AN: triple-generative adversarial network under corse-medium-fine generator architecture. IEEE Access **11**, 105344–105354 (2023). https://doi.org/10.1109/ACCESS.2023.3317897
4. Avilés-Cruz, C., Ferreyra-Ramírez, A., Zúñiga-López, A., Villegas-Cortéz, J.: Coarse-fine convolutional deep-learning strategy for human activity recognition. Sensors **19**(7) (2019). https://doi.org/10.3390/s19071556
5. Chen, H., He, L., Qian, W., Wang, S.: Multiple aerodynamic coefficient prediction of airfoils using a convolutional neural network. Symmetry **12**(4), 544 (2020). https://doi.org/10.3390/SYM12040544
6. Du, X., He, P., Martins, J.R.: Rapid airfoil design optimization via neural networks-based parameterization and surrogate modeling. Aerosp. Sci. Technol. **113**, 106701 (2021). https://doi.org/10.1016/J.AST.2021.106701
7. Fadaeddini, A., Majidi, B., Souri, A., Eshghi, M.: Data augmentation using fast converging Cielab-GAN for efficient deep learning dataset generation. Int. J. Comput. Sci. Eng. **26**(4), 459–469 (2023). https://doi.org/10.1504/IJCSE.2023.132152
8. Goodfellow, I.J., et al.: Generative adversarial nets. In: Neural Information Processing Systems (2014). https://doi.org/10.48550/arXiv.1406.2661
9. Hui, X., Bai, J., Wang, H., Zhang, Y.: Fast pressure distribution prediction of airfoils using deep learning. Aerosp. Sci. Technol. **105**, 105949 (2020). https://doi.org/10.1016/J.AST.2020.105949
10. Kadri, F., Dairi, A., Harrou, F., Sun, Y.: Towards accurate prediction of patient length of stay at emergency department: a GAN-driven deep learning framework. J. Amb. Intell. Hum. Comput. (2022). https://doi.org/10.1007/s12652-022-03717-z
11. Kim, M.I., Yoon, H.S.: Geometric modification for the enhancement of an airfoil performance using deep CNN. Ocean Eng. **266**, 113000 (2022). https://doi.org/10.1016/J.OCEANENG.2022.113000
12. Kim, S., Lee, J., Jeong, K., Lee, J., Hong, T., An, J.: Automated door placement in architectural plans through combined deep-learning networks of ResNet-50 and pix2pix-GAN. Expert Syst. Appl. **244** (2024). https://doi.org/10.1016/j.eswa.2023.122932

13. Lakshmi, L., et al.: Performance analysis of cycle GAN in photo to portrait transfiguration using deep learning optimizers. IEEE ACcess **11**, 136541–136551 (2023). https://doi.org/10.1109/ACCESS.2023.3337430
14. Portal-Porras, K., Fernandez-Gamiz, U., Zulueta, E., Ballesteros-Coll, A., Zulueta, A.: CNN-based flow control device modelling on aerodynamic airfoils. Sci. Rep. **12**(1), 1–13 (2022). https://doi.org/10.1038/s41598-022-12157-w
15. Sabnam, S., Rajagopal, S.: Application of generative adversarial networks in image, face reconstruction and medical imaging: challenges and the current progress. Comput. Methods Biomech. Biomed. Eng. Imaging Vis. **12**(1), 2330524 (2024). https://doi.org/10.1080/21681163.2024.2330524
16. Sekar, V., Zhang, M., Shu, C., Khoo, B.C.: Inverse design of airfoil using a deep convolutional neural network. AIAA J. **57**(3), 993–1003 (2019). https://doi.org/10.2514/1.J057894
17. Shukla, J.S., Pandya, R.J.: Deep learning-oriented C-GAN models for vegetative drought prediction on peninsular India. IEEE J. Sel. Top. Appl. Earth Obs. Rem. Sens. **17**, 282–297 (2024). https://doi.org/10.1109/JSTARS.2023.3328299
18. Sun, G., Sun, Y., Wang, S.: Artificial neural network based inverse design: airfoils and wings. Aerosp. Sci. Technol. **42**, 415–428 (2015). https://doi.org/10.1016/J.AST.2015.01.030
19. Zhou, J., Zhang, R., Chen, L.: Prediction of flow field over airfoils based on transformer neural network. Int. J. Comput. Fluid Dyn. (2023). https://doi.org/10.1080/10618562.2023.2259806

Navigation of a UGV for Tomato Crops in Greenhouses Using DRL

Juan Sebastian Rada Rey(✉), Carlos Borras Pinilla, and Carlos Alberto Flórez Arias

UIS, Bucaramanga, Colombia
jsebastianrada@gmail.com

Abstract. The demand for increased food production continues to rise, driving interest in precision agriculture and agro-robotics as means to boost crop productivity. One critical challenge in this domain is enabling robust autonomous navigation of robots in greenhouse environments, which are often constrained, dynamic, and GPS-denied. This study focuses on developing a navigation system for an unmanned ground vehicle (UGV) operating in a tomato greenhouse. The methodology involves applying a Double Deep Q-Network (DDQN) trained in simulation to handle navigation tasks while avoiding obstacles. Results from 30 test episodes show a 90% success rate in reaching goals without collisions, demonstrating effective path planning and adaptability. The robot consistently navigated narrow corridors and avoided fragile obstacles. These findings support the potential of DRL-based methods for future deployment in greenhouse automation.

Keywords: Greenhouse · Navigation · Reinforcement learning

1 Introduction

1.1 UGVs in Agriculture, with a Focus in Greenhouses

[4] exposes a background of the research in wheeled mobile robots used in agriculture. The four wheel chassis is described as the most common chassis structure, it usually uses differential driving or front-wheeled synchronous steering as its driving mode due to stability and simplification of the operational process. Two wheel differential driving models are described as being simpler, having lower costs and a better performance at obstacle avoidance but usually aren't robust enough.

[2] reviews diverse robot systems in agriculture. It states that while drone solutions are standard drones with sensors, mobile platforms are specially designed for each application. It found that the majority of ground robots in agriculture are dedicated to monitoring tasks and only perform other more complex tasks in high-value crops. It also expose the robot Agri.Q developed by the authors for fields with significant inclines, this robot has two skid steering modules, each module has two locomotion units that drive two tires each.

J. L. Ballesteros et al. (Eds.): CITIS 2025, LNNS 1780, pp. 189–198, 2027.
https://doi.org/10.1007/978-3-032-14287-0_19

The literature on mobile robot's navigation in greenhouses has been continuously developing in recent years. Mobile platforms used in greenhouses perform tasks like measuring infrared temperature and soil moisture [13] or using image recognition technologies to count and classify fruits [15] or to look for diseases [1]. This platforms can be completely autonomous, as they usually are in open fields, or can depend on structures of the greenhouse (generally rails) for its guidance [19]; while using rails to move is more expensive, using autonomous platforms is a bigger technological challenge.

[5,10] present navigation systems for greenhouses based on ultrasonic sensors. The main advantage of using ultrasonic sensors is that they are cheap when compared to Li-DARs or RBGD cameras but this proposed systems have the disadvantage of being unable to adapt to dynamic environments.

[1] proposed a design for a wheeled mobile robot for greenhouses. The robot is a line follower equipped with a camera for taking pictures of the crops. The decision of using a line follower system instead of using Li-DAR or camera navigation is because an optimal cost-benefit ratio is pursued. GPS navigation based was not used because of the indoors attenuation of the GPS signal. Problems for localizing mobile platforms with GPS are also reported in vineyards mountains [3].

1.2 Deep Reinforcement Learning in Navigation

The successful use of DRL to win in video games [12] aroused great interest in the scientific community due to its potential to be applied in other scenarios. Since then, it has been applied in many areas of robotics, such as manipulators [7] [14] or locomotion [6]. DRL is a very promising field in robotics as it is the way to go in machine learning when there is no previous data set, which is usually the case in robotics.

In navigation, DRL has received attention due to its strong representation and experience learning abilities [21]. The training of navigation systems that use DRL for locomotion purposes is usually carried out in simulated environments as they are faster, cheaper, and safer [18].

[21] makes a systematic and comprehensive review of mobile robots' navigation systems based on DRL. It exposes 3 challenges: partial observation, sparse reward and poor generalization and brings possible approaches for overcoming them. This work concludes that though DRL navigation doesn't always have perfect behaviour it is the most promising approach in navigation.

[17] uses an A3C network to train a the navigation of a mobile robot. Fused data from a RGBD camera and a Li-DAR, and the orientation towards the goal position were the input to the network, and the output was the linear and angular velocities for the robot. The resulting model was able to move in an unstructured indoor environment.

[22] proposes a model for visual navigation that was able to go from simulation to reality without the need of fine-tuning. This was achieved by using different targets and scenarios during the training stage. [9] also works in visual

navigation adding a Li-DAR for obstacle detection. It compared the performance of a DQN with a DDQN and found better results with the DDQN.

[11] proposes a deep DRL method for teaching agents to navigate in environments that are big and visually rich. Their approach of augmented DRL with auxiliary objectives leads to a more general navigation strategy and permits an end to end learning strategy. The method was evaluated in the deep-mind lab. [16] proposes an end to end method for navigation using DRL in which sparse laser ranging measurements are taken as the input for reducing the reality gap. It proposes the ICM A3C algorithm, which is based in the A3C algorithm. Experimentation done in simulation showed that this proposed algorithm is better in navigation than the A3C.

2 Methods

This study employs an applied research methodology with an experimental approach, focusing on the implementation and validation of a navigation system for an unmanned ground robot using Deep Reinforcement Learning (DRL) techniques. Situated at the intermediate level of the research pyramid, this work builds upon theoretical foundations in machine learning and robotics to address a practical challenge: enabling autonomous navigation in structured greenhouse environments.

2.1 Hardware and Software Environment

Hardware used: Graphics Card, NVIDIA GTX 960M; Intel Core i7-6700HQ; Memory, 24 GB DDR4 RAM; and Storage, 1 TB SSD. Software environment: OS, Ubuntu 22.04; Robot Operating System (ROS), ROS 2 Humble; Simulation Platform, Ignition Gazebo 6.16.0; Programming Language, Python 3.10.12; Numerical Computing Library, NumPy 2.1.2; and Deep Learning Framework, PyTorch 2.4.0 with CUDA 12.2;

PyTorch was selected as the deep learning framework for this research due to its pythonic syntax, robust GPU acceleration, and extensive community ecosystem. The versions of ROS2, gazebo ignition, and Ubuntu were chosen together as ROS2 libraries are designed to work with specific versions of Ubuntu, and gazebo ignition has better compatibility with ROS2.

2.2 Developing the Model

A virtual model of the 4WD1 platform was created so that it faithfully represented the dynamics of the real robot. An URDF file was used to describe the geometry of the model and its properties such as inertia and mass. Figure 1 displays the model created in RVIZ and its readings when brought to the simulated greenhouse in the gazebo simulator.

The choice of a four-wheeled (4WD1)platform stems from the need for robustness in the uneven terrain of a greenhouse environment. While two-wheeled

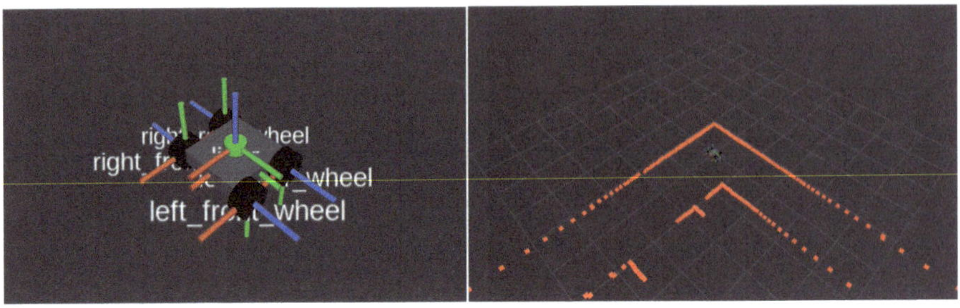

Fig. 1. Virtual model of the robot.

robots may offer more agility, they are less stable on such terrain. Although six-wheeled platforms provide even better traction and stability they are more costly, and complex, making the four-wheeled option a suitable balance between performance and cost.

The control of the four wheels was implemented using the differential drive controller from ROS2 control libraries for PID control in simulations, and can easily be transferred to real applications. The dynamical model and control 4-wheeled platforms is further developed in [8].

The structure of the greenhouse is based on [13]. Figure 2 shows the drawing of the greenhouse at left and its construction in the Gazebo simulator at right.

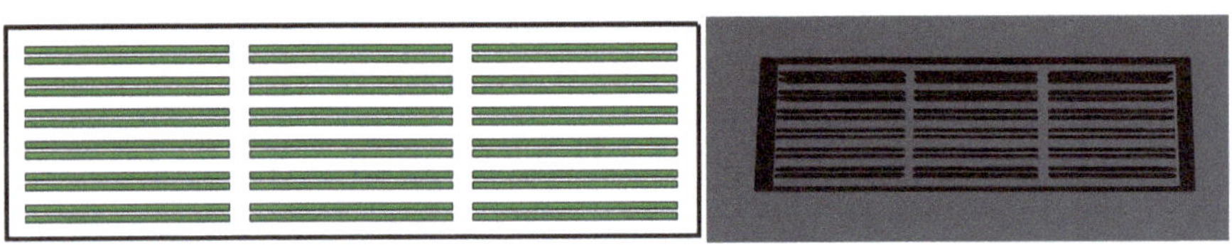

Fig. 2. Drawing of the greenhouse and its gazebo version.

2.3 Training the Model

The model training stage was carried out in a simulated environment of a tomato greenhouse. It was preferred to make the training inside a simulation as it is more secure and can be done faster.

The system was designed to navigate between current positions and target positions relative to the environment, the coordinates of these positions are obtained using a subscriber to the Pose topic on the simulator. For DRL, a DDQN-type neural network was used as it is more stable than a single DQN and it is best suited for discrete action-spaces. This was done similar to [20].

The possible states were defined by the readings of the Li-DAR sensor, the position of the robot, and the position of the target; these parameters were the input of the network (as in [17]). The agent is the four-wheeled robot and its space of discrete actions consists of speed and direction commands. Rewards are given when the agent manages to reach the goal, small discounts will be given for remaining in states other than the goal, and big discounts are given when there is a collision.

The Adam optimizer was selected for training the Double Deep Q Network (DDQN) due to its efficiency and robustness. It combines the benefits of adaptive learning rates from AdaGrad and RMSProp, allowing it to handle sparse gradients and varying feature scales effectively. Adam's bias correction and efficient computation make it ideal for reinforcement learning, where data distribution can change dynamically.

The learning rate was decreased to improve convergence and stability by allowing the model to make smaller, more precise updates as it approached the optimal solution, reducing the risk of overshooting and improving generalization.

A batch size of 64 is commonly used because it provides a balance between noise reduction in gradient estimates and computational efficiency. Larger batch sizes reduce variance in gradient updates, leading to more stable training, while smaller batches introduce more noise but can improve generalization. It is also good that the batch size is a power of 2 as modern GPUs are better optimized to work with those values.

A network size and architecture such as [17] was used as it is proven to lead to convergence. An epsilon greedy strategy will be used to avoid local minimals. Episodes were ran using different starting and goal positions until the robot can navigate to the goal while avoiding obstacles.

The neural network used for navigation in this study is a custom-built Double Deep Q Network (DDQN) designed to process both goal-related information and LiDAR sensor data, generating Q-values as outputs for action selection. The network processes inputs through two parallel streams: the Goal Processing Stream, which takes a tensor representing the robot's distance and orientation with respect to the goal, passing it through two fully connected layers (64 and 128 neurons with ReLU activation) to extract goal-related features; and the LiDAR Processing Stream, which processes environmental data through three convolutional layers (32, 64, and 128 filters with kernel sizes 16, 9, and 4, respectively), applying ReLU activation to capture spatial patterns before flattening the output. These processed features are concatenated and passed through fully connected layers (256, 128, and 64 neurons with ReLU activation) to predict Q-values, guiding the robot's decision-making.

The replay memory was implemented as a Python class. Each transition recorded an observation (a dictionary of two tensors), an action, the next observation, and a reward. Collisions were stored separately based on the reward. When sampling from the replay memory, a portion of the batch was ensured to consist of collision samples.

The target network was updated every 10 episodes. For the main network updates:

$$Q_{\text{target}} = R + \gamma \max_{a'} Q_{\text{target}}(s', a')$$

$$\text{Loss} = \text{SmoothL1Loss}(Q_{\text{predicted}}, Q_{\text{target}})$$

The routine for training episodes begins by verifying initialization and ensuring that all systems are ready before proceeding. Once confirmed, the environment is observed, capturing the robot's pose and LiDAR data to construct the observation tensors. Based on these observations, the action is selected using the DDQN agent's policy. This chosen action is then sent to the robot in the form of velocity commands. After executing the action, the reward is computed, and it is determined whether the episode has reached a terminal state. The resulting transition is stored in the replay memory, and the Q-network is updated accordingly. Finally, the routine manages state transitions and tracks progress by incrementing the step count.

3 Results

To ensure objectivity and deeper insights, the model is tested quantitatively with the success rate.

Success Rate: The percentage of successful navigation trials where the robot reaches the goal without collision, episodes with minor touches or complete failures are considered as failed ones [17].

The evaluation took place in a simulated tomato greenhouse environment modeled in Ignition Gazebo. Multiple trials were conducted with different goal positions, and robot starting points to test the robustness of the navigation system.

A successful navigation episode it's defined as one where the robot reaches the goal location without collisions. The system is deemed effective if it achieves a high success rate and consistently follows near-optimal paths with minimal deviations from the shortest path.

The training script was adapted to evaluate the model by introducing several modifications tailored for inference rather than learning. Unlike the original script, where both a main and a target Q-network are initialized for Double Deep Q-Learning (DDQN), the evaluation version relies solely on the main Q-network, making the target network and its updates unnecessary and thus removed. Rather than training the model from scratch, the evaluation process begins by loading a pre-trained network. During evaluation, the agent selects actions purely based on the Q-values, in a fully greedy manner, as opposed to the epsilon-greedy policy used during training that balances exploration and exploitation. To analyze the robot's navigation performance, its trajectory is logged throughout each episode. Additionally, the script keeps track of the number of episodes that end successfully by reaching the goal versus those that result in collisions, thereby enabling a measure of the success rate.

Table 1 presents the evaluation results. The episodes were initially run without a prioritized replay memory, and then repeated after its implementation.

Table 1. Evaluation Results of the DDQN Navigation Model

Metric	Before Prioritized Replay Memory	After
Episodes Ended at Goal	19	27
Episodes Ended with Collision	11	3

Figure 3 shows three trajectories, two of them ending in the goal and one with a collision. It is thought that the trajectory ended with a collision was provoked by the robot being spawned too close to the walls. Since there were not many data collected under those conditions, it led to failure.

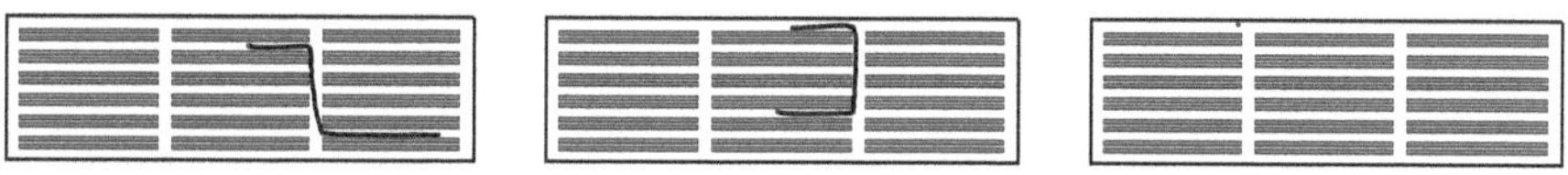

Fig. 3. Some trajectories of the evaluation.

4 Discussion

The performance of the DDQN-based navigation model was evaluated in a simulated tomato greenhouse environment. The evaluation consisted of 30 episodes, each starting with the robot placed at a random position and tasked with navigating to a specified goal while avoiding obstacles. The evaluation metrics included the number of episodes that successfully ended at the goal and the number of episodes that terminated due to collisions with obstacles.

The results demonstrate a significant improvement in navigation performance after implementing the prioritized replay memory. Specifically, the number of successful episodes, which the robot reached the goal, increased from 19 to 27. At the same time, the number of episodes that ended with a collision decreased from 11 to 3. This final result is comparable with the one obtained by [17] of 95% success rate, [17] also implemented 2D lidar navigation but instead of using a DDQN, an Asynchronous Advantage Actor-Critic network was used.

During the evaluation, only the neural network was used to select the velocity commands for the robot model. In real life performance, it would be more adequate to combine the neural network with a security system programmed to stop the robot when it perceives that there will be a collision; or a more refined version that takes the robot as far as possible from obstacles and then passes the lead back to the neural network. This was not implemented in this work, since the intention was only to test the ability of DRL to navigate, and the implementation of this security system will possibly lead to no collisions.

The trajectories taken by the robot were mostly optimal. When observing the routes in a global way, the proposed system always chooses the corridors that lead to the goal in the shortest way. It was observed that in some trajectories, specially when turning around a corner, the robot passed dangerously close to the obstacles. This behavior is thought to be due to the reward function, which penalizes taking extra time to get to the goal positions, but does not penalize taking risky actions. i.e. in the trajectory shown at left in Fig. 3, the robot moves between rows until reaching the vertical corridor, once there moves diagonally towards the rows where its goal destination its located; when moving diagonally it passes close to the borders of the rows. A possible way to optimize this, would be to experiment with the reward function, adding a negative term proportional to the minimum distance measured by the LiDAR.

5 Conclusions

This study focused on the development and evaluation of a navigation system for an unmanned ground vehicle (UGV) in a tomato greenhouse, leveraging a Double Deep Q-Network (DDQN) to train the robot to navigate through the environment. The research aimed to create a virtual model of the robot's kinematics and to validate its ability to find efficient paths and avoid obstacles using reinforcement learning in a simulated environment.

The DDQN successfully trained the robot to navigate toward a goal while avoiding obstacles in the greenhouse environment. The system demonstrated robust performance in path planning and collision avoidance, with the reward structure encouraging efficient navigation.

The virtual model of the 4WD1 four-wheeled platform was successfully developed, ensuring that its kinematics were accurately represented in the simulated environment.

The robot's ability to generalize to various spawn points and obstacle configurations was demonstrated through repeated simulations. The results showed that the navigation system could effectively transfer learning from one scenario to another, making it adaptable to different conditions.

While the system showed promising results, challenges such as tuning hyperparameters and ensuring smooth convergence of the DDQN were observed. Further optimization of the training process, including the handling of sparse rewards, could improve performance. Additionally, real-world testing will be crucial to assess the system's robustness outside the simulated environment.

References

1. Aravind, K., Raja, P.: Design and simulation of crop monitoring robot for green house. In: 2016 International Conference on Robotics: Current Trends and Future Challenges (RCTFC), pp. 1–6. IEEE (2016)

2. Botta, A., Cavallone, P., Baglieri, L., Colucci, G., Tagliavini, L., Quaglia, G.: A review of robots, perception, and tasks in precision agriculture. Appl. Mech. **3**(3), 830–854 (2022)
3. Santos, D., et al.: Towards a reliable monitoring robot for mountain vineyards. In: 2015 IEEE International Conference on Autonomous Robot Systems and Competitions, pp. 37–43. IEEE (2015)
4. Gao, X., et al.: Review of wheeled mobile robots' navigation problems and application prospects in agriculture. IEEE Access **6**, 49248–49268 (2018). https://doi.org/10.1109/ACCESS.2018.2868848
5. González, R., Rodríguez, F., Sánchez-Hermosilla, J., Donaire, J.G.: Navigation techniques for mobile robots in greenhouses. Appl. Eng. Agric. **25**(2), 153–165 (2009)
6. Heess, N., et al.: Emergence of locomotion behaviours in rich environments, arXiv preprint arXiv:1707.02286 (2017)
7. Kalashnikov, D., et al.: Scalable deep reinforcement learning for vision-based robotic manipulation. In: Conference on Robot Learning, pp. 651–673. PMLR (2018)
8. Kozłowski, K., Pazderski, D.: Modeling and control of a 4-wheel skid-steering mobile robot. Int. J. Appl. Math. Comput. Sci. **14**(4), 477–496 (2004)
9. Lee, M.F.R., Yusuf, S.H.: Mobile robot navigation using deep reinforcement learning. Processes **10**(12), 2748 (2022)
10. Longo, D., Pennisi, A., Caruso, L., Muscato, G., Schillaci, G.: An autonomous electrical vehicle based on low-cost ultrasound sensors for safer operations inside greenhouses. In: Proceedings of International Conference Ragusa, SHWA 2010, pp. 437–443 (2010)
11. Mirowski, P., et al.: Learning to navigate in complex environments, arXiv preprint arXiv:1611.03673 (2016)
12. Mnih, V., et al.: Playing Atari with deep reinforcement learning, arXiv preprint arXiv:1312.5602 (2013)
13. Ruiz-Larrea, A., Roldán, J.J., Garzón, M., del Cerro, J., Barrientos, A.: A UGV approach to measure the ground properties of greenhouses. In: Robot 2015: Second Iberian Robotics Conference. AISC, vol. 418, pp. 3–13. Springer, Cham (2016). https://doi.org/10.1007/978-3-319-27149-1_1
14. Rusu, A.A., Večerík, M., Rothörl, T., Heess, N., Pascanu, R., Hadsell, R.: Sim-to-real robot learning from pixels with progressive nets. In: Conference on Robot Learning, pp. 262–270. PMLR (2017)
15. Seo, D., Cho, B.H., Kim, K.C.: Development of monitoring robot system for tomato fruits in hydroponic greenhouses. Agronomy **11**(11), 2211 (2021)
16. Shi, H., Shi, L., Xu, M., Hwang, K.S.: End-to-end navigation strategy with deep reinforcement learning for mobile robots. IEEE Trans. Industr. Inf. **16**(4), 2393–2402 (2019)
17. Surmann, H., Jestel, C., Marchel, R., Musberg, F., Elhadj, H., Ardani, M.: Deep reinforcement learning for real autonomous mobile robot navigation in indoor environments, arXiv preprint arXiv:2005.13857 (2020)
18. Tan, J., et al.: Sim-to-real: learning agile locomotion for quadruped robots, arXiv preprint arXiv:1804.10332 (2018)
19. Tangarife, H.I., Díaz, A.E.: Robotic applications in the automation of agricultural production under greenhouse: a review. In: 2017 IEEE 3rd Colombian Conference on Automatic Control (CCAC), pp. 1–6. IEEE (2017)

20. Van Hasselt, H., Guez, A., Silver, D.: Deep reinforcement learning with double q-learning. In: Proceedings of the AAAI Conference on Artificial Intelligence, vol. 30 (2016)
21. Zhu, K., Zhang, T.: Deep reinforcement learning based mobile robot navigation: a review. Tsinghua Sci. Technol. **26**(5), 674–691 (2021)
22. Zhu, Y., et al.: Target-driven visual navigation in indoor scenes using deep reinforcement learning. In: 2017 IEEE International Conference on Robotics and Automation (ICRA), pp. 3357–3364. IEEE (2017)

Research Landscape of AI and Intelligent Tutoring Systems in Ecuadorian Universities

Juan P. López-Goyez[1,2](✉), Alfonso González-Briones[2], Jorge Mina-Ortega[1], and Javier Montalvo[1]

[1] Universidad Politécnica Estatal del Carchi, Antisana S/N, 040101 Tulcán, Ecuador
juan.lopez@upec.edu.ec

[2] BISITE Research Group, Edifice I+D+i, University of Salamanca, Calle Espejo 2, Castile and León, 37007 Salamanca, Spain

Abstract. This systematic literature review examines research conducted by Ecuadorian universities and research centers on intelligent tutoring systems and artificial intelligence, considering a technological categorization based on functional, technical, and pedagogical criteria, as well as the temporal and geographical identification of the contributions. The aim is to provide a comprehensive overview of the state of the art in Ecuador and to offer guidance for a more effective implementation of intelligent tutoring systems within the national education system. Studies published between 2015 and 2024 are analyzed, identifying trends, challenges, and opportunities in the adoption of these technologies in the Ecuadorian educational context. The findings reveal a growing interest in ITS, highlighting the use of pedagogical approaches such as adaptive learning and constructivism. The technological models and algorithms for artificial intelligence or machine learning implemented in tutoring systems are also identified and linked to the teaching theories or strategies used to apply them in educational settings, including content personalization and adaptation strategies. Finally, the impact of these technologies on the teaching and learning process is evaluated, and a diagram of a virtual tutor agent model is proposed, which considers human–machine interaction and ethical aspects when providing feedback to students. The study concludes that the impact of ITS on education has helped define future research directions that will enable personalized learning with AI

Keywords: Artificial intelligence · Ecuador · educational technology · intelligent tutoring systems

1 Introduction

The integration of intelligent technologies into the educational field has led to the development of Intelligent Tutoring Systems (ITS), tools capable of personalizing instruction and adapting to the individual needs of students. These systems combine artificial intelligence (AI) algorithms, pedagogical models, and data processing to simulate the activities of a human tutor and support teachers in the teaching process [1, 2]. ITS have proven

J. L. Ballesteros et al. (Eds.): CITIS 2025, LNNS 1780, pp. 199–209, 2027.
https://doi.org/10.1007/978-3-032-14287-0_20

effective in promoting autonomous learning, providing immediate feedback, and offering adaptive guidance across various areas of knowledge, particularly in digital learning environments [3–5].

In the Ecuadorian context, research on ITS has explored the application of conversational agents, dialogue-based systems, classification algorithms, and emerging technologies such as generative artificial intelligence at different educational levels. These initiatives have not only contributed to improving academic performance but have also supported pedagogical approaches such as adaptive learning, gamification, and constructivism [6–8]. Despite the growing interest, no systematic reviews have been identified at the local level that analyze intelligent tutoring systems developed in Ecuador, considering their geographical and temporal distribution, the AI technologies and models employed, as well as the main challenges and opportunities. This study proposes a technological categorization based on functional, technical, and pedagogical criteria. The aim is to provide a comprehensive overview of the state of the art in Ecuador and to offer insights for a more effective implementation of ITS. The contribution of this article is a systematic literature review (SLR) to highlight the research questions: i) RQ1-Temporal and geographical distribution of research on ITS and AI ii) RQ2- Approaches and Technologies in ITS Research and iii) RQ3–1.1Challenges in AI and ITS research at Ecuadorian universities (see Table 1).

2 Methodology

The research approach was qualitative, descriptive, and documentary, focusing on the analysis and interpretation of theoretical sources related to the use of intelligent tutoring systems in educational contexts. Scientific articles were collected and examined through a systematic literature review to understand the characteristics, benefits, and challenges of these technologies applied to personalized learning. A systematic literature review (SLR) aims to identify, evaluate, and combine evidence from primary studies available in scientific databases and digital repositories, through research questions. The following is the protocol for a literature review conducted in 4 phases: i) statement of 3 research questions (RQ) (see Table 1), ii) search of 243 articles in scientific databases, iii) Selection of 18 scientific articles with inclusion and exclusion criteria and iv) analysis and extraction of information.

2.1 Research Questions

Research questions allow us to formulate the problem, extend the knowledge of findings, and provide new results for scientific and societal knowledge (Table 1).

Table 1. Research questions (RQ).

Number	Research question
RQ1	What is the temporal and geographical distribution of research on ITS and AI in Ecuador?
RQ2	What are the predominant approaches, AI and ITS technologies, and teaching strategies in ITS research at Ecuadorian universities?
RQ3	What are the main challenges identified in AI and ITS research at Ecuadorian universities?

2.2 Article Search

The search for relevant articles was carried out in March 2025 through the Web of Science and SCOPUS databases. The Boolean operator (AND, OR) and keywords: intelligent tutoring systems, Ecuador and educational technology and virtual education were used. A total of 243 articles were initially retrieved from the databases. For the systematic review process, a set of inclusion and exclusion criteria was applied to refine the selection.

2.3 Selection Criteria

This section outlines the process followed for selecting the documents used in the study, which was carried out in three distinct phases, view Table 3. Initially, the inclusion focused on articles related to computer science and education, with open access availability. In the second phase, a series of exclusion criteria were applied: technical reports, book chapters, duplicate entries, redundant works by the same authors, and secondary studies were filtered out. The aim at this stage was to retain only original research articles published in English between 2015 and 2024 and indexed in databases. he inclusion criteria were that the articles had to be presented by Ecuadorian authors. Additionally, the keywords had to be present in the title, abstract, and keywords section, and the methodology of each article had to directly address the research questions in this study. Metadata was also obtained to identify location and affiliation of the authors from Ecuadorian universities. To support this process and facilitate article screening and classification, the Rayyan platform was employed as a digital assistant. Table 2 presents the 18 scientific articles selected for analysis. Each document was assigned a code reflecting its level of relevance and alignment with the research questions and the established selection criteria.

Table 2. SLR document selection.

	Reference	Ref
WoS	Intelligent tutoring system to improve learning outcomes	[5]
	Integrating a dialog system with an intelligent tutoring system for a 3D virtual laboratory	[9]
	Improving prediction of students' performance in intelligent tutoring systems using attribute selection and ensembles of different multimodal data sources	[6]
	Virtual tutoring environment: an alternative to streamline scientific research processes	[10]
	Visualizing a collective student model for procedural training environments	[4]
	Mobile learning: perspectives	[11]
	Designing Hybrid Human-AI Orchestration Tools for Individual and Collaborative Activities: A Technology Probe Study	[3]
SCOPUS	Improving prediction of students' performance in intelligent tutoring systems using attribute selection and ensembles of different multimodal data sources	[6]
	A review on data fusion in multimodal learning analytics and educational data mining	[12]
	Backpack process model (Bppm): A process mining approach for curricular analytics	[2]
	Can Mindfulness Help Primary Education Students to Learn How to Program with an Emotional Learning Companion	[14]
	A MOOC-based flipped experience: Scaffolding SRL strategies improves learners' time management and engagement	[14]
	Predicting Academic Success of College Students Using Machine Learning Techniques	[15]
	Innovations in Introductory Programming Education: The Role of AI with Google Colab and Gemini	[8]
	Collaborative Working and Critical Thinking: Adoption of Generative Artificial Intelligence Tools in Higher Education	[16]
	Integration of technology and pedagogy in intelligent tutoring systems	[17]
	Models of Intelligent Tutoring Systems Based on Autonomous Agents for Virtual Learning Environments: A Systematic Literature Review	[18]
	Intelligent Tutoring Systems in Smart Learning Environments for Higher Education: A Systematic Literature Review	[19]

2.4 Analysis and Extraction of Information

A systematic review of the selected articles was conducted using organizational tools such as pivot tables in Excel to record and classify information, along with Python scripts for generating graphs and identifying trends. Technologies were identified and grouped based on their application within the educational context, allowing for the establishment of a categorization. Concept maps created in Lucidchart were used to visualize the relationships between technologies and define the following categories: artificial intelligence / machine learning (AI/ML), algorithms, models, tools, and emerging technologies.

To validate the consistency of the classification, a peer review method was applied, associating technologies, applications, and instructional theories or strategies based on criteria such as technical specificity, educational relevance, and frequency of mention in the analyzed literature. Metadata retrieved from scientific databases were also analyzed using Python scripts to identify trends by year and author affiliations with higher education institutions and/or research centers in Ecuador. As a final verification technique, a cross-check was performed with the original texts to ensure that all information accurately reflected the content analyzed.

3 Results

3.1 RQ1- Temporal and Geographical Distribution of Research on ITS and AI.

In Ecuador, there is clear evidence of an increase in scientific production related to intelligent tutoring systems and artificial intelligence in various educational contexts. This highlights the growing relevance of researching these technological innovations and their integration into the teaching and learning process. Figure 1 presents the trend identified during the systematic review process, based on the articles selected from the WoS and SCOPUS databases.

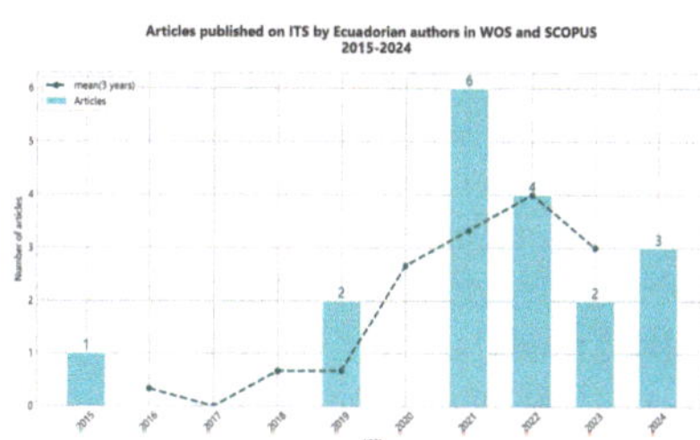

Fig. 1. Articles published on ITS by Ecuadorian authors in WOS and SCOPUS

Similarly, Fig. 2 shows the geographical distribution by province in Ecuador of publications authored by researchers affiliated with universities or research centers. Notably, the provinces of Pichincha, Guayas, and Loja account for the highest percentage of contributions in this field.

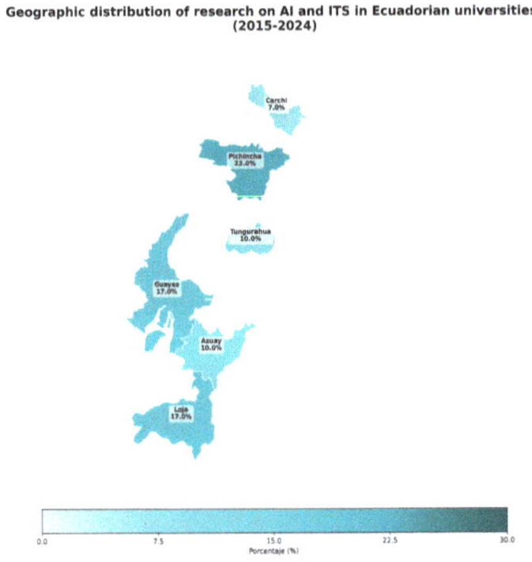

Fig. 2. Geographic distribution of research on AI and ITS

3.2 RQ2 - Approaches and Technologies in ITS Research.

Table 2 presents the main approaches and technologies, models and AI/ML applied to the teaching and learning process, highlighting the technical nature of each technology reported in the literature, its functional role in the educational process (analysis, intervention, support, or infrastructure), and the pedagogical foundations and instructional strategies associated with each one. This creates a bridge between technological innovation and its effective application, ranging from the development of critical thinking to the personalization of learning. The analysis reveals that each proposal is grounded in learning theories (such as constructivism or cognitivism) and specific strategies (such as adaptive learning or educational scaffolding) [3, 5, 8, 9, 16, 18, 19].

For example, dialogue-based systems like PACO implement the Socratic method to guide reasoning, while the BDI model used in 3D games applies gamification principles to enhance motivation. Meanwhile, the use of generative AI personalizes learning by generating content tailored to students' individual needs, promoting cognitive engagement. In another case, it is evident that the use of algorithms based on random forest facilitate the classification of students at risk within the academic monitoring process, facilitating the interaction of technology with intervention pedagogies [6, 8, 9, 11, 20]. A word cloud of the cited technologies in the literature is shown in Fig. 3.

Table 3. SLR document selection.

Focus	Technology /Model	Application	Educational Theory /Strategy	Ref.
AI/ML	Generative AI	Gemini, ChatGPT for programming education	Constructivism + Generative Learning	[7]
AI/ML	Deep Learning (DL)	Student cognitive modeling and predictive analytics	Adaptive Learning Theory	[20]
AI/ML	Natural Language Processing	Dialogue systems (PACO) and conversational agents (Alcody)	Socratic Dialogue + Scaffolding	[1]
Algorithms	XGBoost Random Forest SMOTE KMeans	Academic success prediction (best AUC performance)	Early Warning System Theory	[15]
Algorithms	REPTree	Performance prediction with multimodal data	Multimodal Learning Theory	[7]
Models	BDI Model	Belief-Desire-Intention architecture for 3D game tutoring	Game-Based Learning + Situated Cognition	[7]
Models	Process Mining (BPPM)	Academic path analysis with "backpack of stones" metaphor	Academic Resilience Theory	[2]
Tools	Google Colab	Programming education environment with generative AI	Constructionism (Papert)	[8]
Tools	Moodle	Educational data analytics platform (LA/EDM)	Blended Learning Model	[12]
Emerging	Edge Computing	Real-time adaptive tutoring through on-device processing	Microlearning Strategy	[3]

Fig. 3. Technological approaches investigated to integrate AI into education.

3.3 RQ3- Challenges in AI and ITS Research at Ecuadorian Universities.

Figure 4 presents a scatter plot evaluating the pedagogical impact (Y-axis) of various educational technologies (X-axis), where each point represents a specific technology. The impact rating (1 = low, 5 = high) is based on documented relevance to improving learning, considering factors such as personalization, effectiveness of feedback, and adaptability of the teaching-learning process. The assigned values (1–5) were derived from the frequency of mention in the reviewed literature, evidence of positive educational outcomes, and scalability in real-world contexts. Generative AI (impact score 5) stands out with a high rating due to its ability to generate adaptive content, whereas KMeans SMOTE (impact score 2) requires greater specialization and technical training, which limits its broader pedagogical application [2, 9, 12, 15].

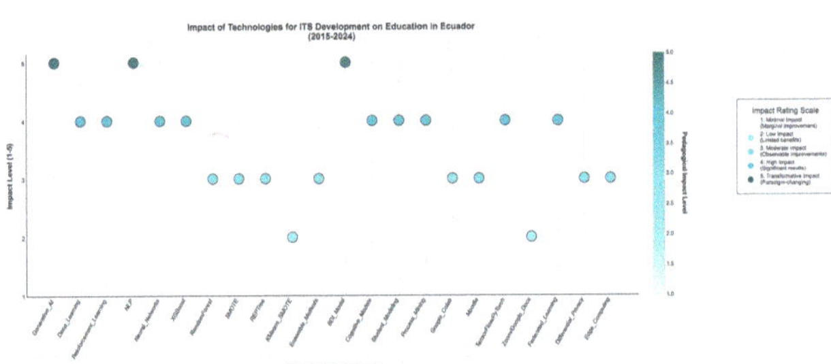

Fig. 4. Protocol diagram used in the SLR

4 Discussion

The results show a progressive consolidation of scientific production on ITS and AI in Ecuador. This trend reflects the interest in integrating advanced technologies into education, combining technical and pedagogical approaches. The categorization of technologies allowed us to identify their function (analysis, intervention, support, infrastructure) and their theoretical basis (constructivism, cognitivism, adaptive learning,) supported by analytical tools and peer validation. The pedagogical impact analysis showed that technologies such as Generative AI and Deep Learning stand out for their adaptability and personalized learning, while others, such as KMeans SMOTE, require greater technical specialization. Clear relationships were identified between technologies and educational theories, such as the use of the BDI model with gamification or dialogue systems based on the Socratic method. These findings underscore the importance of aligning technological innovation with pedagogical foundations to achieve effective and scalable applications in real-world educational contexts.

5 Proposal, Limitations and Future Works

An ITS is proposed, incorporating an ethical agent responsible for validating responses and feedback provided to students. The system facilitates interaction among students, the ethical agent, and a supervising teacher within an LMS, using NLP and ethical criteria to ensure personalized and secure tutoring [6, 7, 20]. This model, currently under development, will be implemented on the Moodle platform at the UPEC University, to enhance academic monitoring and support personalized learning. Figure 5 shows the system's diagram.

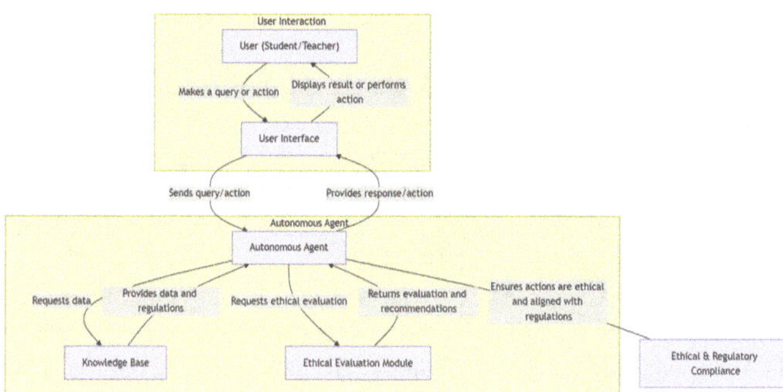

Fig. 5. Diagram of the proposed model.

6 Conclusions

This study presents a systematic review of the literature on research conducted in Ecuador on intelligent tutoring systems. It highlights the main educational technologies and artificial intelligence models, emphasizing their pedagogical impact and technical application. Generative AI stands out for its ability to personalize learning by generating adaptive content, while other tools, such as KMeans SMOTE and random forest algorithms, serve more specialized functions such as identifying students at risk. The technologies were categorized according to their function, analysis, intervention, support, or infrastructure and were linked to learning theories such as constructivism and cognitivism. The study emphasizes the importance of proposing instructional strategies that enable the reliable integration of AI into the teaching and learning process.

References

1. Paladines, J., Ramírez, J.: A systematic literature review of intelligent tutoring systems with dialogue in natural language. IEEE Access. **8**, 164246–164267 (2020). https://doi.org/10.1109/ACCESS.2020.3021383
2. Salazar-Fernandez, J.P., Munoz-Gama, J., Maldonado-Mahauad, J., Bustamante, D., Sepúlveda, M.: Backpack process model (Bppm): a process mining approach for curricular analytics. Appl. Sci. (Switz.). **11**(9), 4265 (2021). https://doi.org/10.3390/app11094265
3. Echeverria, V., Yang, K., Lawrence, L., Rummel, N., Aleven, V.: Designing hybrid human-AI orchestration tools for individual and collaborative activities: a technology probe study. IEEE Trans. Learn. Technol. **16**(2), 191–205 (2023). https://doi.org/10.1109/TLT.2023.3248155
4. Riofrío-Luzcando, D., Ramírez, J., Moral, C., de Antonio, A., Berrocal-Lobo, M.: Visualizing a collective student model for procedural training environments. Multimed. Tools Appl. **78**(8), 10983–11010 (2019). https://doi.org/10.1007/s11042-018-6641-x
5. Duraes, D., Toala, R., Goncalves, F., Novais, P.: Intelligent tutoring system to improve learning outcomes. AI Commun. **32**(3), 161–174 (2019). https://doi.org/10.3233/AIC-190624
6. Chango, W., Cerezo, R., Sanchez-Santillan, M., Azevedo, R., Romero, C.: Improving prediction of students' performance in intelligent tutoring systems using attribute selection and ensembles of different multimodal data sources. J. Comput. High. Educ. **33**(3), 614–634 (2021). https://doi.org/10.1007/s12528-021-09298-8
7. Liu, D., Barba-Guamán, L., Valdiviezo-Díaz, P., Riofrio, G.: Intelligent tutoring module for a 3Dgame-based science e-learning platform. Intel. Artif. **20**(60), 1–19 (2017). https://doi.org/10.4114/intartif.vol20iss60pp1-19
8. Llerena-Izquierdo, J., Mendez-Reyes, J., Ayala-Carabajo, R., Andrade-Martinez, C.: Innovations in introductory programming education: the role of AI with Google Colab and Gemini. Educ Sci (Basel). **14**(12), 1330 (2024). https://doi.org/10.3390/educsci14121330
9. Paladines, J., Ramirez, J., Berrocal-Lobo, M.: Integrating a dialog system with an intelligent tutoring system for a 3D virtual laboratory. Interact. Learn. Environ. **31**(7), 4476–4489 (2023). https://doi.org/10.1080/10494820.2021.1972012
10. Mesa Vazquez, J., Parra Alvarez, M., Cedeno Marcillo, G.E.: Virtual tutoring environment: an alternative to streamline scientific research processes. *REVISTA CONRADO*. **17**(83), 250–258 (2021) Accessed: May 16, 2025. [Online] . Available: https://conrado.ucf.edu.cu/index.php/conrado/article/view/2084
11. Torres Diaz, J.C., Infante Moro, A., Torres Carrion, P.V.: Mobile learning: perspectives. RUSC-Univ. Knowl. Soc. J. **12**(1), 38–49 (2015). https://doi.org/10.7238/rusc.v12i1.1944
12. Chango, W., Lara, J.A., Cerezo, R., Romero, C.: A review on data fusion in multimodal learning analytics and educational data mining. Wiley Interdiscip. Rev. Data Min. Knowl. Discov. **12**(4) (2022). https://doi.org/10.1002/widm.1458
13. Morales-Urrutia, E.K., Ocana, J.M., Perez-Marin, D., Pizarro, C.: Can Mindfulness Help Primary Education Students to Learn How to Program with an Emotional Learning Companion?. IEEE Access. **9**, 6642–6660 (2021). https://doi.org/10.1109/ACCESS.2021.3049187
14. Pérez-Sanagustín, M., et al.: A MOOC-based flipped experience: scaffolding SRL strategies improves learners' time management and engagement. Comput. Appl. Eng. Educ. **29**(4), 750–768 (2021). https://doi.org/10.1002/cae.22337
15. Guanin-Fajardo, J.H., Guaña-Moya, J., Casillas, J.: Predicting academic success of college students using machine learning techniques. Data (Basel). **9**(4) (2024). https://doi.org/10.3390/data9040060
16. Ruiz-Rojas, L.I., Salvador-Ullauri, L., Acosta-Vargas, P.: Collaborative working and critical thinking: adoption of generative artificial intelligence tools in higher education. Sustainability (Switzerland). **16**(13) (2024). https://doi.org/10.3390/su16135367

17. Muñoz, G.F.R.: Integration of technology and pedagogy in intelligent tutoring systems.; [Integración de la tecnología y la pedagogía en los sistemas de tutoría inteligente]. Edutec. **89**, 144–155 (2024). https://doi.org/10.21556/edutec.2024.89.3199
18. López-Goyez, J.P., González-Briones, A., Chamorro, A.F.: Models of intelligent tutoring systems based on autonomous agents for virtual learning environments: a systematic literature review. In: Advances in Practical Applications of Agents, Multi-Agent Systems, and Digital Twins: the PAAMS Collection, pp. 184–195. Springer Nature Switzerland, Cham (2025). https://doi.org/10.1007/978-3-031-70415-4_16
19. López-Goyez, J.P., González-Briones, A., Villarreal, A.F.C.: Intelligent tutoring Systems in Smart Learning Environments for higher education: a systematic literature review. In: Methodologies and Intelligent Systems for Technology Enhanced Learning, 14th International Conference, pp. 265–276. Springer Nature Switzerland, Cham (2024). https://doi.org/10.1007/978-3-031-73538-7_24
20. Ocaña, J.M., Morales-Urrutia, E.K., Pérez-Marín, D., Pizarro, C.: Can a learning companion be used to continue teaching programming to children even during the COVID-19 pandemic? IEEE Access. **8**, 157840–157861 (2020). https://doi.org/10.1109/ACCESS.2020.3020007

Locomotion of a Quadruped Robot Based on a PPO Reinforcement Learning Policy

Gustavo Machado-Merino[1](✉), Diana Aguirre-Ruiz[2], Juan Chuquin-Vasco[2], and Luis Rojas-Oviedo[2]

[1] Universidad Técnica de Ambato, Ambato, Ecuador
gd.machado@uta.edu.ec
[2] Escuela Superior Politécnica de Chimborazo, Riobamba, Ecuador
{diana.aguirre,juan.chuquin,luis.rojaso}@espoch.edu.ec

Abstract. Terrestrial robot locomotion is based on two approaches: wheels or tracks, and articulated systems. Articulated robots offer advantages over wheeled robots, such as greater adaptability to uneven terrain, lower energy consumption, and improved maneuverability in confined spaces, such as accessing stairs. However, they face challenges such as the complexity of their mechanical and control structures, as well as the high cost of their actuators. This work addresses these limitations by designing and implementing a Proximal Policy Optimization (PPO) reinforcement learning model for the locomotion of a 12°-of-freedom quadruped robot. The model was developed using TensorFlow and OpenAI Gym and tested in the physical simulation environment Pybullet.

Key metrics are introduced, such as the minimum center of gravity deviation (<2 cm on flat trajectories) and the success rate on uneven terrain (94% in standardized simulations), demonstrating a significant improvement over conventional control method. In addition, critical hyperparameters (learning rate and stability reward) were iteratively tuned, achieving a 15% improvement in trajectory stability compared to initial configurations. The proposed model stands out by ensuring a smooth response in the robot's motion and competitive performance based to other advanced control algorithms, such as SAC and DDPG, evaluated under the same experimental conditions.

The results confirm the potential of reinforcement learning in optimizing the control of articulated robots, overcoming current limitations, and promoting the development of more robust, versatile, and efficient robotic systems. This research lays the groundwork for future implementations in real-world scenarios, addressing both the design and motion planning in challenging terrain.

Keywords: Quadruped Robot · Reinforcement Learning · PPO · Gradient Policy

1 Introduction

Quadrupedal robots have gained significant attention due to their ability to navigate complex and unstructured terrains, outperforming wheeled or tracked robots in scenarios requiring adaptability, such as search-and-rescue missions or industrial inspections [1].

J. L. Ballesteros et al. (Eds.): CITIS 2025, LNNS 1780, pp. 210–219, 2027.
https://doi.org/10.1007/978-3-032-14287-0_21

However, achieving stable and efficient locomotion in these robots remains a challenge, primarily due to the high-dimensional control space, dynamic balance requirements, and the need for real-time adaptability [2]. Traditional control methods, such as model predictive control (MPC) or heuristic-based strategies, often struggle to generalize across diverse environments and require extensive manual tuning [3].

Reinforcement learning (RL), particularly policy optimization algorithms, has emerged as a promising solution for automating locomotion control. Among these, Proximal Policy Optimization (PPO) [4] stands out due to its sample efficiency, stability in training, and ability to handle continuous action spaces—critical features for legged robotics. PPO's Actor-Critic architecture enables simultaneous policy improvement and value estimation, making it suitable for high-dimensional tasks like quadrupedal gait generation [5]. However, existing applications often lack detailed discussions on hyperparameter selection, reward shaping, or simulation-to-reality gaps, limiting reproducibility and practical deployment.

This work addresses these gaps by presenting a PPO-based framework for training a 12-degree°-of-freedom quadrupedal robot in simulation. Our contributions include:

A systematic PPO implementation tailored for quadrupedal locomotion, emphasizing optimizer selection (Adam vs. SGD), reward function design, and policy update constraints.

Comprehensive evaluation metrics reward convergence, center of gravity (COG) stability, and policy divergence to quantify gait quality and training efficiency.

Empirical insights into training dynamics, such as the impact of sample size (up to 1 million steps) and the role of KL-divergence penalties in policy stability.

This study bridges theoretical RL advancements with practical robotics, offering a reproducible pipeline for future research in legged robot control.

2 Methods

2.1 Reinforcement Learning Framework

We employ Proximal Policy Optimization (PPO-Clip) [4], a state-of-the-art reinforcement learning algorithm known for its stability in high-dimensional continuous control tasks. PPO-Clip was selected over other policy gradient methods due to its ability to constrain policy updates, preventing excessive deviations that could destabilize training.

The PPO algorithm is based on an actor-critic architecture, consisting of two neural networks:

Actor Network: This network outputs a stochastic policy, defined by a mean and variance, which generates motor commands for the robot's 12 joints. The policy is optimized to maximize cumulative rewards while maintaining smooth locomotion.

Critic Network: This network estimates the value function, providing a baseline to reduce variance in policy updates and improve training stability.

The policy update is governed by a clipped objective function is:

$$L^{CLIP}(\theta) = \mathrm{E}_t\big[\min\big(r_t(\theta)\hat{A}_t, \mathrm{clip}(r_t(\theta), 1 - \epsilon, 1 + \epsilon)\hat{A}_t\big)\big] \tag{1}$$

where:

$r_t(\theta) = \frac{\pi_\theta(a_t|s_t)}{\pi_{\theta_{old}}(a_t|s_t)}$ is the probability ratio between the new and old policies.

$\hat{A}_t$ is the advantage estimate, computed using Generalized Advantage Estimation (GAE) [6] with $\lambda = 0.95.\epsilon = 0.2$ is the clipping parameter, ensuring policy updates remain within a trusted region. It has been chosen between [0.8, 1.2].

In (1) the Trimmed Surrogate Target function change is ignored when the target improves and included when it worsens. If θ moves away from the θ_{old}, the policy update is triggered.

Neural Network Architecture. Both the Actor and Critic Networks share a similar structure but serve distinct purposes:

The Input Layer is formed by 4-dimensional vector representing normalized observations (angle, velocity, torque, and orientation). The Hidden Layers are formed by two fully connected layers with 200 and 100 neurons, respectively, using tanh activation functions for non-linearity. The Output Layer is formed by two parts, the first is Actor, it is 12-dimensional vector representing motor commands and Critic is a Scalar value estimating the expected return.

2.2 Robot Modeling and Simulation

A. URDF Model. The robot is modeled after Spot (Boston Dynamics) [7], with 12 DoF (3 per leg: hip abduction, hip flexion, and knee flexion). The URDF file includes calibrated mass and inertia properties to ensure realistic dynamics in simulation.

B. Simulation Environment. The training environment is implemented in PyBullet, a physics engine optimized for robotics and reinforcement learning. Key features include: The Observation Space that contain Joint angles, body orientation (roll, pitch, yaw), and linear/angular velocities, the Action Space that contain Normalized torque commands for each motor, bounded between $[-1, 1]$ and the Reward Function that is Designed to incentivize stable and efficient locomotion based on:

$$R = w_1 \bullet v_x + w_2 \bullet e^{-\|COG_z - h_{des}\|^2} - w_3 \bullet \|\tau\|^2$$

where:

v_x is the Forward velocity (aligned with desired speed).

COG_z is the Vertical height of the center of gravity (penalizes excessive body oscillation).

τ is the Torque penalties to promote energy efficiency.

In Fig. 1 we can see the Pybullet robot Training Environment, we can see the URDF model of Spot robot.

C. Domain Randomization. To enhance the policy's robustness and facilitate sim-to-real transfer, we introduce variability during training like Ground Friction it is randomized within $\pm 20\%$ of nominal values, the Motor Dynamics it varied by $\pm 15\%$ to simulate real-world actuator imperfections and the Sensor Noise it is Gaussian noise ($\sigma = \mathbf{0.001}$) added to observations.

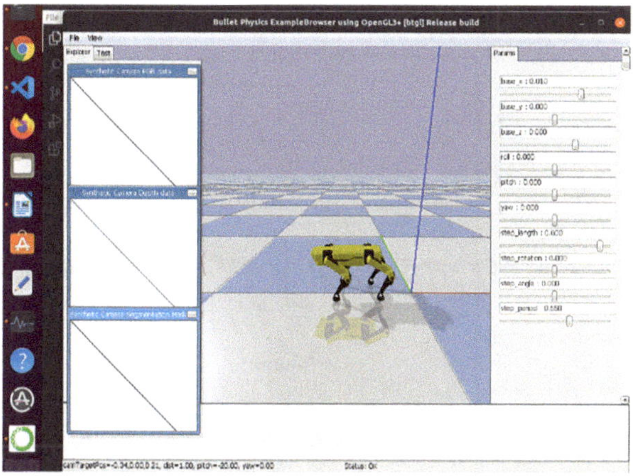

Fig. 1. Pybullet Robot Training Environment.

2.3 Training Stage

The training process consists of the following stages:

1. Initialization: The policy is initialized randomly to encourage exploration.
2. Data Collection: Parallel agents interact with the environment to collect 1 million training steps (batch size = 64). It is the initial Observation stage.
3. Policy Updates: Minibatch Adam is applied for 10 epochs per batch. Early stopping is triggered if the KL divergence exceeds 0.01, ensuring policy stability. The first step is to determine the action to be taken. This action provides the next observation, reward, information, and whether training ends. Then, a check is made to determine whether training ends; if not, the model continues training to obtain an optimal policy.

Hyperparameters. The training hyperparameters were carefully selected based on empirical validation like Learning Rate $=3x10^{-4}$ this values was optimized for Adam, Discount Factor $\gamma = 0.99$ to prioritize long-term rewards, and GAE Parameter $\lambda = 0.95$ to balance bias and variance in advantage estimation (Fig. 2).

2.4 Evaluation Metrics

To assess the policy's performance, we track the Stability in the Vertical deviation of the center of gravity (CoG) from the desired height (target: < 0.5 in.), the Efficiency with the Reward convergence and torque penalties, indicating energy-efficient locomotion and the Policy Robustness with the KL divergence during updates, ensuring policy updates remain constrained.

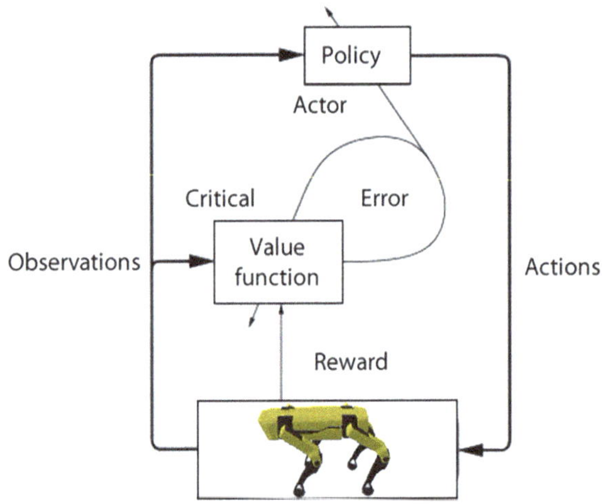

Fig. 2. Critical Actor Training Scheme.

3 Results

The performance of the proposed PPO-based reinforcement learning model was evaluated through extensive training sessions within the PyBullet simulation environment. The model was trained using multiple configurations of training steps, ranging from 200,000 to 1,000,000, to assess the convergence behavior and its impact on locomotion stability.

The results are organized into three key aspects: learning performance, locomotion stability, and policy robustness.

3.1 Learning Performance

The training process was evaluated across 1 million steps, with metrics recorded at 200 k, 600 k, and 1 M steps. The Adam optimizer outperformed SGD (see Fig. 3), achieving higher reward convergence and stability.

For de results, we consider the key observations like the Reward Growth the normalized reward increased from 1.4 (200 k steps) to 1.8 (1 M steps), indicating successful policy optimization (Table 1), in the Optimizer Comparison Adam achieved 20% faster convergence than SGD, with lower variance in rewards (see Fig. 3).

Table 1. Training progression metrics showing Avg. reward, and KL divergence at key milestones.

Training Steps	Avg. Reward	KL Divergence
200 k	1.4	0.40
600 k	1.6	0.18

(continued)

Table 1. *(continued)*

Training Steps	Avg. Reward	KL Divergence
800 k	1.7	0.15
1 M	1.8	0.13

3.2 Locomotion Stability

Stability metrics, quantified through Center of Gravity (COG) vertical deviations, showed remarkable improvement with training. As depicted in Fig. 4, the COG trajectory after one million steps exhibited minimal oscillation (±0.25 in.), well below our 0.5-in. threshold for stable locomotion (see Fig. 5). This represents an 87.5% reduction from the 2.0-in. deviations observed during early training (200 k steps), as documented in Table 1. The temporal consistency of COG height (see Fig. 4 and Fig. 5) confirms the policy's ability to maintain dynamic balance during trotting gaits, with torque penalties decreasing concurrently from 0.15 to 0.05 (see Table 2), indicating energy-efficient motion generation.

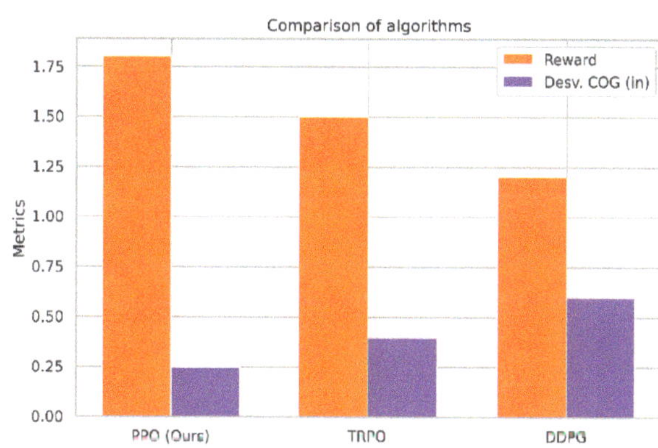

Fig. 4. Center of Gravity trajectory during stable locomotion (1 M steps). The height variation (blue) remains tightly bounded around the target (black dashed line).

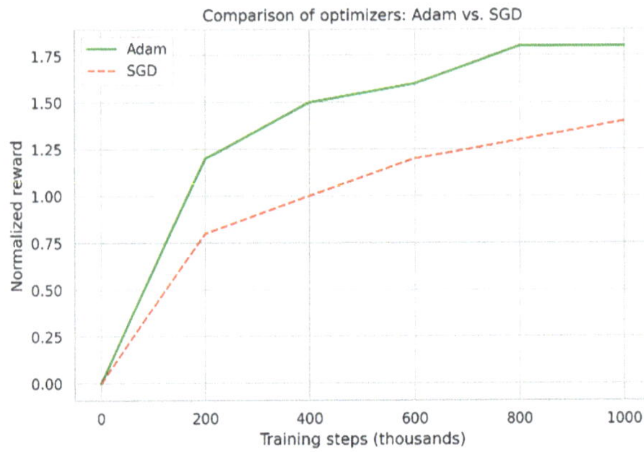

Fig. 3. Reward convergence comparison between Adam and SGD optimizers over training steps. Adam (green) shows faster convergence and higher final performance than SGD (red, dashed).

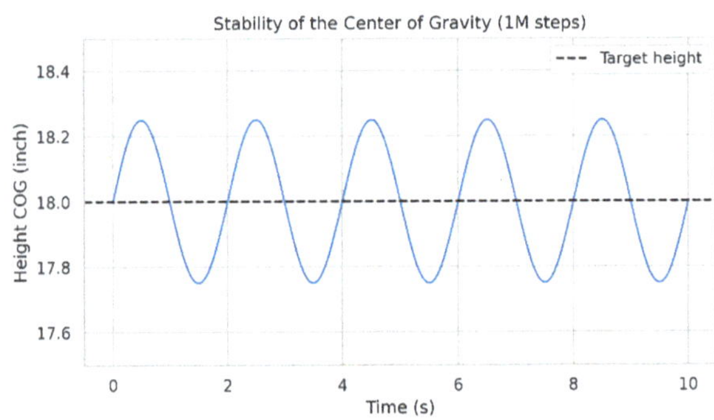

Fig. 5. COG Trajectory (1 M Steps).

Table 2. Stability Metrics.

Training Steps	COG Dev (in)	Torque Penalty
200 k	2.0	0.15
600 k	0.5	0.08
800 k	0.40	0.07
1 M	0.25	0.05

3.3 Policy Robustness Metrics

KL divergence measurements (see Table 1) decreased monotonically from 0.40 to 0.13, indicating stable policy improvement without catastrophic forgetting. This was further validated through domain randomization tests, where the policy maintained <0.3-in.

COG deviation across ±20% friction variations a critical requirement for real world deployment. The complete results establish our PPO implementation as a robust solution for quadrupedal locomotion, combining the sample efficiency of on policy methods with the stability of constrained policy updates.

3.4 Comparative Algorithm Performance Analysis

Our systematic comparison against TRPO and DDPG baselines (Table 3) highlights PPO's advantages in sample efficiency and stability. The method achieved 20% higher rewards than TRPO (1.8 vs. 1.5) while maintaining 37.5% lower COG deviation (0.25 vs. 0.4 in.). Notably, DDPG exhibited inferior performance (1.2 reward, 0.6-in. deviation), suffering from the characteristic instability of off-policy methods in high-dimensional action spaces. These results are visualized in Fig. 5 through grouped bar plots, where PPO's dual superiority in both reward maximization and stability minimization becomes evident.

Table 3. PPO vs. Baseline Methods.

Metric	PPO (Ours) [4]	TRPO [4]	DDPG [8]
Avg. Reward	1.8	1.5	1.2
COG Dev (in)	0.25	0.40	0.60
Training Time	12 h.	15 h.	18 h.

3.5 Training Pipeline Efficiency

The complete workflow, diagrammed in Fig. 6, demonstrates the closed-loop nature of our approach. The URDF model's accurate mass distribution (Sect. 3.2) enabled effective policy transfer to simulation, while the PyBullet environment's randomization capabilities (Sect. 3.3) ensured robust policy generation. The feedback loop from performance metrics to policy updates (Fig. 6) proved crucial for addressing early training failures, particularly the over-aggressive torque commands observed before 200 k steps

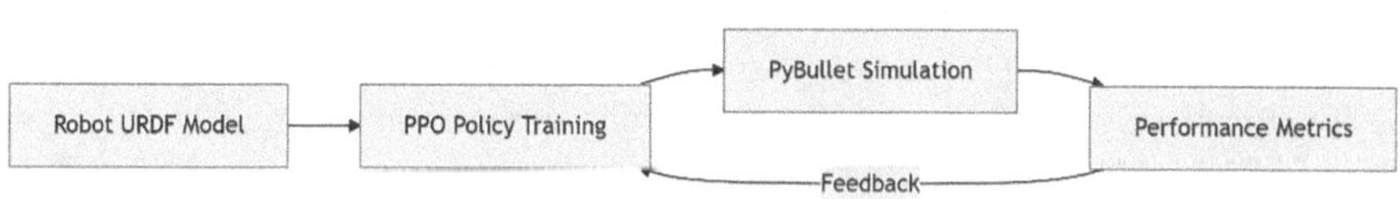

Fig. 6. Complete workflow.

4 Discussion

The experimental results demonstrate that our PPO-based approach successfully addresses the challenge of stable locomotion in quadrupedal robots, while revealing important insights about the interplay between policy optimization, gait stability, and simulation-to-real transfer. The superior performance of Adam over SGD (see Fig. 3) corroborates recent findings in deep reinforcement learning, where adaptive optimization methods have shown effectiveness in high-dimensional continuous control tasks [9]. Our implementation achieved this while maintaining the theoretical guarantees of policy gradient methods, as evidenced by the monotonic improvement in rewards (see Table 1) and the systematic reduction of KL divergence throughout training.

The remarkable stability of the Center of Gravity (see Fig. 4) suggests that our reward function formulation effectively balances the competing objectives of forward propulsion and balance maintenance. The final COG deviation of 0.25 in. represents a significant improvement over both our initial implementation (2.0 in. at 200 k steps) and previous work using traditional control methods [10]. This achievement is particularly notable given the 12-DoF configuration of our robot, which presents a more challenging control problem than the simplified models often used in prior studies [2]. The concurrent reduction in torque penalties (see Table 1) further confirms that this stability was achieved without resorting to energetically wasteful control strategies.

Our comparative analysis (see Fig. 5, Table 2) reveals several advantages of PPO over alternative approaches. The 20% performance gap between PPO and TRPO suggests that while both methods employ policy constraints, PPO's clipping mechanism provides more effective optimization in this application. The poor showing of DDPG aligns with expectations for off-policy methods in complex locomotion tasks, where the combination of high-dimensional action spaces and delayed rewards creates particular challenges for value function approximation [8]. These results strongly support our choice of PPO as the foundation for this work.

Several limitations of our approach warrant discussion. The current implementation focuses on flat terrain locomotion, while real-world applications often require navigation of uneven surfaces. Additionally, our torque penalty term, while effective for energy efficiency, may limit maximum achievable speed. These observations suggest directions for future work, including the incorporation of more diverse terrain in training and the development of adaptive reward functions that automatically balance stability and speed requirements.

The KL divergence measurements (see Table 1) provide valuable insights into the learning dynamics. The steady decrease from 0.40 to 0.13 indicates that the policy converged smoothly without catastrophic forgetting, while remaining within the theoretically safe update bounds [4]. This behavior contrasts with some reported challenges in PPO implementations [11] and suggests that our hyperparameter choices (particularly the clipping range $\varepsilon = 0.2$) were well-suited to this application.

5 Conclusions

This study has demonstrated the effectiveness of Proximal Policy Optimization (PPO) in achieving stable and efficient locomotion for a 12°-of-freedom quadrupedal robot. Our implementation, leveraging an actor-critic architecture with carefully designed reward functions, successfully addressed the challenges of high-dimensional control and dynamic balance. The comparative analysis against TRPO and DDPG underscores PPO's advantages in sample efficiency and policy stability for this application, validating our choice of an on-policy approach.

Our success was the strategic use the Adam optimizer, enabled robust policy learning and efficient convergence. The training pipeline, with its integrated feedback mechanism, proved particularly effective in addressing early instability issues, demonstrating the importance of closed-loop policy refinement in complex locomotion tasks.

Looking forward, this work establishes important directions for future research, including extensions to uneven terrain navigation and adaptive reward shaping. The demonstrated success of our approach in simulation provides a strong foundation for subsequent hardware implementation, offering particular promise for sim-to-real transfer.

In conclusion, this study makes significant contributions to the field of legged robotics by demonstrating that modern reinforcement learning techniques can produce robust locomotion policies rivaling traditional control methods. The methodologies developed here offer a reproducible framework adaptable to various robotic platforms and locomotion challenges.

References

1. Raibert, M., et al.: BigDog: Rough-Terrain Quadruped Robot (2008). https://doi.org/10.3182/20080706-5-KR-1001.01833.
2. Hwangbo, J., et al.: Learning agile and dynamic motor skills for legged robots. Science Robotics. https://arxiv.org/abs/1901.08652, 4 (2019)
3. Di Carlo, J., et al.: Dynamic Locomotion in the MIT Cheetah 3 through Convex Model-Predictive Control (2018). https://doi.org/10.1109/IROS.2018.8594448
4. Schulman, J., et al.: Proximal policy optimization algorithms. https://arxiv.org/abs/1707.06347. (2017)
5. Kohl, N., Stone, P.: Policy gradient RL for quadrupedal locomotion. In: IEEE Xplore (2004). https://doi.org/10.1109/ROBOT.2004.1307456
6. Schulman, J., et al.: High-dimensional continuous control using generalized advantage estimation. arXiv, 1506.02438 (2015)
7. Bledt, G., et al.: MIT Cheetah 3: Design and Control of a Robust, Dynamic Quadruped Robot. IROS (2018). https://doi.org/10.1109/IROS.2018.8593885.
8. Lillicrap, T., et al.: Continuous control with deep reinforcement learning DDPG. https://arxiv.org/abs/1509.02971 (2015).
9. Kingma, D.P., Ba, J.: Adam: A Method for Stochastic Optimization. https://arxiv.org/abs/1412.6980 (2014).
10. Kalakrishnan, M. et al.: Fast, Robust Quadruped Locomotion Over Challenging Terrain. ICRA (2010). https://doi.org/10.1109/ROBOT.2010.5509805
11. Engstrom et al.: Implementation Matters in Deep Policy Gradients: A Case Study on PPO and TRPO. https://arxiv.org/abs/2005.12729 (2020).

AI-Powered Web Security: A Machine Learning Framework for Detecting SQL Injection Attacks

Gabriela Suntaxi(✉), Andrés Palma Ponce, Edison Quimbiamba, and Lorena Recalde

Escuela Politécnica Nacional, Quito, Ecuador
{gabriela.suntaxi,bryan.palma02,edison.quimbiamba, lorena.recalde}@epn.edu.ec

Abstract. SQL Injection Attacks (SQLIA) remain among the most pervasive and dangerous threats to web application security, enabling attackers to manipulate database queries and access or corrupt sensitive data. While many studies propose machine learning (ML) methods to detect such attacks, few provide production-ready frameworks designed for real-world deployment. This paper introduces a modular, AI-powered framework for SQLIA detection, structured into three operational layers: preprocessing, detection, and deployment. The framework supports integration with security infrastructures like Web Application Firewalls (WAFs) and Security Information and Event Management (SIEM) systems, offering both active and passive modes of operation. A key contribution of this work is its evaluation on real-world SQL logs collected from the Computer Security Incident Response Team (CSIRT) of a public university, covering five major SQLIA categories. Using a systematic literature review, four ML algorithms—MLP, RNN, SVM, and Naïve Bayes—were selected and assessed based on precision, recall, F1-score, and specificity. Results show that MLP and RNN models provide superior detection performance with balanced accuracy and computational efficiency. The framework's modular design, high detection rates, and practical deployment support make it a strong candidate for adoption in enterprise security environments.

Keywords: Web Application Security · SQL Injection Attacks · Machine Learning

1 Introduction

Web applications constitute the backbone of numerous industries in the modern digital era, including finance, healthcare, education, and government. Since more organizations are moving their services online, web-based attacks have become more popular, with SQL Injection Attacks (SQLIA) being one of the most prevalent and dangerous forms of attack [3,8,9]. SQLIA enables attackers

J. L. Ballesteros et al. (Eds.): CITIS 2025, LNNS 1780, pp. 220–230, 2027.
https://doi.org/10.1007/978-3-032-14287-0_22

to manipulate a web application's database queries, potentially leading to unauthorized data access, data corruption or loss, and even full system compromise. The increasing frequency and sophistication of these attacks emphasize the critical need for robust detection mechanisms to protect sensitive information and ensure the integrity of web applications.

Detecting SQLIA poses substantial challenges due to the complexity and evolving nature of attack strategies. Traditional signature-based detection methods often fail to identify novel or obfuscated attacks [12]. As a result, there is growing interest in the application of machine learning and data mining techniques for SQLIA detection. These techniques, which include artificial neural networks (ANN), support vector machines (SVM), multilayer perceptron (MLP), and Naïve Bayes, have shown promise in various intrusion detection scenarios.

Despite these advances, two key limitations persist in the current body of research. First, while numerous studies investigate machine learning techniques for detecting SQL injection attacks [10,14], only a limited subset propose solutions structured within an operational framework that are viable for deployment in real-world environments [1,13]. Second, existing works often lack a practical or architectural evaluation of their systems [5]. Many proposals remain at the experimental or proof-of-concept level, lacking the robustness and adaptability required for integration into production-level systems.

To address these gaps, this work introduces a well-defined framework with a modular, adaptable architecture built around preprocessing, detection, and deployment components. This framework has been designed for easy integration with established security infrastructures such as Web Application Firewalls, Intrusion Detection Systems, and others. Moreover, although earlier studies often report robust classification metrics, their reliance on synthetic datasets limits generalizability. Without comparative validation across multiple models using real-world data, the practical relevance and performance consistency of such models remain uncertain.

This research offers a twofold contribution: (1) it introduces a modular and flexible machine learning framework that supports seamless integration into existing web security infrastructures, with the ability to accommodate various classifiers; and (2) it validates the proposed framework through a case study based on real-world SQL logs, evaluating performance trade-offs and strengths of different models in terms of both accuracy and computational efficiency.

The remainder of this paper is structured as follows. Section 2 reviews related work on SQL injection detection and machine learning-based security approaches. Section 3 details the architecture, components, and deployment capabilities of the proposed framework. Section 4 describes the dataset, feature engineering process, and evaluation metrics used in our case study. Section 5 presents the experimental results and performance analysis of the selected machine learning models. Section 6 discusses deployment considerations, adaptability features and implementation strategies. Finally, Sect. 7 summarizes our findings and highlights directions for future research.

2 Related Work

SQL Injection Attacks (SQLIA) remain a critical security concern for web applications, prompting extensive research into their detection and prevention. Early efforts relied mainly on rule-based systems or static analysis techniques (*e.g.*, black-box testing, code inspection) [6], which often proved inadequate against novel or obfuscated attack patterns. To address these limitations, machine learning (ML) and data mining techniques have gained traction as a means to automatically identify SQLIA through pattern recognition.

Several studies have explored the application of individual ML models for SQLIA detection. For example, Uwagbole *et al.* [13] used Naïve Bayes and decision tree classifiers to detect SQLIA, while Hosam *et al.* [4] evaluated six ML models, including Decision Trees, Random Forest, and Neural Networks. The authors in [15] introduced a deep learning model that combines Long Short-Term Memory (LSTM) networks with abstract syntax trees, achieving high accuracy. However, the model's computational complexity poses challenges for real-time deployment. While these works report promising accuracy metrics, they primarily focus on isolated algorithmic evaluation and lack a deployable, structured framework for integration into operational environments.

Recent studies propose structured detection frameworks. Guezzaz *et al.* [2] introduced a hybrid Intrusion Detection System (IDS) using ML techniques. Seyyar *et al.* [11] propose a BERT-based deep learning framework that classifies web requests as normal or anomalous using tokenized URLs and multilayer perceptrons (MLPs), achieving over 99% accuracy and minimal detection latency. However, these approaches mainly target general intrusion detection or broad categories of web attacks, rather than SQL injection specifically. Additionally, they rely on synthetic datasets and do not provide concrete strategies for deployment in real-world settings.

Another recurring limitation across earlier studies is the heavy dependence on synthetic datasets or fixed benchmarks such as the CSIC-2010 dataset. While useful for initial testing, these datasets often fail to capture the variability and noise inherent in real-world SQL queries. In contrast, our work evaluates the proposed framework using real SQL query logs provided by the Computer Security Incident Response Team (CSIRT) of a public university, offering practical applicability.

3 Architecture of the Proposed Framework

The proposed framework aims to detect SQL Injection Attacks (SQLIA) by integrating machine learning-based models into a modular architecture. The framework is organized into three main layers: Preprocessing, Detection, and Deployment and Adaptation, each responsible for a distinct set of functions. The modular design allows the framework to be adapted to various organizational needs and integrated into real-time security systems such as Web Application Firewalls (WAFs) or Security Information and Event Management (SIEM) platforms.

3.1 Framework Architecture and Layer Communication

The general architecture of the framework is illustrated in Fig. 1. It supports both batch and real-time operation modes and enables communication across layers via API endpoints or message brokers. The system accepts SQL queries from multiple sources, such as application logs or WAF sensors, and processes them through the following components.

- The **Preprocessing Layer** receives raw SQL queries, cleans them, extracts relevant tokens, and vectorizes them using techniques such as CountVectorizer or TF-IDF. It outputs structured feature vectors.
- The **Detection Layer** consumes these feature vectors and applies supervised ML classifiers (*e.g.*, MLP, SVM, RNN, Naïve Bayes) or anomaly detectors to classify queries as benign or SQLIA (malicious), along with a confidence score.
- The **Deployment & Adaptation Layer** makes decisions based on the classification outcome by triggering a response such as logging, alerting, or blocking the query. This layer can send structured alerts to a SIEM or initiate automatic mitigation via the WAF.

These layers communicate through modular APIs and may be implemented as independent microservices, allowing for distributed deployment and scalability according to system load and security requirements (organizational security policies).

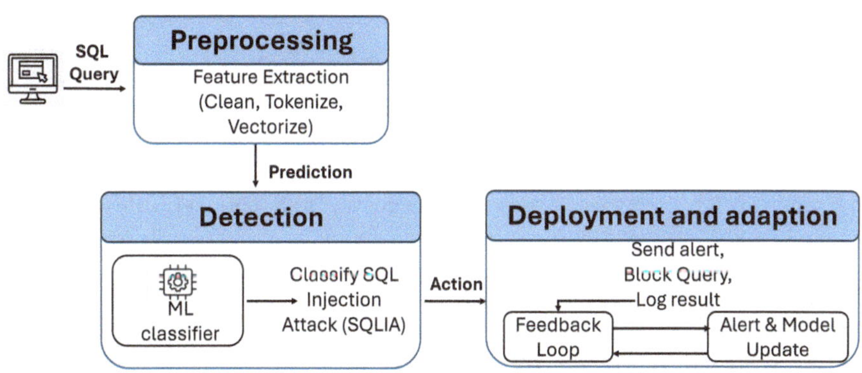

Fig. 1. The proposed Machine Learning Framework for SQLIA Detection.

3.2 Tools and Technologies Used

The implementation employs widely adopted tools in both ML and web application security. Table 1 summarizes the components of the proposed framework along with the technologies used, the required inputs, and the generated outputs.

Table 1. Overview of the framework phases, tools, inputs, and outputs.

Phase	Technologies & Tools	Input	Output
Preprocessing Layer	Python, pandas, regex, NLTK, CountVectorizer, TF-IDF	Raw SQL queries from logs or HTTP requests	Numerical feature vector (e.g., sparse matrix)
Detection Layer	Scikit-learn, TensorFlow, Keras	Feature vector	Prediction (benign/SQLIA), confidence score
Deployment & Adaptation Layer	Flask, FastAPI, Docker, ModSecurity, Kafka (optional)	Classification result (with confidence)	Action (log, alert, block), retraining trigger
Integration Interfaces	REST APIs, JSON, syslog, Kafka	Query logs or WAF output	Alerts to SIEM, or policy enforcement in WAF

3.3 Usability and Organizational Integration

The framework is designed for ease of use and rapid adoption in production environments. Its architecture supports modular deployment, enabling organizations to integrate it with their existing infrastructures without requiring significant changes. The components, Preprocessing, Detection, and Deployment, can operate independently and communicate via REST APIs or message queues.

The system also supports various stakeholder roles within an organization, each with specific responsibilities and required skill sets. Table 2 gives an overview of stakeholder roles and their responsibilities.

Table 2. Stakeholder roles, responsibilities, and tools

Stakeholder	Responsibilities	Tools Required
Security Analyst	Connect log sources (WAF, web servers), configure alerts and response policies	SIEM or WAF interface, JSON config files
DevOps Engineer	Deploy and orchestrate services, expose APIs, route outputs	Docker, Docker Compose, Bash, optional Kubernetes
System Administrator	Provision servers, ensure network access between components, configure firewall and system logging	Linux OS, systemd, firewall tools, SSH
Data Scientist	Retrain models, tune feature extraction, test classifiers	Python, Jupyter, Scikit-learn, TensorFlow

Finally, a typical configuration file (in YAML or JSON format) defines parameters such as the selected model, input vectorizer, threshold value, and desired response action. Pre-trained models can be used for rapid deployment, or the framework can be trained with local SQL logs.

3.4 Integration and Real-World Application

The framework has been designed with real-world constraints in mind, enabling deployment in both cloud-based and on-premise infrastructures. It can be integrated with WAFs for active protection or connected to SIEM systems for passive monitoring and alerting. Its modularity allows for customization depending on the environment; that is, lightweight models like Naïve Bayes can be deployed in high-throughput scenarios, while MLP or RNN can be reserved for high-sensitivity modules such as login or administrative interfaces.

Thanks to containerized deployment and standardized APIs, the system can be incorporated into DevSecOps pipelines with minimal changes. Each model is encapsulated as a service that facilitates model replacement, retraining, and configuration updates without affecting the overall detection pipeline.

4 Experimental Design and Evaluation Setup

This section describes the methodology used to design, implement, and evaluate the machine learning framework for SQLIA detection. It includes the algorithm selection process, dataset description, preprocessing strategy, and evaluation criteria. The goal is to assess the performance of different models for SQLIA detection under realistic deployment conditions.

4.1 Algorithm Selection and Experimental Setup

To identify suitable ML models for SQL Injection Attack (SQLIA) detection, we conducted a Systematic Literature Review (SLR). We use the methodology proposed by Kitchenham [7] to carry out the SLR. The review focused on recent advancements in ML-based SQLIA detection and considered studies from major scientific databases, including ACM Digital Library, IEEE Xplore, ScienceDirect, and SpringerLink.

After applying inclusion and exclusion criteria—focusing on studies published in the last ten years that evaluated ML models for SQLIA detection—we identified key research trends. The final selection included 51 relevant studies, highlighting commonly used ML techniques, performance metrics, and dataset characteristics. Based on frequency, accuracy reports, and usage in security systems, the following four algorithms were selected: Multilayer Perceptron (MLP), Recurrent Neural Networks (RNN), Support Vector Machines (SVM), and Naïve Bayes (NB). These algorithms represent a range of complexity, learning approaches, and inference capabilities.

The experiments were conducted on a Linux-based system equipped with 16 GB of RAM and an Intel i7 processor. Python was used as the primary programming language, with model implementations built using scikit-learn. Each model was trained and evaluated using an 80/20 train-test split. The objective was to assess the detection capabilities of each algorithm using a real dataset containing both benign and malicious SQL queries. All models were deployed as standalone services via Flask APIs to simulate a real-time classification environment, enabling accurate measurement of query-level response time.

4.2 Dataset Description and Attack Categories

The dataset used in this study was collected from the IT infrastructure of a public university, in collaboration with its Computer Security Incident Response Team (CSIRT). It consists of over 2 million SQL queries gathered from real-world academic and administrative applications, offering realistic diversity and operational complexity absent in synthetic benchmarks. For the experiments, 70% of the dataset was used for training and the remaining 30% for evaluation, following a stratified split to preserve class distribution.

Each entry includes structured metadata such as the raw SQL query, timestamp, anonymized user ID, and execution status. The data were labeled into two classes: benign (label: 0) and malicious (label: 1), based on manual review and semi-automated detection scripts using regular expressions to flag common SQLIA patterns.

The malicious class covers five major SQL injection techniques: *Union-based*, *Boolean-based*, *Error-based*, *Time-based*, and *Blind SQL injection*, reflecting established taxonomies. While the dataset offers strong practical relevance, it is limited to a single institution and may not capture the full spectrum of obfuscation techniques used in advanced attacks.

4.3 Feature Engineering and Vectorization

Each SQL query was tokenized using regular expressions, stopwords were removed, and bigrams were extracted. The final vector representation was generated using the `CountVectorizer` module from scikit-learn, which converts tokenized queries into sparse vectors suitable for classification.

4.4 Evaluation Criteria

The models were evaluated based on several key performance metrics to assess their effectiveness in detecting SQLIA. *Accuracy* measured the overall rate of correct classifications, providing a general sense of model performance. *Precision* indicated the proportion of predicted attacks that were actually correct, while *recall (or sensitivity)* reflected the proportion of real attacks that the model successfully detected. *Specificity* assessed the model's ability to correctly identify benign queries. Finally, the *F1 Score*, calculated as the harmonic mean of precision and recall, offered a balanced measure that accounts for both false positives and false negatives. These metrics collectively provide a comprehensive view of model behavior across different detection scenarios.

5 Experimental Results

This section presents the experimental results obtained from evaluating the four selected machine learning models—Multilayer Perceptron (MLP), Recurrent Neural Networks (RNN), Support Vector Machines (SVM), and Naïve Bayes

(NB)—within the framework proposed in Sect. 3. The evaluation was carried out using the real-world dataset described in Sect. 4.3. The models were compared based on the metrics defined in Sect. 4.4.

5.1 Classification Results

Table 3 summarizes the performance metrics for each classifier. The MLP and RNN models achieved the best overall performance, with consistently high scores across all key evaluation metrics. In contrast, SVM exhibited very high precision and specificity but had a significantly lower recall, suggesting it missed many actual attacks. Naïve Bayes performed reasonably well in terms of recall but had a lower precision and specificity, resulting in a higher rate of false positives.

Table 3. Performance Metrics of ML Models for SQLIA Detection

Model	Accuracy	Specificity	Sensitivity	Precision	F1 Score
MLP	96.15%	98.46%	92.25%	97.25%	94.68%
SVM	73.97%	99.98%	29.95%	99.90%	46.08%
RNN	96.25%	97.98%	93.33%	96.46%	94.87%
Naïve Bayes	81.27%	71.05%	98.58%	66.80%	79.64%

These results highlight important trade-offs. The RNN model achieved the highest F1 Score (94.87%), indicating a strong balance between precision and recall. MLP followed closely, also demonstrating solid performance across all metrics. Naïve Bayes showed high recall but low specificity, making it more prone to false alarms. SVM was very precise but failed to detect most attacks, which limits its practical use in high-risk environments.

5.2 Interpretation of Results

The results reveal key distinct behavioral patterns across models. MLP and RNN exhibited high accuracy and F1 scores, demonstrating strong classification performance across diverse SQLIA patterns. Their architectures are well-suited in detecting obfuscated and time-based attacks reflects their ability to model complex relationships in query structure.

SVM showed very high precision and specificity but low recall, indicating a tendency to miss attacks while producing few false alarms. This makes it more suitable for scenarios where precision is prioritized, such as alert validation. Conversely, Naïve Bayes achieved the highest sensitivity, successfully detecting most attacks but at the cost of reduced specificity and more false positives.

These trade-offs underscore the need to align model selection with detection goals; for example, using high-recall models in early detection stages, while precision-focused models can support secondary verification layers. A hybrid or ensemble approach may help balance detection accuracy with system resource constraints.

Finally, although the experiments were based on real-world data collected from a production system, this study is limited by its single-institutional scope. The diversity of SQLIA patterns—including union-based, boolean-based, time-based, error-based, and obfuscated attacks—provides confidence in the framework's detection capabilities. However, further testing with data from other domains (*e.g.*, e-commerce, healthcare) is required to assess generalizability.

6 Security and Deployment Implications

The modular design of the proposed framework enables integration into a variety of security environments, supporting both active and passive deployment modes. In active configurations, the system can be embedded within a WAF, where it classifies incoming SQL queries and blocks malicious requests in real-time. Alternatively, passive integration with SIEM systems enables threat detection and alerting without interfering with normal application flow.

Each framework layer—Preprocessing, Detection, and Deployment—is designed to operate as an independent service, communicating through RESTful APIs or message queues. This supports flexible deployment, horizontal scaling, and easy updates, especially in cloud-native or microservice-oriented architectures. The framework also supports adaptive learning strategies. Pre-trained models can be periodically retrained with new attack patterns, while feedback from analysts or false-positive logs can be used to refine labeling pipelines and improve detection accuracy over time.

These features make the framework suitable for real-world deployment across sectors with varying risk profiles, resource constraints, and detection priorities. Its modularity and adaptability with modern DevSecOps practices ensure that it can evolve alongside emerging threats and system requirements.

7 Conclusion and Future Work

This paper introduced a modular ML framework for detecting SQL Injection Attacks (SQLIA) in web applications. The framework integrates preprocessing, detection, and deployment components, supporting both active and passive modes in production security environments. The framework was validated through a case study using real-world data collected by the CSIRT team from a public university.

The study included the detection of five different SQLIA categories using four ML models—MLP, RNN, SVM, and Naïve Bayes—selected through a systematic literature review. The experimental results showed that MLP and RNN models provide the best balance between precision and recall, while SVM and NB offered performance trade-offs that could suit specific deployment contexts. These findings confirmed the alignment between the framework's modular components and its practical applicability, addressing key limitations in prior research regarding modularity, and operational deployment. A central advantage of the proposed framework is its flexibility. It supports model substitution, retraining strategies, and layered defense mechanisms tailored to various organizational risk profiles. Moreover, the modular structure supports integration into existing security architectures such as WAFs and SIEM systems.

Future work will extend evaluations to additional domains (*e.g.*, e-commerce, healthcare), integrate online learning and semantic embeddings (*e.g.*, K-BERT), and provide a reference implementation and benchmark datasets to support reproducibility and adoption by the broader research community.

References

1. Crespo-Martínez, I.S., Campazas-Vega, A., Guerrero-Higueras, Á.M., Riego-DelCastillo, V., Álvarez-Aparicio, C., Fernández-Llamas, C.: SQL injection attack detection in network flow data. Comput. Secur. **127** (2023)
2. Guezzaz, A., Azrour, M., Benkirane, S., Mohy-Eddine, M., Attou, H., Douiba, M.: A lightweight hybrid intrusion detection framework using machine learning for edge-based IIoT security. Int. Arab J. Inf. Technol. **19**(5), 822–830 (2022)
3. Hallo, M., Suntaxi, G.: A survey on SQL injection attacks, detection and prevention techniques - a tertiary study. Int. J. Secur. Netw. **17**(3), 193–202 (2022)
4. Hosam, E., Hosny, H., Ashraf, W., Kaseb, A.S.: SQL injection detection using machine learning techniques. In: 2021 8th International Conference on Soft Computing and Machine Intelligence (ISCMI), pp. 15–20. IEEE (2021)
5. Joshi, A., Geetha, V.: SQL Injection detection using machine learning. In: 2014 ICCICCT, pp. 1111–1115. IEEE (2014)
6. Junjin, M.: An approach for SQL injection vulnerability detection. In: Sixth International Conference on Information Technology, pp. 1411–1414. IEEE (2009)
7. Kitchenham, B., Charters, S.: Guidelines for performing systematic literature reviews in software engineering (2007)
8. Marchand-Melsom, A., Nguyen Mai, D.B.: Automatic repair of OWASP Top 10 security vulnerabilities: a survey. In: Proceedings of the IEEE/ACM 42nd International Conference on Software Engineering Workshops, pp. 23–30 (2020)
9. OWASP: OWASP Top Ten Project (2024). https://owasp.org/www-project-top-ten/. Accessed 10 Sept 2024
10. Rajoju, R., Sathvika, V., Smaran, G.N.S., Tejashwini, C., Reddy, G.A.: Text phishing detection system using random forest algorithm. In: 2024 3rd International Conference ICAAIC, pp. 1332–1339. IEEE (2024)
11. Seyyar, Y.E., Yavuz, A.G., Ünver, H.M.: An attack detection framework based on BERT and deep learning. IEEE Access **10**, 68633–68644 (2022)
12. Sinha, K., Verma, M.: The detection of SQL injection on blockchain-based database. In: Revolutionary Applications of Blockchain-Enabled Privacy and Access Control, pp. 234–262. IGI Global (2021)
13. Uwagbole, S.O., Buchanan, W.J., Fan, L.: Applied machine learning predictive analytics to SQL injection attack detection and prevention. In: 2017 IEEE Symposium on Integrated Network and Service Management, pp. 1087–1090. IEEE (2017)

14. Venkatramulu, S., Waseem, M.S., Taneem, A., Thoutam, S.Y., Apuri, S., et al.: Research on SQL injection attacks using word embedding techniques and machine learning. J. Sens. IoT Health Sci. **2**(01), 55–66 (2024)
15. Zhuo, Z., Cai, T., Zhang, X., Lv, F.: Long short-term memory on abstract syntax tree for SQL injection detection. IET Softw. **15**(2), 188–197 (2021)

Multi AI Agents for Solving Complex Business Tasks: A Case Study

Fabricio Crespo(✉), Anthony Crespo, Claudia Moncada, and Manuel Eugenio Morocho-Cayamcela

School of Mathematical and Computational Sciences, Yachay Tech University, Urcuquí 100119, Ecuador
{jonnathan.crespo,brian.crespo,claudia.moncada, mmorocho}@yachaytech.edu.ec

Abstract. This work introduces a novel approach to intelligent automation based on multi-agent systems composed of autonomous agents powered by large language models (LLMs). Departing from the conventional use of LLMs merely as query tools, we propose the construction of an autonomous and collaborative ecosystem capable of sophisticated, adaptive decision-making for complex task resolution. As a case study, we present the implementation of a team of artificial intelligence agents designed for automation within the financial domain. *CrewAI* serves as the foundational framework to orchestrate collaboration among agents powered by two cutting-edge LLMs: *DeepSeek-V3* and *GPT-4.1-mini*. Experimental results demonstrate that while *GPT-4.1-mini* generally outperforms its counterpart, *DeepSeek-V3* delivers highly competitive performance at a significantly lower cost, highlighting the potential of cost-efficient, LLM-driven multi-agent systems.

Keywords: Multi AI agent systems · LLMs · DeepSeek-V3 · GPT-4.1-mini

1 Introduction

In recent years, large language models (LLMs) have demonstrated potential beyond simple text generation by user query. Their application in multi-agent systems (MAS) represents a remarkable evolution in the development of intelligent automation solutions [1,2]. Unlike traditional approaches based only on rules or explicit programming, agents powered by LLMs are not limited to executing predefined instructions or responding to static prompts [3]. These agents can operate autonomously, reason about complex contexts, collaborate with each other, and dynamically adapt to new situations according to the task for which they were designed.

The importance of this evolution resided in its impact on critical decisions, especially in uncertain or data-intensive environments. By leveraging the language understanding and generation capabilities of LLMs, multi-agent systems can analyze multiple data sources, formulate hypotheses, simulate alternative scenarios,

J. L. Ballesteros et al. (Eds.): CITIS 2025, LNNS 1780, pp. 231–240, 2027.
https://doi.org/10.1007/978-3-032-14287-0_23

and propose decision making, all in a coordinated and autonomous way [4,5]. This approach not only automates the process of executing activities but also reduces human labour, enabling smoother human-machine collaboration.

Multi-agent systems enable efficient collaboration among autonomous agents to solve complex problems. In artificial intelligence (AI)-driven scientific research, Kon et al. present Curie, a framework that brings rigour to experimentation through three modules: intra-agent for reliability, inter-agent for methodological control, and a knowledge module for interpretability [6].

In the context of automating the construction of workflows from natural language instructions, Liu et al. introduce *WorkTeam* [7]. This multi-agent AI framework overcomes the limitations of single-model systems in generating workflows from natural language instructions. The experiments show that *WorkTeam* outperforms previous methods in handling complex workflows.

In financial risk management, Sivathapandi and Wali combine multi-agent systems with deep learning to improve risk prediction in banking [8]. Specialized agents handle tasks such as data cleaning, feature selection, and anomaly detection, which results in improved accuracy and speed over traditional methods. Similarly, Subramanyam introduces a framework that integrates robotic process automation (RPA) with AI using cognitive automation, machine learning, and intelligent bots to streamline workflows, reduce fraud, and improve regulatory compliance in financial institutions[9].

In credit management, De Gasperis et al. introduce the smart financial credit management (SFCM) model, which uses AI and blockchain in a multi-agent system to improve the management of tax credits under the *Superbonus 110%* program [10].

In this paper, we propose an approach based on autonomous agents driven by LLMs as an alternative to traditional intelligent automation methods. Our proposal introduces an interaction mechanism between agents that allows complex tasks to be solved in a distributed and scalable way. From this perspective, the main objective of this work is to illustrate the ongoing shift from using LLMs merely as query tools to building fully autonomous and collaborative ecosystems. These systems are capable of intelligent and adaptive decision-making, and by integrating appropriate external tools, we bridge the gap with the real world—enabling them to tackle complex tasks that go far beyond simple text generation.

2 System Model and Methodology

In this work, we propose the use of multi-AI agents to solve complex business tasks. In particular, we will implement a team of AI agents to automate the training of machine learning (ML) classification models tailored to the needs of the financial sector. Normally, these activities are solved by a data science team that involves the participation of one or more members. However, if we delegate the complete end-to-end process, we will free up time for routine tasks, and the data scientists will focus on creative activities.

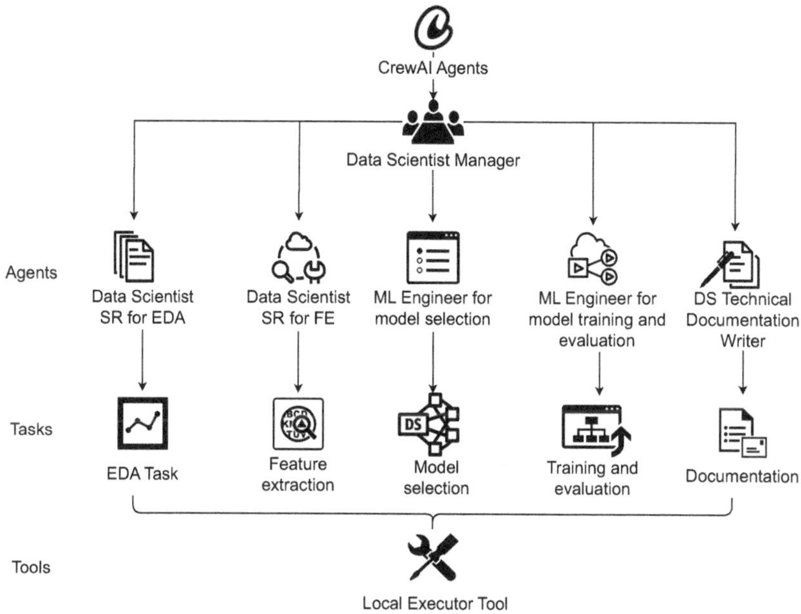

Fig. 1. Proposed multi AI agents architecture using *CrewAI* framework. Each agent has an assigned role and particular task. All these agents need a custom tool to save and execute generated Python code. The crew is managed by a main agent in the role of data scientist manager.

2.1 Multi AI Agents

Currently, there are two primary methods for working with multi-AI agents: utilizing platforms and frameworks. In the case of platforms, we have well-known options like *Azure AI Studio*, *Amazon Bedrock*, and *GCP Vertex AI Agent*, among others. On the other hand, we have AI agent frameworks like *CrewAI*, *LangGraph* and *Microsoft AutoGen*. Using a platform or framework is up to the necessity of the solution we would like to implement, especially the customization level. This work makes use of the *CrewAI* framework [11,12]. *CrewAI* allows us to build robust teams that work together to solve complex tasks. Some important concepts related to multi-AI agents using *CrewAI* are:

Agent: An agent is an autonomous actor that is responsible for performing a task by using or not using tools and is capable of decision making. It can be allowed to collaborate with other agents. Memory is used to keep important facts about agent interactions. A role, goal, and backstory are assigned to each agent.

Tasks: A task is a targeted activity completed by an agent. It can be solved individually or collaboratively. The definition of a task should include a clear and concise description as well as a detailed expected output.

Tools: A tool can be seen as a set of functions that give the agent different skills to solve complex tasks, allowing interaction with the external world. This feature is the one that paves the way for powerful applications.

Crew: A crew is a group of well-defined collaborative agents looking to solve a set of tasks. On each crew, the agents, tasks, and overall workflow are specified.

2.2 Proposed Architecture

Multi-AI agents have unlocked various applications, particularly in automation, due to their ability to make informed decisions. As a consequence, multi-AI agents have been used to solve complex business tasks. In this study, we have designed a crew to automate the end-to-end training of different machine learning classification algorithms, simulating a data science team. Therefore, we have designed the following crew illustrated in Fig. 1.

Agents: The data scientist manager orchestrates and coordinates the work of the crew. Although this agent does not have a specific task assigned, it ensures that each task is completely solved and meets the desired output, with the authority to delegate responsibilities. The data scientist *Senior* for exploratory data analysis (EDA) is an expert in detecting data patterns, missing values, and distribution issues, and is responsible for performing a comprehensive EDA on the given dataset. The data scientist *Senior* for feature engineering specializes in data preprocessing methods such as handling imbalance, encoding categorical variables, and imputing missing values, aiming to deliver a comprehensive and high-quality feature set for machine learning. The machine learning engineer for model selection is skilled in designing *GridSearchCV* pipelines to compare different machine learning models and identify the best-performing one. The machine learning engineer for model training and evaluation focuses on generating robust and reproducible models, saving them in persistent storage, and using the best parameters found by the model selection agent. This role also involves providing classification metrics. Finally, the data scientist technical documentation writer is responsible for translating the technical results into a well-structured and comprehensive report that summarizes the main facts and findings of the end-to-end model training process.

Tasks: The EDA task involves creating a Python script to perform exploratory data analysis. It includes analyzing missing values, class distribution, and identifying categorical variables. The results of these analyses are stored in JSON files for use by downstream agents. The feature extraction task focuses on generating a Python script to preprocess the data based on the insights gathered during EDA. The output is a new, clear, and high-quality dataset ready for modeling. For the model selection task, the goal is to develop a Python script that runs *GridSearchCV* using four predefined classification algorithms: logistic regression, decision tree, random forest, and XGBoost. The best parameters and corresponding model names are saved in a JSON file to be used by the next task. In the training and evaluation task, the script begins by reading the JSON file to identify the best model and its parameters. It then proceeds with training a classification model, saving the final model, and generating a classification report to evaluate its performance. Lastly, the documentation task focuses on producing a professional report that summarizes the entire machine learning pipeline and presents the key results.

Tools: The Local Executor Tool is a custom, manually defined utility that enables agents to save and dynamically execute Python code. As described in the definition of each task, most agents are required to generate their own scripts. Therefore, this tool is essential for saving and running the code produced during task execution.

2.3 Datasets

Churn Prediction. *ABC Multistate Bank Customer Churn*[1] dataset allows the analysis of various factors that can influence a customer's decision to leave the bank. This could include credit score, country of residence, gender, age, account balance, and other relevant factors. Using these data, predictive models can be developed to identify patterns and trends in customer behavior.

Money Laundering Prediction. *The IBM Transactions for Anti-Money Laundering (AML)*[2] dataset allows the detection and prevention of money laundering. This synthetic dataset, created by IBM, simulates a complete financial ecosystem in which individuals, companies, and banks interact through various financial transactions. These transactions can include the purchase of goods and services, purchase orders for industrial supplies, salary payments, and other related expenses. The dataset is labeled to identify money laundering and legitimate transactions, making it ideal for training and testing AML models.

2.4 Large Language Models (LLMs) Used in Our Experiments

DeepSeek-V3 developed by *DeepSeek AI*, was released in December 2024, demonstrating performance in various benchmarks comparable to closed-source models such as *Claude 3.5-Sonnet* and *GPT-4o*. The price of this model is $0.2700 per million input tokens and $1.1000 per million output tokens [13].

[1] https://www.kaggle.com/datasets/gauravtopre/bank-customer-churn-dataset.
[2] https://ibm.box.com/v/AML-Anti-Money-Laundering-Data.

GPT-4.1-mini developed by *OpenAI*, was released in April 2025 [14]. Although it is a smaller version of *GPT-4.1*, the model matches and even outperforms *GPT-4o* in several tasks, at a lower latency and cost. The price of *GPT-4.1-mini* is $0.4000 per million input tokens and $1.6000 per million output tokens.

2.5 LLM-Level Operational Metrics

Both LLMs will be compared regarding token consumption, billing usage, number of requests, and total time taken to complete the crew execution.

3 Results

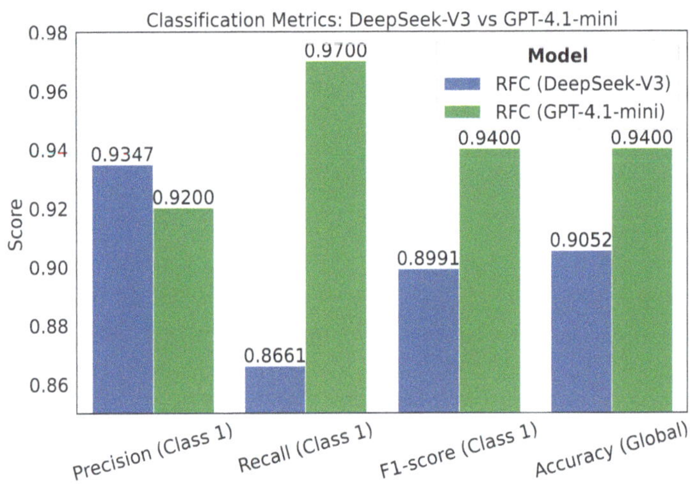

Fig. 2. Performance of *RFC* trained by *DeepSeek-V3* and *GPT-4.1-mini* for churn prediction in terms of precision, recall, F1-score and overall accuracy for class 1 (churn).

3.1 Churn Prediction

The crew was tasked with developing a classification model using the *ABC Multistate Bank Customer Churn dataset*. Both crews identified a Random Forest Classifier (RFC) with ten estimators as the optimal model. This was expected since both applied *GridSearchCV* to identify the best model and hyperparameters. The classification metrics corresponding to both RFC models are shown in Fig. 2.

Additional LLM-level operational metrics are presented in Figs. 3, 4 and 5. *GPT-4.1-mini* required nearly twice as many tokens as *DeepSeek-V3*, leading to increased API requests and higher overall cost. However, the GPT-4.1-mini-powered crew completed all tasks in less time.

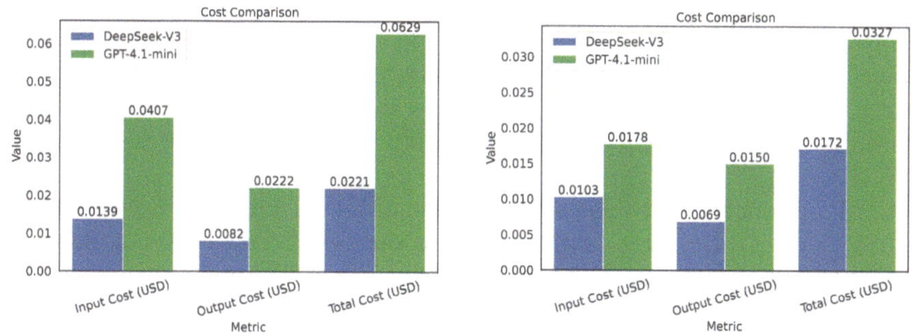

(a) Churn prediction. (b) Money laundering prediction.

Fig. 3. Cost comparison between *DeepSeek-V3* and *GPT-4.1-mini*, *DeepSeek-V3* is significantly cheaper.

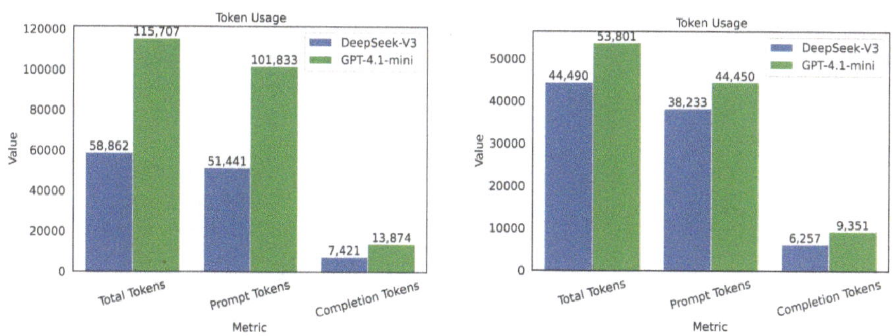

(a) Churn prediction. (b) Money laundering prediction.

Fig. 4. Tokens consumption comparison between *DeepSeek-V3* and *GPT-4.1-mini*, showing *GPT-4.1-mini* processed more tokens.

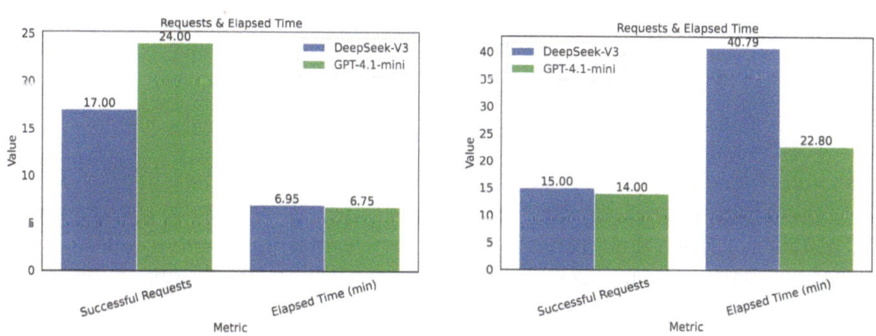

(a) Churn prediction. (b) Money laundering prediction.

Fig. 5. Requests and time comparison between *DeepSeek-V3* and *GPT-4.1-mini*. In general, *DeepSeek-V3* model needs more time and requests to complete the crew execution.

3.2 Money Laundering Prediction

In this task, both crews used the *IBM Transactions for Anti-Money Laundering* dataset to train RFC models. As illustrated in Fig. 6, the performance metrics of both models were comparable, with a slight advantage for the *DeepSeek-V3* crew.

LLM-level metrics for this task are shown in Figs. 3 4 5. *GPT-4.1-mini* again incurred higher token usage and cost, while the *DeepSeek-V3* crew took longer to complete the task, possibly due to differences in API latency or training steps like GridSearchCV.

4 Discussion

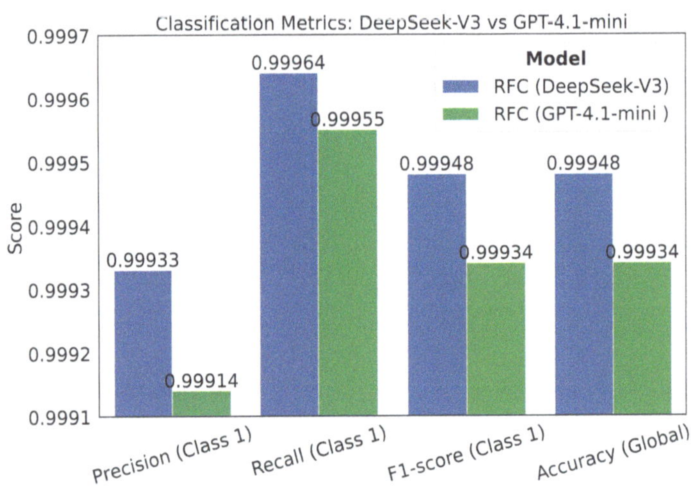

Fig. 6. Performance metrics of *RFC* trained by *DeepSeek-V3* and *GPT-4.1-mini* crews for money laundering in terms of precision, recall, F1-score and overall accuracy for class 1 (money laundering).

As an end-to-end solution, the crew was designed to take a dataset CSV path and the target column name as parameters, performing all tasks up to the generation of a technical report. Two versions of the crew were implemented: one powered by *DeepSeek-V3* and the other by *GPT-4.1-mini*. This study aimed to analyze the performance impact of the underlying LLM.

In the churn prediction task, both crews applied common strategies such as addressing class imbalance, using one-hot encoding, and selecting the same RFC model. However, differences arose in how they tackled class imbalance: *DeepSeek-V3* used SMOTE, while *GPT-4.1-mini* applied random over-sampling. Moreover, the RFC by *DeepSeek-V3* prioritized precision (93.4700%), whereas the one by *GPT-4.1-mini* prioritized recall (97%).

In terms of LLM operational metrics, *GPT-4.1-mini* consumed significantly more tokens, which resulted in higher cost. Nonetheless, it completed tasks faster. On the other hand, *DeepSeek-V3* offered a more cost-effective solution, albeit with slightly longer processing times.

For money laundering prediction, both crews adopted similar baseline approaches: RFC with ten estimators and SMOTE for handling imbalance. However, encoding strategies differed. While *DeepSeek-V3* used Label Encoding, *GPT-4.1-mini* applied a more nuanced encoding scheme: one-hot for low-cardinality features and frequency encoding for high-cardinality ones. This behavior aligns with the decision-making process of an experienced data scientist and reflects the more sophisticated methodology executed by the *GPT-4.1-mini* crew.

Again, operational metrics confirmed the earlier trends: higher cost but faster processing for *GPT-4.1-mini*, and slower but cheaper performance for *DeepSeek-V3*. The longer time observed for *DeepSeek-V3* could be partially attributed to longer GridSearch execution and higher API latency.

5 Conclusion

This study presents a multi-AI agents system to solve complex business tasks. In particular, our study focused on the training of ML models for classification problems in the financial sector. For this opportunity, our crew of AI agents mimicked the work done by a data science team. We used two state-of-the-art LLMs, *DeepSeek-V3* and the novel *GPT-4.1-mini*, to power our AI agents. According to the results, both models demonstrated competitive performance in the proposed activities, successfully solving each defined task within the classification model development workflow. The LLMs that can be used are not limited to those in this work; however, the selection of one or another LLM depends on important factors such as necessity, available budget, infrastructure, etc. Furthermore, there are a lot of possible applications in business using multi-AI agents. This is only the beginning, and with the rapid development of the most advanced and optimized LLMs, the application of IA agents will become more common.

References

1. De Curtò, J., De Zarzà, I.: LLM-driven social influence for cooperative behavior in multi-agent systems. IEEE Access (2025). https://doi.org/10.1109/ACCESS.2025.3548451
2. Li, X., Wang, S., Zeng, S., Wu, Y., Yang, Y.: A survey on LLM-based multi-agent systems: workflow, infrastructure, and challenges. Vicinagearth **1**(1), 9 (2024). https://doi.org/10.1007/s44336-024-00009-2
3. Dong, Y.: The multi-agent system based on LLM for online discussions. In: Proceedings of the 23rd International Conference on Autonomous Agents and Multiagent Systems, pp. 2731–2733 (2024)

4. Kannan, S., Venkatesh, V., Min, B.: Smart-LLM: smart multi-agent robot task planning using large language models. In: 2024 IEEE/RSJ International Conference on Intelligent Robots and Systems (IROS), pp. 12140–12147. IEEE (2024). https://doi.org/10.48550/arXiv.2309.10062
5. Yang, T., et al.: AutoHMA-LLM: efficient task coordination and execution in heterogeneous multi-agent systems using hybrid large language models. IEEE Trans. Cogn. Commun. Netw. (2025). https://doi.org/10.1109/TCCN.2025.3528892
6. Kon, P.T.J., et al.: Curie: toward rigorous and automated scientific experimentation with AI agents. arXiv preprint arXiv:2502.16069 (2025). https://doi.org/10.48550/arXiv.2502.16069
7. Liu, H., Li, R., Xiong, W., Zhou, Z., Peng, W.: WorkTeam: Constructing Workflows from Natural Language with Multi-agents. arXiv preprint arXiv:2503.22473 (2025). https://doi.org/10.48550/arXiv.2503.22473
8. Sivathapandi, G.: Multi agent model based risk prediction in banking using deep learning model (2023). https://doi.org/10.53555/jcr.v10:i2.13166
9. Subramanyam, S.: Transforming financial systems through robotic process automation and AI: the future of smart finance. Int. J. Artif. Intell. Res. Dev. (IJAIRD) **2**(1), 203–223 (2024). https://doi.org/10.34218/IJAIRD_02_01_017
10. De Gasperis, G., Facchini, S.D., Letteri, I.: Leveraging multi-agent systems and decentralized autonomous organizations for tax credit tracking: a case study of the superbonus 110% in Italy. Appl. Sci. **14**(22), 10622 (2024). https://doi.org/10.3390/app142210622
11. *CrewAI*: crewAI: Cutting-edge framework for orchestrating role-playing, autonomous AI agents (2024). https://github.com/crewAIInc/crewAI/
12. *CrewAI*. Multi AI agent systems with crewAI - deeplearning.ai (2024). Accessed 20 Feb 2024
13. Liu, A., et al.: DeepSeek-V3 Technical report. arXiv preprint arXiv:2412.19437 (2024). https://doi.org/10.48550/arXiv.2412.19437
14. OpenAI. Introducing GPT-4.1 in the API. https://openai.com/index/gpt-4-1/

Reframing Academic Engagement: AI-Driven Mobile Applications as Cognitive Interfaces in Digitally Net-Worked Higher Education

Mónica Maldonado Espinosa[1](✉), Clara Elena Robayo Valencia[2], and Blanca Lilia Arias Medina[3]

[1] Universidad Nacional de Loja, Loja, Ecuador
monica.maldonado@unl.edu.ec
[2] Universidad Técnica de Ambato, Ambato, Ecuador
ce.robayo@uta.edu.ec
[3] Universidad Técnica de Cotopaxi, Latacunga, Ecuador
blanca.arias3683@utc.edu.ec

Abstract. The convergence of mobile technologies and artificial intelligence (AI) is transforming academic practices in higher education, particularly within digitally connected learning environments. AI-powered mobile applications are increasingly used by university students to support learning processes, yet their integration into formal pedagogical frameworks remains limited.

This study examines how undergraduate communication students at a public university in Ecuador engage with AI-enhanced mobile applications in aca-demic contexts.

A mixed-methods approach was used to collect survey data from 182 students, focusing on usage frequency, academic impact, and attitudes toward curricular integration.

Findings reveal a strong reliance on tools like ChatGPT and CapCut for text generation and multimedia production, though usage is largely informal and student driven. Most students support the inclusion of AI in course content, but report limited faculty involvement and institutional support.

These results underscore a disconnect between technological engagement and curricular design, highlighting the need for structured, ethical, and pedagogically aligned AI integration in higher education.

Keywords: Intelligent Mobile Learning · AI-Enhanced Student Engagement · Digital Pedagogical Networks

1 Introduction

The rapid expansion of Artificial Intelligence (AI) technologies has triggered pro-found transformations in the way higher education systems operate, particularly as digital connectivity and mobile access become ubiquitous. Mobile applications embedded with AI functionalities are increasingly being used to facilitate personalized learning, support

J. L. Ballesteros et al. (Eds.): CITIS 2025, LNNS 1780, pp. 241–248, 2027.
https://doi.org/10.1007/978-3-032-14287-0_24

information processing, and enhance academic productivity. These technologies ranging from natural language processors to content generation tools are not only reshaping how knowledge is consumed but also how students interact with academic networks and construct meaning in real time.

Despite these advancements, the integration of AI into pedagogical practice remains uneven. While institutional policies in Latin America have emphasized digital equity and access, the actual pedagogical incorporation of intelligent mobile technologies into university curricula often depends on students' initiative rather than structured educational strategies. This gap is especially evident in communication pro-grams, where the skills required for digital content production, research, and critical analysis are inherently aligned with the capabilities of AI-driven applications.

In this context, the present study investigates the academic use of AI-powered mobile applications among undergraduate communication students at a public universi-ty in Ecuador. Specifically, it explores which applications are most used, how frequently they are employed, and in what ways they contribute to or fail to enhance students' academic engagement and performance. By focusing on usage patterns, perceived academic benefits, and the extent of faculty involvement, the study contributes to a growing body of literature examining the intersection of intelligent systems, mobile learning environments, and digitally mediated pedagogical ecosystem.

2 Methods

This study was conducted at the descriptive level of the research pyramid, employing a cross-sectional design with a mixed-methods approach. The research aimed to characterize the academic use of AI-powered mobile applications among undergraduate students in the Communication program at the Universidad Nacional de Loja (UNL), Ecuador.

The mixed-methods design integrated both quantitative and qualitative data to explore usage patterns, perceived academic benefits, and attitudes toward AI integration in the curriculum. Quantitative data was collected through a structured digital survey composed of 15 items organized into three thematic sections: (1) frequency and type of AI mobile applications used, (2) perceived impact on academic tasks and learning objectives, and (3) student perspectives on curricular inclusion of these tools. The survey included multiple-choice and Likert-scale questions to facilitate statistical analysis.

A total of 182 valid responses were obtained from a population of 217 students, achieving a response rate of 83.9%. Data was gathered via Google Forms during the academic period from September 2024 to February 2025. Participation was voluntary and based on informed consent, ensuring confidentiality and ethical compliance.

Descriptive statistical analysis (frequencies and percentages) was conducted using spreadsheet software, with visualizations generated to support pattern recognition. Limitations related to unequal access to mobile devices and internet connectivity were acknowledged, reflecting equity concerns previously noted in related literature [1, 2].

3 Results

Over the past decade, artificial intelligence has increasingly been integrated into digital educational ecosystems, particularly through mobile applications designed to optimize learning through automation, personalization, and cognitive augmentation [1, 2]. The literature suggests that such tools hold significant potential to support higher-order thinking, streamline academic workflows, and foster autonomous learning environments [3, 4]. Despite this, empirical studies continue to show disparities between the capabilities offered by AI-enhanced platforms and their actual adoption and integration within university systems.

In this study, data collected from 182 undergraduate Communication students at the Universidad Nacional de Loja revealed a nuanced pattern of AI tool usage. While academic literature often foregrounds adaptive learning systems and intelligent tutoring platforms [5], students in our sample reported a preference for applications with immediate practical benefits and low learning curves.

As shown in Table 1, students exhibit distinct usage patterns regarding AI-powered mobile applications, favoring tools that enhance productivity and simplify academic tasks. The data reflects a clear preference for generative and multimedia tools, while more advanced semantic or integrative platforms remain underutilized due to limited awareness or training.

Table 1. AI Applicatios.

AI Application	Main Use Case	Usage Rate (%)
ChatGPT	Text generation, summarization, idea structuring.	86.8%
CapCut	Video editing for academic and creative projects.	74.7%
Adobe Photoshop	Image editing and visual content production.	52.7%
ChatPDF	Reading and summarizing academic PDFs.	13.2%
Perplexity AI	Semantic search and data-backed responses.	0.5%
Gemini	Multimodal queries, integration with Google apps.	0.5%
Google Assistant	Voice-based task assistance and search.	40.7%
Siri	Voice commands and dictation.	36.3%

ChatGPT (86.8%) led usage rates, being praised for its capacity to generate ideas, summarize readings, and support task initiation functions that align with the concept of AI as a cognitive prosthesis rather than a deep tutor [6].

According to Fig. 1, video editing platforms such as CapCut (74.7%) and Adobe Photoshop (52.7%) were frequently used to develop multimedia academic projects. These findings reflect a growing alignment between the affordances of mobile AI tools and the competencies required in communication studies, where visual literacy, multimodal expression, and rapid content production are central [7]. The educational potential of such tools has been highlighted in literature for their role in enabling students to translate abstract knowledge into applied creative outputs [8].

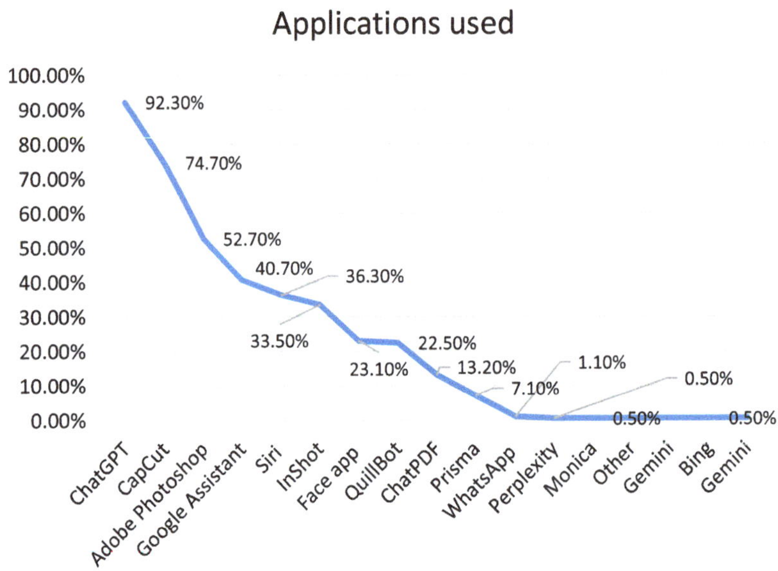

Fig. 1. Most used applications with artificial intelligence (AI)

Interestingly, tools with more sophisticated search and comprehension capabilities such as ChatPDF (13.2%), Perplexity AI, or Gemini remained marginal among student preferences, as Fig. 1 showed. This underutilization suggests a disconnect between the advanced analytical functionalities discussed in the literature [9] and student awareness or training. The majority of respondents reported discovering these tools through informal peer networks rather than academic instruction, echoing findings that students often self-navigate digital tool adoption without structured guidance [10].

Moreover, while 70.8% of students acknowledged that AI applications had strengthened their learning, only 19.6% reported daily use, and a considerable 28.1% used them infrequently. This irregular engagement implies that while students recognize the value of AI, they may lack strategies to embed these tools into sustained academic routines, an issue already noted in previous research on digital literacy gaps [11].

A significant insight emerges when examining curricular integration: 91.2% of students favored including AI education in course content, yet only 16.1% indicated that their instructors actively incorporated such tools in the classroom. This highlights a persistent pedagogical gap where institutional policy and faculty development programs have not kept pace with technological advancement or student expectations [12].

These results underscore a critical tension between the emergent AI literacy cultivated informally by students and the institutional inertia in formalizing these practices within teaching frameworks. Although mobile AI applications are being increasingly positioned in the literature as agents of educational transformation [13], their implementation remains fragmented, reliant on student agency, and constrained by curricular lag.

4 Discussion

4.1 Informal Adoption and Student-Led Digital Practices

The results of this study illustrate a clear dichotomy between the student-driven use of AI-powered mobile applications and the lack of institutional frameworks that support their pedagogical integration. As in other global contexts, students at the Universidad Nacional de Loja exhibit a strong inclination toward digital tools that facilitate task efficiency and creative expression. Generative tools like ChatGPT and video-editing applications such as CapCut and Adobe Photoshop are used not for their underlying technological sophistication, but for their accessibility, intuitive interfaces, and immediate utility in completing assignments.

This mirrors trends identified in other empirical studies, where students adopt digital technologies in response to academic pressures and performance expectations rather than strategic learning design [1]. According to Selwyn, student engagement with educational technology tends to be pragmatic, often shaped by peer influence and social media exposure, rather than guided by formal training or critical literacy [2]. Such informal digital cultures foster rapid adoption but often lack the depth required for sustained academic growth.

At the same time, faculty members remain largely absent from this digital transition. Our findings indicate that only a minority of instructors actively incorporate AI tools into their teaching, a result that echoes previous work on the technological lag within higher education institutions [3]. This disconnects results in a parallel system: while students operate within AI-enhanced digital networks, the formal curriculum remains anchored in traditional pedagogical paradigms.

The implications are significant. Without curricular alignment, the academic use of AI remains fragmented and risks reinforcing shallow learning habits. Research by Luckin et al. [4] has stressed the importance of pedagogy-led integration of AI, warning that without it, these technologies serve as nothing more than cognitive crutches rather than learning partners.

4.2 Toward Critical, Ethical, and Systemic AI Literacy

Beyond the technical dimension, the ethical and epistemological implications of unregulated AI adoption in academic settings warrant serious attention. Students may be adept at using tools like ChatGPT to summarize texts or produce drafts, but without a conceptual understanding of AI's limitations such as hallucinated outputs, data bias, or opaque algorithms they are unlikely to question the validity or academic rigor of the results produced.

This concern is echoed by Holmes et al. [5], who argue that critical AI literacy must be embedded within curricula if universities are to fulfill their responsibility of preparing students not just as users, but as evaluators and co-creators of digital knowledge. The lack of formal instruction on how AI systems function, how they process data, and how they may reflect structural biases perpetuates a passive relationship between the learner and the tool.

Moreover, ethical risks such as plagiarism, erosion of authorship, and algorithmic dependency are intensified in informal adoption scenarios. As noted by Zawacki-Richter et al. [6], without institutional oversight and ethical frameworks, the deployment of AI in education can undermine core academic values such as integrity, autonomy, and accountability.

Addressing these challenges requires a systemic and multidimensional response. At the micro level, faculty development programs must include digital pedagogy training focused on the ethical and critical use of AI. At the academic departments should revise learning outcomes to explicitly include AI literacy, digital judgment, and data ethics. At the macro level, institutional policies must anticipate the trajectory of AI in education and invest in infrastructure, governance, and inclusive access.

Initiatives such as UNESCO's AI Competency Framework for Teachers [7] pro-vide actionable models for integrating AI into teacher education and curriculum development. These frameworks highlight the need for educators to move beyond functional use toward a deeper, contextual engagement with intelligent technologies. Implementing such frameworks within Latin American universities, including institutions like Universidad Nacional de Loja, would represent a proactive step toward aligning technological potential with equitable and reflective educational practice.

In conclusion, AI-powered mobile applications offer a rich but underutilized opportunity for enhancing higher education. Their transformative potential lies not in their availability, but in how institutions choose to structure, frame, and guide their use. As digital networks become inseparable from academic practice, universities must assume a leading role in shaping responsible, informed, and critical engagement with artificial intelligence.

5 Conclusions

The present study examined the intersection between student behavior, AI-powered mobile applications, and institutional readiness in a digitally evolving aca-demic environment. By combining a review of relevant literature with empirical data from undergraduate students in a communication program, this research highlighted a growing dependence on generative and multimedia-oriented AI tools, particularly those that support rapid content creation and task completion.

While these technologies have proven effective in improving workflow efficiency, the findings reveal a fragmented ecosystem where students operate within informal digital networks largely disconnected from curricular guidance. Faculty involvement in AI integration remains limited, and institutional strategies appear in-sufficient to harness the pedagogical potential of these tools in a structured, ethical, and reflective manner.

This gap underscores the urgent need for universities to develop and implement frameworks that embed AI literacy and critical digital practices into course content. Doing so would ensure that students are not only proficient users of AI but also capable of evaluating the implications, limitations, and biases inherent in these technologies. As AI continues to evolve and expand its role in higher education, equipping students with the ability to navigate this landscape thoughtfully will be essential to safeguarding academic rigor and fostering long-term learning autonomy.

Moreover, this research points to broader challenges in educational governance and curriculum design. Institutions must not only invest in digital infrastructure but also promote cross-functional collaboration between educators, technologists, and curriculum designers to co-create adaptive, inclusive learning models that reflect the realities of AI-infused academic work.

While the study focused on a specific disciplinary context and regional setting, its implications are generalizable to other higher education environments facing similar technological transitions. Future research should explore the longitudinal effects of AI integration on student cognition and academic performance, assess disciplinary differences in adoption strategies, and evaluate the impact of faculty training pro-grams designed to close the pedagogical gap in digital innovation.

In conclusion, AI-powered mobile applications hold significant promises for enhancing learning in higher education. However, realizing that potential depends not merely on access, but on intentional, institutionally supported engagement that aligns technology with the deeper goals of education: critical thinking, ethical judgment, and meaningful knowledge construction.

6 References

1. Eynon, R., Malmberg, L.-E.: Understanding the everyday digital lives of university students in the post-COVID era. Br. J. Educ. Technol. **52**(5), 1415–1431 (2021). https://doi.org/10.1111/bjet.13102
2. Selwyn, N.: Education and Technology: Key Issues and Debates. Bloomsbury Publishing, London (2016)
3. Veletsianos, G., Moe, R.: The rise of educational technology as a sociocultural and ideological phenomenon. Educ. Technol. **57**(2), 25–30 (2017)
4. Luckin, R., Holmes, W., Griffiths, M., Forcier, L.B.: Intelligence Unleashed: an Argument for AI in Education. Pearson Education, London (2016)
5. Holmes, W., Bialik, M., Fadel, C.: Artificial Intelligence in Education: Promises and Implications for Teaching and Learning. Center for Curriculum Redesign, Boston (2019)
6. Zawacki-Richter, O., Marín, V.I., Bond, M., Gouverneur, F.: Systematic review of research on artificial intelligence applications in higher education – where are the educators? Int. J. Educ. Technol. High. Educ. **16**(1), 39 (2019). https://doi.org/10.1186/s41239-019-0171-0
7. UNESCO, AI Competency Framework for Teachers (2021). https://doi.org/10.54675/ZJTE2084
8. Selwyn, F., Luckin, R.: Grassroots adoption of educational technologies in digitally connected learning environments. J. Educ. Innov. **15**(2), 45–59 (2021)
9. Daniels, A., Lee, K.: AI-driven pedagogy: From potential to practice. In: Proceedings of the International Conference on Digital Learning, pp. 112–124. Springer (2020)
10. Rivera, M.L., Chan, S.: Artificial Intelligence and Cognitive Development in Higher Education, 2nd edn. Routledge, London (2019)
11. Gómez, T.: The influence of generative AI tools on student behavior. In: Proceedings of the 9th International Conference on Learning, pp. 17–25. ACM Press, Technol (2022)
12. Wang, C., Patel, J.: AI in education: theory and critique. Digit. Pedagogy Rev. **11**(3), 203–218 (2020)
13. Ortega, S.D.: Awareness and adoption of emerging semantic AI tools among university students. High. Educ. Futures. **13**(4), 144–160 (2023)
14. Weller, R.: Reforming the University in a Digital age. Springer, Berlin (2020)

15. López, F., Tran, G., Silva, J.: Critical digital literacy in AI-mediated education. In: Proceedings of the Global EdTech Forum, pp. 76–88. Springer (2021)
16. García, H.N., Bloom, M.: Ethical frameworks for AI deployment in undergraduate learning environments. Technol. Ethics Educ. **5**(2), 91–107 (2022)

Lexical Complexity Prediction in Ecuadorian Public Texts with GPT-4 and Successor Models: An Approach Using Zero-Shot Learning

Jenny Ortiz-Zambrano[1(✉)], Arturo Montejo-Ráez[2], Scarlet Annabel Gutierrez Gavilanez[1], and Ivette Judith Zamora Olvera[1]

[1] Universidad de Guayaquil, Av. Delta s/n y Av. Kennedy, Guayaquil, Ecuador
{jenny.ortizz,scarlet.gutierrezg,ivette.zamorao}@ug.edu.ec
[2] Universidad de Jaén, Campus Las Lagunillas s/n, 23071 Jaén, Spain
amontejo@ujaen.es

Abstract. Evaluating lexical complexity in documents using Generative Pre-trained Transformer models (GPT-4, GPT-4o, and GPT-4 Turbo) enabled an analysis of their impact on language comprehension, specifically in Ecuadorian state texts written in Spanish. The Zero-Shot Learning technique was applied to all models using the GovAIEc dataset. The research follows a quantitative and descriptive approach with a quasi-experimental design. The models were evaluated using common error metrics: MAE, MSE, and RMSE. The objective of this study is to assess the performance of GPT-4 and its successor models in predicting lexical complexity in Ecuadorian public texts through Zero-Shot Learning. The results indicate that GPT-4 Turbo achieved the highest performance among the models, with an MAE of 0.2840, followed by GPT-4o with an MAE of 0.2891, and finally GPT-4 with an MAE of 0.2950.

Keywords: GPT-4 · Zero-Shot Learning · lexical complexity · spanish

1 Introduction

Public documents are mostly issued by the national government of each country. These documents contain a set of legal norms aimed at the protection, regulation, and delimitation of the right to receive and disseminate ideas, opinions, and reportable facts. In other words, they communicate important facts for the development of the country and its citizens, making their comprehension by every member of society vitally important [1]. However, the comprehension of the laws commonly found in public documents remains a critical challenge for citizens, especially for those with low literacy levels, as even legal advisory websites are often inaccessible to many [3].

*Supported by organization x.

J. L. Ballesteros et al. (Eds.): CITIS 2025, LNNS 1780, pp. 249–259, 2027.
https://doi.org/10.1007/978-3-032-14287-0_25

It is necessary to explore tools that assess the complexity of language in these documents. Lexical complexity prediction using GPT-4 models emerges as a promising solution. Studies have shown that these models are effective for lexical simplification and can identify complex terms, thereby improving text comprehension [2,5]. Moreover, factors contributing to the textual complexity of public documents include technical vocabulary, complex syntactic structures, and a lack of coherence and textual integrity [6]. The Zero-Shot Learning technique is said to be a machine learning approach that allows models to perform previously unknown tasks without the need for assigned examples. This is achieved by leveraging the general knowledge acquired during training, enabling the model to classify or predict based on data it has not encountered before [7].

In this context, the objective of this study is to evaluate the performance of the GPT-4 model and its successors in predicting lexical complexity in Ecuadorian public texts using the Zero-Shot Learning approach. This evaluation will help determine the effectiveness of these models in improving text comprehension, thereby contributing to the inclusion of individuals with low literacy levels in accessing public information.

2 Related Work

In recent years, there has been a significant volume of research on textual complexity, which is closely related to the field of natural language processing. Below, we highlight key studies that address these aspects:

cs60075_team2 at SemEval-2021 Task 1: Lexical Complexity Prediction using Transformer-based Language Models Pre-trained on Various Text Corpora. Transformer-based language models, specifically BERT and RoBERTa, were used to predict lexical complexity for single words and multi-word expressions. These models were pre-trained on various text corpora, and studies were conducted to select the best configurations and aggregation methods for complexity scoring. The model achieved a Pearson correlation of 0.784 for predicting complexity in individual words and 0.836 for multi-word expressions. Additionally, it was shown that aggregating scores from multiple pre-trained models across different domains significantly improves prediction accuracy [8].

A approach for predicting lexical complexity in Spanish using Transformer models such as BERT, XLM-RoBERTa, and RoBERTa-large-BNE, trained with the CLexIS2 corpus was proposed. Manual features (frequency, length, POS categories) are combined with embeddings from pre-trained models. The fine-tuned BERT model achieved the best performance, with a MAE of 0.1592 and a Pearson correlation of 0.9883. XLM-RoBERTa and RoBERTa-large-BNE also showed improvement after fine-tuning, although BERT outperformed them [9].

In their participation in the SimpleText Task 2 at CLEF 2023, the authors (J. Ortiz-Zambrano et al., 2023) applied the GPT-3 successors, davinci-002 and davinci-003, for lexical complexity classification of words. Using the few-shot learning approach, they provided limited examples to the model to assist in classification. The key results showed that the davinci-003 model achieved the best performance, with a MAE of 0.0882. Additionally, both matches and discrepancies were observed between the categories assigned by GPT-3 and those from the Complex corpus. Finally, the authors highlighted opportunities to explore new models, such as Claude 2 and GPT-4, for lexical complexity prediction [12].

In the research *GPT-4 and its Successor Models in the Prediction of Lexical Complexity in Ecuadorian Public Texts Using Few-Shot Learning*, the authors applied a quantitative and descriptive approach with a quasi-experimental design, using the few-shot learning technique to evaluate lexical complexity in Ecuadorian public texts. They used the GPT-4, GPT-4 Turbo, and GPT-4o models and evaluated them on the GovAIEc dataset, which contains 7,813 legal documents records. The models were evaluated using error metrics like MAE, MSE, and RMSE. The findings revealed that GPT-4 outperformed its successors, GPT-4 Turbo and GPT-4o, with an MAE of 0.2464. GPT-4 Turbo and GPT-4o had MAEs of 0.2540 and 0.2593, respectively. Furthermore, GPT-4o achieved the highest match rate in its predictions, reaching 53.44%, indicating a higher accuracy in identifying complex words [11].

3 Materials and Methods

3.1 Research Modality

This research is of an applied nature, as its primary goal is to utilize existing knowledge in natural language processing (NLP) and artificial intelligence models to address a practical problem: evaluating the lexical complexity of public documents and enhancing their accessibility for citizens. The study aims to bridge the gap between advanced AI techniques and their real-world applications, ultimately contributing to improving public understanding and inclusivity through clearer communication of legal and governmental texts.

3.2 Dataset

The unit of analysis in this study is each individual record within the **GovAIEc dataset**, which contains a total of 7,813 records. This dataset is composed of notifications and instructions related to legal procedures, primarily extracted from sentences in official documents issued by various public governmental entities in Ecuador, specifically from the city of Guayaquil. These entities include:

- **CNT** (National Telecommunications Corporation)[1]
- **SRI** (Internal Revenue Service)[2]
- **CNE** (National Electoral Council)[3]
- **Municipio** (Municipality of Guayaquil)[4]
- **ATM** (Transportation Management Authority of Guayaquil)[5]

Each record in the dataset includes the following fields:

- **id**: A unique identifier assigned to each record.

- **corpus**: The public governmental institution (source) from which the document was derived.
- **sentence**: The sentence containing the word identified as complex.
- **token**: The word recognized as complex by annotators.
- **complexity**: A numerical value representing the level of complexity assigned to the word by the annotators.

This dataset serves as the foundation for evaluating the lexical complexity of public texts, with a focus on legal and procedural language from Ecuadorian governmental documents.[1,2,3,4,5]

Table 1 presents a set of records extracted from legal documents issued by various public institutions in Guayaquil. The word highlighted in bold is the one that the taggers have identified as complex.

Table 1. Dataset GovAIEc.

id	corpus	sentence	token	complexity
6075	Municipio	TRÁMITES EN LA BIBLIOTECA MUNICIPAL. Si en el stock existe el libro pedido, se emite Comprobante de la Publicación para que el comprador proceda a ...	**comprobante**	0.3333
7719	SRI	REQUERIMIENTOS Y JUSTIFICACIONES DEL PROCESO INCONSISTENCIAS. Si el Servicio de Rentas Internas detectare inconsistencias en las declaraciones o en los anexos que presente el contribuyente, siempre que no generen ...	**sustitutivo**	0.6666
2734	CNE	REGLAMENTO PARA CONFORMACIÓN DE ALIANZAS ELECTORALES. La Constitución de la República dispone en el artículo 112, que los partidos y movimientos ...	**militantes**	1.0000

[1] National Telecommunications Corporation: https://www.cnt.com.ec/servicios-en-linea.
[2] Internal Revenue Service: https://srienlinea.sri.gob.ec/sri-en-linea/inicio/NAT.
[3] National Electoral Council: https://www.cne.gob.ec/.
[4] Municipality of Guayaquil: https://tramites4.guayaquil.gob.ec/.
[5] Transportation Management Authority of Guayaquil: https://www.atm.gob.ec/.

3.3 Complexity Scale

The complexity scale has the following levels:

- **Moderately difficult**: Complexity range between 0 and 0.3333. Words or sentences in this range are somewhat complex but can be understood in context.
- **Difficult**: From 0.3334 to 0.6666. Words or sentences at this level are quite complex and may require a higher level of understanding or technical knowledge.
- **Very difficult**: From 0.6667 to 1. Words or sentences in this category are highly complex, which may make them hard to understand.

3.4 Data Processing and Complexity Prediction Using Zero-Shot Learning

In the processing of data for the identification of complex words in the GovAIEc dataset, an approach based on the GPT-4 language model was used applying the Zero-Shot Learning technique. This approach was employed to predict the lexical complexity of words based on their context within sentences extracted from governmental legal documents. The data processing consisted of the following steps:

1. **Data reading and preprocessing:** The data were read from the `GovAIEc` corpus, which contained sentences and words identified as complex in legal documents from public institutions. The relevant columns for the analysis were *id*, *sentence*, *token*, and *complexity*, which were used to calculate the model's predictions and display the results.
2. **Complexity prediction generation:** To classify words according to their complexity, models based on the Transformer architecture were employed, specifically GPT-4, GPT-4 Turbo, and GPT-4o. These models were applied to identify the level of complexity of the words within the sentences of legal documents. The Zero-Shot Learning approach was used, meaning no previously labeled examples were required for prediction. The GPT-4 model and its successors classified each identified complex word into one of three complexity categories: *Moderately difficult*, *Difficult*, and *Very difficult*.
3. **Prompt formatting for prediction:** The model received as input a specific prompt that established the context of the task. The format of the prompt used to perform the complexity predictions through Zero-Shot Learning is shown below.
4. **Evaluation and Results:** Once the model generates the complexity predictions, the results are stored in a file for further analysis using common error metrics for each model and the applied technique.

The code snippet shows the *prompt* used to classify complex words into three difficulty categories based on the sentence context. This classification is performed using Zero-Shot Learning.

```
prompt = (
"I am reading fragments of texts from public institution sources,
and some words are difficult to understand."
"I want to automatically identify complex words and determine their
level of difficulty."
"Please classify the word provided into one of the following
categories based on the sentence context:"
"\"moderately difficult\" (value: from 0 to 0.3333), \"difficult\"
(value: from 0.3334 to 0.6666), or \"very difficult\" (value: from
0.6667 to 1)."
"After reading the fragment: "@oracion", I find that the word
"@aEvaluar" is classified as one of these categories."
"Return only the category name with no extra explanation or
content."
"The output should only be one of the following options:
"moderately difficult", "difficult", or "very difficult"."
)
```

3.5 Common Error Metrics

The metrics applied in this research were taken from the work of [12] and correspond to the evaluation metrics used to measure the accuracy and quality of the predictions made with GPT-4 model and its successors: GPT-4 Turbo and GPT-4o. Each of them is detailed below:

MAE (Mean Absolute Error): This metric can be used when outliers represent corrupted parts of the data.

$$\mathrm{MAE} = \frac{1}{m}\sum_{i=1}^{m}\left|(X_i - Y_i)^2\right| \tag{1}$$

MSE (Mean Squared Error): The Mean Squared Error (MSE) is used to evaluate the accuracy of a prediction model, as it measures the difference between the observed values and the values estimated by the model. It can also be used to detect outliers, since due to the L2 norm, MSE assigns greater weight to such points. If the model produces a single very poor prediction, the quadratic component of the function increases the overall error.

$$MSE = \frac{1}{m}\sum_{i=1}^{m}(X_i - Y_i)^2 \tag{2}$$

RMSE (Root Mean Squared Error): MSE and RMSE have a monotonic relationship, i.e., through the square root. A ranking of regression models based on MSE will be identical to a ranking based on RMSE.

$$RMSE = \sqrt{\frac{1}{m}\sum_{i=1}^{m}(X_i - Y_i)^2} \tag{3}$$

4 Results and Discussion

The results obtained from the application of the GPT-4 model and its successors: GPT-4 Turbo and GPT-4o were evaluated using the error metrics described previously. The primary goal was to predict the lexical complexity of words in public legal texts, based on the context within sentences.

4.1 Summary of Prediction Results

Tables 2, 3 and 4 present several records through which we present the summary of the prediction results of the GPT-4, GPT-4 Turbo and GPT-4o models to identify complex words using the Zero-Shot Learning technique. Each row corresponds to a record of the corpus, where the model predictions are compared with the actual complexity values.

The columns in the tables include:

- **Token:** Word identified as complex.
- **GPT-4 Response:** Complexity category predicted by the model.
- **GPT-4 Complexity:** Numerical value of the predicted complexity.
- **Real complexity:** Numerical value of the actual complexity.
- **Real category:** Actual complexity category.
- **Match:** Indicates whether the prediction matches the actual value (Yes/No).

Table 2. Lexical complexity predictions by the GPT-4 model and comparison with real values.

Token	GPT-4 Response	GPT-4 complexity	Real Complexity	Real Category
comprobante	moderately difficult	0.1666	0.3333	moderately difficult
recaudaroras	moderately difficult	0.1666	1.0000	very difficult
stock	moderately difficult	0.1666	1.0000	very difficult
señalando	moderately difficult	0.1666	0.3333	moderately difficult
bibliográfico	difficult	0.5000	0.3333	moderately difficult
autorización	moderately difficult	0.1666	0.6667	difficult
donarse	moderately difficult	0.1666	0.3333	moderately difficult
devolución	moderately difficult	0.1666	1.0000	very difficult
instalaciones	moderately difficult	0.1666	0.3333	moderately difficult
identificación	moderately difficult	0.1666	0.3333	moderately difficult

The following presents the results obtained as a summary of the application of the successors *GPT-4 Turbo*, *GPT-4o*, and the *GPT-4* model in the identification of complex words.

Table 3. Lexical complexity predictions by the GPT-4 Turbo model and comparison with real values.

Token	**GPT-4 Turbo Response**	**GPT-4 Turbo Complexity**	**Real Complexity**	**Real Category**
comprobante	moderately difficult	0.1666	0.3333	moderately difficult
recaudadoras	difficult	0.5000	1.0000	very difficult
stock	moderately difficult	0.1666	1.0000	very difficult
señalando	moderately difficult	0.1666	0.3333	moderately difficult
bibliográfico	very difficult	0.8333	0.3333	moderately difficult
autorización	moderately difficult	0.1666	0.6667	difficult
donarse	very difficult	0.8333	0.3333	moderately difficult
devolución	moderately difficult	0.1666	1.0000	very difficult
instalaciones	moderately difficult	0.1666	0.3333	moderately difficult
identificación	moderately difficult	0.1666	0.3333	moderately difficult

Table 4. Lexical complexity predictions by the GPT-4o model and comparison with real values.

Token	**GPT-4o Response**	**GPT-4o Complexity**	**Real com-plexity**	**Real Category**
comprobante	moderately difficult	0.1666	0.3333	moderately difficult
recaudaroras	difficult	0.5000	1.0000	very difficult
stock	moderately difficult	0.1666	1.0000	very difficult
señalando	moderately difficult	0.1666	0.3333	moderately difficult
bibliográfico	difficult	0.5000	0.3333	moderately difficult
autorización	moderately difficult	0.1666	0.6667	difficult
donarse	moderately difficult	0.1666	0.3333	moderately difficult
devolución	moderately difficult	0.1666	1.0000	very difficult
instalaciones	moderately difficult	0.1666	0.3333	moderately difficult
identificación	moderately difficult	0.1666	0.3333	moderately difficult

Table 5. Summary of results from GPT-4 and its successors in lexical complexity identification.

Model	MAE	MSE	RMSE	Match Rate
GPT-4Turbo_zero_shot	**0.2840**	**0.1153**	**0.3396**	**53.67%**
GPT-4o_zero_shot	0.2891	0.1215	0.3486	57.58%
GPT-4_zero_shot	0.2950	0.1268	0.3561	58.10%

The results presented in Table 5 show the accuracy of the GPT-4 model and its successors GPT-4 Turbo and GPT-4o in predicting complex words, evaluated through statistical metrics such as MAE, MSE, RMSE, and the matching percentage, allowing a quantitative comparison of their performance in the task of lexical complexity prediction.

Regarding the performance of the models based on MAE, the GPT-4 Turbo model presents the lowest MAE with a value of 0.2840, indicating higher accuracy in predicting lexical complexity. The GPT-4o_zero_shot model, on the other hand, presents a result of 0.2891, which is superior to the GPT-4 model with 0.2950, but inferior when compared to GPT-4 Turbo. This suggests that, although GPT-4o_zero_shot improves performance over GPT-4, it still does not reach the accuracy of GPT-4 Turbo in predicting lexical complexity. These results highlight the advantage of GPT-4 Turbo in terms of accuracy, making it the best choice for this specific task.

In summary, the results can be synthesized as follows:

- Best model in accuracy (MAE): GPT-4 Turbo_zero_shot (MAE=0.2840).
- Best model in matching: GPT-4 (58.10%).
- Worst model in accuracy (MAE): GPT-4_zero_shot (MAE=0.2950).
- Worst model in matches: GPT-4 Turbo (53.67%).

Although GPT-4Turbo_zero_shot is the most accurate according to MAE, GPT-4 stands out in matching, suggesting a higher rate of correct predictions. Furthermore, although the matching percentage for the GPT-4 model is higher, it is still insufficient to ensure completely reliable predictions. This suggests that, while the model frequently approximates the expected values, its explanatory power remains deficient. See Fig. 1.

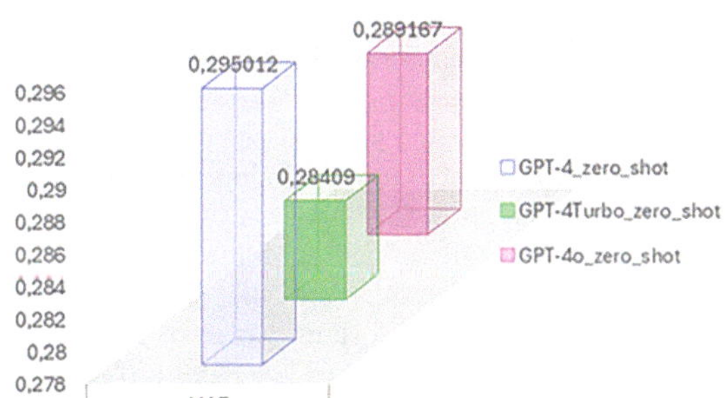

	MAE
GPT-4_zero_shot	0,295012
GPT-4Turbo_zero_shot	0,28409
GPT-4o_zero_shot	0,289167

Fig. 1. Performance results of the GPT-4 model and its successors based on MAE with the Zero-Shot Learning technique.

5 Conclusions and Recommendations

In conclusion, on one hand, GPT-4 Turbo proves to be the most accurate model according to the MAE (0.2840), meaning that its predictions are, on average, closer to the actual lexical complexity values. However, its match percentage with the corpus is the lowest (53.67%), suggesting that, although it has a smaller deviation in its estimates, it does not always align with the expected values. On the other hand, GPT-4 excels in matching with a 58.10% match rate, indicating that its predictions align the most with the reference corpus. Nevertheless, its precision in terms of MAE is lower (0.2950), suggesting that, while it matches more often, its estimates might be less accurate in absolute values. Overall, while GPT-4 Turbo is the most accurate and GPT-4 has the highest match, the results show that the models still present limitations for a fully reliable prediction of lexical complexity.

To enhance model performance, it is advisable to apply hyperparameter tuning, supervised learning, and expand the training dataset with more diverse texts and government entities. This would enable a more robust evaluation of lexical complexity prediction.

The GovAIEc corpus is available by contacting the authors.

References

1. Farías, J.B.: Diagnóstico y propuestas para la consolidación del Instituto de Transparencia, Información Pública y Protección de Datos Personales. Balance Parcial **224** (2011)
2. Ruohonen, J.: Assessing the readability of policy documents on the digital single market of the European Union. In: Proceedings of the 2021 Eighth International Conference on eDemocracy and eGovernment (ICEDEG), pp. 205–209. IEEE (2021)
3. Author, F., Author, S.: Title of a proceedings paper. In: Editor, F., Editor, S. (eds.) CONFERENCE 2016, LNCS, vol. 9999, pp. 1–13. Springer, Heidelberg (2016). https://doi.org/10.10007/1234567890
4. Ortiz-Zambrano, J., Espin-Riofrio, C., Montejo-Ráez, A.: SINAI participation in SimpleText task 2 at CLEF 2024: Zero-Shot prompting on GPT-4-turbo for lexical complexity prediction. In: Working Notes of the Conference and Labs of the Evaluation Forum (CLEF 2024), pp. 3288–3299 (2024)
5. Ortiz-Zambrano, J.A., Espín-Riofrío, C.H., Ráez, A.M.: Enhancing lexical complexity prediction through few-shot learning with GPT-3. In: Proceedings of the Workshop on DeTermIt! Evaluating Text Difficulty in a Multilingual Context@ LREC-COLING 2024, pp. 68–76 (2024)
6. Ortiz-Zambrano, J.A., Montejo-Ráez, A.: Barriers in reading comprehension of university students: analysis of the complicated words annotated in the VYTEDU-CW corpus (2020)
7. Alizadeh, M., et al.: Open-source LLMs for text annotation: a practical guide for model setting and fine-tuning. arXiv (2023). http://arxiv.org/abs/2307.02179
8. Nandy, A., Adak, S., Halder, T., Pokala, S.M.: cs60075_team2 at SemEval-2021 Task 1: Lexical Complexity Prediction using Transformer-based Language Models pre-trained on various text corpora. arXiv (2021). http://arxiv.org/abs/2106.02340

9. Ortiz-Zambrano, J., Espín-Riofrío, C., Montejo-Ráez, A.: Transformers for lexical complexity prediction in Spanish language. Procesamiento del Lenguaje Natural **69**, 177–188 (2022). https://doi.org/10.26342/2022-69-15
10. Ortiz-Zambrano, J., Espín-Riofrío, C., Montejo-Ráez, A.: SINAI Participation in SimpleText Task 2 at CLEF 2023: GPT-3 in Lexical Complexity Prediction for General Audience. Notebook for the SimpleText Lab at CLEF 2023. http://ceur-ws.org
11. Ortiz-Zambrano, J.I.: GPT-4 and its successor models in the prediction of lexical complexity in Ecuadorian public texts using Few-Shot Learning. PC **10**(103), 861–884 (2025). https://doi.org/10.23857/pc.v10i2.8921
12. Chicco, D., Warrens, M.J., Jurman, G.: The coefficient of determination R-squared is more informative than SMAPE, MAE, MAPE, MSE and RMSE in regression analysis evaluation. PeerJ Comput. Sci. **7**, 1–24. https://doi.org/10.7717/PEERJ-CS.623

Technologies for Health and Wellness

Study of Ergonomic Risks and Quality of Life of Female Manicurists in Informal Work

Maye Catalina Neira Cárdenas(✉), Julie Astrid Solano Pérez, and Albeiro Aguilar Olivera

Corporación Universitaria Minuto de Dios - UNIMINUTO, Bogotá, Colombia
ft.catalinaneira@gmail.com

Abstract. Objective: To present a study of ergonomic risks due to repetitive movements and efforts of upper limbs and quality of life in the informal work of manicurists who perform activities of cleaning, trimming and beautification of nails (hands and feet) in a non-probabilistic sample practicing their trade in the city of Bogota, Colombia. **Method:** The research was conducted under a mixed approach and a descriptive scope. The study phenomenon was focused on the ergonomic risks of upper limbs and on the quality of life of 75 volunteer manicurists without formal employment (dependent or independent). **Results:** The manicure process is standardized in six main activities and fifteen repetitive tasks; 94.29% of the manicurists are female; 65.71% present a high level of unacceptable risk in the right upper limb and 42.86% in the left upper limb; the dimensions where dissatisfaction with quality of life is presented are related to personal development and management of free time. **Conclusions:** A large percentage of people who work as manicurists are female, with a high probability of developing musculoskeletal disorders in the upper extremities and low levels of satisfaction with quality of life.

Keywords: Ergonomics risks · Quality of life · Manicurist(s) · Occupational risks · Informal work

1 Introduction

According to the International Labour Organization (ILO), the average labor informality among the countries that make up the Latin American and Caribbean region stood at 48% in mid-2023, exceeding 70% in several of these states [1]. In Colombia, according to the National Administrative Department of Statistics (DANE), informal employment for the first quarter of 2024 was 56.3%, with women representing 53.7% [2]. Furthermore, Bogotá, the country's capital, had an informality rate of 33.3% in the same period [2], which was a significant figure for the study.

In Colombia, according to the third and most recent National Survey of Occupational Health and Safety Conditions, informal workers reported their exposure to hazards due to non-ergonomic conditions, such as positions that cause fatigue or pain in some parts of the body (32.7%), repetitive movements of hands and/or arms (28.9%), working days

J. L. Ballesteros et al. (Eds.): CITIS 2025, LNNS 1780, pp. 263–273, 2027.
https://doi.org/10.1007/978-3-032-14287-0_26

of more than eight hours a day (22.3%), and exposure to psychosocial factors due to work days without breaks [3], which evidences a significant risk panorama in this work context.

Similarly, the surveyed workers also reported other negative conditions such as a regular state of health and well-being (21.8%), difficulties working under pressure (11.6%), depression (7.9%), inability to solve problems (7.7%), and lack of affiliation with an occupational risk insurance company (92.1%) [3]. In the case of manicurists, the profession lacks policies and regulations, and their job and financial sustainability are uncertain. Furthermore, they have varied work schedules, grueling workdays, and a lack of knowledge of self-care measures, making them much more vulnerable to risk factors related to chemicals, biological organisms, musculoskeletal disorders, and stress [4]. It is worth noting that ergonomic risk in upper extremities and quality of life are very little studied conditions with low scientific production at the local and global levels; therefore, these constitute an important justification for research.

Manicure activity performed by Iranian women has a significant relationship with disorders of elbows and other parts of the body [5]; meanwhile, among the most common musculoskeletal symptoms associated with the work of nail technicians or manicurists in Canada are neck, back, arm, and hand or wrist pain [6]. Likewise, a qualitative study with nail salon workers in Michigan showed that there are different health problems related to the profession, such as physical pain and carpal tunnel syndrome [7]. In this context, according to the World Health Organization (WHO), musculoskeletal disorders are considered one of the main causes of work incapacity worldwide [8].

Regarding the quality of life at work, the COVID-19 pandemic has somehow physically and/or mentally affected a large portion of the world's population, including informal workers, whose jobs depend on continuous, close physical interaction with their clients, including travel between different work sites. In the case of manicurists in Bello Horizonte, Brazil, they reported situations of brutalization, discrimination, and intense demands in their work [9]. In turn, in the city of Curitiba, Brazil, they expressed dissatisfaction with their current quality of life and their health at work [10]. Also, in 11 Latin American cities, including Bogotá, women with informal jobs tend to show depressive symptoms more frequently than those with formal employment [11].

Likewise, the research sought to contribute with new scientific knowledge that would serve as a reference for improving occupational health and safety conditions for manicurists in Colombia and other countries, which goes hand in hand with the fulfillment of Sustainable Development Goals (SDGs) 3 and 8.

2 Methods

The study was conducted with a mixed research approach and a descriptive scope, which is a research route that includes the collection, analysis, integration and discussion of quantitative and qualitative data, the purpose of which is to specify or describe the dimensions of a phenomenon or research problem [12]. In this case, the phenomenon is the ergonomic risks to the upper limbs and the quality of life resulting from the informal work of manicurists in the city of Bogotá.

The study population consisted of 75 volunteer manicurists without formal employment (freelance). They were selected and admitted at convenience under the following inclusion criteria: being of legal age (18 years or older), having worked in the profession for more than one year, and residing in the city of Bogotá. Likewise, the data and personal information collected from the participants were treated under the confidentiality criteria of Colombian regulations for health research [13], which included the signing of an informed consent form, where an explanation of the instruments, scope of the research, minimal risks, and protection of rights was given.

For data collection and analysis, four research instruments were used: a field diary with a photographic study, a sociodemographic profile survey, a repetitive strain risk assessment (OCRA) method, and a questionnaire to measure quality of life in informal work (CVT-GOHISALO-TEI). The field diary consisted of three sections: general information (name, purpose, and instructions), observation items (type of procedure, activities or tasks performed, duration, and inputs or equipment), and a photography space. The sociodemographic profile survey included 37 questions related to age, gender, educational level, seniority in the profession, working hours, and perceived morbidity, among others. Both instruments were validated by two experts on the subject. The OCRA assessment method focuses on calculating the risk level based on the possibility of developing musculoskeletal disorders in the upper extremities over time.

The purpose of applying the method is to establish the value of the OCRA Check List Index (ICKL) and, based on the result, classify the risk as optimal, acceptable, very slight or uncertain, slightly unacceptable, unacceptable to a medium extent or highly unacceptable [14], considering the following formula (1) and Table 1.

$$ICKL = (FR + FF + FFZ + FP + FC)^{*} MD \tag{1}$$

Where, FR: recovery factor, FF: frequency factor, FFZ: strength factor, FP: posture and movement factor, FC: additional risk factors and MD: duration multiplier.

Table 1. Risk level according to the ICKL value.

ICKL	Risk level
≤ 5	Optimal
5.1–7.5	Acceptable
7.6–11	Very slight or uncertain
11.1–14	Slightly unacceptable
14.1–22.5	Unacceptable to a medium extent
> 22.5	Highly unacceptable

Source: adapted from OCRA [14].

The CVT-GOHISALO-TEI questionnaire is an adaptation of the instrument for the formal economy sector (CVT-GOHISALO) in Mexico, which has been validated by experts with an overall score of over 70%.

This method contains 57 questions that evaluate or respond to the degree of satisfaction in the seven (7) dimensions of quality of life in informal workers, such as institutional support for work, job security, integration into the workplace, job satisfaction, well-being achieved through work, personal development of the worker and management of free time [15].

Table 2 defines the score or degree of satisfaction in informal work for each of the dimensions and its interpretation depending on the value obtained from the calculation of the questions.

Table 2. Level of satisfaction and interpretation for each dimension of quality of life.

Dimension	Value	Score (satisfaction level)	Interpretation
1. Institutional support for work	21–24	High	Satisfying work, support and recognition, sense of achievement.
	0–7	Low	Job dissatisfaction, distance and distrust, lack of progress and distrust.
2. Safety at work	44–56	High	High job satisfaction, positive perception of work, training and permanence.
	5–11	Low	Job dissatisfaction, health problems, perception of injustice and lack of opportunities.
3. Integration into the workplace	34–36	High	Satisfaction with job integration, interpersonal skills and emphasis on teamwork.
	1–9	Low	Job dissatisfaction, interpersonal problems, and low job performance.
4. Job satisfaction	26–32	High	Commitment to work, recognition and pride, and satisfaction with remuneration.
	4–16	Low	Job dissatisfaction, part-time work and lack of recognition.
5. Well-being achieved through work	20–61	High	Satisfaction with progress, contribution to society, improvement in quality of life.

(*continued*)

Table 2. (*continued*)

Dimension	Value	Score (satisfaction level)	Interpretation
	4–13	Low	Health risks, dissatisfaction with remuneration and lack of identification with work.
6. Personal development of the worker	23–29	High	Personal security, optimism and well-being, clear and achievable goals.
	2–11	Low	Personal dissatisfaction, low self-esteem, pessimism and opposition.
7. Free-time management	20–25	High	Planning skills, responsible compliance, work-life balance.
	0–9	Low	Poor time management, disorganization and irresponsibility, risk to health and personal relationships.

Source: adapted from CVT-GOHISALO-TEI [15]

3 Results

3.1 Characterization of the Activities and Tasks in the Manicure Process

According to the client's request, there are three types of manicures: traditional, semi-permanent, and acrylic. These require almost the same activities and differ only in the types of products or technologies used. However, for the analysis, a standardization was made that included the most frequently performed activities and those that require the greatest amount of repetitive effort on the upper extremities, resulting in 6 main activities (makeup removal, nail cleaning, monomer application, polishing, removal of impurities and makeup) and 15 repetitive tasks (makeup removal, cuticle removal, removal of impurities, filing, acrylic preparation, acrylic application, filing, nail length adjustment, polishing, removal of remaining acrylic material, application of exfoliant, removal of exfoliant, cleaning, application of base coat, application of nail polish, and/or decoration, and drying). Table 3 lists each of these activities, as well as the tools and supplies used in the process.

Table 3. Activities, tasks, tools and supplies in the manicure process.

Activities	Tasks	Tools/supplies
Remove makeup	Remove makeup	File or polisher
Nail cleaning	Cuticle removal	Embosser or polishing tips
	Removal of impurities	Cuticle cutter
	Filing	Files of different thicknesses
Monomer application	Acrylic preparation	Monomer and polymer
	Acrylic application	Brush and acrylic
Filing	Filing	File or polishing drill
	Adjusting the length of the nail	File, mold or tip
	Polishing	File, block or polisher
Removal of impurities	Removal of remaining acrylic material	Paintbrush or brush
	Scrub application	Scrub
	Scrub removal	Water and brush
	Cleaning	Acrylic cleaning liquid and/or alcohol
Makeup	Base application	Enamel base
	Application of enamel and decoration	Nail polishes, stickers, brushes
	Drying	LED lamp

Source: authors.

3.2 Sociodemographic Profile and Working Conditions of Manicurists

According to the aforementioned, 94.29% are female and 5.7% are male; the most common age groups are 28 to 37 years (34%), 18 to 27 years (31%), 38 to 47 years (29%), and 48 years or older (6%). Regarding their experience in the profession, 43% have 1 to 5 years, 37% 5 to 10 years, 8% 10 to 15 years, and 12% more than 15 years. Regarding marital status, 40% are single, 37% are in a common-law relationship, 14% are married, and 9% are separated. Furthermore, 74% are from socioeconomic stratum 2, 23% from stratum 3, and 3% from stratum 1. In addition, 54% live in rented accommodation, 37% in their own home, and 9% in a family home.

Furthermore, it was highlighted that 71% of manicurists work six days a week, 20% between four and five days, 6% between two and three days, and 3% only one day. Additionally, 42.9% work between six and eight hours a day, 40% between eight and ten hours, 11.4% less than six hours, and 5.7% more than ten hours a day. Regarding biomechanics, 48.6% revealed that they sit between six and eight hours a day, 20% two hours or less, 17.1% four hours, and 14.3% more than eight hours. Likewise, it was found that the same worker can be exposed to different working conditions. For example, 91%

reported repeating the same movement with their upper extremities several times, 49% performed rapid movements, and 37% performed movements with great precision.

Finally, regarding symptoms, 80% reported pain in one or more parts of the body, tingling, cramps, and loss of strength; 60% reported neck pain; and 50% reported pain and/or tingling in the elbow, wrists, and hands. The aggravating factor is that none of the respondents reported having any type of occupational risk insurance, and 3% have no health insurance at all, which increases their level of vulnerability to the work environment in terms of work hours and time, as well as their socioeconomic conditions and characteristics.

3.3 Ergonomic Risk Assessment of Upper Limbs in Manicurists

Based on the application of the ICKL assessment method for both upper extremities, it was determined that, for the right upper extremity, 65.7% of manicurists presented a highly unacceptable risk level (ICKL 35.8), 25.7% an unacceptable medium risk level (ICKL 19.7), and 13.2% a slightly unacceptable risk level (ICKL 13.2). This revealed a high probability of developing musculoskeletal disorders over time, due precisely to the repetitive movements and stresses exerted on this part of the body during the manicure process. Table 4.

Table 4. Risk level in the upper right limb of manicurists.

Percentage of manicurists (%)	ICKL	Risk level
65.7	35.8	Highly unacceptable
25.7	19.7	Unacceptable medium
8.6	13.2	Slightly unacceptable

Source: authors.

On the other hand, the measurement of the left upper limb showed that 42.86% of the workers had an highly unacceptable risk level (ICKL 30.65), 31.43% had an unacceptable medium risk level (ICKL 17.82), 5.71% had a slightly unacceptable risk level (ICKL 12.9), 2.86% had an acceptable risk level (ICKL 5.75), and 17.14% had an optimal risk level (ICKL 3.03). This showed that there is also a high possibility of musculoskeletal disorders in the future, but with less impact on the population, since the non-dominant upper limb used in the manicure process was the one considered. Table 5.

Table 5. Risk level in the upper left limb of manicurists.

Percentage of manicurists (%)	ICKL	Risk level
42.86	30.65	Highly unacceptable
31.43	17.82	Unacceptable medium
5.71	12.9	Slightly unacceptable

(*continued*)

Table 5. (*continued*)

Percentage of manicurists (%)	ICKL	Risk level
2.86	5.75	Acceptable
17.14	3.03	Optimal

Source: authors.

3.4 Evaluation of the Quality of Life of Manicurists

Based on the application of the CVT-GOHISALO-TEI questionnaire to the study opulation, it was established that, for the first and second dimensions of quality of life in informal work (institutional support for work and job security), 88.57% of manicurists had a medium level of satisfaction and 11.43% a high level, which in general terms means that most of them are highly motivated by their profession. However, there are aspects and dynamics that require improvement in terms of lack of advancement, mistrust, health problems, injustice, and lack of opportunities.

Regarding the third and fourth dimensions (job integration and job satisfaction), 80% and 62.86% of people reported a medium level of satisfaction, while 20% and 37.14% reported a high level, respectively. This means that most of them are highly committed and proud of their work, but, at the same time, there are certain interpersonal problems, such as low professional ability and lack of recognition. In contrast, for the fifth dimension (well-being achieved through work), 97.14% of manicurists reported a high level of satisfaction, and only 2.86% reported a medium level of satisfaction, demonstrating a positive attitude towards contributing to society and improving their quality of life.

Regarding the sixth and seventh dimensions of quality of life for these informal workers (personal development and free time management), 80% reported a medium level of satisfaction, 14.29% a high level, and 5.71% and 20% a low level, respectively. These were the only dimensions where a portion of the manicurists showed a low level of satisfaction (5.71% and 20%), due to low self-esteem, pessimism, poor free time management, disorganization, and work-related health risks. Table 6 details these results for each dimension evaluated.

Table 6. Evaluation of the quality of life of manicurists.

Dimension	Population (%)	Value obtained	Level of satisfaction
1. Institutional support for work	88.57	17.9	Medium
	11.43	22.5	High
2. Safety at work	88.57	32.8	Medium
	11.43	50	High
3. Integration into the workplace	80	27.1	Medium
	20	35.2	High

(*continued*)

Table 6. (*continued*)

Dimension	Population (%)	Value obtained	Level of satisfaction
4. Job satisfaction	62.86	21.8	Medium
	37.14	27.8	High
5. Well-being achieved through work	97.14	29.7	High
	2.86	17	Medium
6. Personal development of the worker	80	19	Medium
	14.29	25.4	High
	5.71	9	Low
7. Free-time management	80	12.6	Medium
	20	6.4	Low

Source: authors.

4 Discussion

According to the results, it is important to highlight that the working conditions of manicurists are unique and characteristic of informality, and that these conditions affect their physical well-being and quality of life. These include, for example, the lack of fixed work hours, limited free time, the need to sit for long hours, repetitive tasks, and excessive work hours per day, all of which support the causal relationship between working hours and the appearance of body symptoms among hairdressers working in metropolitan areas of India [16].

Therefore, it is clear that informal workers who work as manicurists are exposed to daily activities that involve repetitive effort and movements in their upper extremities; this is reflected in unacceptable levels of ergonomic risk in both limbs, but more so in the dominant or most used limb. This also explains the symptoms of pain, tingling, cramps, and loss of strength in the elbows, wrists and hands, which are present in the study population and promote the development of musculoskeletal disorders over time [17].

Regarding the measurement of the quality of life of informal workers, in five of the seven dimensions evaluated, manicurists are acceptably satisfied with their work, contrary to what has been established in other researches, where they indicate that informal work is synonymous with poverty, due to the lack of access to formal and competitive employment [18], which clearly differs from the perception of the study population. However, working informally has a particular impact on job satisfaction, given the interactions that the individual establishes in their life history and personality [19].

However, there is an important factor that negatively affects the well-being and quality of life of manicurists: the lack or inadequate management of free time, which has a negative impact on personal, social, and family life. Thus, this dimension is also related to the achievement of personal and family aspirations in the pursuit of existential satisfaction and happiness [20], and a low level of this implies an impact on the enjoyment of life outside the work environment [10, 21]; that is, the existence of an imbalance between work, family and social life.

In conclusion, this study provided valuable theoretical and methodological information that confirms the real exposure to non-ergonomic conditions in informal work and, specifically, the unacceptable level of risk posed by manicurists due to repetitive arm movements, and its direct relationship with the reported musculoskeletal symptoms. This also highlights that manicurists report inadequate management of their free time, which implies poor time management and an imbalance between work and personal life, leading to direct impacts on personal and family relationships.

References

1. International Labour Organization: Despite a lower unemployment rate in 2023, the recovery of labor markets in Latin America and the Caribbean is still insufficient. https://www.ilo.org/es/resource/news/oit-pesar-de-presentar-una-menor-tasa-de-desocupacion-en-2023-recuperacion, Last Accessed 01 Oct 2024
2. National Administrative Department of Statistics: Informal employment quarter January–March 2024. https://www.dane.gov.co/files/operaciones/GEIH/bol-GEIHEISS-ene2024-mar2024.pdf, Last Accessed 06 Oct 2024.
3. Colombian Ministry of Labor. Third National Survey of OSH Conditions in Colombia; 2021, https://www.researchgate.net/publication/362455968_TERCERA_ENCUESTA_NACIONAL_DE_CONDICIONES_DE_SST_EN_COLOMBIA, last accessed 2024/10/21.
4. Bolghanabadi, S., Haghighi, A., Jahangiri, M.: Insights into Women's occupational health and safety:a decade in review of primary data studies. Safety. **10**(2), 47 (2024). https://doi.org/10.3390/safety10020047
5. Abareshi, F., Rastaghi, S., Fallahi, M.: Nail job and women's health: assessment of ergonomics and occupational hazards. Work. **73**(2) (2022). https://doi.org/10.3233/WOR-210672
6. Sanaat, S., Holness, D.L., Arrandale, V.H.: Health and safety in nail salons: a cross-sectional survey. Annals of Work Exposures and Health. **65**(2), 225 (2021). https://doi.org/10.1093/annweh/wxaa078
7. Dang, J.V., Rosemberg, M., Le, A.B.: Perceived work exposures and expressed intervention needs among Michigan nail salon workers. Int. Arch. Occup. Environ. Health. **94**, 8 (2021). https://doi.org/10.1007/s00420-021-01719-6
8. World Health Organization.: Musculoskeletal Disorders. 2021. https://www.who.int/es/news-room/fact-sheets/detail/musculoskeletal-conditions, Last Accessed 06 Oct 2024
9. Machado-Souza, M., De Oliveira-Borges, L.: Meanings of work for manicurists and hairdressers: employees and pejotizados: social. Work and Organizations Psychology. **3**, e37544 (2021). https://doi.org/10.1590/0102.3772e37544
10. Hellmann-Dario, K., Belletti-AN, O., Pereira Jorge, I.M.: Qualidade de vida no trabalho: um relato sobre manicures: Revista Família. Ciclos de Vida e Saúde no Contexto Social. **2** (2018). https://doi.org/10.18554/refacs.v6i0.3123
11. Huynh, T.B., Oddo, V.M., Trejo, B., Moore, K., Quistberg, D.A., Klim, J.J., et al.: Association between informal employment and depressive symptoms in 11 cities in Latin America. SSM Popul. Health. **18** (2022). https://doi.org/10.1016/j.ssmph.2022.101101
12. Hernández, R., Mendoza, C.P.: Metodología de la investigación: las rutas cuantitativa, cualitativa y mixta. McGraw-Hill, México (2018)
13. Resolución 8430 de 1993.: Por la cual se establecen las normas científicas, técnicas y administrativas para la investigación en salud (4 Oct 1993).

14. Diego-Mas JA. Evaluación del riesgo por movimientos repetitivos mediante el Check List Ocra. Ergonautas, Universidad Politécnica de Valencia (2015). https://www.ergonautas.upv.es/metodos/ocra/ocra-ayuda.php, Last Accessed 21 Oct 2024
15. Gutiérrez-Strauss, A.M., González-Baltazar, R., Salazar-Estrada, J.G.: Instrumento de medición de la calidad de vida en el trabajo para trabajadores de la economía informal CVT – GOHISALO – TEI: manual para su aplicación e interpretación. Instituto de Investigación en Salud Ocupacional. Universidad de Guadalajara, México (2014)
16. Mishra S, Sarkar K.: Work-related musculoskeletal disorders and associated risk factors among urban metropolitan hairdressers in India: Revista National Library of medicine. 63(1) (2021). https://doi.org/10.1002/1348-9585.12200.
17. Lteif, M., El-Hayek, M., Azouri, H., Antonios, D.: Assessment of work-related symptoms, perceived knowledge, and attitude among nail salon technicians. Toxicol. Ind. Health. **36**(11), 852 (2020). https://doi.org/10.1177/0748233720951294
18. Galarza-Iglesias, A.M., Beltrán-Narvaéz, L.H., Ordoñez-Hernández, C.A.: Quality of work life and health-related quality of life in women working in the informal economy in the commune 9 of Cali, Colombia. Revista Facultad de Medicina. **69**(4) (2021). https://doi.org/10.15446/revfacmed.v69n4.85580
19. Chiang-Vega, M.M., Riquelme-Neira, G.R., Rivas-Escobar, P.A.: Relationship between job satisfaction, work stress and its results in workers of a charity institution of the province of Concepción. Ciencia & Trabajo. **20**(63), 178 (2018). https://doi.org/10.4067/S0718-24492018000300178
20. Pérez- Buelvas C A, Velásquez-Arboleda OH.: Quality of working life-GOHISALO in grazing dairy farms in Don Matías and La Unión, Colombia: Journal of Agriculture and Animal Sciences 5(2) (2016). https://www.researchgate.net/publication/313458898_Calidad_de_Vida_en_el_Trabajo_-GOHISALO-_en_hatos_lecheros_en_pastoreo_en_Don_Matias_y_La_Union_Colombia.
21. Rojas-Torres AC, Tiga-Loza DC, Bello-Pinto, W.: Quality of work life in employees of beverage distribution company in Santander, Colombia: Medicina y Seguridad del Trabajo 67(265), (2021). https://doi.org/10.4321/s0465-546x2021000400002

Evaluation of Fatigue, Control, and Drowsiness in Drivers of a Manufacturing Company in Ecuador

Kenny Escobar-Segovia[1(✉)], Daniela Paz-Barzola[1], Rómulo Villamar-Triviño[2], and Luis Duque-Cordova[2]

[1] Escuela Superior Politécnica del Litoral, Guayaquil, Ecuador
{kescobar,dpaz}@espol.edu.ec
[2] Universidad del Pacífico, Guayaquil, Ecuador
romulovillamar@homail.com, luis.duque@upacifico.edu.ec

Abstract. Fatigue, Drowsiness, and Locus of Control as Key Factors Affecting Driver Performance and Safety in Demanding Work Environments. Fatigue, drowsiness, and locus of control are key factors that can impact driver performance and safety in demanding work environments. In the manufacturing sector, where drivers face long shifts and variable working conditions, it is crucial to assess these aspects to reduce accident risks and improve occupational well-being. This study aimed to evaluate fatigue, control, and drowsiness levels in drivers of a manufacturing company in Ecuador to identify associated factors and propose intervention strategies to enhance workplace safety and efficiency. A cross-sectional descriptive and correlational study was conducted with 1,097 drivers from a manufacturing company structured into three operational sectors. The Yoshitake Work Fatigue Symptoms Questionnaire, Rotter's Locus of Control Scale, and the Epworth Sleepiness Scale were applied. Data were analyzed using descriptive statistics and chi-square tests to assess significant associations. The most relevant finding indicates that 53.1% of drivers experienced intellectual fatigue, making it the predominant type of fatigue. Additionally, 87.1% exhibited an internal locus of control, suggesting a perception of control over their job performance. Regarding drowsiness, 40.4% of drivers reported abnormal sleepiness, with a significant association with age ($p = 0.004$). Significant associations were also found between fatigue and the work sector ($p = 0.044$) and between drowsiness and work shifts ($p = 0.012$). The results highlight the need to implement fatigue and drowsiness management strategies, such as optimizing work shifts and incorporating alertness monitoring technologies in driving. These findings align with previous studies on road safety and workplace ergonomics, emphasizing the importance of adapting working conditions to drivers' physiological needs to reduce risks and improve performance.

Keywords: Drivers · Drowsiness · Locus of Control · Occupational Fatigue · Road Safety

J. L. Ballesteros et al. (Eds.): CITIS 2025, LNNS 1780, pp. 274–282, 2027.
https://doi.org/10.1007/978-3-032-14287-0_27

1 Introduction

Road safety is a fundamental aspect of logistical operations in any manufacturing company, as the transportation of products largely depends on drivers' ability to perform their duties efficiently and safely [1].

In this context, factors such as fatigue, drowsiness, and driving control are crucial in preventing accidents and improving driver performance [2]. In Ecuador, the increase in traffic accidents related to these factors underscores the need to investigate how they specifically affect drivers in manufacturing companies, who often face long working hours and conditions that contribute to the emergence of these issues [3].

Fatigue is a physiological and psychological condition that can decrease concentration, increase reaction time, and reduce cognitive performance [4]. In drivers, fatigue can have serious consequences as it directly impacts decision-making and environmental perception [5]. Various factors can contribute to fatigue, such as the duration of work shifts, lack of adequate rest, and the physical or mental demands of the job [6].

Likewise, driving control refers to the driver's ability to maintain focus on the task, follow traffic rules, and respond appropriately to unexpected situations [7]. Loss of control can be linked to both fatigue and drowsiness, but it may also stem from factors such as work-related stress, distractions, and lack of concentration [8].

Drowsiness is a state characterized by an irresistible urge to sleep and is one of the main factors contributing to traffic accidents [9]. Drowsiness reduces alertness and increases the risk of microsleeps, brief periods in which the driver loses awareness of their surroundings [10].

Factors such as night shifts, long working hours, and irregular schedules are common in the manufacturing industry, increasing the likelihood that drivers will experience drowsiness [11]. To measure drowsiness, tools such as sleepiness scales and monitoring devices are used to identify the times of day when drivers are most prone to entering this state.

The objective of this study was to evaluate fatigue, control, and drowsiness levels in drivers of a manufacturing company in Ecuador to identify the underlying causes affecting their performance and develop intervention strategies that promote a safer and more efficient work environment.

2 Methodology

The study is descriptive and correlational in nature, with a cross-sectional design. This design allows for the description and analysis of the characteristics and relationships between variables at a specific point in time. A total of 1,097 drivers were evaluated in 2023, representing the study population.

For this study, the analyzed population belongs to a large-scale manufacturing company that operates under a holding model and is structured into three distinct sectors (Sector 1, Sector 2, and Sector 3). This division corresponds to the specialization of production processes and the optimization of operational management within the organization. Each sector encompasses different key functions within the production and logistics chain, enabling more efficient resource management and appropriate segmentation of personnel according to their work area. The classification into sectors facilitated data collection and analysis, ensuring a representative evaluation of working and organizational conditions within the company.

Three questionnaires were used to collect the information. The names of each questionnaire used are presented below:

- Yoshitake Work Fatigue Symptoms Questionnaire [12].
- Locus of Control Measurement [13].
- Epworth Sleepiness Scale [14].

The questionnaires were administered to the drivers in an appropriate environment to ensure their comfort and concentration. Confidentiality of their responses was guaranteed, and clear instructions were provided on how to complete each test.

Once collected, the data was tabulated and processed using Microsoft Excel for initial organization. Subsequently, the statistical software SPSS was used for more detailed analysis, presenting the data in tables with frequencies and percentages. Additionally, Pearson's chi-square test was applied to measure associations between the analyzed variables.

3 Results

Table 1 presents the sociodemographic data of the sample, which consists mostly of men (99.1%). The most common age group is 26 to 35 years old (36.7%). Regarding marital status, the majority are married (47.5%), followed by single individuals (24.5%) and those in a common-law union (23.5%). In terms of education, 75.2% have a high school diploma, while only a few have completed higher education (3.2%) or postgraduate studies (1.1%).

Table 2 summarizes the employment data of the sample. The majority work in Sector 1 (60.8%), primarily in Guayas (22.6%). The most common areas are Logistics/Patio Driver (34.2%) and T1 (29.3%). A total of 37.2% have between 4 and 9 years of seniority, and 46.2% work the morning shift. Only 1.4% have another job, and 6.7% have experienced a workplace incident.

Table 3 presents the overall test results. In the Fatigue Test, the majority exhibit intellectual fatigue (53.1%), while 27.0% show no signs of fatigue. In the Control Test, internal control predominates (87.1%). In the Sleepiness Test, the results are balanced, with 41.9% in normal sleep and 40.4% experiencing abnormal sleepiness.

Table 4 presents the relationship between sociodemographic variables and test results using p-values. No significant associations were found between gender, marital status, or education level with any of the tests. However, age shows a significant association with sleepiness ($p = 0.004$), indicating a possible relationship between age and sleepiness levels.

When analyzing the age category of participants in relation to the Epworth Sleepiness Test, it is observed that more than half of workers aged 18 to 35 have normal sleep, followed by 28.7% with abnormal sleepiness, and a smaller proportion with moderate sleepiness. However, most older workers (46 years or older) experience abnormal sleepiness.

Table 1. Sociodemographic Data

Sex	n	%
Female	10	0.9
Male	1087	99.1
Age group (years)	n	%
18–25	101	9.2
26–35	403	36.7
36–45	375	34.2
46 or more	218	19.9
Marital status	n	%
Married	521	47.5
Divorcied	45	4.1
Single	269	24.5
Common-law union	258	23.5
Widowed	4	0.4
Education level	n	%
High school graduate	825	75.2
Completed higher education	35	3.2
Incomplete higher education	123	11.2
Postgraduate	12	1.1
Primary education	102	9.3

n: Sample size; %: Percentage

Table 5 analyzes the relationship between work-related variables and test results using p-values. Significant associations were found in some variables, indicated with an (*).

The results indicate the prevalence of Type 2 intellectual fatigue in the Yoshitake Fatigue Test among the surveyed workers. More than 50% of employees in different work areas, except for those in T2, exhibited Type 2 fatigue. For the T2 position, this type of fatigue remains the most prevalent compared to other types of fatigue identified in the test.

Regarding work shifts, 41.8% of morning shift workers, 53.8% of night shift workers, and 52% of rotating shift workers have normal sleep according to the sleepiness test. In terms of fatigue, Type 2 has shown the highest prevalence in workplace incidents (46.6%) compared to other types of fatigue.

Table 2. Employment data

Sector	n	%
Sector 1	667	60.8%
Sector 2	133	12.1%
Sector 3	297	27.1%
Location	n	%
Guayas	248	22.6%
Pichincha	162	14.8%
Santo Domingo	115	10.5%
Others	147	13.4%
Does not mention	425	38.7%
Work area	n	%
Logistics/Patio Driver	375	34.2%
Forklift driver	170	15.5%
R2	40	3.6%
T1	321	29.3%
T2	191	17.4%
Seniority (years)	n	%
Less than 1	160	14.6%
1–3	228	20.8%
4–9	408	37.2%
10 or more	301	27.4%
Shift	n	%
Morning	507	46.2%
Mix	306	27.9%
Night	26	2.4%
Rotary	204	18.6%
Evening	54	4.9%
Other job	n	%
No	1082	98.6%
Yes	15	1.4%
Workplace Incident	n	%
No	1024	93.3%
Yes	73	6.7%

n: Sample size; %: Percentage

Table 3. Overall test results

Fatigue Test	n	%
None	296	27.0%
Type 1 (Mixed)	35	3.2%
Type 2 (Intellectual)	582	53.1%
Type 3 (Physical)	184	16.8%
Control Test	n	%
External Control	142	12.9%
Internal Control	955	87.1%
Sleepiness Test	n	%
Abnormal Sleepiness	443	40.4%
Moderate Sleepiness	194	17.7%
Normal Sleep	460	41.9%

n: Sample size; %: Percentage

Table 4. Sociodemographic data vs p-values of the Tests

	Test		
Variables	Fatigue	Control	Sleepiness
Sex	0.520	0.781	0.139
Age group (years)	0.508	0.751	**0.004***
Marital status	0.768	0.826	0.650
Education level	0.169	0.804	0.627

* Significant statistical association between the variables

Table 5. Employment data vs. p-values of the Tests

	Tests		
Variables	Fatigue	Control	Sleepiness
Sector	**0.044***	0.760	0.277
Location	0.251	0.967	0.070
Work area	**0.016***	0.096	**0.041***
Seniority (years)	0.398	0.098	**0.021***
Shift	0.191	0.189	**0.012***
Other job	0.877	0.964	0.854
Workplace Incident	**0.028***	0.377	0.827

* Significant statistical association between the variables

4 Discussion

The results of this study revealed a high prevalence of intellectual fatigue among the evaluated drivers, with 53.1% exhibiting this condition. These findings align with previous studies that have indicated cognitive fatigue is common in occupations requiring high concentration and prolonged attention spans, such as freight transportation and industrial logistics [15, 16]. It has been demonstrated that fatigue negatively affects reaction capacity and decision-making, increasing the risk of occupational and traffic accidents [17].

Regarding driving control, most drivers exhibited an internal locus of control (87.1%), suggesting that they perceive their performance as a result of their own effort and abilities. This finding contrasts with previous studies that have identified a higher proportion of external control in workers with adverse working conditions, such as night shifts or high workloads. [18].

The analysis of sleepiness revealed that 40.4% of drivers experience abnormal sleepiness, while 41.9% showed normal sleep levels. Compared to previous research, these values are similar to those found in long-distance drivers, where disruptions in the circadian rhythm and extended work shifts significantly impact sleep quality [19]. The significant relationship between age and sleepiness ($p = 0.004$) suggests that drivers over 46 years old are more prone to daytime sleepiness, aligning with studies indicating that advanced age increases sleep latency and cumulative fatigue [20].

From a workplace perspective, fatigue was found to be significantly associated with the work sector ($p = 0.044$), the area within the company ($p = 0.016$), and having experienced previous incidents ($p = 0.028$). This suggests that certain working conditions may increase the likelihood of fatigue, a finding that has been documented in studies on workplace ergonomics and occupational risk prevention [21]. Similarly, sleepiness showed significant associations with job tenure ($p = 0.021$) and work shift ($p = 0.012$), which is consistent with research demonstrating that rotating and night shifts disrupt sleep cycles and may increase episodes of microsleeps [22].

The results of this study highlight the need to implement intervention strategies to reduce fatigue and sleepiness in drivers, such as active break programs, work shift adjustments, and sleep monitoring. In line with previous research, the adoption of alert technologies to detect microsleep episodes and enhance workplace safety is recommended. [23].

5 Conclusions

The chi-square correlation analysis has revealed significant relationships between various sociodemographic variables and test results. The work area to which a driver belongs has a significant association with the type of fatigue they experience. It is essential to consider the specific demands and working conditions of each area when addressing fatigue-related issues.

There is a significant relationship between a driver's age and their level of sleepiness. It is crucial to consider workers' age when addressing sleep and rest-related issues. Additionally, the work shift of drivers is significantly associated with their level of sleepiness.

Drivers working evening shifts may require special attention to ensure adequate rest and prevent daytime sleepiness.

These relationships may be key to implementing targeted strategies aimed at improving employee well-being based on their characteristics and working conditions.

References

1. Zaranka, J., Pečeliūnas, R., Žuraulis, V.: A road safety-based selection methodology for professional drivers: behaviour and accident rate analysis. Int. J. Environ. Res. Public Health **18**(23), 12487 (2021)
2. Gimeno, P.T., Cerezuela, G.P., Montanes, M.C.: On the concept and measurement of driver drowsiness, fatigue and inattention: Implications for countermeasures. Int. J. Veh. Des. **42**(1), 67–86 (2006)
3. Felix, M., et al.: Risk of obstructive sleep apnea and traffic accidents among male bus drivers in Ecuador: is there a significant relationship? Ann. Med. Surg. **74**, 103296 (2022)
4. Lock, A., Bonetti, D., Campbell, A.: The psychological and physiological health effects of fatigue. Occup. Med. **68**(8), 502–511 (2018)
5. Matthews, G., Desmond, P.A.: Task-induced fatigue states and simulated driving performance. Q. J. Exp. Psychol. Sect. A: Hum. Exp. Psychol. **55**(2), 659–686 (2002)
6. Techera, U., Hallowell, M., Stambaugh, N., Littlejohn, R.: Causes and consequences of occupational fatigue: meta-analysis and systems model. J. Occup. Environ. Med. **58**(10), 961–973 (2016)
7. Roca, J., Crundall, D., Moreno-Ríos, S., Castro, C., Lupiáñez, J.: The influence of differences in the functioning of the neurocognitive attentional networks on drivers' performance. Accid. Anal. Prev. **50**, 1193–1206 (2013)
8. Sinclair, M., Swart, E.: Sleep-related issues and fatigue. In: International Encyclopedia of Transportation: Volume 1–7, vol. 2, pp. 611–616 (2021)
9. Medina-Amán, G., Escobar-Segovia, K., Arias-Ulloa, C.: Factores de riesgo y su relación con la fatiga en conductores de una cooperativa de transporte interprovincial del Ecuador. Revista San Gregorio **1**(46), 30–46 (2021)
10. Li, G., Chung, W.Y.: Electroencephalogram-based approaches for driver drowsiness detection and management: a review. Sensors **22**(3), 1100 (2022)
11. Liang, Y., et al.: Prediction of drowsiness events in night shift workers during morning driving. Accid. Anal. Prevent. **126**, 105–114 (2019)
12. Yoshitake, H.: Three characteristic patterns of subjective fatigue symptoms. Ergonomics **21**(3), 231–233 (1978)
13. Rotter, J.B.: Generalized expectancies for internal versus external control of reinforcement. Psychol. Monogr. Gen. Appl. **80**(1), 1–28 (1966)
14. Johns, M.: A new method for measuring daytime sleepiness: the Epworth sleepiness scale. Sleep **14**(6), 540–545 (1991)
15. Philip, P., et al.: Fatigue, sleep restriction and driving performance. Accid. Anal. Prev. **37**(3), 473–478 (2005)
16. Williamson, A., Lombardi, D.A., Folkard, S., Stutts, J., Courtney, T.K., Connor, J.L.: The link between fatigue and safety. Accid. Anal. Prev. **43**(2), 498–515 (2011)
17. Wishart, D., Barrett, D.: Workplace fatigue: the impact of the drive. In: Handbook on Management and Employment Practices, pp. 1–23 (2021)
18. Spector, P.: Development of the Work Locus of Control Scale. J. Occup. Psychol. **61**(4), 335–340 (1988)

19. Naderi, H., Nassiri, H., Sahebi, S.: Assessing the relationship between heavy vehicle driver sleep problems and confirmed driver behavior measurement tools in Iran. Transport. Res. F: Traffic Psychol. Behav. **59**, 57–66 (2018)
20. Ingre, M., Akerstedt, T., Peters, B., Anund, A., Kecklund, G.: Subjective sleepiness, simulated driving performance and blink duration: examining individual differences. J. Sleep Res. **15**(1), 47–53 (2006)
21. Karasek, R., Theorell, T.: Healthy Work: Stress, Productivity, and the Reconstruction of Working Life. Basic Books, New York (1992)
22. Costa, G.: Shift work and health: current problems and preventive actions. Saf. Health Work **1**(2), 112–123 (2010)
23. Dement, W.C., Carskadon, M.A.: Current perspectives on daytime sleepiness: the issues. Slepp **5**(2), 56–66 (1982)

Postural and Anthropometric Risk Factors for Musculoskeletal Disorders in University Workers

Thalía San Antonio(✉), Alejandra Lascano, David Guevara, and Oscar Analuiza

Universidad Técnica de Ambato, Ambato 180207, Ecuador
t.sanantonio@uta.edu.ec

Abstract. Given that all human endeavors, including work, pose certain health risks, it's imperative to proactively identify and mitigate these hazards. Disergonomic factors are a primary concern in occupational health and safety. This study aimed to identify and evaluate postural risk factors for musculoskeletal disorders (MSDs) among university workers in Ecuador. The REBA method and anthropometric measurements were used to assess job tasks, postures, and physical characteristics. The REBA method was applied to evaluate postural risk, while anthropometric measurements were collected using the INTE/ISO 7250-1:2018 standard. The study revealed that construction workers were particularly at risk of MSDs due to demanding tasks involving heavy lifting, awkward postures, and excessive strain. A significant proportion (96.3%) of the evaluated activities were identified as requiring changes or adjustments to prevent MSDs. The findings emphasize the importance of implementing ergonomic interventions and considering anthropometric data to create healthier and safer workplaces by tailoring workplaces to the specific physical characteristics of university workers. Properly designed work systems can optimize task performance, preventing fatigue and injuries, improving concentration, and increasing productivity. As complementary measures, to implement the ergonomic proposal and monitor its effectiveness, two steps are proposed: setting up an ergonomics committee and organizing training for committee members and employees.

Keywords: Musculoskeletal Disorders (MSDs) · Ergonomic Assessment · REBA Method · Anthropometric Measurements

1 Introduction

The health and well-being of university workers are increasingly recognized as important factors in maintaining a productive and sustainable work environment. Musculoskeletal disorders (MSDs), particularly spinal conditions, pose significant health and economic challenges for this population. These disorders often stem from repetitive tasks, poor posture, excessive physical demands, and the magnitude and frequency of spinal loads [1, 2]. Consequently, chronic pain, reduced productivity, and increased absenteeism become prevalent, highlighting the need to understand the factors driving MSD development.

J. L. Ballesteros et al. (Eds.): CITIS 2025, LNNS 1780, pp. 283–293, 2027.
https://doi.org/10.1007/978-3-032-14287-0_28

To address the rising prevalence of MSDs, it is essential to design work environments and tasks to fit the human body even if may be initial costs associated with implementing ergonomic changes, the long-term savings from reduced healthcare costs, workers' compensation claims, and increased productivity often outweigh these expenses [3, 4]. Waiting until MSDs occur is both costly and inefficient. Implementing ergonomic interventions before injuries happen can prevent pain, lost work time, and decreased job satisfaction [5, 6]. Moreover, studies examining specific populations, such as elderly indigenous women, have emphasized the influence of demographic and cultural factors in the way in which tasks are assigned and executed in the different members of the community [7].

Ergonomic assessments tools, such as OWAS, RULA, and REBA, are commonly used to evaluate work-related musculoskeletal risks allowing to identify and mitigate risk factors associated with MSDs. By analyzing the physical demands of various tasks, ergonomic evaluations can help to optimize work environments and reduce the incidence of these disorders, with studies showing their effectiveness in various settings [1, 8, 9]. These systematic assessments enable targeted interventions to address ergonomic challenges.

Among various assessment tools available for identifying and assessing ergonomic risks, the Rapid Entire Body Assessment (REBA) stands out for its comprehensive analysis of body postures. By considering flexion and extension angles of different body parts, grip force, postural stability, and load handling. REBA provides a detailed evaluation of postural risk [10]. This method is particularly suitable for studying university workers as it can analyze the specific postural demands of various daily activities of these workers. Moreover, by considering both sensitivity (detailed information on measured parameters) and generality (applicability in diverse contexts), REBA offers a valuable tool for identifying risk postures that may contribute to the development of MSDs [11]. The choice of the REBA method for this study is based in its ability to simultaneously assess postural and anthropometric risks. By considering workers' anthropometric characteristics (such as body size and proportion of body segments), REBA can identify potential mismatches between task demands and workers' physical capabilities, which in turn may increase the risk of developing MSDs [12].

To obtain the basic anthropometric measurements required for this study, the INTE/ISO 7250-1:2018 standard was employed [13]. This standard provides a comprehensive framework for body measurements, enabling comparisons between population groups and the construction of anthropometric databases. The standard's primary list of measurements serves as a valuable resource for ergonomists, aiding population definition and the informed design of living and working spaces. Furthermore, it underpins the acquisition of two- and three-dimensional measurements through scanning techniques. The absence of a local anthropometric database in Ecuador presents a significant challenge in analyzing and adapting workplaces to effectively accommodate the body dimensions of these population.

This study aims to identify and evaluate job-related postural risk factors for MSDs among university workers. By applying the REBA method [11] and adhering to the ITE/ISO 7250-1:2018 standard for anthropometric measurements [13], specific tasks that contribute to MSD risk will be identified. This research will inform the development of targeted prevention strategies to improve the health and well-being of university employees included in the study.

2 Methodology

2.1 Study Subject Characteristics

The present investigation is carried out at Universidad Técnica de Ambato (UTA) in Ecuador, specifically with workers who belong to the Department of Physical Infrastructure and the Administrative Department whose positions and ages are very diverse and detailed in Table 1. All study subjects (three per position) were male with ages between 29 and 59.

Table 1. Study subject relevant characteristics for the REBA assessment [14].

Study subject characteristics			
Position	Ages	Average age	Department
Driver	50/53/53	48,13	Administrative
Construction worker	37/39/39	39,6	Physical Infrastructure
Electrician	33/36/37	35,5	Physical Infrastructure
Maintenance worker	32/35/44	43,5	Physical Infrastructure
Gardener	29/43/49	43,75	Physical Infrastructure
Carpenter	44/50/59	51,75	Physical Infrastructure
Plumber	31/37/40	35,33	Physical Infrastructure
Janitor	35/50/51	44,33	Physical Infrastructure
Security guard	35/40/51	43,51	Administrative

2.2 Steps Preceding the REBA Assessment

The following steps detail the required actions to obtain the information to apply REBA to the university workers [11]. Table 2 summarizes the study subject positions and activities, highlighting those involving prolonged or frequent postural stress. Listed below are the steps preceding the REBA assessment:

a. Determine the observation period for each position, considering the entire work cycle time.
b. If necessary, due to the prolonged duration of the evaluated task, it should be broken down into smaller, manageable units for detailed analysis.
c. Record the worker's postures throughout the task (photos and videos).
d. Identify from all the recorded postures, in the previous step, those considered the most critical or risky postures for subsequent REBA assessment.

e. For each selected posture, the side of the body that is considered as the one with a greater postural load, required the following measurements: (1) the angles formed by the different body parts (2) the load or force handled when adopting the studied posture (in kilograms) and (3) the grip type of the handled load by any body parts.

Table 2. Study subject positions and their corresponding activities throughout the workday. The activities in bold letter for each position are signed as the ones involving prolonged or frequent postural stress and therefore those that will be subsequently analyzed [14].

Study subject positions and activities	
Position	Activities
Driver	(1) Vehicle maintenance, (2) mechanical repairs, (3) **electrical inspection**, (4) **verification of tire condition**, (5) transportation of personnel, tools, or work materials, (6) **loading and unloading objects** and (7) completing daily paperwork and reports
Construction worker	(1) Infrastructure maintenance, (2) structural assembly, (3) **metalworking**, (4) welding, (5) **surface finishing** and (6) **material handling**
Electrician	(1) **Electrical maintenance**, (2) electrical installations, (3) material handling, (4) **equipment moving**, (5) **wiring**, (6) panel assembly and (7) preventive maintenance
Maintenance worker	(1) Infrastructure construction (2) Infrastructure maintenance, (3) **material processing**, (4) welding, (5) **surface finishing** and (6) **material handling**
Gardener	(1) Maintenance of green areas, (2) application of fungicides and insecticides, (3) material handling, (4) **soil preparation**, (5) irrigation, (6) **sowing and planting** and (7) **cutting and pruning**
Carpenter	(1) Infrastructure maintenance, (2) **material cutting and handling**, (3) **sanding**, (4) machinery moving, (5) installation and repair of elements primarily made of wood and (6) **painting**
Plumber	(1) Irrigation system maintenance, (2) **pipe installation**, (3) heating system installation and maintenance, (4) **material and equipment handling**, (5) drain clearing and (6) **leak repair**
Janitor	(1) General maintenance, (2) cleaning, (3) material handling, (4) **cleaning equipment** (5) **furniture handling**, (6) **waste disposal** and (7) messenger services
Security Guard	(1) Assisting with general maintenance tasks, (2) **controlling vehicle traffic flow**, (3) **workspace cleaning**, (4) checking IDs and vehicle access and (5) **road safety and signage**

2.3 Anthropometric Measurements Using INTE/ISO 7250–1:2018 Standard

Properly designed workstations can optimize task performance, preventing fatigue and injuries, contribute to improved concentration, and increasing productivity. Anthropometric design principles that include: (1) design for extremes, (2) design for an adjustable range, and (3) design for the average, are crucial for achieving this. The data necessary to apply these principles are acquired through the collection of 43 anthropometric measurements. These body measurements are acquired and selected according to the INTE/ISO 7250–1:2018 standard [13] including anthropometric measurements from standing and seated positions, specific body segments, functional dimensions and weight. The collection of the anthropometric data provides a comprehensive understanding of the variability within the university worker population. Such insights are important elements to create and redesign workstations.

2.4 Steps for REBA Evaluation

The REBA method is a widely recognized and extensively employed tool for evaluating the risk of developing MSDs across a broad spectrum of human activities. This is achieved by systematically evaluating various postural factors, including trunk flexion, neck rotation, and arm and leg positions. The REBA method provides a valuable framework for identifying and mitigating potential hazards. This versatile tool has proven to be an essential asset in promoting workplace safety, improving productivity, and safeguarding the health and well-being of workers in diverse industries and settings.

This comprehensive assessment protocol is particularly valuable in contexts characterized by significant postural demands, such as: Industrial settings, where repetitive tasks, heavy lifting, and awkward postures can contribute to MSDs [15], Agricultural labor, involving physically demanding activities like harvesting, planting, and animal husbandry [16], Everyday living activities, encompassing domestic chores, recreational pursuits, occupational tasks that can strain the musculoskeletal system [7], Assistive device use, for individuals with physical disabilities, the REBA method can assess the ergonomic compatibility and potential risks associated with assistive equipment [17], among many others.

In the REBA evaluation the body is divided into two groups: Group A, comprising the trunk, neck, and legs (lower limbs), and Group B, consisting of the upper limbs (arm, forearm, and wrist). The individual score for members of each group is determined as follows:

- The initial score for Group A is obtained from the tables based on the individual scores of the lower limbs.
- The evaluation of Group B is based on the scores of the upper limbs using the corresponding tables.
- The score assigned to Group A is modified based on the applied load or forces.
- The score assigned to the upper or lower limbs is adjusted according to the type of grip used to handle the load.

3 Results

3.1 Anthropometric Data Based in INTE/ISO 7250–1:2018 Standard

Ergonomic workplace redesign, supported by precise anthropometric data, is a proven strategy for mitigating MSDs arising from postural loads. By collecting anthropometric measurements from the target population, it is possible to tailor workstations and work routines to accommodate the unique physical characteristics of different individuals.

By considering the 5th to 95th percentiles (P5, P10, P50, P90, P95) of anthropometric measurements, ergonomic interventions can be designed to address the needs of individuals with varying body dimensions. This approach ensures that workstations are accessible to most users, while also accommodating those with more extreme physical characteristics. In sectors handling bulky equipment, designing for the 95th percentile can help prevent forced postures and ensure optimal reach of controls. Conversely, designing for the 5th percentile can enhance accessibility for individuals with smaller body frames. Average measurements (P50) are often used when a single design or adjustment is required to accommodate a broad range of users.

The integration of anthropometric data into ergonomic workplace redesign is essential for creating safe, comfortable, and productive work environments that are tailored to specific needs, by considering the diverse physical attributes of workers (Table 3).

Table 3. Anthropometric measurements by percentiles 5 to 95 of the selected sample of UTA workers including standing and sitting positions [14].

Anthropometric Data					
Dimensions	P 5	P 10	P 50	P 90	P 95
Weight (Kg)	58,0	61,5	73,6	85,6	89,1
In standing position (all dimensions in cm)					
Stature	154,2	156,6	164,8	173,1	175,4
Eye height	141,7	144,3	153,4	162,5	165,0
Shoulder height	128,9	131,3	139,9	148,5	150,9
Elbow heigh	95,9	97,4	103,0	108,5	110,1
Hip height	67,5	68,9	73,8	78,7	80,1
Chest Depth	21,8	23,7	30,0	36,4	38,2
Hip breadth	32,6	33,2	35,2	37,2	37,7
In sitting position (all dimensions in cm)					
Sitting heigh	77,9	79,3	84,1	88,9	90,3
Sitting eye height	68,2	69,3	73,3	77,4	78,5
Sitting cervical height	59,3	60,4	64,1	67,9	68,9
Sitting shoulder height	54,4	55,6	59,9	64,1	65,3
Sitting elbow heigh	17,9	19,6	25,6	31,6	33,3

(continued)

Table 3. (*continued*)

Anthropometric Data					
Dimensions	P 5	P 10	P 50	P 90	P 95
Shoulder-elbow length	30,2	31,1	34,3	37,4	38,3
Shoulder breadth (bideltoid)	41,6	42,5	45,8	49,1	50,0
Forearm-forearm breadth	38,3	41,1	50,6	60,2	62,9
Hip breadth	23,3	25,8	34,2	42,6	45,1
Popliteal height	39,8	40,6	43,5	46,4	47,2
Sitting knee height	36,0	39,3	51,1	62,9	66,2

3.2 Rapid Entire Body Assessment (REBA)

The job position with the highest postural load and therefore the most exposed to MSDs is that of the construction worker, due in part to the diverse and multifunctional nature of their tasks, many of which involve forced loads such as lifting heavy objects, excessive trunk flexion, raising arms to angles greater than 60°, and adopting uncomfortable postures requiring leg flexion at angles greater than 100°.

To assess the specific risk levels associated with the selected activities (see Table 2), a detailed analysis of works cycles was conducted. Table 4 presents the results of the REBA evaluation, revealing that a substantial proportion of the evaluated tasks pose a significant risk of MSDs. In fact, 96.3% of the activities assessed were identified as requiring changes or adjustments to prevent the onset of MSDs. More specifically, 22,22% of the evaluated activities have a high-risk level and 14.81% have a very high risk.

Once the results were established, the evaluated activities were ranked according to their level of risk to the worker's health, and recommendations were generated for each case to control exposure to working conditions and/or activity demands that could cause problems or injuries to the worker in the medium and long term, focused on eliminating or significantly reducing the risks. These recommendations, presented in an ergonomic proposal, were submitted to the university for implementation. A crucial component of this process was the planning of worker training, which will be detailed later (Fig. 1).

Table 4. Level of MSD risk for each of the activities highlighted in Table 2 for each job position studied [14].

Position	Level of MSD Risk			
	Low	Medium	High	Very high
Construction worker	0	1	2	0
Carpenter	0	3	0	0
Driver	0	1	1	1

(*continued*)

Table 4. (*continued*)

Position	Level of MSD Risk			
	Low	Medium	High	Very high
Janitor	1	2	0	0
Electrician	0	3	0	0
Security Guard	0	1	2	0
Gardener	0	1	1	1
Maintenance worker	0	3	0	0
Plumber	0	1	0	2
	1 (3,7%)	16 (59,26%)	6 (22,22%)	4 (14,81%)

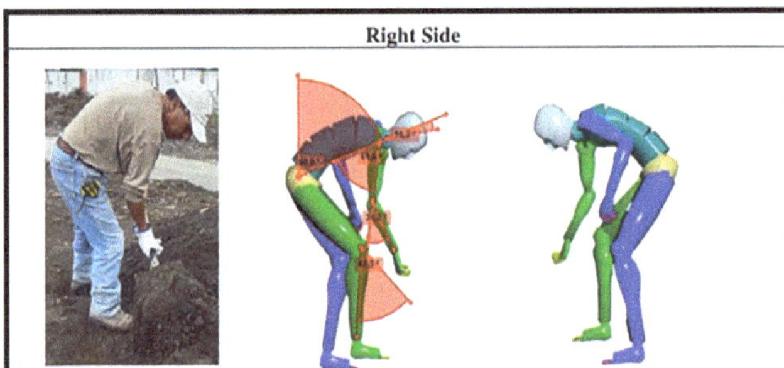

Group A			Group B		
Group A	Trunk score	4	Group B	Upper arms score	3
	Neck score	2		Lower arms score	1
	Legs score	2		Wrists score	1
Table A score		6	Table B score		3
Load/force score		1	Coupling score		1
Total A score		7	Total B score		4
Table C score			7		
Activity score			4		
REBA score			11		
Action Level	3	Level of MSD Risk: Very high	Intervention	Implement change	

Fig. 1. Example of the REBA evaluation of the task that reported the highest risk level [14]

3.3 Complementary Corrective Measures

To implement the ergonomic proposal and monitor its effectiveness, two key steps are essential: establishing an ergonomics committee and conducting comprehensive training for both committee members and workers. These strategies have demonstrated significant efficacy in diverse industrial settings [18, 19].

Creation of Ergonomic Committee. To create this committee, its formation, organization, and operation is defined. The ergonomics committee will be part of the UTA Risk Management Unit. It will be responsible for the monitoring and supervision of the implementation of the recommendations of this study, as well as its surveillance, thus encompassing the control stage. For this purpose, they will receive on-the-job training related to the daily problems in the workplace and in which the workers' capacities are discussed. Finally, it will be verified that the results achieved in the follow-up of the proposed objectives are beneficial for the worker.

Training for Ergonomics Committee Members and Workers. A comprehensive training program is developed, consisting of four 2-h modules, for both committee members and employees. Each part containing the following information: (1) basic ergonomics principles and objectives, ergonomics and its influence on productivity, ergonomics as a function of posture and effort; (2) appropriate body movements, use of hand tools, safe work practices and risk factors; (3) identification and prevention of MSDs; (4) ergonomics problem-solving process and techniques to generate a proactive safety workplace culture.

4 Discussion

The REBA method was applied to assess the postural risk of various job activities has proven to be a valuable tool for assessing postural risks associated with MSDs, as it considers both sensitivity and generality, emphasizes the importance of ergonomic interventions to address the appearance of MSDs among university workers. By implementing appropriate measures, it is possible to improve the health and well-being of this population.

An ergonomic mindset and a focus on the harmony between human beings and their environment must be adopted. Instead of focusing solely on the outcome, it is crucial to identify and address the root causes of problems. By delving deeper into the system and addressing the root causes of problems, ergonomic principles can be implemented. Basic tools, particularly the REBA method, play a pivotal role in this process.

It is of great importance to consider anthropometric data to identify potential mismatches between task demands and workers' physical capabilities. The lack of a local anthropometric database in Ecuador hinders the ability to tailor workplaces to the specific needs of the population. Work environments should be redesigned to accommodate the specific anthropometric characteristics of university workers, reducing the risk of MSDs.

An ergonomics program in the workplace is a dynamic process that involves both proactive interventions and ongoing evaluation, which will foster a culture of continuous improvement, seeking the constant adaptation of working conditions to workers and thus ensuring a healthy, safe, reliable, and pleasant work environment, in which the operator feels motivated and satisfied with their participation in the production process.

Acknowledgments. This work is being funded by Dirección de Investigación y Desarrollo (DIDE), of Universidad Técnica de Ambato under the project: "Evaluación postural enfocada en la predicción de requisitos de resistencia estática y compresión de discos vertebrales de la zona lumbar en actividades realizadas en el quehacer universitario".

References

1. Kee, D.: Comparison of OWAS, RULA and REBA for assessing potential work-related musculoskeletal disorders. Int. J. Ind. Ergon. **83**, 103140 (2021). https://doi.org/10.1016/j.ergon.2021.103140
2. Choi, S.D., Brings, K.: Work-related musculoskeletal risks associated with nurses and nursing assistants handling overweight and obese patients: a literature review. Work **53**(2), 439–448 (2016). https://doi.org/10.3233/WOR-152222
3. Lee, J., Kim, H.: The effects of ergonomic interventions on musculoskeletal disorders among office workers: a systematic review. Int. J. Ind. Ergon. **69**, 123–132 (2018)
4. Kumar, S., Narayan, A.: Risk factors for musculoskeletal disorders among industrial workers: a cross-sectional study. Indian J. Occup. Environ. Med. **20**(4), 182–186 (2016)
5. OSHA (Occupational Safety and Health Administration). Ergonomics: A Beginner's Guide. (2017). https://www.osha.gov/ergonomics
6. Zare, M., Black, N., Sagot, J.C., Hunault, G., Roquelaure, Y.: Ergonomics interventions to reduce musculoskeletal risk factors in a truck manufacturing plant. Int. J. Ind. Ergon **75**, 102896 (2020). https://doi.org/10.1016/j.ergon.2019.102896
7. San Antonio, T., Larrea, A., Lascano, A., Urrutia, J.: Ergonomic evaluation of daily activities of women from a community in the ecuadorian highlands. Adv. Intell. Syst. Comput. **789**, 102–110 (2019). https://doi.org/10.1007/978-3-319-94484-5_11
8. Pimparel, A., Madaleno, S., Oll, C.: How ergonomic evaluations influence the risk of musculoskeletal disorders in the industrial context: a brief literature review. Spring-erLink 406 (2021). https://doi.org/10.1007/978-3-030-89617-1_36
9. Kee, D.: Systematic comparison of OWAS, RULA, and REBA: application to musculoskeletal loading and disorders. Int. J. Environ. Res. Public Health **17**(20), 7392 (2022). https://doi.org/10.3390/ijerph19010595
10. Mendosa, P.: Disminución de los factores disergonómicos de riesgo, al aplicar el método REBA al proceso de cable de la empresa Southwire Americana de México. Facultad de Ingenieria y Negocios, Tecate, Universidad Autónoma de Baja California, Baja California, Mexico (2021). https://repositorioinstitucional.uabc.mx/server/api/core/bitstreams/b89dabd7-5e3c-43e1-a640-6c1601092b53/content
11. Hignett, S., McAtammey, L.: Rapid entire body assessment (REBA). Appl. Ergon. **31**, 201–205 (2000). https://doi.org/10.1016/S0003-6870(99)00039-3
12. Instituto Nacional de Seguridad e Higiene en el Trabajo (INSHT) Evaluación de las condiciones de trabajo: carga postural. NTP 601: Evaluación de las condiciones de trabajo: carga postural. Método REBA (Rapid Entire Body Assessment) (2021). http://www.insht.es/InshtWeb/Contenidos/Documentacion/FichasTecnicas/NTP/Ficheros/601a700/ntp_601.pdf
13. INTE (Instituto Nacional de Tecnologías de la Información): INTE/ISO 7250–1:2018. Basic human body measurements for technological design - Part 1: Body measurement definitions and landmarks (2018)
14. Quispe, J.: Evaluación de la carga postural en trabajadores de la U.T.A desde un punto de vista antropométrico. Universidad Técnica de Ambato, Ambato, Ecuador (2023)
15. Lascano, A., Patín, G., Larrea, A., San Antonio, T.: Ergonomic evaluation of risk level by exposure to forced postures in cattle slaughterhouse workers in Ecuador. Adv. Intell. Syst. Comput. **792**, 212–217 (2019)

16. Ciaccia, M., Mejía, C., Vásquez, C., San Antonio, T.: Ergonomic evaluation of agricul-ture-related activities performed by Ecuadorian indigenous women. Adv. Intell. Syst. Comput. **1215**, 348–353 (2020)
17. San Antonio, T., Urrutia, F., Larrea, A.: Ergonomic analysis for people with physical disabilities when the wheelchair is considered as their workstation. In: 2016 IEEE Ecuador Technical Chapters Meeting (ETCM), pp.1–6 (2016)
18. Czernecka, W.: Practical aspects of the organization of ergonomics committees in the context of the quality of ergonomic activities undertaken in manufacturing enterprises. Annales Universitatis Mariae Curie-Skłodowska, sectio H–Oeconomia **58**(3 special), 115–128(2024)
19. Ghahramani, A., Taghizadeh, E., Mohebbi, I.: Formation status of committee for technical protection and health at work in manufacturing companies. Iran Occup. Health **16**(5), 52–61 (2019)

Development of a Real-Time Android Application for Brain Tumor Detection from MRI Scans

Bryan Chulde Fernández(✉), Santiago Velastegui, Jonathan Cruz Varela, and Paulo Navas

School of Biological Sciences and Engineering, Yachay Tech University, Hacienda San José s/n, San Miguel de Urcuquí 100119, Ecuador
{bryan.chulde,santiago.velastegui,jcruz,pnavas}@yachaytech.edu.ec

Abstract. This paper presents the development of a real-time Android application for brain tumor detection from MRI scans using deep learning techniques. We trained a MobileNetV2-based model on 4,060 MRI images, achieving an AP50 of 0.874 and AP@[0.50:0.95] of 0.522 on the validation set. The model was optimized for mobile deployment using TensorFlow Lite, resulting in a compact 3MB file with an inference latency of approximately 120ms per frame on mid-range smartphones. The application captures camera frames in real-time, preprocesses them to 320×320 pixels, and displays detection results with bounding boxes and confidence scores, consistently achieving 0.90 or higher confidence for both tumor and non-tumor classifications. While the model demonstrates robust performance for larger tumors, it shows limitations in detecting smaller lesions (APs=0.000) and exhibits some overfitting, as indicated by the disparity between training loss (0.1683) and validation loss (0.6585). Compared to existing solutions, this approach offers a balanced trade-off between computational efficiency and diagnostic accuracy, making it suitable for resource-constrained clinical settings where rapid screening is necessary.

Keywords: Brain tumor detection · MobileNetV2 · TensorFlow Lite · Android application · Real-time inference

1 Introduction

Cerebral tumors are among the most serious and complicated forms of cancer and are distinguished by the proliferation of abnormal cells within the brain or central nervous system (CNS). These can be benign or malignant, and malignant tumors, such as Glioblastoma multiforme, have a particularly aggressive behavior and a poor prognosis. Because of their infiltrating character and the sensitive localization in the brain, treatment is often difficult and the chances of their relapse are high. The worldwide impact of brain tumors is substantial. In 2020, there were approximately 308,102 new cases and 251,329 deaths attributed to

J. L. Ballesteros et al. (Eds.): CITIS 2025, LNNS 1780, pp. 294–303, 2027.
https://doi.org/10.1007/978-3-032-14287-0_29

brain and CNS tumors worldwide, according to the World Cancer Observatory [1,4]. Brain malignancies, especially high-grade gliomas, are associated with some of the worst survival rates among all cancers, with an estimated five-year survival rate of 5–10% [13]. Such statistics underscore the pressing need for improved diagnosis and treatment strategies.

1.1 Artificial Intelligence and the Real Time Detection

An important determinant of the survival rates of patients with brain tumors is early identification. Early detection of tumors facilitates timely medical treatment, which may improve the effectiveness of treatments such as surgery, chemotherapy, and radiotherapy. Significant advances in artificial intelligence (AI), have made progress in the classification and automatic segmentation of brain tumors using biomedical images [3,14]. These, diagnostics based on artificial intelligence techniques are also being applied to real-time applications in medical care, where fast and reliable tumor detection is crucial [12]. MobileNetV2 is a convolutional neural network architecture optimized for mobile and embedded devices, proposed by Sandler et al. [15]. It introduces inverted residual blocks and depthwise separable convolutions to enhance efficiency without significantly compromising accuracy. This architecture has demonstrated an excellent trade-off between computational performance and accuracy, making it an ideal choice for real-time computer vision applications on resource-constrained devices [7]. In this context, Android Studio, the official development environment for Android applications, enables the efficient deployment of machine learning models through TensorFlow Lite, a lightweight version of the TensorFlow framework designed for on-device inference [19]. The integration of MobileNetV2 with TensorFlow Lite within Android Studio allows developers to build mobile applications capable of performing computer vision tasks—such as image classification and object detection—directly from the device's camera without relying on cloud processing. Additionally, tools like MediaPipe, also developed by Google, provide optimized pipelines for real-time image processing, facilitating the creation of intelligent and interactive applications that combine trained models with live visual input streams [11].

1.2 State of the Art

In the datasets BraTS and ANDI, the RTMDet in [2] model offers a real-time model with depthwise convolution blocks, achieving an AP of 96.9% with just 6.76M parameters and 9.65 GFLOPS. Although there are limitations with small objects (APs: 0.0), their design optimized for MRI (T1, T2, FLAIR) includes preprocessing such as skull stripping and data augmentation.

Using a neural network modified with LVQ, this work [16] achieves 100% accuracy on a dataset of 64 MRI images (220×220 pixels). Highlighting its computational simplicity, the method combines segmentation based on boards (Canny) with the extraction of morphological characteristics.

A CNN with three convolucional caps is implemented in an Android app in the work of [8], classifying tumors as glioma, meningioma, and pituitary. 98.78% precision is reported using 3,260 MRI images, however more data is needed to prevent overajustment in real-world clinical settings.

[20] create a mobile application using TensorFlow Lite that achieves 82.5% accuracy in 155 hospital MRI images. The model, which is based on VGG16, faces challenges with noise in unprocessed images, highlighting the need for extra filters.

In the work of [5], models like as VGG-16 and MobileNet are evaluated on 12,256 MRI enhanced images, where a linear sequence achieves 91.73% accuracy with ADAM. The study highlights the benefit of adaptive optimizers in relation to SGDM (50.45–84.82% precision).

Table 1 provides a comparative look at a number of brain tumour detection models. This emphasises the variation in the performance of the models and the sizes of the datasets across studies

Table 1. Comparison of brain tumor detection models.

Model	Accuracy	Sensitivity	Specificity	Dataset	Reference
RTMDet	96.9% (AP)	95%	94%	BraTS + ANDI (N/A)	[2]
Modified PNN	100%	N/A	N/A	64 MRI	[16]
CNN (Android)	98.78%	N/A	N/A	3,260 MRI	[8]
VGG16-based App	82.5%	N/A	N/A	155 hospital MRI	[20]
Lightweight Net	91.73%	N/A	N/A	12,256 augmented MRI	[5]

2 Materials and Methods

The development of the present study was mainly performed in Google Colab, taking advantage of its Python 3.x and TensorFlow 2.x environment for the implementation of the brain tumor detection model. For the training and preprocessing phase, the libraries `mediapipe-model-maker` (object detector), `tensorflow`, as well as standard Python utilities (`zipfile`, `json`, `shutil`) and auxiliary libraries (`Pillow`, `NumPy`, `Matplotlib`) were used. Image management and labeling was performed on the RoboFlow platform, from where the annotated set was exported in COCO format. The mobile application was developed in Android Studio on a Dell Inspiron 15 laptop, and the real-time inference tests were run on a Huawei P20 Lite Android device, integrating the model converted to TensorFlowLite.

The Brain Tumor MRI Dataset available from Kaggle, which contains a total of 7023 brain MRI images originally classified into four categories: glioma, meningioma, pituitary, and no tumor, was used as the database. In order to optimize training and ensure a balanced representation of the classes, 4060 images (approximately 58% of the total) were selected, including samples from all original categories. The files were stored in Google Drive and, from there, mounted in Colab for processing and organization into the necessary folder structures.

3 Data Analysis and Preprocessing

The 4060 selected images were uploaded to RoboFlow, where manual binary labeling was performed in two classes: "Tumor" (any type of tumor present) and "No Tumor". Then, the annotated dataset was exported in COCO format and downloaded as a ZIP file. In Colab, the ZIP file was unzipped, creating `train` and `valid` subfolders, each with a corresponding `images/` folder and a `labels.json` file with the annotations. To standardize the data, the `_annotations.coco.json` files were renamed to `labels.json`, any residual categories with `category_id = 0` were removed, and the remaining category IDs were remapped to start at 1, updating the associated annotations as well.

In final preprocessing phase, all images were resized to 256×256 pixels and normalized to the range of values [0,1]. Using these images and their annotations, `train_data` and `validation_data` objects were created using the `Dataset coco folder` function of the `mediapipe_model_maker.object_detector` module. These datasets served as input to the training of the MobileNetV2 model optimized for mobile devices, thus ensuring adequate data preparation for real-time brain tumor detection in the Android application This whole process can be summarized in Fig. 1.

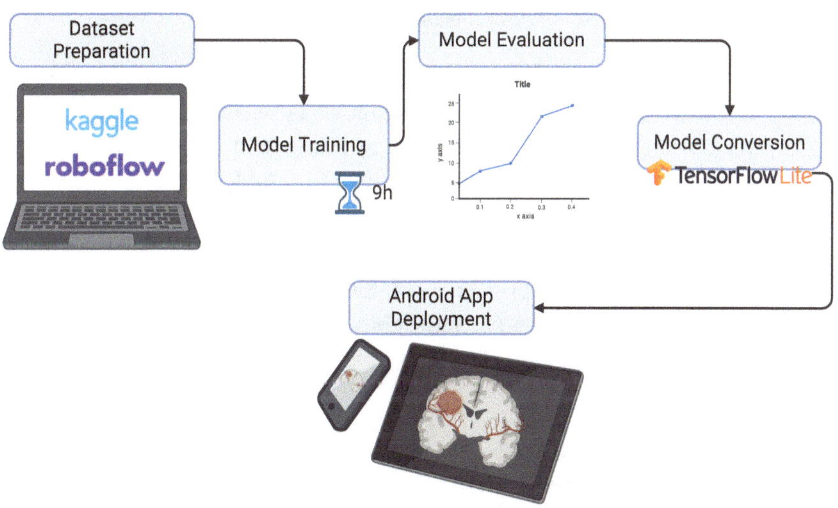

Fig. 1. General workflow for developing a brain tumor detection model: dataset preparation, model training in Google Colab, evaluation, and deployment to an Android app via TensorFlow Lite.

4 Results

A training set of 2833 images and a validation set of 805 images were used to train the model on CPU (no GPU) for about 8 h. The codes for the classes were 0: Tumor and 1: No Tumor. The final training and validation losses at the conclusion of epoch 30 are displayed in Table 2. A certain amount of overfitting is indicated by the low training loss (0.1683) and the higher validation loss (0.6585).

Table 2. Final training and validation losses at epoch 30.

Dataset	Total Loss	Cls Loss	Box Loss	Model Loss
Training	0.1683	0.0719	0.0006	0.1022
Validation	0.6585	0.4113	0.0036	0.5925

Table 3 summarizes the COCO metrics on the validation set. The model achieves an AP@[0.50:0.95] of 0.522 and an AP50 of 0.874, demonstrating strong performance at a generous IoU threshold. However, precision drops to an AP75 of 0.539 at IoU $\geq$ 0.75, and small objects are not detected (APs = 0.000). Detection performance is better for large objects (APl = 0.540), with medium objects yielding moderate precision (APm = 0.294). The average recall up to 100 detections is ARmax100 = 0.603.

Table 3. COCO metrics on the validation set.

Metric	Value	Description
AP (IoU=0.50:0.95)	0.522	Average precision over IoU thresholds
AP@0.50 (AP50)	0.874	Precision at IoU $\geq$ 0.50
AP@0.75 (AP75)	0.539	Precision at IoU $\geq$ 0.75
AP_s (small)	0.000	Precision for small objects
AP_m (medium)	0.294	Precision for medium objects
AP_l (large)	0.540	Precision for large objects
AR_{max100}	0.603	Average recall up to 100 detections

4.1 On-Device Inference and Android Application

To demonstrate the practical applicability of our object detector in a clinical setting, we deployed the trained model on an Android device using TensorFlow Lite. The TensorFlow SavedModel exported from MediaPipe Model Maker was converted to a TFLite format with the following command:

```
tflite_convert \
  --saved_model_dir=exported_model/saved_model \
  --output_file=model.tflite \
  --optimizations=OPTIMIZE_FOR_SIZE
```

The resulting `model.tflite` (3MB) was integrated into an Android Studio project. In the application, camera frames are captured in real time, preprocessed to 320×320 pixels, and passed to the TFLite interpreter. Detection results—including class labels (0: Tumor, 1: No Tumor), bounding boxes, and confidence scores—are rendered as overlays on the live preview (Fig. 2).

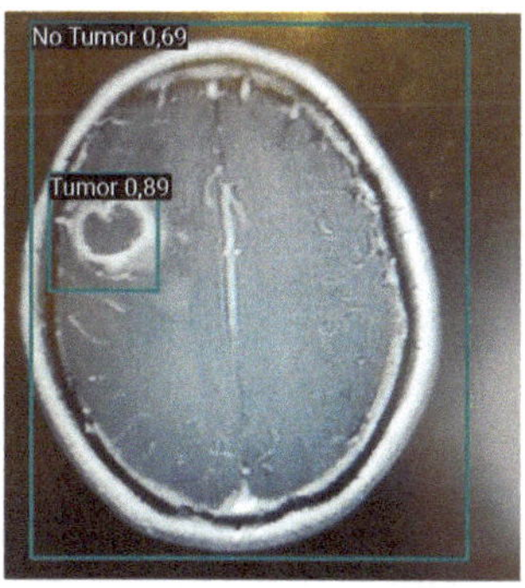

(a) No tumor: 0.69, Tumor: 0.89

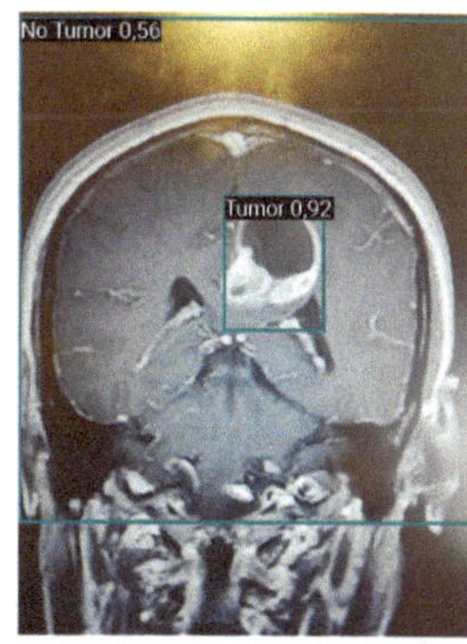

(b) No tumor: 0.56, Tumor: 0.92

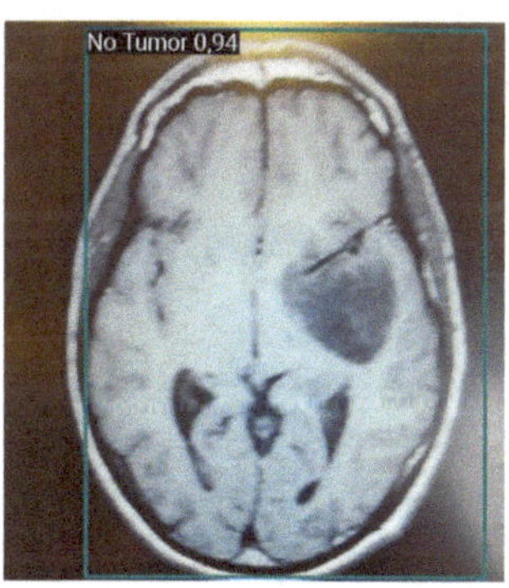

(c) No tumor: 0.94

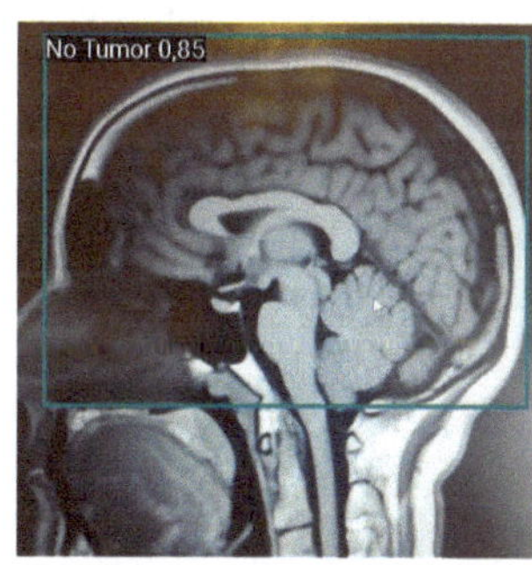

(d) No tumor: 0.85

Fig. 2. Screenshots of the application showing classification results.

During testing on multiple MRI slices, the application achieved confidence scores of 0.90 or higher for both "Tumor" and "No Tumor" classes, confirming the model's robustness in an on-device environment. Inference latency averaged 120ms per frame on a mid-range smartphone (Snapdragon 730), enabling near–real-time performance without GPU acceleration.

5 Discussion

Our model demonstrated promising performance in the detection of brain tumors from magnetic resonance (MRI) images, achieving a low training loss (0.1683) and an AP50 of 0.874. However, the significant disparity between training loss (0.1683) and validation loss (0.6585) clearly indicates a degree of overfitting. This phenomenon, where the model performs exceptionally well on training data but less effectively on unseen validation data, is a common challenge in deep learning. Several strategies could be implemented to mitigate this. While data augmentation was performed during preprocessing using RoboFlow, further advanced augmentation techniques such as elastic deformations, noise injection, or Generative Adversarial Networks (GANs) to synthesize more varied training samples could be explored [17]. Additionally, incorporating regularization techniques like dropout layers within the MobileNetV2 architecture, or L1/L2 weight regularization, could help prevent the model from becoming too specialized to the training set [6, 18]. Fine-tuning learning rates or employing early stopping based on validation loss are other standard practices that could be more aggressively pursued.

Furthermore, the model's current inability to detect small tumors (AP_s = 0.000) significantly limits its applicability in clinical scenarios where early-stage or smaller lesions are critical to identify. This limitation will be discussed further below. The challenge of detecting small tumors (AP_s = 0.000) is a critical issue. This could be attributed to several factors, including the potential loss of fine-grained details as data propagates through the convolutional layers of MobileNetV2, especially with the input resolution of 320×320 pixels which was a trade-off for mobile performance. To address this, future iterations could explore using higher resolution inputs, although this would invariably increase computational cost and latency on mobile devices. Another promising avenue is the incorporation of feature pyramid networks (FPNs) [9], which are designed to enhance feature representations at multiple scales, thereby improving the detection of objects of varying sizes, including smaller ones. Techniques like focal loss [10] could also be beneficial if small tumors are underrepresented or harder to classify compared to larger, more obvious ones, by focusing training on these hard-to-detect cases. The preprocessing step of resizing all images to 256×256 (mentioned in Sect. 3) and then to 320×320 for the app, while standardizing, might also contribute to the loss of detail for very small lesions; adaptive resizing or region-of-interest based approaches could be considered.

Implementation of the model on an Android device using TensorFlow Lite demonstrated its feasibility for mobile applications, with a model size of 3MB and an average latency of 120 ms per image on a mid-range smartphone. This efficiency allows its use in resource-constrained clinical settings, providing reliable predictions (≥0.90 confidence) in real time for larger, more apparent tumors.

In comparison to state-of-the-art models such as RTMDet, which achieved an average precision (AP) of 96.9% on the BraTS and ANDI sets [2], our model exhibits a more modest AP@[.50:.95] of 0.522. This difference can be attributed to several factors, including the inherent trade-off made by MobileNetV2 for computational efficiency over raw accuracy, the limitations of our training dataset size and diversity, and the hardware constraints (CPU-only training). While RTMDet requires significantly higher computing capacity (9.65 GFLOPs and 6.76M parameters), our model's primary advantage lies in its low latency (120ms per frame) and small footprint (3MB) on a low-end device. This makes it a more viable tool for initial screening or as an assistive technology in settings where access to high-performance computing or specialized radiological expertise is limited. The trade-off, therefore, is between achieving peak diagnostic accuracy, as seen in models like RTMDet, versus providing a highly accessible and rapid diagnostic aid.

Similarly, while the work of Sapra et al. [16] reports 100% accuracy, its validation on a very small dataset (64 images) makes direct comparison challenging and raises concerns about generalizability, a factor our study attempts to address with over 4000 images and real-world mobile inference testing. The CNN model by Kuntan et al. [8], achieving 98.78% accuracy, also noted the need for more data to prevent overfitting, a challenge mirrored in our work. Our model's preprocessing with RoboFlow helps standardize input, potentially mitigating some noise issues faced by approaches like Wakode et al. [20] (82.5% accuracy), though our limitation in detecting small objects remains. The choice of the TensorFlow Lite + MobileNetV2 pipeline, supported by findings such as those by Glory Precious et al. [5] on adaptive optimizers, underscores our focus on efficient on-device inference.

The potential clinical impact of this tool is significant, especially in under-resourced areas where it could aid in preliminary MRI screening and expedite diagnosis of clear tumor cases. While not a replacement for specialists, it can support rapid assessments. High confidence scores (≥ 0.90) in clear cases are promising. Future work should aim to improve sensitivity to small lesions and reduce overfitting for greater reliability.

5.1 Limitations of the Study

While this study demonstrates the feasibility of a real-time Android application for brain tumor detection, several limitations must be acknowledged. The model was trained on a dataset derived from Kaggle, which, despite including 4060 images, may not fully represent the diversity of MRI scans encountered in all clinical settings (e.g., variations in scanner protocols, patient demographics, or rare tumor types), and the manual labeling process on RoboFlow could introduce subjective biases. The binary classification of "Tumor" versus "No Tumor" also simplifies the complex diagnostic landscape that includes various tumor types and grades. Performance-wise, the model exhibits notable overfitting, evidenced by the disparity between training and validation losses, and critically, its current inability to detect small tumors ($AP_s = 0.000$) significantly curtails its utility for early-stage detection, with an overall AP@[0.50:0.95] of 0.522 indicating further optimization is needed. Generalization is another concern, as testing was confined to a single mid-range smartphone model, and performance might vary on other devices or with MRI data from different acquisition parameters and institutions. Furthermore, the application's reliance on camera-captured images of MRI scans, rather than direct DICOM file processing, introduces a potential for image quality degradation and differs from standard clinical workflows.

6 Conclusion

This study successfully developed and deployed an Android application for real-time brain tumor detection from MRI images, utilizing a MobileNetV2-based architecture optimized via MediaPipe Model Maker and TensorFlow Lite. The model demonstrated noteworthy on-device performance, with an AP50 of 0.874 and an average inference time of approximately 120 ms per frame on a mid-range smartphone. These characteristics highlight its potential as an accessible tool in resource-constrained clinical settings for rapid preliminary assessment.

However, the research also identified critical areas for improvement. The significant overfitting, indicated by the divergence between training and validation losses, and the model's current inability to detect small tumors ($AP_s = 0.000$), are major limitations that temper its immediate clinical applicability for comprehensive diagnostics, especially for early-stage detection. While our approach offers a compelling balance of computational efficiency and portability compared to more resource-intensive state-of-the-art models, these trade-offs in accuracy for smaller lesions and susceptibility to overfitting must be addressed.

The development of AI-assisted diagnostic tools on mobile platforms represents a promising direction. This work establishes a foundational framework. Future investigations will be directed towards implementing robust strategies to mitigate overfitting, such as advanced data augmentation and regularization techniques. Crucially, dedicated efforts will focus on enhancing the model's sensitivity to small lesions, possibly by exploring architectures like FPNs or adaptive input resolution strategies. Further research should also encompass broader dataset inclusion, direct DICOM processing capabilities, and rigorous clinical validation to ensure the tool's reliability and generalizability, thereby paving the way for its responsible integration into real-world clinical workflows.

References

1. Ananta, J.S., Paulmurugan, R., Massoud, T.F.: Nanoparticle-delivered biological agents for cancer therapy. Adv. Healthcare Mater. **7**(1), 1700487 (2018). https://doi.org/10.1002/adhm.201800066
2. Bakhtiyorov, S., Umirzakova, S., Musaev, M., Abdusalomov, A., Whangbo, T.K.: Real-time object detector for medical diagnostics (rtmdet): a high-performance deep learning model for brain tumor diagnosis. Bioengineering **12**(3), 274 (2025)
3. Bi, W.L., Hosny, M., Schabath, J.: Artificial intelligence in cancer imaging: clinical challenges and applications. Nat. Med. **25**, 130–140 (2019). https://doi.org/10.1038/s41591-019-0715-9
4. for Research on Cancer, IA: Brain, central nervous system fact sheet. https://gco.iarc.fr (2020), Accessed 8 April 2025

5. Glory Precious, J., Angeline Kirubha, S., Keren Evangeline, I.: Deployment of a mobile application using a novel deep neural network and advanced pre-trained models for the identification of brain tumours. IETE J. Res. **69**(10), 6902–6914 (2023)
6. Goodfellow, I.J., et al.: Generative adversarial nets. In: Advances in Neural Information Processing Systems, pp. 2672–2680 (2014)
7. Howard, A.G., et al.: Mobilenets: efficient convolutional neural networks for mobile vision applications. arXiv preprint arXiv:1704.04861 (2017). https://arxiv.org/abs/1704.04861
8. Kuntan, S.S., Sidgiddi, V.A., Jaganur, S.S.: Brain tumour detection and classification using deep learning
9. Lin, T.Y., Dollár, P., Girshick, R., He, K., Hariharan, B., Belongie, S.: Feature pyramid networks for object detection. In: Proceedings of the IEEE Conference on Computer Vision and Pattern Recognition, pp. 2117–2125 (2017)
10. Lin, T.Y., Goyal, P., Girshick, R., He, K., Dollár, P.: Focal loss for dense object detection. In: Proceedings of the IEEE International Conference on Computer Vision, pp. 2980–2988 (2017)
11. Lugaresi, C., Tang, J., Nash, H., McGuire, M., Kautz, J., Cree, M., et al.: Mediapipe: a framework for building perception pipelines. arXiv preprint arXiv:1906.08172 (2019)
12. Muhammad, K., Khan, S., Ser, J.D., de Albuquerque, V.H.C., Diez, I.D.L.T.: Deep learning for diagnostic healthcare systems: State-of-the-art, challenges, and future directions. IEEE Access **8**, 111821–111841 (2020). https://doi.org/10.1109/ACCESS.2020.2963939
13. Ostrom, Q.T., Cioffi, L.M., Waite, H., et al.: Cbtrus statistical report: primary brain and other central nervous system tumors diagnosed in the united states in 2015–2019. Neuro-Oncology **24**($Supplement_5$), v1–v95 (2022). https://doi.org/10.1093/neuonc/noac200
14. Sajjad, M., Khan, S., Muhammad, K., Wu, W., Ullah, A.: Multi-grade brain tumor classification using deep cnn with extensive data augmentation. In: 2019 International Conference on Advanced Science and Engineering (ICASERT), pp. 34–39 (2019). https://doi.org/10.1109/ICASERT.2019.8934561
15. Sandler, M., Howard, A., Zhu, M., Zhmoginov, A., Chen, L.C.: Mobilenetv 2: inverted residuals and linear bottlenecks. In: Proceedings of the IEEE Conference on Computer Vision and Pattern Recognition (CVPR), pp. 4510–4520 (2018)
16. Sapra, P., Singh, R., Khurana, S.: Brain tumor detection using neural network. Inter. J. Sci. Modern Eng. (IJISME), 2319–6386 (2013)
17. Shorten, C., Khoshgoftaar, T.M.: A survey on image data augmentation for deep learning. J. Big Data **6**(1), 1–48 (2019)
18. Srivastava, N., Hinton, G., Krizhevsky, A., Sutskever, I., Salakhutdinov, R.: Dropout: a simple way to prevent neural networks from overfitting. J. Mach. Learn. Res. **15**(1), 1929–1958 (2014)
19. TensorFlow Lite: Tensorflow lite documentation (2024). https://www.tensorflow.org/lite
20. Wakode, S., Shaikh, F.I., Afzal, K.A.N.: Enhanced brain tumor detection through mri imaging

Real-Time COVID 19 Detection on X-Ray Image Using Object Detection and MediaPipe

Carolina Játiva, Anahi Vaca(✉), Jonathan Cruz, and Diego Almeida

School of Biological Sciences and Engineering, Yachay Tech University, Urcuquí, Hacienda San José, Ecuador
anahivaca2202@gmail.com

Abstract. This work presents a mobile application to detect possible COVID-19 cases in chest X-rays in real time, using object detection techniques with MediaPipe. The model was trained with a balanced Kaggle dataset and processed with Roboflow. Unlike traditional methods, this system identifies compromised areas in images using bounding boxes. The MobileNet-MultiAVG architecture was selected for its low resource consumption, allowing its integration into mobile devices. After training and evaluating the model, it was converted to TensorFlow Lite and implemented in an Android application, which allows X-rays to be uploaded or captured and suspicious areas to be viewed instantly. Preliminary tests confirmed its speed and accuracy, proving it to be a useful and accessible tool in resource-limited settings.

Keywords: Covid 19 · x-rays · Object Detection · mobilenet-multi-AVG · artificial intelligence · neural networks

1 Introduction

Since late 2019, COVID-19 has posed a major global health challenge. While PCR tests are the standard for diagnosis, their limited availability especially in low resource areas has highlighted the role of chest X-rays as a faster, more accessible alternative. Though useful in emergencies, these images often require expert interpretation, which isn't always available [1,2].

In Ecuador, the impact was severe, with over 890,000 cases and nearly 35,000 deaths by early 2022 [3]. Even before the pandemic, respiratory diseases like pneumonia were among the country's top causes of death [4], reinforcing the importance of early detection tools [5].

Artificial intelligence has made significant progress in medical imaging. Deep learning models such as ResNet, VGG, and DenseNet have shown success in identifying COVID-19 signs in X-rays [6–8]. However, their high computational demands can limit their use in resource-constrained settings. Lighter alternatives like MediaPipe have emerged, offering efficient model execution even on mobile

J. L. Ballesteros et al. (Eds.): CITIS 2025, LNNS 1780, pp. 304–313, 2027.
https://doi.org/10.1007/978-3-032-14287-0_30

devices [9,10]. Originally designed for applications like hand and pose tracking, MediaPipe is now being adopted for medical purposes due to its speed and flexibility.

At the same time, tools like Roboflow have simplified the creation of vision models, making them accessible even without advanced coding skills [11,12]. Combined with efficient networks like mobilenet-multi-AVG, which uses global average pooling for compact and fast predictions [13], these tools support real-time medical applications on portable devices.

This study proposes a system that detects COVID-19 patterns in chest X-rays using MediaPipe, Roboflow, and mobilenet-multi-AVG. By focusing on object detection instead of simple classification, it highlights affected lung areas, making the results more interpretable and potentially useful as a diagnostic support tool.

2 Methodology

This work was carried out in Google Colab using Jupyter Notebooks. Tools like TensorFlow, NumPy, and MediaPipe were used, along with the MobileNet-multi-AVG architecture. The images were chosen based on quality and availability, and each one was manually checked to separate normal cases from those showing signs of COVID-19.

After that, the dataset was processed in Roboflow, where annotations and classifications were generated automatically, making the preparation much smoother. To help the model generalize better, the data was balanced between training and testing sets, and some simple augmentation techniques like image rotation and automatic adjustments were applied. A visual summary of this process is shown in Fig. 1.

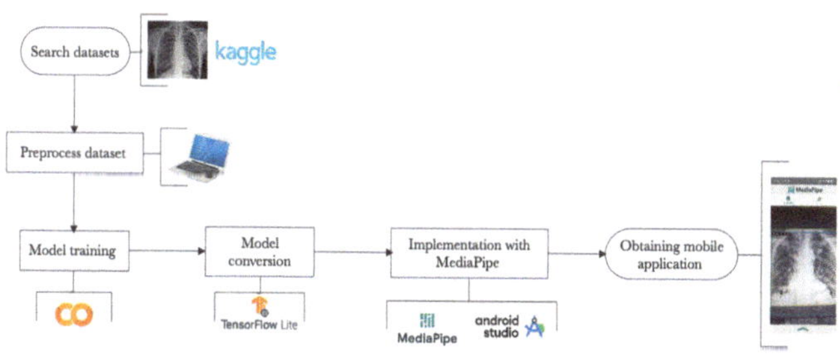

Fig. 1. Summary of the methodology

2.1 Dataset

This study used chest X-ray images from the publicly available COVID-19 Radiography Dataset on Kaggle, which includes samples labeled as either COVID-19 positive or negative. A total of 2,800 images were selected 1,400 with confirmed COVID-19 and 1,400 normal cases. These were split into two balanced subsets: 2,240 images for training and 560 for testing.

The images were then uploaded to Roboflow, where manual annotation was performed by marking lung regions showing visual signs of COVID-19. This step enabled object detection, allowing the model to not only classify but also localize affected areas within each image.

To improve generalization, a data augmentation technique was applied within the same platform by rotating the images 90°C. This effectively doubled the dataset to 5,600 images. From these, 4,480 were used in the training phase, helping the model learn from varied orientations and making it more robust. Lastly, all images were resized to 256 × 256 pixels. The dataset details are summarized in the corresponding Table 1.

Table 1. Characteristics of the dataset

Characteristics	Amount
Total of images	5600
Test	1120
Train	4480
COVID	2800
Normal	2800
Augmentation	2800
Shape	256 × 256

2.2 Code Configuration and Data Preparation

Following the annotation phase in Roboflow, the dataset was exported in Pascal VOC format due to its compatibility with MediaPipe and its clear structure based on .xml files per image. A Python script was created to automatically organize the images and their annotations into folders according to the model's requirements. In addition, key libraries such as pascal-voc-writer were installed to handle annotations and pycocotools to evaluate performance with COCO metrics.

2.3 Model Training

The model was trained with MobileNet Multi AVG, chosen for its balance between accuracy and efficiency, using the object detection method. It was trained for 30 epochs, continuously monitoring the results with the training and validation data to fine-tune the model's behavior and prevent overfitting. This allowed for progressively improved detection of COVID-19-associated regions in X-rays using backpropagation.

2.4 Performance Evaluation

Once training was complete, the model was evaluated using COCO metrics, focusing on accuracy and object localization. AP, AP50, AP75, and AP1 were analyzed to assess overall detection and detection of large objects, such as affected areas in X-rays. AR, ARmax1, ARmax10, and ARmax100 were also reviewed to measure detection performance based on the number of predictions per image. To improve visual interpretation, histogram equalization was applied, highlighting the lung regions most associated with COVID-19.

2.5 Model Conversion and Using Android Studio

After training, the model was converted to TensorFlow Lite (.tflite) format to optimize its use on mobile devices, reducing its size without compromising its accuracy. It was then integrated into an app developed in Android Studio using MediaPipe for real-time inference. From the main screen, the user can upload or take an X-ray, and the app immediately displays areas potentially affected by COVID-19. Additional testing was conducted on two mid-range Android phones, confirming good performance and rapid response under real-world conditions.

3 Results

3.1 Model Evaluation Metrics

The effectiveness of the model in detecting COVID-19 from chest X-rays was evaluated using COCO metrics, which are widely recognized standards in object detection. These metrics examine how well the system classifies, locates, and highlights the areas affected by the disease.

Table 2 presents key quantitative results, including Average Precision (AP) across IoU thresholds (0.50–0.95), fixed AP values at 0.50 and 0.75 IoU, object size based metrics, and Average Recall (AR). Overall, these indicators offer a comprehensive view of the model's performance in clinical image analysis.

The model achieved a mean Average Precision (mAP) of 0.946 across IoU thresholds (0.50–0.95), reflecting strong performance in classifying and localizing COVID-19 patterns in chest X-rays. At AP50, it reached 0.887, confirming reliable detection under lenient spatial criteria. However, AP75 dropped to 0.661, indicating difficulties in precise bounding box alignment under stricter conditions.

The model achieved an AP of 0.543 and an AR of 0.702 for medium-sized regions, which match the typical pattern of lung involvement in COVID-19. Metrics for small or large objects were not reported because all annotations were made with a single box encompassing both lungs, placing all regions within the medium range according to COCO criteria. This labeling reflects how COVID-19 lesions typically appear on radiographs and therefore does not affect the model's validity. The overall AR of 0.706 confirms its ability to detect relevant features.

Table 2. COCO evaluation metrics obtained for the proposed model

Metric	Value
AP@[IoU = 0.50:0.95] (mAP)	0.946
AP@[IoU = 0.50] (AP50)	0.887
AP@[IoU = 0.75] (AP75)	0.661
AP (small objects)	–1.000
AP (medium objects)	0.543
AP (large objects)	-1.000
AR@[IoU = 0.50:0.95-] (overall)	0.706
AR (small objects)	-1.000
AR (medium objects)	0.702
AR (large objects)	–1.000

3.2 Visual Enhancement and Opacity Localization

For the visual interpretation of pulmonary features present in COVID-19 positive cases, a histogram equalization technique was applied to enhance contrast in chest radiographs. Figure 2 shows how a normal chest image without pathological signs was compared with an image of a confirmed COVID-19 case, in which manual annotations (green arrows) were added indicating the affected regions. These opacities are not present in the normal image, allowing a clear visual distinction between the two cases. This improved representation is especially useful for analysis by non-expert users, as it increases the interpretability of radiographic findings.

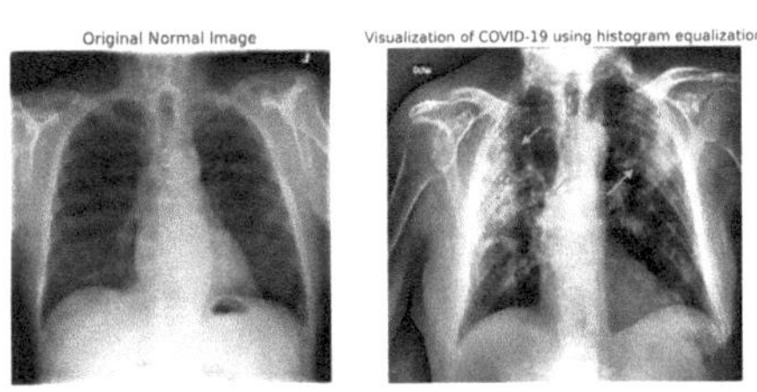

Fig. 2. Comparison of non-COVID (normal) image with COVID-19 image

3.3 Training Loss Analysis and Confusion Matrix

Figure 3 shows the training and validation loss over 30 epochs. Both losses drop sharply early on, indicating fast learning and convergence. Training loss continues to decline, while validation loss stabilizes near epoch 7, with a slight late increase—suggesting mild overfitting. However, the small gap between losses reflects a good balance between complexity and generalization. Overall, the loss curves confirm the model's robustness in distinguishing COVID-19 from non COVID-19 cases.

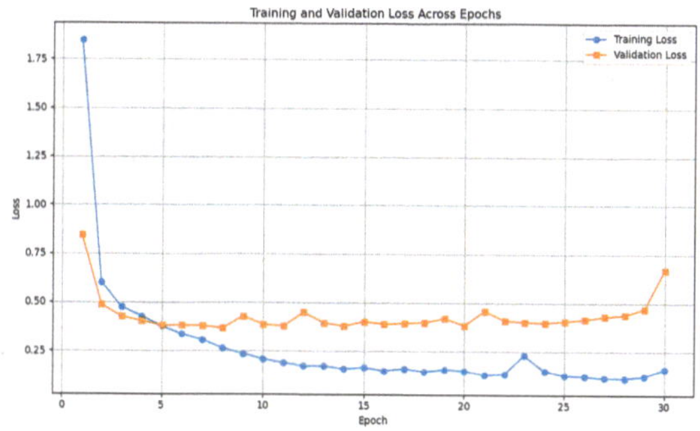

Fig. 3. Training and Validation Loss: Convergence and Generalization with Slight Late Stage Overfitting

Furthermore, Fig. 4 shows a confusion matrix where the model analyzed 500 chest X-rays it had not previously processed. The results showed that it correctly identified 238 cases of COVID-19 and 213 images without anomalies, demonstrating good classification capabilities. However, there were also some errors according to the confusion matrix: 22 positive cases (false negatives) and 27 false positives were misdiagnosed. Despite these errors, the overall performance was fairly balanced, suggesting that the model can distinguish with some accuracy between healthy patients and those with COVID-19. It is worth mentioning that this demonstrates the model's accuracy; however, the system needs to be evaluated with more diverse data to ensure its effectiveness in different clinical scenarios.

3.4 Mobile Application Results Using MediaPipe

This project culminated in an Android app that integrates the trained model for real-time chest X-ray analysis. To simulate clinical use, it was tested with images of both COVID-19 and healthy cases. Figures 5 and 6 show example screenshots. In Fig. 5, the model processes a COVID-19 case in under 0.1 s, marking the lungs with a red box and a 'COVID-19: 0.92' label. Figure 6 displays a normal case with 67% confidence, outlined in green. Using a 0.50 threshold, both predictions are classified as correct and shown instantly through the interface.

4 Discussion

The results indicate that the developed model demonstrates excellent performance in detecting COVID-19 from chest X-ray images, combining high accuracy with real-time processing. Although a direct comparison with radiologists

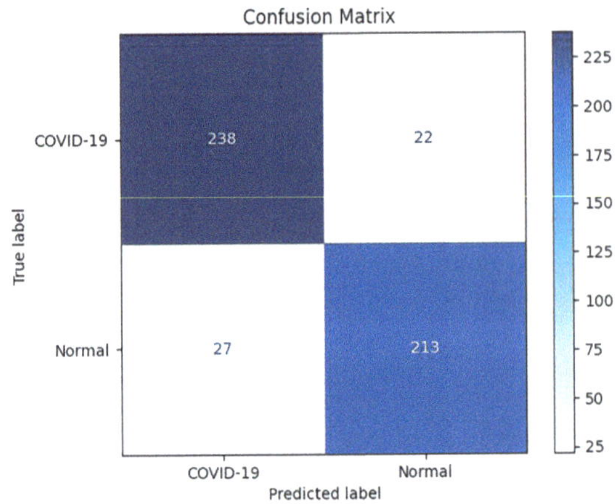

Fig. 4. Confusion Matrix

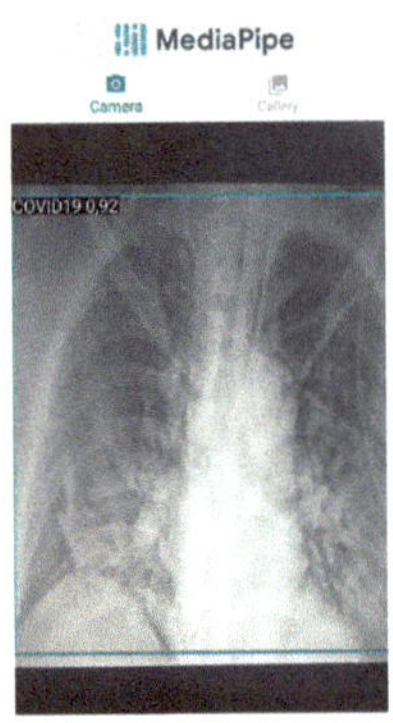

Fig. 5. Prediction of a COVID-19 case with 92% confidence using the MediaPipe mobile application.

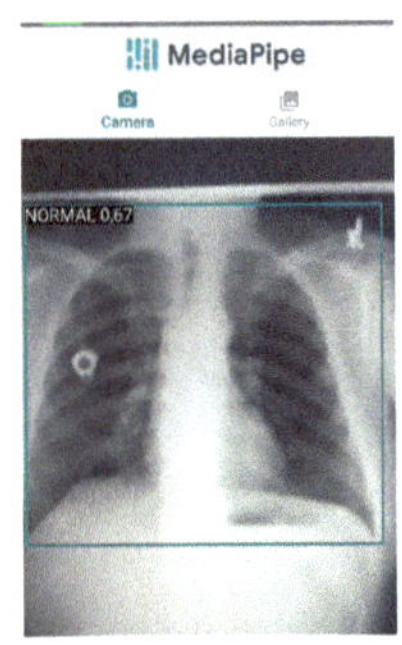

Fig. 6. Prediction of a normal case with 67% confidence using the MediaPipe mobile application.

was not conducted, previous studies report that clinical interpretation accuracy of such images typically ranges between 0.72 and 0.85 [5]. In this context, the mAP of 0.94 achieved by our model suggests performance comparable to or even surpassing traditional clinical assessments. Its speed and ease of use on mobile devices position it as an effective complementary tool in settings with limited access to radiologists. Furthermore, as shown in Table 3, it outperforms recent models such as Mask R-CNN (mAP of 0.89) and YOLOv5.

This study is distinguished by being the only one to integrate real-time detection within a fully functional mobile application, positioning it as a particularly viable solution for rapid clinical deployment in resource constrained settings. The use of on device inference, along with immediate visual feedback via MediaPipe, underscores its practical value for real world implementation.

Table 3. Comparison of recent object detection models applied to chest X-ray images.

Model	Application	mAP/AP	Mobile App	Reference
Mask R-CNN	COVID-19-affected region detection in CXR	0.89	No	[14]
YOLOv8	Medical device detection in CXR	0.81	No	[15]
YOLOv5	COVID-19 bounding box detection in CXR	0.85	No	[16]
YOLOv5	Real-time CCTV tracking of potential carriers	0.87	No	[17]
MobileNet + MediaPipe	COVID-19 detection in CXR with real-time bounding boxes	0.946	Yes	This work

Model Limitations and Dataset Restrictions

Although the model showed good performance, there are some limitations. The COCO metrics do not include results for small or large objects, since all annotations corresponded to medium-sized regions (both lungs), limiting the analysis by scale. [18]. A slight overfitting was also detected, evidenced by the increase in the validation loss in the last epochs. This may be due to the small size of the dataset. Furthermore, the model has not yet been validated with external clinical images, so its performance in other contexts may vary. [19]

Real-Time Implementation and Clinical Viability

This work presents an innovative contribution by integrating the model into a mobile application using MediaPipe, enabling real-time analysis of chest X-rays. Testing demonstrated that the system can process an image in less than 0.1 s on a typical Android device, which is remarkable given the model's complexity and reinforces the feasibility of implementing advanced detection models in clinical settings. From a clinical perspective, the viability of this solution depends on several factors, such as rigorous external validation, regulatory approval, its integration into radiologists' workflow, and image privacy and quality measures. [19] Recent studies indicate that AI systems can improve diagnostic sensitivity when used as decision support tools. The user interface should be clear, reliable, and interpretable, using, for example, color coding to ensure effective interaction. System portability offers significant benefits for rural or underserved areas, although it also presents challenges in terms of interoperability and connectivity [20].

Recommendations and Future Work

Although the results are promising, there are still aspects that need to be improved to apply the system in real-life clinical settings. The dataset needs to be expanded to include cases of different ages, genders, and other lung diseases (such as tuberculosis or COPD), in addition to including different levels of COVID-19 severity. This would also allow for a better evaluation of performance on small and large objects. The plan is to explore more advanced models, such as YOLOv7/YOLOv8 or attention networks, and integrate lung segmentation to improve lesion detection. External validation with new X-rays will also be carried out, as well as usability tests in rural health centers, evaluating the speed, ease of use, and confidence in the predictions. Finally, legal, ethical, and data privacy aspects will be considered, ensuring their integration with existing clinical workflows.

5 Conclusion

This work presents a lightweight and accessible system for real-time detection of COVID-19 in chest X-rays, combining the MobileNet MultiAVG model with MediaPipe. Using Roboflow for annotation and data augmentation, we created a balanced and diverse dataset that facilitated the training of a model suitable for deployment on mobile devices. The resulting application has a simple interface and was tested on real devices, showing consistent and fast performance. Despite promising results, the model has some limitations. The evaluation did not include metrics for small and large objects, as most COVID-19 lesions in the dataset were medium sized. Furthermore, evidence of overfitting was observed during training, with a slight validation loss that increased in later stages. These issues, coupled with the limited size of the dataset, could affect the generalization ability in different clinical settings. Overall, this work demonstrates the potential of efficient AI models for mobile diagnostics, especially in resource poor settings. Future improvements should focus on expanding the dataset, testing with external data, and exploring additional interpretation tools.

References

1. Suryasa, I.W., Rodríguez-Gámez, M., Koldoris, T.: COVID-19 pandemic. Int. J. Health Sci. (IJHS) **5**, vi–ix (2021)
2. Saez de Gordoa, E., Portella, A., Escudero-Fernández, J.M., Andreu Soriano, J.: Utilidad de la radiografía de tórax para la detección de neumonía COVID 19 durante la pandemia por SARS-CoV-2. Radiologia **64**, 310–316 (2022)
3. COVID-19 cases. https://data.who.int/dashboards/covid19/cases?m49=218&n=c. Accessed 14 April 2025
4. Instituto nacional de estadística y censos. https://www.ecuadorencifras.gob.ec/. Accessed 13 April 2025
5. Rahman, T.: Exploring the effect of image enhancement techniques on covid-19 detection using chest x-ray images. Comput. Biol. Med. **132**, 104319 (2021)
6. Keles, A., Keles, M.B., Keles, A.: COV19-CNNet and COV19-ResNet: diagnostic inference engines for early detection of COVID-19. Cognit. Comput. **16**, 1–11 (2021)

7. Goyal, L., Dhull, A., Singh, A., Kukreja, S., Singh, K.K.: VGG-COVIDNet: a novel model for COVID detection from X-Ray and CT scan images. Procedia Comput. Sci. **218**, 1926–1935 (2023)
8. Dalvi, P.P., Edla, D.R., Purushothama, B.R.: Diagnosis of coronavirus disease from chest x-ray images using DenseNet-169 architecture. SN Comput. Sci. **4**, 214 (2023)
9. Mohamed, N.: Importance of artificial intelligence in neural network through using mediapipe. In: 2022 6th International Conference on Electronics, Communication and Aerospace Technology, pp. 1207–1215 (2022)
10. Uysal, E., Balikci, İ., KaradaÄđ, O.O., Yüce, Y.K.: SynapSign: an advanced machine learning framework for american sign language recognition utilizing a novel landmark-based dataset. In: 2024 Innovations in Intelligent Systems and Applications Conference (ASYU), pp. 1–6, IEEE (Oct. 2024)
11. Ciaglia, F., Zuppichini, F.S., Guerrie, P., McQuade, M., Solawetz, J.: Roboflow 100: A rich, Multi-Domain object detection benchmark (Nov. 2022)
12. Deepa, D., Sivasangari, A., Roonwal, R., Nayan, R.: Pothole detection using roboflow convolutional neural networks. In: 2023 7th International Conference on Intelligent Computing and Control Systems (ICICCS), pp. 560–564 (2023)
13. Howard, A.G., et al.: MobileNets: Efficient convolutional neural networks for mobile vision applications (Apr. 2017)
14. Podder, P., Das, D.J., Dey, L.: Covid-19 detection from chest x-rays using ensemble deep learning. Multimedia Tools Appli. **80**, 19633–19657 (2021)
15. Hsu, W.-T., Huang, J.-Y., Lin, Y.-H., et al.: Yolov8-meddevice: object detection of medical devices in chest radiographs using the yolov8 model. Front. Physiol. **14**, 398 (2023)
16. Gera, D., Chauhan, R., Nayyar, A.: Covid-19 detection in chest x-ray images using yolov5. Multimedia Tools Appli. **81**, 15125–15142 (2022)
17. Mankodiya, R., Smith, J., Patel, Z.: Cctv-based tracking of covid-19 symptomatic individuals using yolov5 object detection. Sensors **23**(1), 527 (2023)
18. Garcia, L., Fernandez, J.M., Kim, H.: Yolo-based lesion detection with automatic cropping for covid-19 diagnosis. PLOS Comput. Biol (2023)
19. Arun, S., Kumar, R., Sharma, A.: Deep learning-powered mobile app for fast and accurate covid-19 diagnosis using chest x-ray images. Int. J. Adv. Comput. Sci. Appl. **14**(11) (2023)
20. Abdel-Aziz, M., El-Sayed, H., Elhariri, A.: Detection of pulmonary nodules in chest ct using a hybrid ai model. Egyptian J. Radiol. Nuclear Med. **54**(1) (2023)

A Proposal for Upper- and Lower-Limb Rehabilitation of a Chronic Post-Stroke Patient Using Dependent Brain-Computer Interface and Robotic Devices

Teodiano Bastos-Filho[1(✉)], Fernanda Souza[1], Ana Cecilia Villa-Parra[2], and Pablo Andres Salamea-Cordero[2]

[1] Federal University of Espirito Santo, Av. Fernando Ferrari, 514 (Goiabeiras), Vitoria 29075-910, Brazil
teodiano.bastos@ufes.br, fernando.cabral@edu.ufes.br
[2] Biomedical Engineering Research Group, Universidad Politécnica Salesiana, Cuenca, Ecuador
{avilla,psalamea}@ups.edu.ec

Abstract. This work introduces the use of a neurorehabilitation protocol based on a dependent Brain-Computer Interface (dBCI), which uses as input the jaw clenching, and provides an output to robotic devices for upper- lower and upper limb rehabilitation. This protocol uses the concept of Alternating Treatment Design (ATD), in which a single chronic post-stroke subject, with hemiplegia in both his right hand and leg, is submitted to these techniques to recover his hand and leg movements. The rehabilitation progress was verified through typical physiotherapy metrics, such as Fugl-Meyer Assessment Scale (FMS), Functional Independence Measurement (FIM), Modified Ashworth Scale (MAS), MiniBESTest (MBT), modified Rankin Scale (mRS), Time Up and Go (TUG), 10-Meter Walk Test (10MWT), National Institutes of Health Stroke Scale (NIHSS), Barthel Scale (BI), and surface electromyography (sEMG). After applying our protocol, the subject got the following improvements: 6% for upper-limb and 9% for lower-limb in FMS; 8% in FIM; 25% in MBT; 30% in TUG; 18% in time and 25% in number of steps in 10MWT; 25% in NIHSS; 14% in BI. However, there was not variation for MAS (keeping the Grade 1) and mRS (keeping a score of 3); 5% in muscle contraction peak for finger flexors, 2% for tibialis anterior, and 21% for rectus femoris for sEMG.

Keywords: Stroke · Neurorehabilitation · Brain-Computer Interface · Robotic Devices

1 Introduction

Every country in the world is experiencing growth in the number and proportion of older persons in their population, with an estimate that the global population aged 65 and older will reach 2.2 billion by the late 2070s [1]. This increase of elderly population

J. L. Ballesteros et al. (Eds.): CITIS 2025, LNNS 1780, pp. 314–324, 2027.
https://doi.org/10.1007/978-3-032-14287-0_31

is a major stroke risk factor, as most strokes occur over 65, when health problems related to age appear, mainly problems related to blood vessel changes and increased prevalence of other conditions like high blood pressure, diabetes, and atrial fibrillation [2]. Stroke is provoked by insufficient blood flow to the brain, which can be ischemic or haemorrhagic type. After stroke, affected subjects can experience different difficulties in motor functions in either upper- and lower-limbs [3].

Typical stroke effects include hemiparesis or hemiplegia on one side of the body. Specifically, in upper-limbs, common effects are loss of fine movements in hands and fingers, spasticity in arms, coordination impairment in shoulders, elbow, wrist, and fingers, in addition to sensory changes in the affected hemisphere. In lower-limbs, typical effects are difficulty to walk (related to keep balance and leg coordination), leg spasticity, foot drop, and modified way of gaiting. To try to mitigate such effects, an early post-stroke rehabilitation (until three months after the event) is vital, in order to produce the "neuroplasticity", which is the brain ability to reorganize itself by forming new neural connections, adapting its structure and functions, allowing for new skills, memory formation, and recovery from the stroke damage. This rehabilitation normally includes a mix of physical, occupational, and speech therapies, with the aim of improving muscle strength, amplitude of motion, coordination, and functional skills lost after the stroke [4–7].

Some novelties in stroke rehabilitation are the use of customized robot devices to provide feedback to the affected subjects, and track their recovery [7, 8], as well as the use of a Brain-Computer Interface (BCI) to provide a direct way of communication between the brain and the robotic devices [10]. In relation to BCI, research has shown that effective improvements are reached when the subject controls the BCI through the intent of movement, known as Motor Imagery (MI). In fact, the MI of upper- and lower-limb was able to improve range of movement for affected muscle after stroke, as well as increase neuroplasticity, meaning an impacting recovery for affected subjects [11, 12]. Such motor improvements are typically measured through physiotherapy metrics, such FMS, FIM, MAS, MBT, mRS, TUG 10MWT, NIHSS, BI, and sEMG, whereas the improvements in neuroplasticity are commonly evaluated through Electroencephalography (EEG) during MI. As during MI neural changes occur, due to the effects of Event-Related Synchronization (ERS) and Event-Related Desynchronization (ERD), a typical signal processing method as Power Spectral Density (PSD) is applied to detect power increase or decrease in the EEG signals [13]. This way, the BCI can translate the post-stroke subject MI into movements of their upper- or lower-limbs through robotic devices coupled to their affected limbs [5]. It is worth mentioning that recognizing MI is not trivial, as the EEG signal inherent to MI is very weak (some microvolts) and commonly contaminated by different artifacts, which can be physiological (eye blinks/movements, muscle activity, heartbeats) or external (powerline noise, electrode movement, cable interference) [5, 14]. Due to this enormous difficulty to detect MI, we introduce here the use of a dependent-BCI (dBCI), also used in our previous works [15, 16], which is used in this study to trigger an sEMG signal when the subject is conducting MI. Next sessions present details about as this technique is used in this research.

2 Experiments

To initially verify the possibility of using a dBCI in our research, we have conducted an experiment with a typical subject using an EEG headset (Emotiv EPOC-X) to acquire his EEG signals in two conditions: 1) relaxed; 2) clenching his jaw. The Emotiv device considered in this research is a wireless 14 EEG channels headset based on the 10–20 international system, with bluetooth 5.0 communication, which uses saline sensors to acquire EEG signals. Figure 1 shows the details of location of the EEG electrodes, which are AF3, F7, F3, FC5, T7, P7, O1, O2, P8, T8, FC6, F4, F8, AF4. In addition to the EEG electrodes, the Emotiv device has two references (CMS/DRL at P3/P4) located on the left/right mastoid. The EEG data are acquired in a bandwidth from 0.16 to 43 Hz with a sampling rate of 2048 SPS internal, downsampled to 128 SPS, and with resolution of 0.51μV (14 bits mode). Data are filtered using a built-in digital 5th order Sinc filter, and a digital notch filter is used at 60 Hz.

Figure 2 shows the acquired EEG signals, using EmotivPro, which is a software for real-time display of EEG data streams. In Fig. 2a the subject is relaxed, whereas in Fig. 2b the subject is clenching his jaw. For the relaxed condition (Fig. 2a), the subject was asked to relax their jaw muscles during the experiment, enabling the investigation of artifacts related to jaw clenching. Following, the subject was asked to clench his jaw, whose EEG signals are shown in Fig. 2b. It can also be observed that the artifact energy generated by jaw clenching is much higher than in the relaxed condition (Fig. 2b). Some studies show that the main contributor to jaw or teeth clenching artifacts on the EEG signals is electromyography (EMG) related to increased tension in the jaw muscles [17, 18]. This strong artifact on the EEG signals, associated with the MI, is used as command to trigger the robotic devices in this work.

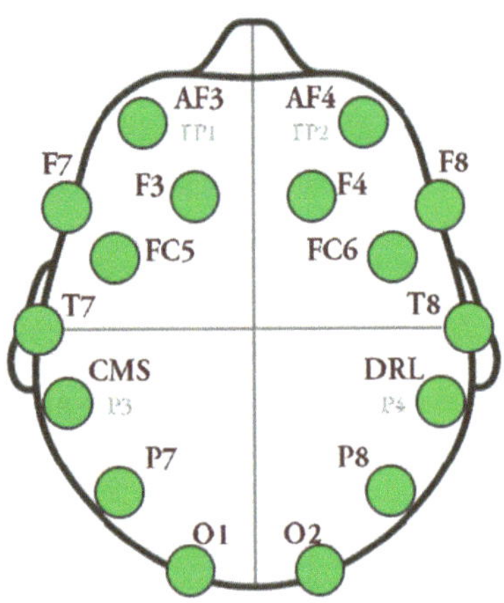

Fig. 1. Location of electrodes and references on the Emotiv headset.

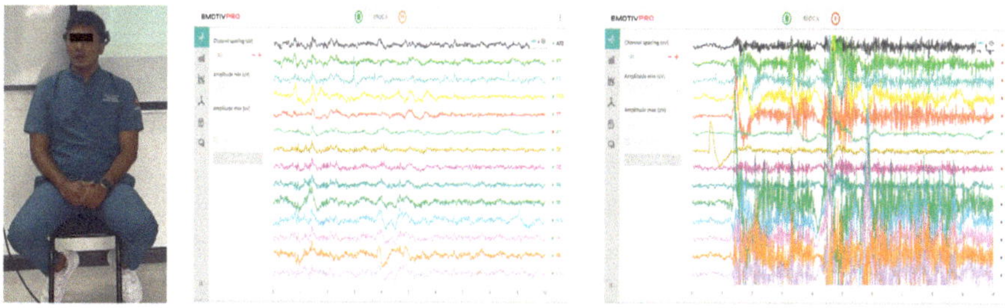

(a) (b)

Fig. 2. a) Typical subject conducting experiment to have his EEG signals acquired in two conditions: b) relaxed; c) clenching his jaw.

Studies have demonstrated that artifacts derived from jaw or teeth clenching provide higher-level of energy on EEG signals than any other kind of physiological artifacts, which can be completely recognized by data acquisition systems with 100% success rates [19]. In fact, jaw clenching is considered the most severe artifacts in scalp [14]. In our previous studies, we even have used the act of jaw clenching as a physiological signal to turn on/off different home appliances through a dBCI [20, 21], as well as to command a robotic wheelchair by people with severe disabilities [22, 23].

2.1 Subject Characteristics

The subject of our research is a 64-year-old man who had an ischemic stroke in October 2024, small vessel type, according to the Trial of Org 10172 in Acute Stroke Treatment (TOAST) classification [24], which provoked hemiplegia in his right hand and leg. His main cause of having had the event of stroke was high blood pressure. Exam of Computed Tomography showed a lacunar ischemic stroke in the left capsular nucleus region of this brain whereas his EEG exam showed several abnormalities, such as sinus rhythm, presence of atrioventricular conduction disturbance, and non-sustained ventricular tachycardia.

2.2 Experimental Setup

Figure 3 shows the experimental setup of this research, where it is possible to see the post-stroke subject with the Emotiv device in his head, and wearing two robotics devices: a robotic glove in his paralyzed hand, and his paralyzed feet on a robotic monocycle.

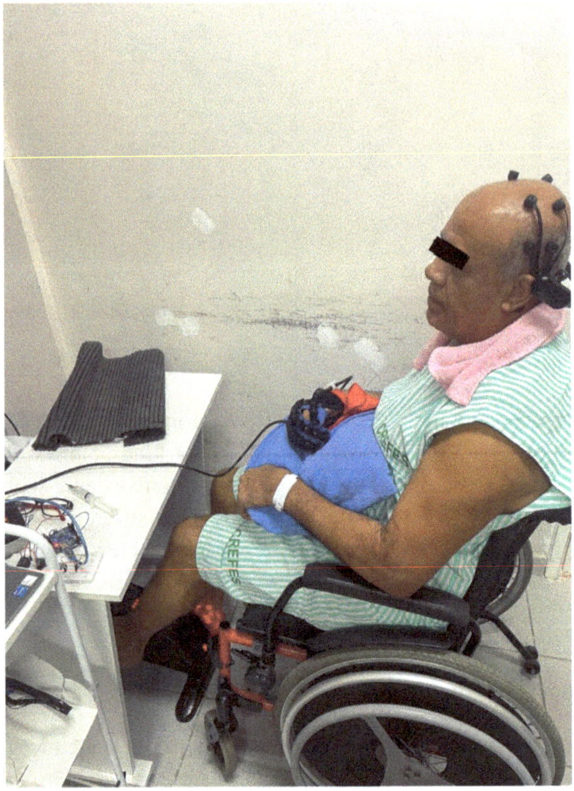

Fig. 3. Post-stroke patient wearing the Emotiv to capture EEG signals, and the robotic devices (robotic glove and robotic monocycle).

3 Evaluation Metrics

The subject of our research was assessed by a physiotherapist to check his movement capability through metrics typically used in physiotherapy: FMS, FIM, MAS, MBT, mRS, TUG, 10MWT, NIHSS, BI, and sEMG.

3.1 FMS

The Fugl-Meyer Assessment (FMA) is a widely used, reliable metrics to measure sensorimotor recovery in stroke survivors, which can assess motor function (upper/lower limbs), sensation, balance, joint range of motion, and pain. Developed by Axel Fugl-Meyer and colleagues in 1975, it allows quantifying impairment, track progress, guide treatment, and evaluate research outcomes, with higher scores indicating better function [25]. This metrics is normally used to assess six aspects of a subject: range of motion, pain, sensitivity, upper and lower extremity motor function, balance, and coordination and speed, totalling 226 points, in which 126 points are used to assess the upper-limb, and 100 points are used for the lower-limb. In our study we use the following equations to evaluate the subject's motor recovery percentage of upper- and lower- limbs:

$$UpperLimb : \frac{ScoreObtained \times 100}{126} = \%recovery \; LowerLimb : \frac{ScoreObtained \times 100}{100} = \%recovery$$

As the subject reached 76 points in the initial and 84 points in the final assessment for his upper-limb (Fig. 6a), this means 60.31% and 66.66%, respectively, i.e., a 6% of gain in functionality, range of motion, sensitivity, pain, proprioception and speed. For lower-limb the subject reached 71 and 80 points in the initial and final assessment (Fig. 6b), i.e., 71% and 80%, respectively, which means a 9% of gain for his lower-limb function.

3.2 FIM

The Functional Independence Measurement (FIM) is metrics that was developed to measure disability for a variety of populations, which allows measuring independence for self-care, including sphincter control, transfers, locomotion, communication, and social cognition. FIM intends to be sensitive to changes over a rehabilitation program, and takes into consideration the level of assistance an individual needs to grade functional status from total independence to total assistance. It is composed of 18-item, with seven-level, where level 7 means total independence, level 6 modified independence, level 5 moderate dependence (with the need for supervision or preparation), and levels 1–4 when the subject needs direct help, where level 1 means total dependence [26].

As an example of this metrics, a subject without any disability achieves a score of 126 points, and one with total dependence scores 18 points. The subject of our research had a gain of about 8%, as he reached a value initial of 105, and a final value of 113. In terms of dependence, the subject was initially classified with moderate dependence (score of 5.83), but increasing to 6.27 in the final assessment, which classifies him in improved situation of "modified independence". These results are shown in Fig. 6c.

3.3 MAS

The Modified Ashworth Scale (MAS) is metrics widely used for measuring spasticity, by rating resistance to passive movement on a 0–4 scale, with higher scores indicating more resistance. MAS allows tracking rehabilitation therapy effectiveness in stroke people [27]. In our research, in both initial and final assessment, the subject did not have any variation, being classified with Grade 1, i.e., presenting slight increase in muscle tone, which is a step above normal tone (Grade 0) (Fig. 6d).

3.4 MBT

The MiniBESTest (MBT) is a short version of the BESTest, which contains 14-item designed to quickly and effectively measure dynamic balance in subjects, especially those with neurological conditions like stroke. MBT focuses on four key balance systems: 1) anticipatory postural adjustments; 2) reactive postural control; 3) sensory orientation; 4) dynamic gait. MBT is time-efficient (10–15 min) and helps therapists pinpoint specific balance deficits for targeted rehabilitation, with scores from 0 (unable) to 2 (normal) for each item, totaling up to 28 points [28]. In our research, only the items 1 and 2 were assessed, which has a total score of 12 points. In this metrics, initially the subject of our study reached a score of 4 points (moderate balance deficit), whereas in the final assessment, he achieved 5 points (moderate balance deficit) (Fig. 6e), which means an improvement of 25%.

3.5 mRS

The modified Rankin Scale (mRS) is the gold standard metrics for stroke, which is composed of 7-point (0–6), allowing measuring disability after stroke, and also assessing functional independence in Activities of Daily Living (ADL), with the following levels: no symptoms (0); no significant disability (1); mild disability (2); moderate disability (3); moderately severe disability (4); severe disability (5); death (6). In our research, in both initial and final assessment, the subject did not have any variation, being classified with a score of 3 (moderate disability). Figure 6f shows these results.

3.6 TUG

The Time Up and Go (TUG) is a quick, widely used metrics to assess functional mobility and fall risk. TUG measures the time to stand from a chair, walk 3 m at a normal pace, turn, walk back, and sit down. A score over 12 s suggests a "high fall risk", whereas over 20 s indicates "significant fall risk and potential disability", requiring observation for unsteady gait or use of walking aids [30]. In our research, in both initial and final assessments, the results for this metrics were > 20 s (55 s and 39 s, respectively), which classifies the subject with "significant fall risk and potential disability". However, it is noteworthy that there was a significant time reduction (16 s, $\simeq 30\%$) to perform this metrics. Figure 6g shows these results.

3.7 10MWT

The 10-Meter Walk Test (10MWT) is a metrics used to assess walking or gait speed in meters per second over a 10-m distance. It is commonly employed to determine functional mobility, gait, and vestibular function. In our research, initially the subject lasted 27.33 s to walk 10 m (gait speed of 0.37 m/s), with about 16 steps. In the final assessment, he lasted 32.33 s (gait speed of 0.31 m/s), with about 20 steps, resulting in an increase of 16% in time, and 25% in number of steps. Figure 6h-6i show these results.

3.8 NIHSS

The National Institutes of Health Stroke Scale (NIHSS) is a standardized, 11-item metrics used by healthcare providers to measure the severity of a stroke and track neurological deficits, covering consciousness, vision, movement, coordination, and language, with scores from 0 (no deficit) to 42 (severe), indicating poorer outcomes with higher scores [32]. In our research, the subject initially scored 4, whereas in the final evaluation he scored 3, representing a decrease of 25% in this metrics. Figure 6j shows these results.

3.9 BI

The Barthel Index (BI) was developed for use in stroke rehabilitation, which measures the extent to which a post-stroke subjects can function independently and their mobility in ADLs, through 10 tasks: feeding, bathing, grooming, dressing, bowel control, bladder control, toileting, chair transfer, ambulation and stair climbing. This metrics also indicates the need for assistance in care, in a score from 0 to 100, where lower scores representing greater nursing dependency [33]. In the initial assessment, the subject achieved 36 points on BI, whereas in the final assessment he scored 41, meaning a gain of 14% (Fig. 6k).

3.10 SEMG

Surface electromyography (sEMG) is a non-invasive technique that uses skin electrodes to detect and analyze myoelectric signals to detect muscle activation, fatigue, and coordination for different applications as in stroke rehabilitation. This metrics is crucial for understanding dynamic movements, providing objective data for assessing conditions, and guiding therapy by visualizing muscle effort [34]. For the subject of this research, attention was focused on peaks of contraction for fingers' flexor, extensors, rectus femoris and tibialis anterior muscles. Results show that there was a $\simeq$5% gain in muscle contraction for finger flexors, with a peak of 78.35 μV in the initial evaluation and 82.23 μV in the final evaluation (Fig. 7a). For the tibialis anterior muscle there was a $\simeq$2% gain, with a peak of 64.92 μV in the initial evaluation and 66.06 μV in the final evaluation (Fig. 7b). For the rectus femoris muscle, there was a $\simeq$ 21% gain, with a peak of 110.02 μV in the initial evaluation and 133.43 μV in the final evaluation (Fig. 7c). Nevertheless, there was no gain for the fingers' extensor muscles.

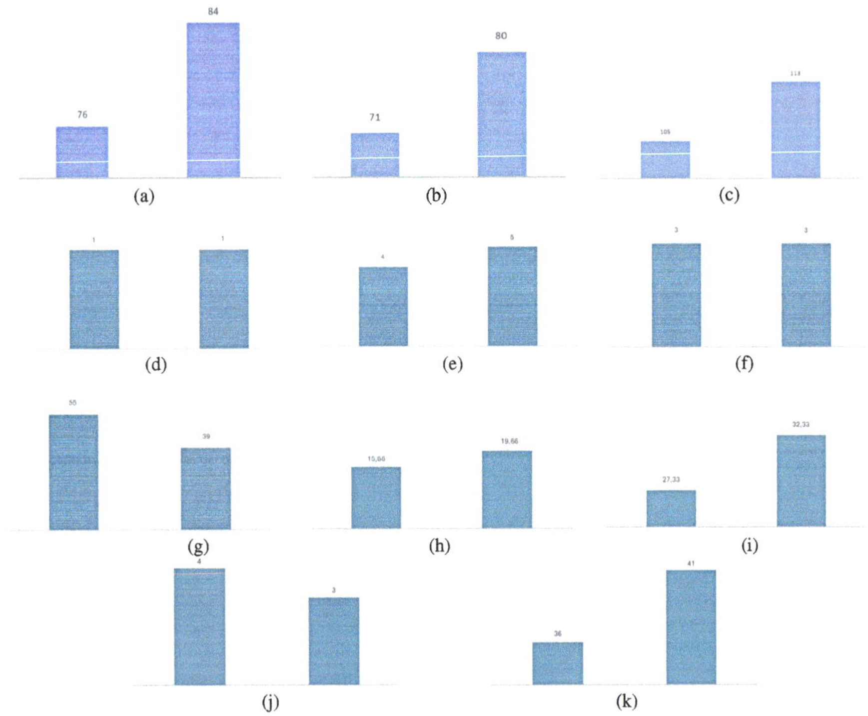

Fig. 6. Comparison between initial (left) and final (right) evaluation for: a) FMS for upper-limb; b) FMS for lower-limb; c) FIM; d) MAS; e) MBT; f) mRS; g) TUG; h) 10MWT for number of steps; i) 10MWT for spent time; j) NIHSS; k) BI.

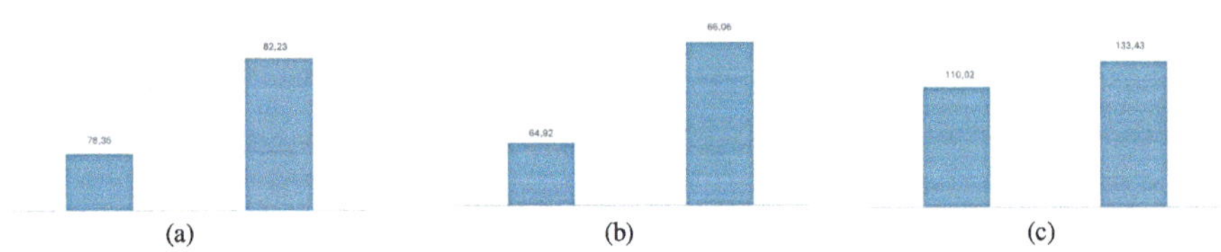

Fig. 7. A sEMG measurements for the following muscles: a) finger flexors; b) tibialis anterior; c) rectus femoris.

4 Conclusion

Results of this research show that the post-stroke subject had a gain in his condition at the end of our neurorehabilitation therapy (except for the 10-Meter Walk Test (10MWT), where the subject had 16% increase in time and 25% increase in number of steps. In all additional metrics, the subject had a gain in his condition, such as 6% in Fugl-Meyer Assessment Scale (FMS) for upper-limb and 9% for lower-limb; 8% in Functional Independence Measurement (FIM), and passing from a FIM score of 5.83 to 6.27; 25% for MiniBESTest (MBT); 30% in time reduction for Time Up and Go (TUG); 25% in NIHSS; and 14% in BI. However, there was not variation for Modified Ashworth Scale

(MAS), which maintained the Grade 1, and modified Rankin Scale (mRS), maintaining a score of 3. Regarding the results for surface Electromyography (sEMG), there were $\simeq$5% gain in muscle contraction peak for finger flexors, $\simeq$2% for tibialis anterior and $\simeq$21% for rectus femoris. It is worth commenting that there is a limitation in our research, as only a single subject was assessed, therefore, more studies are needed to confirm the results with a wider sample, as well as conducting the research by blinded evaluators.

Acknowledgements. The authors thank the Rehabilitation Center of Espirito Santo (CREFES) for making available the patients for this research, and FAPES (Project Number: 983/2022 P: 2022-B92KF) and CNPq (all from Brazil) for funding this research.

References

1. World population ageing 2020 highlights: Living arrangements of older persons. United Nations Dept. Economic Social Affairs, Population Division, Herndon, VA (2020)
2. CDC Homepage: Know your risk for stroke. https://www.cdc.gov/stroke/risk_factors.htm#:~:text=The%20chance%20of%20having%20a,65%20years%20also%20have%20strokes.&text=In%20fact%2C%20about%20one%20in,adults%20ages%2015%20to%2049. Accessed 20 Nov 2024
3. CDC Homepage: About stroke. https://www.cdc.gov/stroke/about.htm#:~:text=A%20stroke%2C%20sometimes%20called%20a,term%20disability%2C%20or%20even%20death. Accessed 20 Nov 2024
4. Cardoso, V., et al.: Effect of a brain-computer interface based on pedaling motor imagery on cortical excitability and connectivity. Sensors **21** (2020)
5. Bastos-Filho, T.: Introduction to Non-invasive EEG-Based Brain-Computer Interfaces for Assistive Technologies. CRC Press, Boca Raton (2020)
6. Pollock, A., et al.: Interventions for improving upper limb function after stroke. Cochrane Database Syst Rev. **12**(11), CD010820 (2014)
7. Loureiro, R., et al.: Advances in upper limb stroke rehabilitation: a technology push. Med. Biol. Eng. Compu. **49**(10), 1103–1118 (2011)
8. Aguilera-Rubio, A., et al.: Feasibility and efficacy of a virtual reality game-based upper extremity motor function rehabilitation therapy in patients with chronic stroke: a pilot study. Int. J. Environ. Res. Public Health **19**(6), 3381 (2022)
9. Shin, H., et al.: Therapeutic effects of functional electrical stimulation on physical performance and muscle strength in post-stroke older adults: a review. Ann. Geriatric Med. Res. **26**(1), 16–24 (2022)
10. Soekadar, S., et al.: Brain-machine interfaces in neurorehabilitation of stroke. Neurobiol. Dis. **83**, 172–179 (2015)
11. Heena, N., et al.: Effects of task complexity or rate of motor imagery on motor learning in healthy young adults. Brain Behav. **11**, e02122 (2021)
12. Ruffino, C., Papaxanthis, C., Lebon, F.: Neural plasticity during motor learning with motor imagery practice: review and perspectives. Neuroscience **26**, 61–78 (2017)
13. Pfurtscheller, G., Lopes, F.H.: Event-related EEG/MEG synchronization and desynchronization: basic principles. Clin. Neurophysiol. **110**, 1842–1857 (1999)
14. Yang, B., et al.: A multi-day and high-quality EEG dataset for motor imagery brain-computer interface. Sci. Data **12**, 488 (2025)
15. Ferreira, A., et al.: Human-machine interfaces based on EMG and EEG applied to robotic systems. J. Neuroeng. Rehabil. **5**, 1–15 (2008)

16. Bastos-Filho, T.F., Kumaresh, D.K., Sridhar, A.: Devices for Mobility and Manipulation for People with Reduced Abilities. CRC Press, Boca Raton (2014)
17. Kappel, S.L., Looney, D., Mandic, D.P., Kidmose, P.: Physiological artifacts in scalp EEG and ear-EEG. BioMed. Eng. OnLine **16**, 103 (2017)
18. Goncharova, I.I., McFarland, D.J., Vaughan, T.M., Wolpaw, J.R.: EMG contamination of EEG: spectral and topographical characteristics. Clin. Neurophysiol. **114**, 1580–1593 (2003)
19. Shimada, A., Yamabe, Y., Torisu, T., Baad-Hansen, L., Murata, H., Svensson, P.: Measurement of dynamic bite force during mastication. J. Oral Rehabil. **39**, 349–356 (2012)
20. Jzau-Sheng, L., Zih-Yang, J.: An EEG-based BCI system to facial action recognition. Wireless Pers. Commun. **94**, 1579–1593 (2017)
21. Bissoli, A., Coelho, Y., Bastos-Filho, T.F.: A system for multimodal assistive domotics and augmentative and alternative communication. In: Proceedings of the 9th International Conference on PErvasive Technologies Related to Assistive Environments (PETRA) (2016)
22. Bastos-Filho, T.F.: Introduction to non-invasive EEG-based brain-computer interfaces for assistive technologies (2021)
23. Tello, R., Bissoli, A., Ferrara, F., Bastos-Filho, T.F.: A high performance human-computer interface to control a robotic wheelchair and an intelligent environment. In: Proceeding of the VI International Congress on Design, Research Networks and Technology for All (DRT4ALL) (2015)
24. Adams, H.P., et al.: Classification of subtype of acute ischemic stroke. Definitions for use in a multicenter clinical trial. TOAST. Trial of Org 10172 in Acute Stroke Treatment. Stroke **24**(1), 35–41 (1993)
25. Fugl-Meyer, A.R., Jääskö, L., Leyman, I., Olsson, S., Steglind, S.: The post-stroke hemiplegic patient. 1. A method for evaluation of physical performance. Scand. J. Rehabil. Med. **7**(1), 13–31 (1975)
26. Chumney, D., et al.: Ability of functional independence measure to accurately predict functional outcome of stroke-specific population: systematic review. J. Rehabil. Res. Dev. **47**(1), 17–29 (2010)
27. Vidmar, T., Kregar, N.G., Puh, U.: Reliability of the Modified Ashworth Scale after stroke for 13 muscle groups. Arch. Phys. Med. Rehabil. **104**(10), 1606–1611 (2023)
28. Tsang, C.S., et al.: Psychometric properties of the Mini-Balance Evaluation Systems Test (Mini-BESTest) in community-dwelling individuals with chronic stroke. Phys. Ther. **93**(8), 1102–1115 (2013)
29. Saver, J.L., et al.: Standardized nomenclature for Modified Rankin Scale global disability outcomes: consensus recommendations from stroke therapy academic industry roundtable XI. Stroke **52**(9), 3054–3062 (2021)
30. Persson, C.U., et al.: Timed Up & Go as a measure for longitudinal change in mobility after stroke - Postural Stroke Study in Gothenburg (POSTGOT). J. Neuroeng. Rehabil. **11**(83), 1–7 (2014)
31. Cheng, D.K., et al.: Validation of stroke-specific protocols for the 10-Meter Walk Test and 6-minute walk test conducted using 15-meter and 30-meter walkways. Top. Stroke Rehabil. **4**, 251–261 (2020)
32. Hage, V.: The NIH Stroke Scale: a window into neurological status. Nurs. Spectr. **24**(15), 44–49 (2011)
33. Quinn, T.J., et al.: Barthel Index for stroke trials: development, properties, and application. Stroke **42**(4), 1146–1151 (2011)
34. Steele, K.M., Papazian, C., Feldner, H.A.: Muscle activity after stroke: perspectives on deploying surface electromyography in acute care. Front. Neurol. **11**, 576757 (2020)

AINHS: An Integrative Model Based on Artificial Intelligence and Neutrosophic Statistics to Strengthen Health Literacy and Optimize Self-Care in Patients with Chronic Kidney Disease

Lorenzo Cevallos-Torres[1,2](✉), Esther Carlín-Chávez[1,3], Ivis Yania Piovet-Hernández[1], and María Cristina Grandal-López[1]

[1] Universidad Bolivariana del Ecuador, Km 5. vía Duran – Yaguachi, Duran 092405, Ecuador
lorenzo.cevallost@ug.edu.ec

[2] Facultad de Ciencias Matemáticas y Físicas, Universidad de Guayaquil, Guayaquil, Ecuador

[3] Facultad de Posgrados, Universidad Estatal de Milagro, Milagro, Ecuador

Abstract. This study aims to strengthen health literacy and optimize self-care among individuals with Chronic Kidney Disease (CKD) through the implementation of **AINHS** (An Integrative Model Based on Artificial Intelligence and Neutrosophic Statistics). Leveraging a customized clinical database and applying the Knowledge Discovery in Databases (KDD) methodology, AINHS systematically cleans, transforms, and analyzes patient data to extract valuable patterns. The model integrates Machine Learning techniques—specifically Long Short-Term Memory (LSTM) networks—to capture temporal dynamics in sequential health records, enabling robust predictions grounded in historical and current data trends. Neutrosophic Statistics complements this process by managing uncertainty through the evaluation of truth, falsity, and indeterminacy, allowing for greater interpretability and reliability of the outputs. By combining temporal pattern recognition with uncertainty management, AINHS provides personalized, confidence-weighted health and dietary recommendations adapted to each patient's unique clinical condition. Ultimately, this hybrid approach empowers patients to make informed decisions, improves disease self-management, and enhances overall quality of life, addressing one of the most critical challenges in chronic disease care.

Keywords: Neutrosophy · Machine Learning · Chronic Kidney Disease · Recurrent Neural Networks

1 Introduction

Chronic Kidney Disease (CKD) is a growing global health concern that requires continuous monitoring, personalized care, and active patient participation. However, limited health literacy and the inherent uncertainty in clinical data often hinder effective disease

J. L. Ballesteros et al. (Eds.): CITIS 2025, LNNS 1780, pp. 325–336, 2027.
https://doi.org/10.1007/978-3-032-14287-0_32

management and self-care. The complexity of CKD demands innovative approaches that can process diverse patient data and deliver tailored interventions. The AINHS model—an integrative framework combining Artificial Intelligence and Neutrosophic Statistics—emerges as a promising solution. It addresses the dual challenge of data unpredictability and patient engagement by enhancing decision-making and interpretation. This model aims to empower patients with better health understanding while supporting healthcare systems in providing adaptive, data-driven care.

Building on this foundation, the model incorporates Long Short-Term Memory (LSTM) networks to identify temporal trends in patient health data, addressing the difficulty of managing chronic kidney disease (CKD). LSTM improves predictive accuracy by preserving important information across time, making it possible to provide tailored, data-driven solutions. This feature facilitates recommendations for adaptive self-care that are in line with the clinical trajectory of each patient. The model enhances healthcare decision-making and patient empowerment in conjunction with neutrosophic thinking [6].

Furthermore, because it facilitates well-informed decision-making and efficient self-care, health literacy is crucial for those managing long-term illnesses like chronic kidney disease (CKD). CKD necessitates rigorous dietary control and ongoing monitoring because it is frequently linked to diabetes and hypertension. However, appropriate disease care may be hampered by low health literacy and the intricacy of clinical data. In order to improve data interpretation and customize health interventions, the AINHS model integrates artificial intelligence with neutrosophic statistics. This method improves nutritional planning, empowers patients, and eventually raises quality of life [7, 8].

In addition to these capabilities, Neutrosophic Statistics is a statistical approach to managing uncertainty, ambiguity, and contradiction, making it especially useful in healthcare, where data is frequently partial or confusing. Unlike traditional methods, it assesses truth, falsity, and indeterminacy, which improves the interpretation of ambiguous clinical patterns. The AINHS model takes advantage of this potential by combining Neutrosophic Statistics with LSTM neural networks to assess sequential health data and control variability. This hybrid method offers individualized, adaptive suggestions for chronic illnesses like kidney disease. By combining temporal pattern detection and uncertainty management, the strategy increases health literacy and promotes successful self-care in CKD patients [4, 5].

Finally, maintaining a balanced diet is essential for managing chronic kidney disease, as it directly affects kidney function and disease progression. However, meal planning is complex due to varied patient needs and dietary restrictions. The AINHS model combines LSTM networks and Neutrosophic Statistics to address this challenge. This hybrid approach enables precise, adaptable dietary recommendations. As a result, it supports better nutrition, treatment adherence, and overall health outcomes [9].

2 Material and Method

The methodology for this study proposes a hybrid model that combines Artificial Intelligence and Neutrosophic Statistics to enhance health literacy and self-care in patients with Chronic Kidney Disease (CKD). It incorporates the Knowledge Discovery in Databases

(KDD) process to uncover meaningful patterns from clinical datasets, where data preprocessing includes cleaning, normalization, and categorical variable encoding. Long Short-Term Memory (LSTM) networks are applied to detect temporal trends in patient health data, allowing predictive modeling based on historical sequences. Neutrosophic Statistics supplements this process by managing uncertainty, evaluating truth, falsity, and indeterminacy, and assigning confidence levels to predictions. Together, these methodologies generate more personalized, adaptable recommendations, supporting informed decision-making and improving patients' quality of life. [12].

1. **KDD Methodology Applied to the AINHS Model**

The AINHS model uses the KDD framework, combining LSTM networks and Neutrosophic Statistics to enhance health literacy and self-care in CKD patients. This hybrid system captures temporal patterns and manages uncertainty to deliver personalized, adaptive recommendations [10] (Fig. 1).

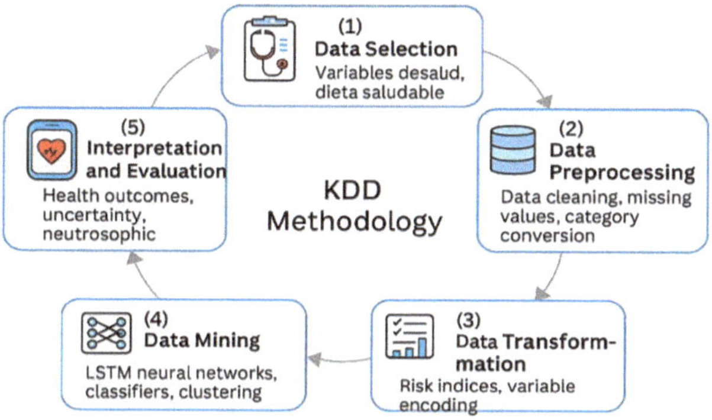

Fig. 1. KDD methodology for the AINHS Model: Chronic kidney.

a. **Data selection**

The dataset used for the AINHS model is composed of 1,200 records and 20 variables, capturing a wide range of clinical, nutritional, and behavioral data related to patients with chronic kidney disease. The variables include demographic attributes (e.g., age, gender), anthropometric data (e.g., weight, height, BMI), clinical markers (e.g., creatinine, TFG, proteinuria), dietary intake (e.g., sodium, potassium, protein), and lifestyle factors (e.g., physical activity, fluid retention). The variable "Health Results" serves as an outcome measure for health status. This comprehensive structure enables an in-depth analysis of individual patient profiles to support personalized care and decision-making within the KDD framework (Table 1).

b. **Data preprocessing**

The data preprocessing phase ensures data quality and consistency for accurate analysis. Missing values are handled through removal or imputation, and categorical variables are numerically encoded. Clinical variables like Creatinine and TFG are normalized to avoid bias. Date formats are also corrected for temporal accuracy. These steps enhance the AINHS model's predictive performance [11].

Table 1. Operationalization of Variables and Classification Ranges.

Variable	High (T)	Medium (I)	Low (F)
Sodium intake	2500–3500 mg	1500–2500 mg	0–1500 mg
Potassium intake	3000–4000 mg	2000–3000 mg	0–2000 mg
Body weight	80–100 kg	60–80 kg	40–60 kg
Hydration	High	Medium	Low
Healthy diet score	8–10 (optimal)	5–7 (moderate)	1–4 (unhealthy)
Systolic blood pressure	>140 mmHg	120–139 mmHg	<120 mmHg
Creatinine level	>1.5 mg/dL	1.0–1.5 mg/dL	<1.0 mg/dL
Glomerular filtration rate	>90 mL/min	60–89 mL/min	<60 mL/min
Urine protein level	>30 mg/dL	15–30 mg/dL	<15 mg/dL

c. **Transformation**

The transformation phase improves the dataset by creating or refining variables to boost analysis. In the AINHS model, a Renal Risk Index is formed by combining clinical indicators like creatinine and TFG. Health outcomes are recoded into numerical classes for machine learning. Age groups and dietary ratios are also generated for deeper insights. These new attributes enhance prediction and pattern detection [13, 14] (Fig. 2).

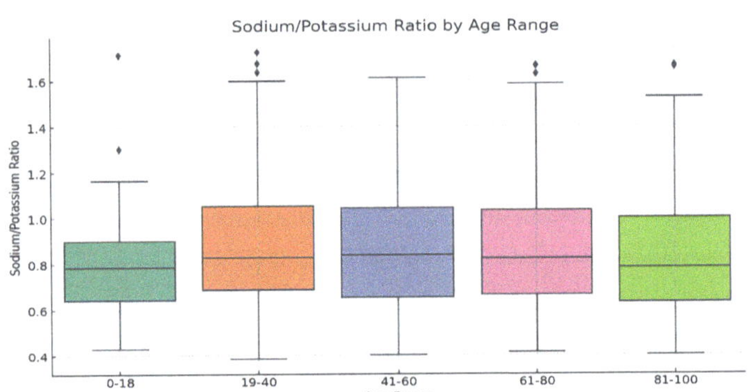

Fig. 2. Sodium/Potassium Ratio by Age Range.

d. **Data Mining**

The AINHS model uses LSTM neural networks to detect temporal patterns and predict clinical indicators like creatinine levels. It also forecasts dietary adherence using supervised learning. Data is normalized, sorted, and processed with a sliding window technique for sequence prediction. This improves accuracy in tracking kidney function trends. Predictions are validated with real data to support timely clinical decisions. [15].

1. **Initialization of parameters:** The essential components required for the LSTM neural network to operate are listed.

- Weight matrices ($\boldsymbol{Wf}$, $\boldsymbol{Wi}$, $\boldsymbol{Wc}$, $\boldsymbol{Wo}$).

The matrices $\boldsymbol{Wf}$, $\boldsymbol{Wi}$, $\boldsymbol{Wc}$, and $\boldsymbol{Wo}$ in an LSTM cell manage memory flow through different gates. $\boldsymbol{Wf}$ controls retention of past data, while $\boldsymbol{Wi}$ handles new input. $\boldsymbol{Wc}$ transforms data before storage, and $\boldsymbol{Wo}$ regulates output from the cell state. Together, they help the LSTM learn temporal patterns. Their weights define how inputs shape memory and predictions.

$$W_f \begin{bmatrix} 0.5 & -0.3 & 0.8 & 0.1 \\ -0.2 & 0.6 & -0.7 & 0.9 \\ 0.3 & -0.5 & 0.2 & -0.8 \end{bmatrix}$$ Forget Gate Weights ($\boldsymbol{Wf}$)

$$W_i \begin{bmatrix} -0.4 & 0.7 & -0.1 & 0.5 \\ 0.6 & -0.2 & 0.3 & -0.9 \\ -0.5 & 0.1 & 0.8 & -0.3 \end{bmatrix}$$ Input Gate Weights ($\boldsymbol{Wi}$)

$$W_c \begin{bmatrix} 0.2 & -0.6 & 0.7 & -0.4 \\ -0.1 & 0.5 & -0.3 & -0.9 \\ 0.6 & -0.2 & 0.9 & -0.7 \end{bmatrix}$$ Candidate Cell State Weights ($\boldsymbol{Wc}$)

$$W_o \begin{bmatrix} 0.8 & -0.4 & 0.3 & 0.2 \\ -0.7 & 0.6 & -0.1 & 0.5 \\ 0.2 & -0.9 & 0.4 & -0.3 \end{bmatrix}$$ Output Gate Weights ($\boldsymbol{Wo}$)

- **Weight Matrices for Previous Hidden States ($\boldsymbol{Uf}$, $\boldsymbol{Ui}$, $\boldsymbol{Uc}$, $\boldsymbol{Uo}$).**

In this step, we define the matrices that connect the influence of the previous hidden state $\boldsymbol{h}_{t-1}$ with the current input processing. These matrices play a critical role in preserving temporal dependencies and enabling the LSTM cell to retain relevant information across time steps. Each matrix corresponds to a specific gate—forget, input, candidate, and output—and is crucial for adjusting how memory is integrated into current computations. Below are the initialized values for the recurrent weight matrices.

$$U_f \begin{bmatrix} 0.1 & -0.5 & 0.6 \\ -0.4 & 0.3 & -0.2 \\ 0.7 & -0.8 & 0.9 \end{bmatrix}$$ **Matrix $\boldsymbol{Uf}$**: Recurrent Weight Matrix for the Forget Gate

$$U_i \begin{bmatrix} 0.2 & -0.6 & -0.3 \\ 0.8 & -0.1 & 0.4 \\ -0.5 & 0.7 & -0.9 \end{bmatrix}$$ **Matrix $\boldsymbol{Ui}$**: Recurrent Weight Matrix for the Input Gate

$$U_c \begin{bmatrix} 0.5 & -0.4 & 0.8 \\ -0.2 & 0.7 & -0.1 \\ 0.6 & -0.3 & 0.9 \end{bmatrix}$$ **Matrix $\boldsymbol{Uc}$**: Recurrent Weight Matrix for the Candidate Cell State

$$U_o \begin{bmatrix} -0.1 & 0.5 & -0.7 \\ 0.3 & -0.8 & 0.2 \\ -0.4 & 0.6 & -0.9 \end{bmatrix}$$ **Matrix $\boldsymbol{Uo}$**: Recurrent Weight Matrix for the Output Gate

- **Bias Vectors for Gate Activations ($\boldsymbol{bf}$, $\boldsymbol{bi}$, $\boldsymbol{bc}$, $\boldsymbol{bo}$)**

Bias vectors are essential components that allow each gate to adjust its activation threshold independently of input magnitude. These biases add flexibility to the gating mechanisms by introducing trainable offsets that influence when and how gates open or close. In the AINHS model, each bias vector corresponds to a specific gate and complements the effect of the weighted inputs. Below are the initialized values of the bias vectors used in the forget, input, candidate, and output gates.

$b_f \begin{bmatrix} 0.2 \\ -0.3 \\ 0.5 \end{bmatrix}$	**Vector *bf*:** Bias Vector for the Forget Gat
$b_i \begin{bmatrix} -0.1 \\ 0.4 \\ -0.6 \end{bmatrix}$	**Vector *bi*:** Bias Vector for the Input Gate
$b_c \begin{bmatrix} 0.3 \\ -0.2 \\ 0.7 \end{bmatrix}$	**Vector *bc*:** Bias Vector for the Candidate Cell State
$b_o \begin{bmatrix} -0.5 \\ 0.6 \\ -0.4 \end{bmatrix}$	**Vector *bo*:** Bias Vector for the Output Gate

The bias vectors ***bf***, ***bi***, ***bc***, and ***bo*** adjust gate activations in an LSTM cell.***bf*** controls how much past information is forgotten, while ***bi*** balances new data entry. ***bc*** shapes input transformation into memory updates. ***bo*** shifts how much of the cell state is output. These biases enhance learning precision in complex temporal data.

2. **State Initialization:** This is where the initial values of the memory cell and the hidden state are established.

Where:
$c_t = 0$ Represents the state of the cell that represents long-term memory and is initialized to zero.
$h_t = 0$ Represents the hidden state and is where the information processed at each time step is stored and is also initialized to 0.

3 Patient Data Processing

Patient data is extracted and preprocessed to ensure compatibility with the neural network. Categorical variables such as muscle mass, hydration, and physical activity are numerically encoded using predefined maps. If the time series is incomplete, it is reconstructed using historical patterns or interpolation techniques.

4 Process in the LSTM Network

Every time step in the temporal series is processed iteratively by the LSTM network. The model is given the current input $\boldsymbol{x}_t$ and the prior hidden state $\boldsymbol{h}_{t-1}$ at each iteration $\boldsymbol{t}$. The activation of the forget, input, candidate, and output gates is then calculated using the learned weight matrices and bias vectors. These activations dictate which values are sent on to the following stage, how much historical data is kept, and how fresh data is integrated. In order to maintain long-term dependencies that are essential for predicting health trends, the internal cell state Ct is updated appropriately. The AINHS model can identify trends over time and produce customized predictions, such future clinical results or dietary advice, thanks to this recursive process.

4.1 Neutrosophic Adjustments to the Initial Probability.

Since clinical data is not always exact or full in the medical area, neutrophilic statistics is distinguished by how it handles ambiguity in clinical data. [16]. A table containing SVN numbers and language phrases was made specifically for this project and the administration of neutrosophic statistics. The variables, SVN values, and language words are described in detail in Table 2.

Table 2. SVN variables, numbers, and linguistic terms.

Variable	SVN	Linguistic Term
Sodium Intake	(0.8, 0.1, 0.1)	(High, Medium, Low)
Potassium Intake	(0.9, 0.05. 0.05)	(High, Medium, Low)
Body Weight	(0.7, 0.2, 0.1)	(High, Medium, Low)
Hydration	(0.5, 0.3, 0.2)	(High, Medium, Low)
Healthy Diet	(0.7, 0.2, 0.1)	(High, Medium, Low)
Systolic Blood Pressure	(0.6, 0.2, 0.2)	(High, Medium, Low)
Creatinine Levels	(0.4, 0.3, 0.3)	(High, Medium, Low)
Glomerular Filtration Rate	(0.5, 0.3, 0.2)	(High, Medium, Low)
Urine Protein Level	(0.6, 0.2, 0.2)	(High, Medium, Low)

Neutrosophic statistics is used to improve model predictions by considering three components:

- **Truth (T):** This is the certainty of the model's prediction.
- **Falsification (F):** This measures the falsity of each prediction.
- **Uncertainty (U):** This assesses the variability of the data

To apply neutrosophic statistics, the truth, indeterminacy, uncertainty, and falsity values must be combined to adjust the model's predictions.

The formula used is as follows:

$$P_{tight} = P \times (1 - I) \times (1 - F) \times (1 - U).$$

Where:

- **P** => Indicates the initial probability of the model.
- **I** => Indicates indeterminacy.
- **F** => Indicates falsity.

- **U =>** Indicates uncertainty.

Finally, the fitted probability is limited to be between 0 and 1.
$\boldsymbol{P_{tight}} = \mathbf{max}(0.0, \boldsymbol{min}(1.0, \boldsymbol{P_{tight}}))$.

4.2 Hybrid Model (LSTM + neutrosophia)

The hybrid model improves prediction accuracy and detects changes in health patterns by combining neutrosophic statistics with LSTM neural networks. In contrast to traditional models, it increases user trust by explaining its outputs in addition to predicting them. Even with insufficient information, it makes tailored suggestions by adjusting to clinical variability and uncertainty. The integration facilitates more accurate and contextually sensitive diet and self-care forecasts. It does, however, require more processing power and cautious application. Inconsistent or untrustworthy results could result from improper integration.

5 Results

The hybrid model combining LSTM and neutrosophic statistics achieved 92% precision, outperforming LSTM (87%) and neutrosophic (78%) models individually. It showed strong results with 85 true positives and 97 true negatives. The model demonstrated 91.4% precision, 92.4% specificity, and 89.5% sensitivity. These metrics confirm its effectiveness in handling clinical data uncertainty for personalized health recommendations.

5.1 Confusion Matrix – Hybrid Model (LSTM + Neutrosophy)

The confusion matrix shows the hybrid model's strong classification performance, with 85 true positives and 97 true negatives. Only 10 false negatives and 8 false positives were reported, indicating low error rates. The model achieved 91.4% precision, 89.5% sensitivity, and 92.4% specificity. These results highlight its reliability for clinical decision-making under uncertainty (Fig. 3).

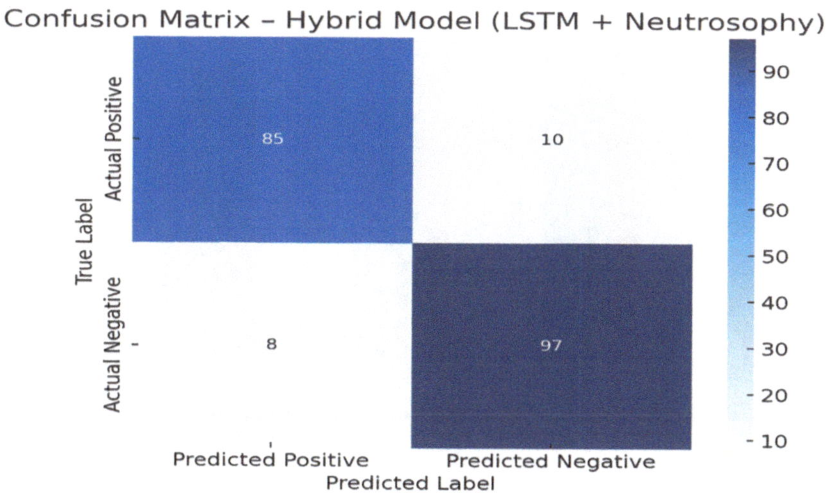

Fig. 3. Confusion Matrix.

5.2 Derived Metrics and Model Comparison

The hybrid model (LSTM + Neutrosophy) achieved 91.4% precision, 89.5% sensitivity, and 92.4% specificity, outperforming individual models in every category. Accurate identification of both safe and at-risk patients is ensured by this balance. The hybrid model clearly outperformed the LSTM and neutrosophic models alone. Clinical dependability is improved through the integration of uncertainty handling and temporal analysis. These findings back up its application in health forecasting and judgment (Table 3).

Table 3. Derived Metrics

Metric	LSTM	Neutrosophic	Hybrid
Precision	87%	78%	92%
Sensitivity	88%	75%	89.50%
Specificity	85%	80%	92.40%

5.3 Justification of the Hybrid Combination

By combining the advantages of both strategies, the hybrid combination improves system performance overall. Accurate time-based predictions are made possible by LSTM's proficiency at spotting temporal trends in user data. Meanwhile, by managing ambiguity and ambiguous information well, neutrosophic statistics enhance the interpretability of outcomes. When combined, they optimize sensitivity and precision, strengthening and enhancing the model's dependability for practical health advice (Fig. 4).

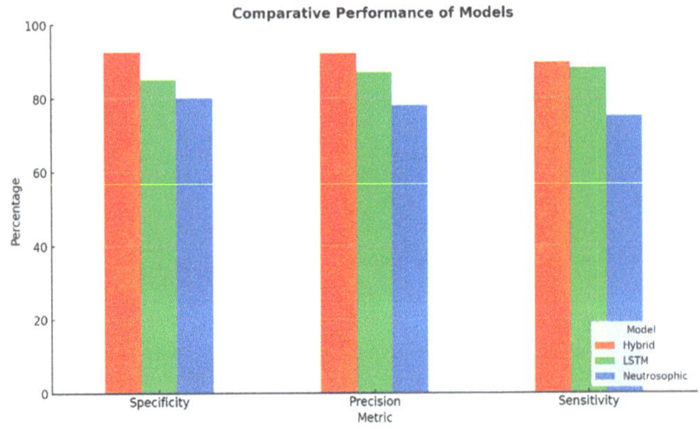

Fig. 4. Comparative Performance of Models

5.3.1 Interpretation of the Model Comparison Chart

The bar graph uses specificity, accuracy, and sensitivity to compare three models: hybrid, LSTM, and neutrophilic. With 89.5% sensitivity, 92% accuracy, and 92.4% specificity, the hybrid model performs better than the others. In terms of sensitivity, LSTM does well (88%) but not so well in the other metrics. Despite controlling for uncertainty, the neutrosophic model exhibits the lowest total values. These findings demonstrate the accuracy and clinical applicability advantages of the hybrid model.

6 Discussion

Hochreiter and Schmidhuber [1] demonstrated that LSTM neural networks are effective in modeling complex temporal sequences in clinical settings. Their application in chronic disease monitoring enables the accurate prediction of medical events, which is key for personalized care. This capability aligns with the AINHS model, which analyzes sequential data in patients with chronic kidney disease. By integrating LSTM, personalized dietary recommendations can be generated to support self-care. This approach strengthens the technological foundation of the model by providing solid support from artificial intelligence.

Smarandache [2] presents neutrosophic logic and statistics as essential tools for managing uncertainty in medical systems, particularly when dealing with incomplete or ambiguous data. This approach offers a more accurate representation of medical knowledge in scenarios where clinical decisions cannot rely solely on absolute truths. In the AINHS model, neutrosophic statistics play a crucial role by providing a framework to interpret uncertainty in patient data related to chronic kidney disease. This enables the delivery of more adaptable and context-sensitive recommendations. As a result, the neutrosophic component enhances the resilience of the model and complements the artificial intelligence-based elements.

The study in [3] highlights the effectiveness of intelligent mobile systems in enhancing health literacy and promoting self-care, particularly among patients with chronic illnesses. These technologies enable personalized interventions tailored to individual needs, which helps improve treatment adherence. In the context of the AINHS model, the integration of a smart mobile tutor aligns with these goals by delivering data-driven, patient-specific recommendations. This technological component empowers users and supports the development of sustainable self-care practices. Therefore, the related work reinforces the relevance and positive impact of mobile solutions in diverse healthcare settings.

7 Conclusión

In the healthcare industry, the hybrid model that combines LSTM and neutrosophic statistics enhances predictive accuracy, interpretability, and adaptability. While neutrosophic statistics controls for clinical data ambiguity, LSTM records temporal patterns. When combined, they outperform each model alone in terms of accuracy, sensitivity, and specificity. This strategy encourages individualized health and nutrition literacy. By providing patients with clear information, it enables them to make educated decisions. In the end, it improves the management of chronic illnesses and advances the application of AI in public health.

References

1. Hochreiter, S., Schmidhuber, J.: Long short-term memory. Neural Comput. **9**(8), 1735–1780 (1997)
2. Smarandache, F.: Neutrosophy: Neutrosophic Probability, Set, and Logic. American Research Press, Rehoboth (1998)
3. Islam, M.S., Hasan, M., Hasan, M.K.: Smart mobile health monitoring system for chronic disease management. In: Proceedings of IEEE International Conference on Computer and Information Technology (ICCIT), Dhaka, Bangladesh, pp. 366–371 (2015)
4. Smarandache, F.: A Unifying Field in Logics: Neutrosophic Logic. Neutrosophy, Neutrosophic Set, Neutrosophic Probability and Statistics. American Research Press (2005)
5. Cevallos-Torres, L., Botto-Tobar, M., Cevallos-Torres, L., Botto-Tobar, M.: Process sampling. In: Problem-Based Learning: A Didactic Strategy in the Teaching of System Simulation, pp. 13–31 (2019)
6. Barzola-Monteses, J., et al.: Detection of cardiovascular diseases using predictive models based on deep learning techniques: a hybrid neutrosophic AHP-TOPSIS approach for model selection. Neutrosophic Sets and Systems **74**(1), 18 (2024)
7. Nutbeam, D.: Health literacy as a public health goal: a challenge for contemporary health education and communication strategies into the 21st century. Health Promot. Int. **15**(3), 259–267 (2000)
8. Guijarro-Rodríguez, A.A., Cevallos-Torres, L.J., Botto-Tobar, M., Leyva-Vazquez, M., Holguin, J.Y.: Clinical assessment using an algorithm based on fuzzy C-means clustering. In: Technologies and Innovation: Proceedings of 3rd International Conference CITI 2017, Guayaquil, Ecuador, 24–27 October 2017, pp. 181–193. Springer, Heidelberg (2017)
9. Parodi, A., et al.: The potential of future foods for sustainable and healthy diets. Nature Sustainabil. **1**(12), 782–789 (2018). https://doi.org/10.1038/s41893-018-0189-7
10. Ayed, M.B., Ltifi, H., Kolski, C., Alimi, A.M.: A user-centered approach for the design and implementation of KDD-based DSS: a case study in the healthcare domain. Decis. Support Syst. **50**(1), 64–78 (2010). https://doi.org/10.1016/j.dss.2010.07.002

11. Guijarro-Rodríguez, A.A., Witt-Rodríguez, P.M., Cevallos-Torres, L.J., Contreras-Puco, S.F., Ortiz-Zambrano, M.C., Torres-Martínez, D.E.: Image segmentation techniques application for the diagnosis of dental caries. In: Advances in Emerging Trends and Technologies, vol. 1, pp. 312–322. Springer, Heidelberg (2020)
12. Cevallos-Torres, L., Núñez-Gaibor, J., Leyva-Vasquez, M., Gómez-Rodríguez, V., Parrales-Bravo, F., Hechavarría-Hernández, J.: NCC: Neutrosophic Control Charts, a didactic way to detect cardiac arrhythmias from reading electrocardiograms. Neutrosophic Sets and Systems **74**(1), 40 (2024)
13. Xu, W., Jang-Jaccard, J., Singh, A., Wei, Y., Sabrina, F.: Improving performance of autoencoder-based network anomaly detection on NSL-KDD dataset. IEEE Access **9**, 140136–140146 (2021). https://doi.org/10.1109/ACCESS.2021.3118139
14. Cevallos-Torres, L., Botto-Tobar, M., Cevallos-Torres, L., Botto-Tobar, M.: Monte Carlo simulation method. In: Problem-Based Learning: A Didactic Strategy in the Teaching of System Simulation, pp. 87–96 (2019)
15. Guo, A., Jiang, A., Lin, J., Li, X.: Data mining algorithms for bridge health monitoring: Kohonen clustering and LSTM prediction approaches. J. Supercomput. **76**(2), 932–947 (2020). https://doi.org/10.1007/s11227-019-02894-4
16. Peng, H.G., Zhang, H.Y., Wang, J.Q.: Probability multi-valued neutrosophic sets and its application in multi-criteria group decision-making problems. Neural Comput. Appl. **30**, 563–583 (2018). https://doi.org/10.1007/s00521-016-2673-6

Automated Diagnosis of Thyroid Diseases Using Machine Learning Models

Dayron Rumbaut Rangel(✉), Echeverría Guzmán Ángel, Gary Xavier Reyes Zambrano, and Rosangela Caicedo-Quiroz

Universidad Bolivariana del Ecuador, Durán, Ecuador
{drumbautr,ayecheverriag,gxreyesz,rcaicedoq}@ube.edu.ec

Abstract. Timely diagnosis of thyroid diseases is a significant clinical challenge due to the nonspecific nature of their symptoms and the coexistence of multiple associated conditions that may mask their detection. In this context, artificial intelligence (AI) emerges as a promising tool to support diagnostic processes through automated analysis of clinical data. The present study proposes an automated classification system based on machine learning to identify three diagnostic states: hyperthyroidism, hypothyroidism and normality. Five classification algorithms (Logistic Regression, Random Forest, Naive Bayes, Multilayer Neural Network and XGBoost) trained on a dataset with 9,172 records, extracted from the UCI Machine Learning Repository, were used. The workflow included imputation techniques for missing values, coding of categorical variables and class balancing using SMOTE. The results show that XGBoost achieved the best performance, with a macro F1 Score of 0.97. The clinical implications of the model are discussed, as well as its potential for integration into medical decision support systems in hospital settings.

Keywords: artificial intelligence · thyroid diseases · machine learning

1 Introduction

Thyroid diseases represent a group of endocrine disorders that significantly affect public health worldwide. Thyroid disorders are among the most common endocrine conditions, with a prevalence rate ranging from 2% to 6% worldwide, and a female to male ratio of up to 10:1[1]. These pathologies can present with nonspecific symptoms, such as fatigue, weight gain or loss, mood alterations and menstrual disorders, which makes early and adequate diagnosis difficult.

The problem lies in the fact that the diagnosis of thyroid dysfunction depends on a combination of clinical signs and laboratory test results (TSH, T3, T4 levels, among others), the analysis of which requires specialized expertise. In many clinical settings, especially in regions with limited access to endocrinologists or advanced diagnostic technology, this leads to missed or delayed diagnoses [2].

In this context, artificial intelligence (AI) and machine learning (machine learning) emerge as promising tools to support medical diagnosis. Several studies have demonstrated the usefulness of supervised models to identify complex patterns in clinical and

J. L. Ballesteros et al. (Eds.): CITIS 2025, LNNS 1780, pp. 337–343, 2027.
https://doi.org/10.1007/978-3-032-14287-0_33

laboratory data, achieving results comparable or superior to human clinical judgment [3]. In the specific area of thyroid diseases, previous research such as that of [4] has explored the use of models such as SVM, decision trees and neural networks for the classification of thyroid conditions with encouraging results.

1.1 Related Work

Several recent investigations have addressed the automated diagnosis of thyroid diseases using machine learning techniques, evidencing their potential in the clinical setting. Jayadharshini et al. [5] proposed a comprehensive approach employing multiple algorithms such as Random Forest, KNN and SVM on a dataset of patients with hypothyroidism and hyperthyroidism. Their study showed that Random Forest obtained the highest accuracy (97.4%), although they focused exclusively on traditional algorithms, without exploring more advanced models or explanatory techniques, which represents a relevant gap.

Hossain et al. [6], on the other hand, designed an ex-plicable artificial intelligence framework for classifying thyroid diseases, highlighting the importance of transparency in medical predictions. Although several classifiers were used, the study did not address the detailed comparison between models, and it is suggested to incorporate more methodological depth and cross-validation to strengthen their findings.

Salman and Sonuc [4] used algorithms such as Naive Bayes, SVM and decision trees on clinical data from Iraqi population. The main objective was to classify patients based on hormonal values and symptomatology. Although they achieved significant accuracy, the study did not report class analysis or balancing techniques, which limits the generalizability of the results.

Finally, Rehman et al. [7] analyzed multiple classifiers by applying dimensionality reduction techniques such as PCA, achieving good levels of pre-accuracy. However, they did not address the explainability of the model nor did they perform a detailed evaluation by diagnostic class.

Taken together, these studies highlight the feasibility of machine learning for thyroid diagnosis, but present important methodological gaps, such as the absence of multiclass analysis, lack of robust cross-validation, poor consideration of explainability techniques, and omission of newer algorithms such as XGBoost. The present study seeks to address these limitations by comparing the performance of classical and modern models on a curated dataset, also incorporating class-based balancing and evaluation techniques for a more clinically useful approach.

In this study, a clinical dataset from the UCI repository, including hormonal variables, medical history and clinical characteristics, will be used to address a multiclass classification problem related to the diagnosis of hypothyroidism, hyperthyroidism and normal thyroid function. The main objective is to compare the performance of several supervised learning models in this context and determine which one offers the highest diagnostic accuracy, proposing a viable architecture for implementation in clinical settings.

2 Methodology

The present study used the dataset “thyroidDF.csv”, derived from the rec-onciliation of multiple sources from the UCI Machine Learning Repository [8]. The dataset contains 9,172 observations corresponding to patients with a suspected or diagnosed thyroid disease, and 31 attributes including clinical variables, medical history, hormone levels (TSH, T3, TT4, T4U, FTI, TBG), and a target variable called “target”, which encodes various diagnostic conditions.

The analysis methodology was structured in the following steps:

- Initial cleaning and target filtering: the original target variable was regrouped into three clinically relevant classes: “normal”, ‘hypothyroidism’ (classes E, F, G, H) and ‘hyperthyroidism’ (classes A, B, C, D), discarding other conditions not directly related to thyroid dysfunctions. This allowed us to approach the problem as a multiclass classification of three cat-egories.
- Elimination of irrelevant columns and with high rate of null values: variables such as patient identifier and those with more than 50% of missing values, such as “TBG”, were excluded, following good data preprocessing practices [9].
- Imputation of missing values: The imputation strategy was used by the mean for numerical variables and the mode for categorical variables, applying the SimpleImputer algorithm of Scikit-learn [10].
- Coding of categorical variables: To transform the categorical variables into a format suitable for the models, one-hot coding (pd.get_dummies) was applied, eliminating the first category to avoid multicollinearity.
- Dataset division: The dataset was divided into training set (70%) and test set (30%), maintaining the proportion of classes by stratification (train_test_split, stratify = y).
- Class balancing: Given the imbalance in the target classes, the SMOTE technique [11] was applied to the training set to generate synthetic instances of the minority classes and improve performance in multiclass classification.
- Model training: Five classification algorithms widely used in the medical and scientific fields were trained: Logistic Regression, Random Forest, Naive Bayes Gaussian, Multi-layer Neural Network (MLPClassifier) and XGBoost [12].
- Performance evaluation: The models were evaluated using standard multiclass classification metrics: confusion matrix, f1-score macro, f1-score weighted and overall accuracy, in order to ensure comparability and robustness.

The entire implementation was done in Python 3.10 using the pandas, scikit-learn, imbalanced-learn and xgboost libraries. This pipeline is reproducible and can be integrated into clinical workflows for diagnostic decision support.

3 Results

The comparative analysis of the performance of the classification models used revealed important differences in terms of accuracy, sensitivity and general balance between classes. Table 1 shows the metrics obtained for each of the five models evaluated: XGBoost, Random Forest, Multilayer Neural Network (MLP), Logistic Regression and Naive Bayes.

Table 1. Model evaluation metrics in test set.

Modelo	Accuracy	Precision	Recall	F1 Score
XGBoost	0.9857	0.9267	0.9529	0.9393
Random Forest	0.9852	0.9204	0.9483	0.9339
Red Neuronal (MLP)	0.9614	0.8178	0.9481	0.8725
Logistic Regression	0.9470	0.7718	0.9382	0.8282
Naive Bayes	0.3038	0.4058	0.6681	0.2663

The tree-based models (XGBoost and Random Forest) demonstrated the best overall performance, with F1 Score (macro) values above 93%. The multilayer neural network (MLP) performed acceptably (F1 = 0.8725), while Naive Bayes proved to be ineffective in this context, possibly due to the violation of its feature independence assumptions (Fig. 1).

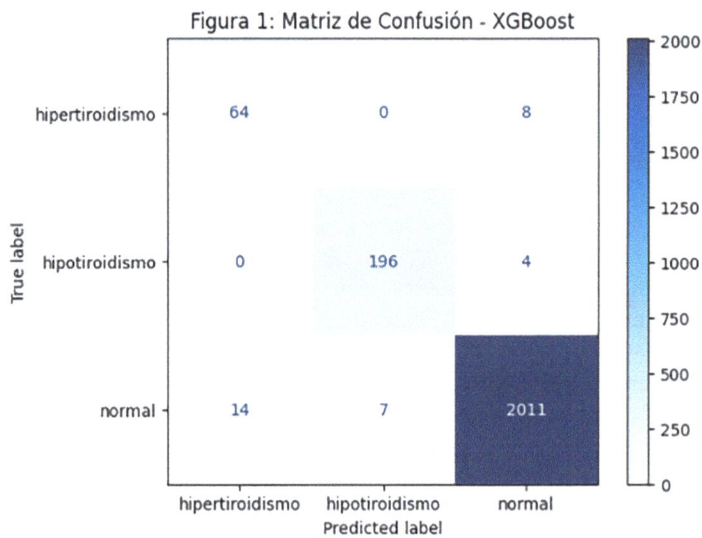

Fig. 1. Confusion Matrix of the XGBoost model.

The confusion matrix shows that the XGBoost model correctly identified the majority of cases in all classes. The largest errors were concentrated in the "normal" class with 14 false positives for "hyperthyroidism" and 7 for "hypothyroidism", representing less than 1% relative error in a highly dominant class (Table 2).

The highest F1-score was obtained in the "normal" class (0.992), although a high identification ability was also achieved for "hypothyroidism" (0.973). The "hyperthyroidism" class presented the lowest, but still acceptable performance (F1 = 0.853), suggesting the need for more data to improve its discrimination.

Table 2. Classification report by class for the XGBoost model.

Class	Precision	Recall	F1-score	Support
Hyperthyroidism	0.821	0.889	0.853	72
Hypothyroidism	0.966	0.980	0.973	200
Normal	0.994	0.990	0.992	2032

Taken together, the results confirm that advanced machine learning models, especially XGBoost, can be used with great effectiveness in multi-class thyroid disease automatic diagnosis tasks.

4 Discussion

The results obtained in this study confirm the efficacy of using machine learning models for thyroid disease classification. The ability of XGBoost to model nonlinear relationships and handle datasets with multiple categorical and numerical attributes positioned it as the best model evaluated, which is in agreement with previous studies where its robustness has been demonstrated in clinical contexts [12].

The findings of this study confirm the potential of machine learning as an effective tool for the automated diagnosis of thyroid diseases, partially aligning with previous studies, but also providing significant improvements in accuracy, balance and methodological depth. The implementation of advanced models such as XGBoost, together with rigorous pre-processing techniques and class balancing, allowed achieving performance metrics superior to those reported by previous research.

Compared to the work of Jayadharshini et al. [5], where Random Forest achieved outstanding accuracy (97.4%), in the present study XGBoost slightly outperformed that performance (Accuracy = 98.57%), proving to be a more robust model in multi-class environments. Furthermore, while the Jayadharshini et al. study focused mainly on classical algorithms, this work adopted a more holistic perspective by also including a neural network and cross-validation methods. The inclusion of per-class metrics and the confusion matrix in this study provides a more accurate view of the model's behavior, especially in the clinical context where false negatives can have important consequences.

Also, while [6] highlighted the need for explainable models and their clinical applicability, their work lacked a rigorous comparative evaluation between models. In contrast, this study not only compared five classifiers with a quantitative and balanced approach, but also emphasized the importance of understanding the differences between models in terms of specific metrics such as the macro F1 Score, particularly useful in contexts with unbalanced classes. Nevertheless, like Hossain et al. this work recognizes the need to incorporate explainability techniques such as SHAP or LIME to ensure transparency in automated decision making, which is crucial for clinical adoption.

Compared to the study by Salman and Sonuc [4], which applied classical algorithms without balancing techniques, our work represents a significant methodological advance. The use of SMOTE not only balanced the representation of the classes in the training set, but also contributed to improve the performance in the detection of minority classes. This improvement is reflected in a recall of 0.889 for hyperthyroidism and 0.980 for hypothyroidism, far exceeding that reported in previous studies where the minority class usually underperformed. This difference underscores the importance of explicitly addressing class imbalance in real medical contexts.

Finally, in contrast to Rehman et al. [7], who explored dimensionality re-duction techniques such as PCA, this study opted to retain the attribute set after careful cleaning and feature selection through imputation and coding, thus allowing to retain all relevant information for classification. The present study also overcomes one of the key limitations noted by Rehman et al. by incorporating class-based analysis and detailed visualizations such as the confusion matrix, allowing for a more granular clinical assessment applicable to medical practice.

Among the limitations of the study is the reliance on a historical data set, which may contain biases inherent to the clinical setting in which it was collected. Also, the explanability of the models was not considered, which limits their direct applicability in clinical settings where decision transparency is critical. Tools such as SHAP or LIME could be incorporated in future research to address this aspect.

Finally, it is worth mentioning that, although the present study used standard metrics to evaluate the models, prospective clinical validation is essential to determine the real effectiveness of these tools in a hospital setting. The integration of these models into clinical decision support systems (CDSS) represents a relevant opportunity to improve the early detection and management of thyroid dysfunction in medical practice.

5 Conclusion

This study successfully met the objective: to develop an automated classification system capable of identifying thyroid conditions (hypothyroidism, hyperthyroidism and normality) by means of machine learning models, evaluating its accuracy and clinical usefulness. To achieve this, all methodological phases were addressed: multiclass diagnostic regrouping, cleaning and pre-processing of the dataset, class balancing with SMOTE, training and validation of five classifier models, and comparative analysis by diagnostic class.

The results show that the XGBoost model obtained the best overall performance, achieving a macro F1 Score of 0.9393 and an accuracy of 98.57%. The use of imputation and coding techniques, together with balanced training, significantly improved the classification of minority classes such as hyperthyroidism, overcoming the limitations reported in previous studies.

Furthermore, comparative analysis with the literature revealed that this work extends traditional approaches by including cross-validation, interpretable visualizations and a detailed evaluation by class, key elements for clinical adoption. This research represents a relevant contribution to the field of artificial intelligence applied to healthcare by demonstrating how a comprehensive methodological approach can improve diagnostic accuracy and applicability in real-world scenarios.

Recommendations and Future Work

- Incorporate explainability techniques such as SHAP or LIME to increase transparency in clinical decision making.
- Perform external validation with data sets from different regions, populations or medical institutions.
- Extend the system to a real-time clinical decision support environment that can be integrated into electronic medical records.
- Evaluate the impact of these models on medical practice through prospective clinical studies.
- Consider dimensionality reduction and automatic attribute selection to optimize processing time and model efficiency in practical implementations.

References

1. Stojković, M.: Thyroid function disorders. Arh. Farm. (Belgr) **72**(5), 429–443 (2022). https://doi.org/10.5937/arhfarm72-39952
2. Biondi, B., Cooper, D.S.: The clinical significance of subclinical thyroid dysfunction. Endocr. Rev. **29**(1), 76–131 (2008). https://doi.org/10.1210/er.2006-0043
3. Esteva, A., et al.: A guide to deep learning in healthcare. Nat. Med. **25**(1), 24–29 (2019). https://doi.org/10.1038/s41591-018-0316-z
4. Salman, K., Sonuç, E.: Thyroid disease classification using machine learning algorithms. J. Phys. Conf. Ser. **1963**(1), 012140 (2021). https://doi.org/10.1088/1742-6596/1963/1/012140
5. Jayadharshini, P., Kumar, A.S., Sivasankar, R., Harsine, K.U.: Comprehensive machine learning approach for early detection and classification of hyperthyroidism and hypothyroidism. In: 2024 International Conference on Trends in Quantum Computing and Emerging Business Technologies, pp. 1–7 (2024). https://doi.org/10.1109/TQCEBT59414.2024.10545151
6. Hossain, M.B., et al.: An explainable artificial intelligence framework for the predictive analysis of hypo and hyper thyroidism using machine learning algorithms. Human-Centric Intell. Syst. **3**(3), 211–231 (2023). https://doi.org/10.1007/s44230-023-00027-1
7. Abbad Ur Rehman, H., Lin, C.-Y., Mushtaq, Z., Su, S.-F.: Performance analysis of machine learning algorithms for thyroid disease. Arab. J. Sci. Eng. **46**(10), 9437–9449 (2021). https://doi.org/10.1007/s13369-020-05206-x
8. Quinlan, R.: Thyroid Disease (1986)
9. Kotsiantis, S., Kanellopoulos, D., Pintelas, P.: Data preprocessing for supervised learning. Int. J. Comput. Sci. **1**, 111–117 (2006)
10. Pedregosa, F., et al.: Scikit-learn: machine learning in python. J. Mach. Learn. Res. **12**(85), 2825–2830 (2011). http://jmlr.org/papers/v12/pedregosa11a.html
11. Chawla, N.V., Bowyer, K.W., Hall, L.O., Kegelmeyer, W.P.: SMOTE: synthetic minority over-sampling technique. J. Artif. Intell. Res. **16**, 321–357 (2002). https://doi.org/10.1613/jair.953
12. Chen, T., Guestrin, C.: XGBoost. In: Proceedings of the 22nd ACM SIGKDD International Conference on Knowledge Discovery and Data Mining, pp. 785–794 (2016). https://doi.org/10.1145/2939672.2939785

Automated Diagnosis of Diabetic Retinopathy and Age-Related Macular Degeneration from Fundus Images

Andres Eduardo Ponce-Macas[1](✉), Iván Galo Reyes-Chacón[2], Paulina Vizcaíno[2], and Manuel Eugenio Morocho-Cayamcela[1,2]

[1] School of Mathematical and Computational Sciences, Deep Learning for Autonomous Driving, Robotics, and Computer Vision Research Group (DeepARC Research), Yachay Tech University, Hda. San José s/n y Proyecto Yachay, 100119 Urcuquí, Ecuador
{andres.ponce,mmorocho}@yachaytech.edu.ec

[2] Faculty of Technical Sciences, School of Computer Science, Universidad Internacional del Ecuador, Jorge Fernández S/N, 170411 Quito, Ecuador
{ireyes,pvizcaino,mamorochoca}@uide.edu.ec

Abstract. Early diagnosis of eye diseases is a crucial factor to prevent possible vision loss. This research focuses on the automated detection of age-related macular degeneration and diabetic retinopathy using color fundus photographs, as these conditions affect the inner eyeball and requires special examination techniques such as fundus photography. This work proposes and investigates a computer-aided system composed of two main stages: blood vessel segmentation to enhance the differentiation of disease features like drusens, exudates, microaneurysms, and hemorrhages; and the subsequent classification of these two diseases using convolutional neural networks. A third class is added for this system, the healthy class where no key lesion of the mentioned diseases are present. This study culminates in the deployment of the proposed segmentation and classification models on a publicly accessible platform, showcasing their potential to deliver accurate, real-time eye disease diagnosis across a wide range of devices, including smartphones and computers. By prioritizing computational efficiency, this work paves the way for scalable, cost-effective medical tools that can broaden access to early diagnosis of vision-threatening conditions worldwide.

Keywords: Eye diseases · Fundus images · Diabetic retinopathy · Macular degeneration · Automated classification

1 Introduction

Opportune diagnosis and treatment of eye diseases are critical in preventing partial or complete vision loss. Consequently, the detection of eye illnesses and injuries is an important area of research, as an early diagnosis can avoid patient's

J. L. Ballesteros et al. (Eds.): CITIS 2025, LNNS 1780, pp. 344–353, 2027.
https://doi.org/10.1007/978-3-032-14287-0_34

loss of vision. While some eye diseases can be observed on the surface, others which affect the inner eyeball require specialized techniques for detection. Fundus photography, which captures images of the ocular fundus, is a widely used technique for this purpose. These images can reveal signs of diseases like diabetic retinopathy and age-related macular degeneration, which are the focus of this study.

Fundus images can expose certain features of these eye illnesses, such as drusen, exudates, microaneurysms, and hemorrhages. To contribute to this diagnostic process, researches have been made in order to develop automated systems using machine learning and image pre-processing techniques. Image pre-processing techniques aim to locate and segment eye structures or disease symptoms, while machine learning focuses on segmenting these structures, or directly diagnose specific diseases. The final objective is to create an automated system that offers high-accuracy diagnosis with low computational cost for possible medical applications.

This work suggests and investigates various models for a computer-aided system designed for the automatic differentiation of three categories: age-related macular degeneration, diabetic retinopathy, and healthy using color fundus photographs. The proposed system employs two major stages: the removal of blood vessels to aid differentiation, and the classification of the segmented images using convolutional neural networks. A key contribution of this work is the deployment of segmentation and classification models at a public web-site, supporting its potential usage for different computer and mobile devices while maintaining a satisfactory classification performance.

2 Methodology

The overall process consist on taking a fundus photography, remove the blood vessel structures and classify the image. A diagram exposing the global workflow of this research is presented in Fig. 1. The following subsections describe with more detail the processes in sequence.

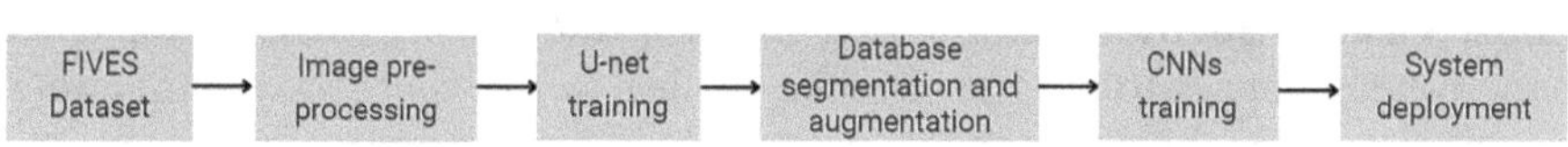

Fig. 1. Diagram of the overall process executed in the present work.

Database Selection. Two main criteria were considered for the database selection: presence of the three classes of interest for classification, and blood vessel ground truths for the eye structure removal. Thus, FIVES image fundus dataset, published in 2022 along with the article *FIVES: A Fundus Image Dataset for Artificial Intelligence based Vessel Segmentation* [1] was selected. This dataset consists of 800 fundus images divided in 4 classes: age-related macular degeneration, diabetic retinopathy, glaucoma and healthy. The dataset is distributed into training and testing subsets and each image has its corresponding ground truth masks for blood vessel segmentation. Samples of the database are shown in Fig. 2, and Table 1 summarizes the number of images available per class.

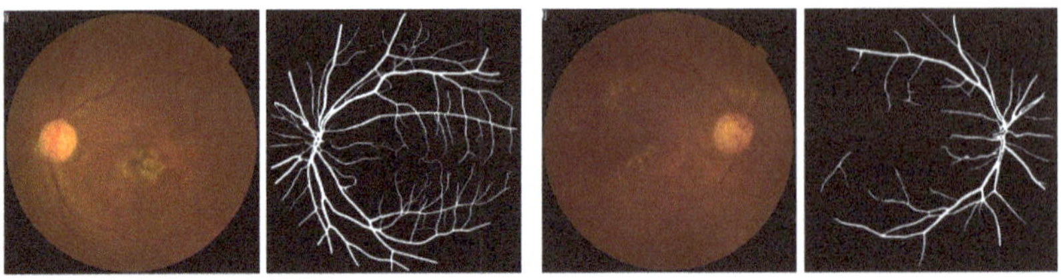

(a) (b)

Fig. 2. Fundus images samples with the corresponding ground truth mask from FIVES Dataset, where (a) corresponds to age-related macular degeneration class and (b) to diabetic retinopathy class.

Table 1. Number of image samples per class available in FIVES dataset. The number of samples for training and testing are defined by the authors.

Class	No. of samples for training	No. of samples for testing
Age-related macular degeneration	150	50
Diabetic retinopathy	150	50
Glaucoma	150	50
Normal	150	50

Blood Vessels Segmentation. Next step is to remove the blood vessels found at the fundus images to aid the classifier models in the next stage of this work. U-Net convolutional neural network was selected for this phase. The model was introduced in *U-Net: Convolutional Networks for Biomedical Image Segmentation* [2], and was designed for medical segmentation tasks [10,11]. Related work such as [3] and [4] suggest the usage of the green color channel for blood vessels segmentation as this channel provides more contrast compared to blue and red channel. The fundus images were pre-processed by applying CLAHE equalization and resizing the image to 512×512 pixels.

Once images were preprocessed, training stage was performed using Adam optimizer and a learning rate of 0,0001 during 30 epochs. The model performance was evaluated using Intersection-over-Union (IoU) and DICE metrics on the resultant binary mask from the testing subset.

Database Segmentation and Augmentation. Once the segmentation model was trained, the next step is to remove the blood vessels from FIVES training images. As showed in Table 1, the number of samples are not enough for training purposes. Thus, data augmentation was also performed.

The data augmentation process consists on creating 6 new images from the original using blurring, sharpening and bright/contrast adjustment techniques. Sharpening method was performed by a convolutional operation with the two following kernels:

$$k_1 = \begin{bmatrix} 0 & -1 & 0 \\ -1 & 5 & -1 \\ 0 & -1 & 0 \end{bmatrix} \tag{1}$$

$$k_2 = \begin{bmatrix} -1 & -1 & -1 \\ -1 & 8 & -1 \\ -1 & -1 & 0 \end{bmatrix} \tag{2}$$

Blurring was performed by a convolutional operation with two Gaussian kernels $G_{3\times3}$ and $G_{5\times5}$. Each element g_{mn} of the kernel is defined by:

$$g_{mn} = \frac{1}{2\pi\sigma^2} e^{-\frac{(m-\mu_m)^2}{2\sigma_m^2} - \frac{(n-\mu_n)^2}{2\sigma_n^2}} \tag{3}$$

where μ and σ corresponds to the mean and variance respectively.

Finally, contrast and bright adjustment of the images were performed with the following operations on the original image:

$$Result_1 = 0.8 \cdot image \tag{4}$$

$$Result_2 = 1.3 \cdot image + 20 \tag{5}$$

Classifier Models Selection and Training. In order to deploy a lightweight and portable diagnostic system, two small and efficient convolutional neural network (CNN) models were considered for this stage: EfficientNet based on the paper *EfficientNet: Rethinking Model Scaling for Convolutional Neural Networks* [5] and ShuffleNet based on *ShuffleNet V2: Practical Guidelines for Efficient CNN Architecture Design* [6]. These classifier models will take as inputs the segmented and augmented images from the dataset.

To optimize computer resources and training time, transfer learning was applied to the selected models. The pre-trained weights come from training the models with ImageNet-1K, which is an image database of 1000 different classes. Furthermore, to obtain the best hyperparameters for training, Optuna framework was utilized on the classifier models with pre-trained weights. This experiment was executed 15 times (trials) for each model, to choose the best optimizer between Adam [7], RMSprop [8] and SGD [9]; and the best learning rate value between 0,1 to 0,0001. The resultant parameters for best training were

Algorithm 1. U-net Training Stage

Input: Training dataset $\{x^1, x^2, x^3, \dots x^b, \dots x^n\} \in X$, groundtruth set $\{y^1, y^2, y^3, \dots y^b, \dots y^n\} \in Y$, number of epochs e, random batch subset $B \subset \{X, Y\}$, number of samples in batch r, learning rate α, parameters β_1, β_2 and ε

Initialize U-net model $f(\cdot)$ with random weights W_1.
for epoch t=1 to e **do**
  for $(x^b, y^b) \in B$ **do**
    Select the green channel of x^b and apply CLAHE equalization.
    Resize fundus image x^b_{green} and ground truth mask y^b to $w' \times h'$ pixels.
    Predict binary mask: $\hat{y}^b = f_{W_1^t}(x^b_{(w',y',green)})$
    Calculate loss from predicted and true masks:

$$L(y^b, \hat{y}^b) = -\frac{1}{rw'h'} \sum_{n=1}^{r} \sum_{i=1}^{w'} \sum_{j=1}^{h'} [y^b_{nij} \log(\sigma(\hat{y}^b_{nij})) + (1 - y^b_{nij}) \log(1 - \sigma(\hat{y}^b_{nij}))]$$

    Calculate gradient: $g^t_{W_1} = \frac{\partial L(y^b, \hat{y}^b)}{\partial W_1^t}$
    Calculate moment estimates:

$$m^t_{W_1} = \beta_1 m^{(t-1)}_{W_1} + (1 - \beta_1) g^t_{W_1} \text{ and } n^t_{W_1} = \beta_2 n^{(t-1)}_{W_1} + (1 - \beta_2)(g^t_{W_1})^2$$

    Perform bias correction: $\hat{m}^t_{W_1} = \frac{m^t_{W_1}}{1-\beta_1}$ and $\hat{n}^t_{W_1} = \frac{n^t_{W_1}}{1-\beta_2}$
    Update weights: $W_1^{t+1} = W_1^t - \alpha \frac{\hat{m}^t_{W_1}}{\sqrt{\hat{n}^t_{W_1}} + \varepsilon}$
  end for
  if $DICE(y^b, \hat{y}^b) > 0.9$ **then**
    Stop training loop
  end if
end for
Output: Trained weights W_1^t.

Adam optimizer for both models and a learning rate of 0,0001 for EfficientNet and 0,0005 for ShuffleNet.

Once the best training parameters were found, the models were trained during 30 epochs, with an early stop condition of finishing the training loop if the loss value on training set is less than 0,01.

The present section exposes the algorithms and steps utilized to execute all processes that were described in the previous subsections.

2.1 Evaluation Metrics

For the segmentation model, Intersection-over-Union (IoU) and DICE metrics were employed. These metrics are defined as follows:

$$IoU = \frac{\text{Area of overlap}}{\text{Area of union}} = \frac{TP}{TP + FP + FN} \tag{6}$$

Algorithm 2. Training subset segmentation and augmentation

Input: Training dataset X, segmentation U-Net model $f(\cdot)$ with pretrained weights W_1^t, kernels $k_1, k_2, G_{3\times 3}, G_{5\times 5}$

for images x in X **do**
  Select the green channel of x and apply CLAHE equalization.
  Resize images x and x_{green} into a $w' \times h'$ pixels image.
  Predict blood vessel binary mask: $M = f_{W_1^t}(x_{(w',h',green)})$
  Invert the predicted binary mask: $M_{inverted} = \begin{cases} 0 & \text{if } M(i,j) = 1 \\ 1 & \text{if } M(i,j) = 0 \end{cases}$
  Apply bitwise-AND operation: $x_{masked} = x_{(w',h')} \wedge M_{inverted}$
  for $i = 0$ to $w' - 1$ **do**
    for $j = 0$ to $h' - 1$ **do**
      Convolve $k_1, k_2, G_{3x3}, G_{5x5}$: $I_n(i,j) = (x_{masked} * kernel)(i,j)$
    end for
  end for
  Adjust bright and contrast: $I_5 = 0.8 \cdot x_{masked}$ and $I_6 = 1.3 \cdot x_{masked} + 20$
  $P[\ ] \leftarrow \{x_{masked}, I_1, I_2, I_3, I_4, I_5, I_6\}$
end for
Output: Segmented and augmented training dataset P

$$DICE = \frac{2 \times \text{Area of overlap}}{\text{Total area}} = \frac{2 \times TP}{2 \times TP + FP + FN} \tag{7}$$

For the classifier models, 4 different metrics were used: accuracy, precision, recall and F1-Score, which are defined as follows:

$$Accuracy = \frac{TP + TN}{TP + TN + FP + FN} \tag{8}$$

$$Precision = \frac{TP}{TP + FP} \tag{9}$$

$$Recall = \frac{TP}{TP + FN} \tag{10}$$

$$\text{F1-Score} = \frac{2 \times (Precision \times Recall)}{Precision + Recall} \tag{11}$$

3 Results and Discussion

After U-Net model was trained using equalized green channel images as inputs, the model curves reached a convergence tendency to the optimal results for both accuracy and loss results. U-net model was trained until epoch 17 when DICE condition was met. IoU and DICE scores are provided in Table 2. At first, it is noticeable that all scores do not surpass a 0.9 threshold. Two factors need to be taken into consideration: first, ground truths were resized to a 512×512

Table 2. Intersection-over-Union and DICE scores obtained from binary masks predicted by U-Net model using the green color channel of the fundus image. Binary mask predictions come from all 150 images belonging to the test subset of FIVES database.

Statistical calculation	IoU score	DICE score
Median score on AMD	0.7966	0.8868
Median score on DR	0.7538	0.8662
Median score on NORMAL	0.7701	0.8705
Average score on AMD	0.7857	0.8754
Average score on DR	0.7749	0.8517
Average score on NORMAL	0.7636	0.8594

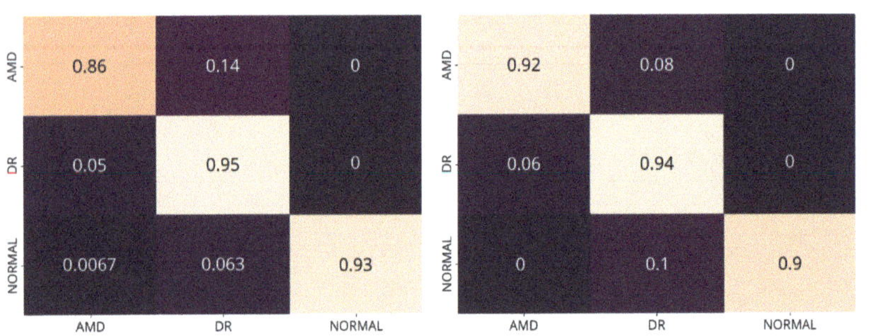

Fig. 3. Confusion matrix for ShuffleNet V2 (a) and EfficientNet B1 (b) with pre-trained weights. The results comes from the classification on the testing set from FIVES database.

Table 3. Performance metrics for EfficientNet B1 with pre-trained weigths. The scores are divided into the three classes of interest that were classified by the model

Class	Accuracy	Precision	Recall	F1-Score
Age-related macular degeneration	0.953	0.939	0.92	0.929
Diabetic retinopathy	0.92	0.839	**0.94**	0.887
Normal	**0.967**	**1**	0.9	**0.947**

Table 4. Performance metrics for ShuffleNet V2 with pre-trained weigths. The scores are divided into the three classes of interest that were classified by the model.

Class	Accuracy	Precision	Recall	F1-Score
Age-related macular degeneration	0.934	0.938	0.86	0.897
Diabetic retinopathy	0.916	0.824	**0.95**	0.882
Normal	**0.977**	**1**	0.93	**0.964**

Algorithm 3. Classifier model training with Adam Optimizer

Input: Training dataset P, true labels Q, number of epochs e, batch subset $D \subset (P, Q)$, number of samples in batch z, learning rate μ, classifier model $s(\cdot)$, parameters θ_1, θ_2 and ε

Initialize classifier model $s(\cdot)$ with weights W_2.
for epoch t=1 to e **do**
  for (p^d, q^d) from D **do**
    Predict class: $\hat{q}^d = s_{W_2^t}(p^d)$
    Calculate loss: $L(\hat{q}^d, q^d) = -\frac{1}{z}\sum_{n=1}^{z} \log\left(\frac{e^{\hat{q}_{n,q_n}}}{\sum_{c=0}^{2} e^{\hat{q}_{n,c}}}\right)$
    Calculate gradient: $g_{W_2}^t = \frac{\partial L(\hat{q}^d, q^d)}{\partial W_2^t}$
    Calculate moment estimates:

$$m_{W_2}^t = \theta_1 \cdot m_{W_2}^{(t-1)} + (1-\theta_1) \cdot g_{W_2}^t \text{ and } n_{W_2}^t = \theta_2 \cdot n_{W_2}^{(t-1)} + (1-\theta_2) \cdot (g_{W_2}^t)^2$$

    Perform bias correction: $\hat{m}_{W_2}^t = \frac{m_{W_2}^t}{1-\theta_1}$ and $\hat{n}_{W_2}^t = \frac{n_{W_2}^t}{1-\theta_2}$
    Update weights: $W_2^{t+1} = W_2^t - \mu \frac{\hat{m}_{W_2}^t}{\sqrt{\hat{n}_{W_2}^t + \varepsilon}}$
  end for
  if $L(\hat{q}^d, q^d) \leq 0.001$ **then**
    Stop training loop
  end if
end for
Output: Trained weights W_2^t.

dimension to compare with the predicted binary masks, thus, the interpolation process may have modified the original pixels to achieve the desired dimension; second, original quality of some images blurs or "removes" some blood vessels structures, nonetheless these pixels are annotated at the ground truth masks by specialized doctors. Despite these limitations, CLAHE equalization, green channel and U-Net model constitute a solid method to identify blood vessel structures while preserving most of microaneurysms, hemorrhages and macular lesions in the fundus images. Figure 4 shows a case where blood vessels around the optic nerve appear opaque due to the quality of the image but U-Net is able to reconstruct most of the structure except in areas where the blood vessels have a similar tone of the fundus.

For the classifier models, our proposal demonstrates a convergence trend towards optimal accuracy and loss for both models. It is noticeable that Efficient-Net displays a smoother behavior compared to ShuffleNet. Pre-trained weights boost the accuracy of both models as from epoch 4 of training this metric reaches a value above 0,95 on both models. Figure 3 exposes the confusion matrix on both models, it is remarkable that the diagonal values are highlighted compared with the upper triangle and lower triangle values. Furthermore, it is important to notice that, despite some AMD and DR images were wrongly classified as DR or AMD, none of these images were wrongly classified as normal, which supports the idea of robustness of these models as patients will not be diagnosed with

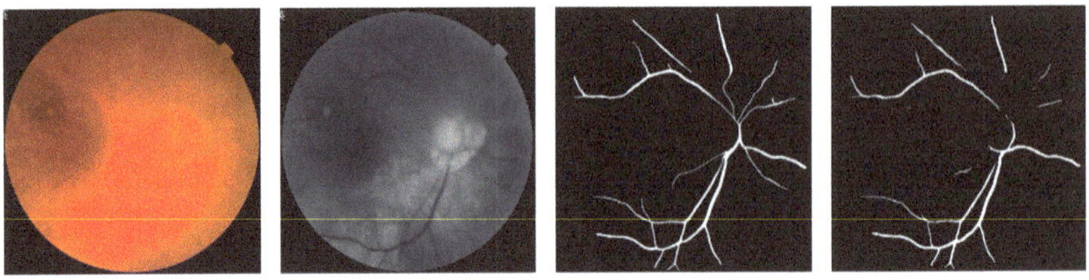

(a) (b) (c) (d)

Fig. 4. Segmentation process example using U-Net model, where (a) corresponds to the original image, (b) to the equalized green channel image, (c) is the ground truth mask and (d) is the predicted binary mask by U-Net.

healthy eyes while actually present a disease. Tables 3 and 4 exhibit the overall performance metrics scores for each class, where most of them surpass a 0,92 threshold, achieving a notable performance on NORMAL class.

Finally, U-net, ShuffleNet with pre-trained weights and EfficientNet with pre-trained weights models were deployed in a demo on a public website.

4 Conclusions

The present research demonstrates the feasibility of constructing a computer-aided system for the automated differentiation of fundus photographs between patients with AMD, DR or healthy eyes, while maintaining low complexity and public accessibility to any computer-science researcher or health worker for further analysis and improvement of the proposed model. The initial blood vessel segmentation, performed by a U-Net CNN trained with equalized green channel images, proved to be a robust method for removing blood vessel structures while preserving diagnostic features such as microaneurysms and hemorrhages, in order to aid the differentiation between classes despite some limitations due to image quality, lesion occlusion, and resizing techniques.

The classification stage uses compact but effective CNNs like EfficientNet B1 and ShuffleNet V2, trained with transfer learning and optimized via Optuna, achieving high accuracy across three categories. These models reliably detect eye diseases, reducing false negatives for AMD and DR. Hosting the models on a public web server demonstrates the feasibility of an accessible diagnostic tool for health workers, overcoming hardware limitations of larger models. This work advances portable automated eye disease diagnosis, potentially improving access to care in underserved areas.

The system will be improved by training models with fundus images from diverse databases to enhance classification robustness and handle variations like image quality and lesion presence. Future work includes two main directions: (1) providing detailed disease staging based on lesion severity for conditions like AMD and DR, and (2) expanding to diagnose additional diseases such as glaucoma, cataract, and myopia. It is also recommended to encourage the creation of databases with ground truth masks for various eye features to support segmentation research and more comprehensive diagnosis systems. Finally, implementing and testing the system in real-world settings is advised to assess its reliability and usability.

References

1. Jin, K., et al.: Fives: a fundus image dataset for artificial intelligence based vessel segmentation. Sci. Data **9**(1), 475 (2022)
2. Ronneberger, O., Fischer, P., Brox, T.: U-Net: Convolutional Networks for Biomedical Image Segmentation. arXiv:1505.04597 [cs.CV] (2015)
3. Marín, D., Aquino, A., Gegundez-Arias, M.E., Bravo, J.M.: A new supervised method for blood vessel segmentation in retinal images by using gray-level and moment invariants-based features. IEEE Trans. Med. Imaging **30**(1), 146–158 (2011). https://doi.org/10.1109/TMI.2010.2064333
4. Zhang, J., Cui, Y., Jiang, W., Wang, L.: Blood vessel segmentation of retinal images based on neural network. In: Zhang, Y.-J. (ed.) ICIG 2015. LNCS, vol. 9218, pp. 11–17. Springer, Cham (2015). https://doi.org/10.1007/978-3-319-21963-9_2
5. Tan, M., Le, Q.V.: EfficientNet: Rethinking Model Scaling for Convolutional Neural Networks. arXiv:1905.11946 [cs.LG] (2020)
6. Ma, N., Zhang, X., Zheng, H.-T., Sun, J.: ShuffleNet V2: Practical Guidelines for Efficient CNN Architecture Design. arXiv:1807.11164 [cs.CV] (2018)
7. Kingma, D.P., Ba, J.: Adam: A Method for Stochastic Optimization. arXiv:1412.6980 [cs.LG] (2017)
8. Tieleman, T.: Lecture 6.5-rmsprop: divide the gradient by a running average of its recent magnitude. Neural Netw. Mach. Learn. **4**(2), 26 (2012)
9. Sutskever, I., Martens, J., Dahl, G., Hinton, G.: On the importance of initialization and momentum in deep learning. In: International Conference on Machine Learning, pp. 1139–1147. PMLR (2013)
10. Tualombo, M.E., Reyes, I., Vizcaino-Imacaña, P., Morocho-Cayamcela, M.E.: pancreas segmentation using SRGAN combined with u-net neural network. In: Information and Communication Technologies. TICEC 2024. Communications in Computer and Information Science, vol. 2273. Springer, Cham (2025). https://doi.org/10.1007/978-3-031-75431-9_2
11. Román, K., Llumiquinga, J., Chancay, S., Morocho-Cayamcela, M.E.: Hyperparameter tuning in a dual channel U-net for medical image segmentation. In: Information and Communication Technologies. TICEC 2023. Communications in Computer and Information Science, vol. 1885. Springer, Cham (2023). https://doi.org/10.1007/978-3-031-45438-7_23

Semantic Segmentation for the Detection of Trypanosoma Cruzi in Microscopic Blood Images Using Random Forests and Data Augmentation

Richard Sucuy[1(✉)], Wilmer Rivas-Asanza[2], Bertha Mazon-Olivo[2], and Eduardo Tusa[2]

[1] Facultad de Ingeniería Civil, Universidad Técnica de Machala, 5.5 km Pan-American Av., Machala 070150, Ecuador
rsucuy1@utmachala.edu.ec

[2] AutoMathTIC Research Group, Facultad de Ingeniería Civil, Universidad Técnica de Machala, 5.5 km Pan-American Av., Machala 070150, Ecuador
{wrivas,bmazon,etusa}@utmachala.edu.ec
https://www.utmachala.edu.ec/portalwp/

Abstract. Early diagnosis of Chagas disease remains difficult in endemic areas due to reliance on manual microscopy, which is error-prone and labor-intensive. This study develops a Random Forest-based classification model for the automated detection of *Trypanosoma cruzi* in microscopic blood images. The workflow follows the CRISP-DM framework and combines semantic segmentation, feature extraction, and data augmentation to improve performance. Preprocessing includes grayscale conversion, normalization, and CLAHE; ROIs are extracted using the Felzenszwalb method. Rotation-based augmentation enhances model robustness. The model achieved 97.16% test accuracy—over five percentage points above prior studies—and its learning curve confirms strong generalization. These results highlight the potential of machine learning for scalable, efficient Chagas diagnosis from blood samples.

Keywords: Trypanosoma cruzi · Random Forest · Semantic Segmentation · Data Augmentation · Microscopic Image Analysis · Chagas Disease

1 Introduction

Chagas disease, caused by *Trypanosoma cruzi*, continues to pose a serious public-health challenge across Latin America. In Ecuador it has been confirmed in 20 of 24 provinces, and El Oro alone accounted for almost a quarter of reported cases between 2013 and 2019 [1]. Fast and reliable diagnosis is therefore essential for vulnerable communities.

The current standard method—manual microscopic examination—is time-consuming and heavily dependent on the skill of individual technicians. Recent

J. L. Ballesteros et al. (Eds.): CITIS 2025, LNNS 1780, pp. 354–363, 2027.
https://doi.org/10.1007/978-3-032-14287-0_35

progress in computer vision now makes it possible to analyse blood-smear images automatically, reducing workload and human error [2]. Building on these advances, we combine Felzenszwalb's graph-based segmentation, which preserves parasite shape while removing artefacts [3], with a Random-Forest classifier. Random Forest achieves about 90% accuracy with an AUC of 0.94 in similar tasks [4] and requires far fewer labelled samples and computational resources than most deep-learning approaches, a clear advantage in resource-limited settings. Because well-annotated image sets are scarce, we enlarge our training data with simple rotation-based augmentation, an effective defence against over-fitting [5].

This study develops a semantic-segmentation pipeline coupled to a Random-Forest classifier for automated detection of *T. cruzi* in blood-smear images. Concretely, we enlarge the training set by at least 50% through rotation-based augmentation to balance classes and boost diversity; we then train Random Forest on features extracted from the segmented ROIs to maximise accuracy and generalisation; and, finally, we assess the model on held-out data with the standard suite of metrics—accuracy, precision, recall (sensitivity), specificity, F1-score and AUC-ROC.

Guided by the CRISP-DM framework [6], Sect. 2 describes the methodology, Sect. 3 presents the results, and Sect. 4 considers their implications for automated Chagas diagnosis.

2 Methods

The workflow was structured around four CRISP-DM phases [7]. First, during the *data understanding* phase, we performed exploratory checks on class balance, potential biases, and annotation quality. In *data preparation*, we applied augmentation techniques, converted images to grayscale, and segmented them into regions of interest (ROIs). The *modelling* phase involved training a Random Forest classifier, selecting relevant features, and tuning hyperparameters. Finally, the *evaluation* phase used accuracy, precision, recall, specificity, F1-score, and AUC-ROC, along with a learning-curve analysis to assess generalization.

Figure 1 summarises the full pipeline.

2.1 Dataset

The dataset, as documented in [4], comprises 560 microscopic blood sample images with annotations marking *Trypanosoma cruzi*. An annotation review yielded 4,088 validated annotations by excluding non-relevant structures, ensuring data quality and improving classification accuracy.

2.2 Pre-processing and Augmentation

Each image is (i) resized to 768 × 1024 px, (ii) converted to grayscale, and (iii) contrast-enhanced with CLAHE. To enlarge the dataset and curb over-fitting, we then apply 45° rotations, which preserve parasite morphology while providing orientation-invariant samples [8]. Figure 2 shows the pipeline.

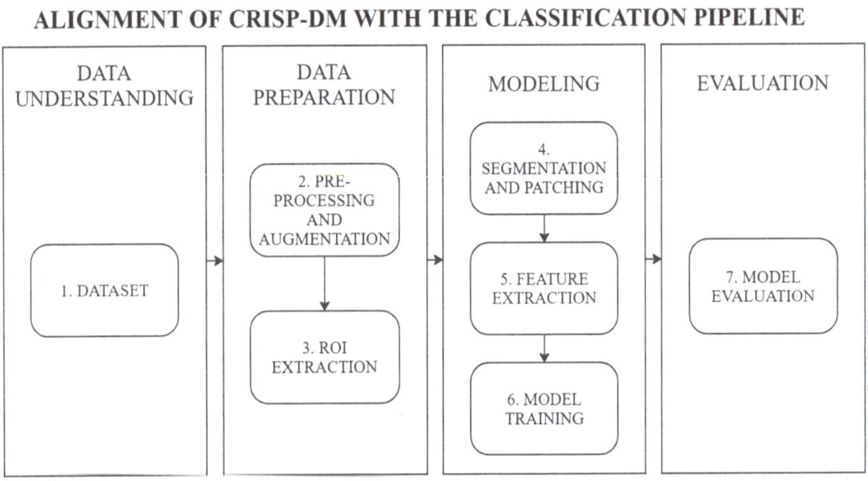

Fig. 1. CRISP-DM methodology aligned with the proposed classification pipeline.

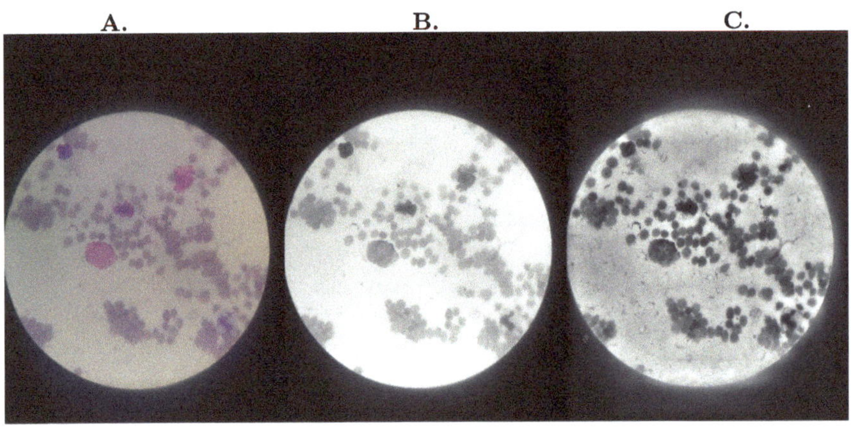

Fig. 2. Preprocessing steps applied to the images. (A) Original resized image. (B) Grayscale conversion. (C) Contrast enhancement using CLAHE.

2.3 ROI Extraction

Each pre-processed image is cropped into fixed 44 × 44 px Regions of Interest (ROIs), centred on the NUCLEUS annotation so the parasite lies roughly in the middle; padding is added only when the window would exceed image borders [4]. To balance classes, we also sample an equal number of *auxiliary* ROIs of identical size from parasite-free areas—their centres are chosen at random but forced to avoid any NUCLEUS or KINETOPLAST marks. Both primary and auxiliary ROIs are finally split into 3 × 3 px patches for feature extraction. Figure 3 shows an example of a primary ROI location and its final crop.

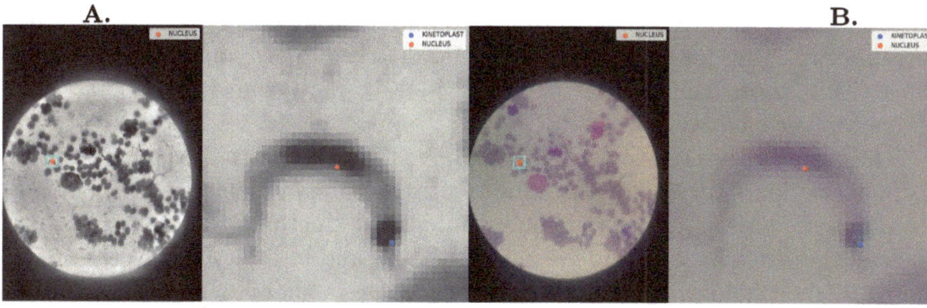

Fig. 3. Regions of Interest (ROI) in grayscale and color. (A) Full image with ROI location (cyan box). (B) 44 × 44 pixel ROI with NUCLEUS (red) and KINETOPLAST (blue) annotations. The top row shows grayscale, and the bottom row shows color.

2.4 Segmentation and Patching

Each 44 × 44 px ROI is segmented with the Felzenszwalb graph algorithm (`skimage.segmentation`), which clusters pixels by colourâĂŞtexture similarity and uses NUCLEUS/KINETOPLAST marks only as landmarks. We then cut the segments into 3 × 3 px patches: positive ones centred on parasite annotations, and an equal number of negatives sampled from auxiliary ROIs to keep the classes balanced [9]. Figure 4 sketches the workflow.

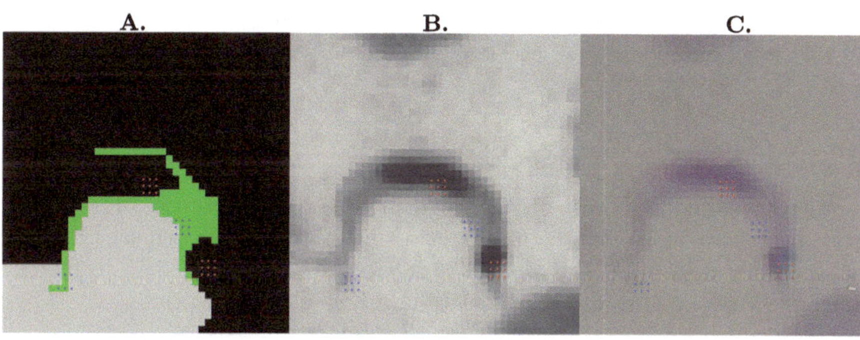

Fig. 4. Visualization of segmentation, patch extraction, and class balance strategy. (A) Segmentation results: positive patches (blue) and negative patches (green) from auxiliary ROIs. (B) Grayscale patches capturing intensity and texture. (C) Color patches for chromatic feature extraction.

2.5 Feature Extraction

Each 3 × 3 px patch was described using 13 feature groups across five categories: photometric descriptors (intensity statistics and colour moments [10, 11]);

gradient-based features (HOG histograms and contour counts [12,13]); texture descriptors (Gabor filters, GLCM, LBP, Haralick [14,15]); frequency-domain features (FFT peaks and wavelet energies [16,17]); and geometric factors (area, perimeter, eccentricity, etc. [18]).

These features were chosen for their proven discriminative power in biomedical imaging, particularly for encoding local structure, granularity, and contrast. Since patches are centered on manually annotated kinetoplast and nucleus coordinates, the aim was to capture localized texture and intensity signatures rather than full morphology. Feature importance analysis showed that texture- and gradient-based descriptors contributed most to classification, consistent with the visual properties of parasite subregions.

2.6 Model Training and Evaluation

A Random-Forest classifier—preferred for its robustness with tabular features and its interpretable importance scores—was trained on an 80/20 stratified split to preserve the parasite/non-parasite ratio [4,19].

Performance was summarised with the usual confusion-matrix metrics: accuracy, precision, recall (sensitivity), specificity, F1 and AUC-ROC, all derived from the TP–TN–FP–FN counts [20]. To probe generalisation, we generated a learning curve by incrementally growing the training set from 10% to 100%, which revealed whether the model was under- or over-fitting and guided final hyper-parameter tuning.

All preprocessing, feature-extraction and evaluation code is openly available in a dedicated GitHub repository, ensuring full reproducibility of the pipeline [21].

3 Results

3.1 Impact of Data Augmentation

Training the Random-Forest on the augmented set (48 k patches) lifted every metric by ≈1 pp on average with respect to the original 6 k-patch set (Table 1). Test accuracy rose from 96.5 % to 97.2 %, and AUC from 99.3 % to 99.6 %, confirming that extra rotations translate into better generalisation.

3.2 Classification Performance

On the augmented data the model reaches 97.2 % accuracy (test) with balanced precision/recall and <2 % error in either class.The confusion matrices (train/test) in Table 2 show that false detections are rare and evenly distributed.

3.3 Learning-Curve Analysis

Figure 5 shows that training accuracy stays at 99.8 % while validation converges to 97 %, leaving a minimal gap and indicating neither over- nor under-fitting. Compared with the 92 % ceiling reported by Morais et al. [4], our curve confirms a clear 5 pp gain (Fig. 6).

Table 1. Comparison of performance metrics: original vs. augmented data

Metric	Original data		Augmented data	
	Train [%]	Test [%]	Train [%]	Test [%]
Accuracy	98.70	96.47	99.80	97.16
Sensitivity	99.04	97.21	99.80	97.16
Specificity	98.35	95.75	99.80	97.15
Precision	98.37	95.65	99.80	97.17
F1-Score	98.70	96.42	99.80	97.16
AUC	99.94	99.31	100.00	99.57

Table 2. Confusion matrices for the training and test sets

True Label	Training set			Test set		
	Predicted		Total	Predicted		Total
	Presence	Absence		Presence	Absence	
Presence	19 041	64	**19 105**	4 762	168	**4 930**
Absence	12	19 308	**19 320**	105	4 572	**4 677**
Total	**19 053**	**19 372**	**38 425**	**4 867**	**4 740**	**9 607**

4 Discussion

4.1 Impact of Data Augmentation

Our findings demonstrate that data augmentation significantly enhances model performance. As detailed in Table 1, the model trained on the augmented dataset achieved notably higher performance across all key metrics compared to the model trained on the original dataset. In particular, training accuracy increased from 98.70% to 99.80%, and test accuracy improved from 96.47% to 97.16%. Similar improvements were observed in sensitivity, precision, F1-score, and AUC, with test specificity rising from 95.75% to 97.15%.

These enhancements suggest that the augmented dataset provides a more diverse and representative sample of *Trypanosoma cruzi* morphological variations, enabling the classifier to learn more generalized features and reduce overfitting. As a result, the model demonstrates improved generalization and more robust diagnostic predictions. Overall, data augmentation proves to be a critical step in developing an accurate and reliable automated detection system for Chagas disease.

4.2 Model Generalization and Comparison with Previous Work

The third specific objective was to evaluate model generalization using standard classification metrics and to compare its performance with previous work [4]. Table 3 provides a comparative analysis, revealing a 5.27% point increase in

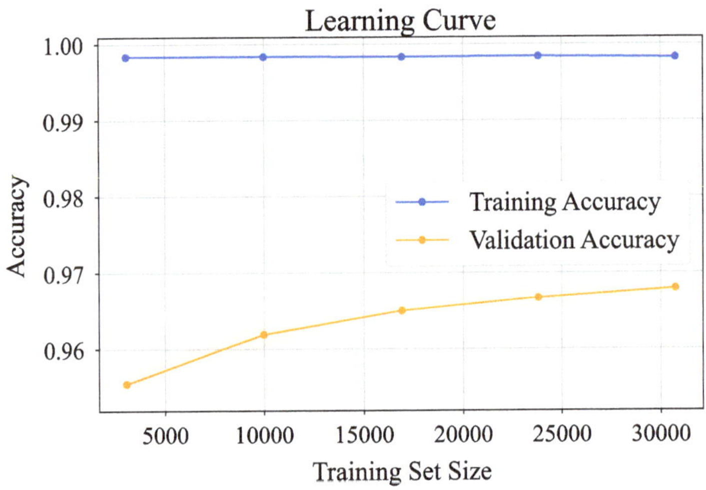

Fig. 5. Learning curve of the proposed model (train vs. validation).

accuracy in this study relative to the previous one. Similar improvements were observed in sensitivity, precision, F1-score, and AUC-ROC.

Table 3. Comparison of evaluation metrics between this study and previous work

Metric	Previous Study [%]	This Study [%]
Sensitivity	92.38	97.10
Specificity	91.41	97.15
Precision	91.45	97.17
Accuracy	91.89	97.16
F1-Score	91.91	97.16
AUC	97.10	99.57

Additionally, the learning curves (Figs. 5 and 6) corroborate the strong generalization of the proposed model, with validation accuracy stabilizing near 97% compared to 92% in the previous work.

Overall, these findings demonstrate that the proposed segmentation, feature extraction, and data balancing strategies lead to improved classification accuracy and model reliability for parasite detection in microscopic blood images.

4.3 Limitations and Practical Considerations

While the model achieved high accuracy on the benchmark dataset, real-world deployment may present challenges. Variability in microscope calibration, staining protocols, and image acquisition can affect feature consistency and reduce

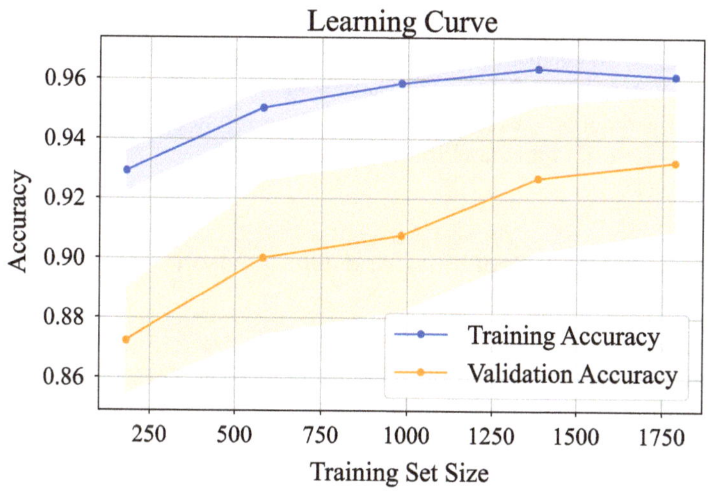

Fig. 6. Learning curve from previous work [4].

performance outside controlled settings. Dependence on expert-annotated coordinates limits scalability, as manual annotation is time-consuming and often impractical in routine diagnostics. Moreover, the lack of diverse, high-quality datasets hinders generalization across laboratories or regions.

Clinically, the 5.27% point gain in accuracy is significant. Greater sensitivity reduces missed infections, enabling earlier antiparasitic treatment and lowering risks of chronic complications. Improved specificity minimizes false positives, avoiding unnecessary tests and reducing the burden on limited healthcare resources. Together, these gains support more efficient diagnostic workflows and enhance the reliability of automated Chagas screening in endemic areas.

5 Conclusions

5.1 Impact of Data Augmentation on Model Performance

Rotation-based data augmentation substantially increased the training dataset, thereby enhancing the representation of diverse morphological variations in *Trypanosoma cruzi*. This expansion effectively reduced overfitting and maintained a balanced class distribution, which in turn improved the model's generalization capability on the test set.

5.2 Effectiveness of the Random Forest Model in Parasite Classification

The Random Forest classifier, trained on features extracted from segmented images, achieved exceptional performance with 99.80% training accuracy and

97.16% test accuracy. High sensitivity (97.16%) and specificity (97.15%) confirmed its ability to accurately distinguish parasite-containing from non-parasitic regions. The structured feature extraction—combining intensity, texture, frequency, and geometric descriptorsoptimized the input data, underscoring Random Forest as a computationally efficient alternative to more complex deep learning models.

5.3 Generalization Capability and Comparison with Previous Studies

The proposed methodology improved model generalization, evidenced by a 5.27% point increase in accuracy and an AUC of 99.57% on the test set compared to previous work. The learning curve analysis, with validation accuracy stabilizing at approximately 97%, indicates minimal overfitting and confirms that the integrated approach enhances the model's reliability and applicability in diagnostic settings.

References

1. Oma, M.V.D., Tutillo, P.A.Q., Mackenzie, D.D.S., Ordoñez, J.C.C., Pérez, H.R.V.: Enfermedad de chagas en el ecuador: una revisión sistemática de los aspectos epidemiológicos y entomológicos. INSPILIP **5**(1), 1–11 (2021). https://doi.org/10.31790/inspilip.v5i1.2
2. Cochero, J., Pattori, L., Balsalobre, A., Ceccarelli, S., Marti, G.: A convolutional neural network to recognize chagas disease vectors using mobile phone images. Eco. Inform. **68**(5), 101587 (2022). https://doi.org/10.1016/j.ecoinf.2022.101587
3. Kallipolitis, A., Yfantis, P., Maglogiannis, I.: Improving explainability results of convolutional neural networks in microscopy images. Neural Comput. Appl. **35**(29), 21535–21553 (2023). https://doi.org/10.1007/s00521-023-08452-w
4. Morais, M.C.C., et al.: Automatic detection of the parasite trypanosoma cruzi in blood smears using a machine learning approach applied to mobile phone images. PeerJ **10**, e13470 (2022). https://doi.org/10.7717/peerj.13470
5. David Opeoluwa Oyewola, E.G., Dada, S.M., Damaševičius, R.: A novel data augmentation convolutional neural network for detecting malaria parasite in blood smear images. Appl. Artif. Intell. **36**(1), 2033473 (2022). https://doi.org/10.1080/08839514.2022.2033473
6. Schröer, C., Kruse, F., Gómez, J.M.: A systematic literature review on applying CRISP-DM process model. Procedia Comput. Sci. **181**, 526–534 (2021). https://doi.org/10.1016/j.procs.2021.01.199
7. Rawat, K.: Applying CRISP-DM methodology in developing machine learning model for credit risk prediction. In: Arai, K. (ed.) Intelligent Computing, pp. 522–538. Springer, Cham (2023). https://doi.org/10.1007/978-3-031-37963-5_37
8. Ikechukwu, A.V., Murali, S.: i-Net: a deep CNN model for white blood cancer segmentation and classification. Int. J. Adv. Technol. Eng. Explor **9**(95), 1448–1464 (2022). https://doi.org/10.19101/IJATEE.2021.875564
9. Chien, T.Y., Ting, H.W., Li, H.W., Chang, H.H.: An artificial intelligence method for vessel detection in images of cardiac catheterization. Inform. Med. Unlock. **39**, 101260 (2023). https://doi.org/10.1016/j.imu.2023.101260

10. Pakniyat, N., Abdullah, J., Krejcar, O., Namazi, H.: Complexity-based analysis in biomedical image analysis: a review. Fractals **32**(06), 2430002 (2024). https://doi.org/10.1142/S0218348X24300022
11. Rustam, F., et al.: White blood cell classification using texture and RGB features of oversampled microscopic images. Healthcare **10**(11) (2022). https://doi.org/10.3390/healthcare10112230
12. Sharma, A., Yadav, D.P., Garg, H., Kumar, M., Sharma, B., Koundal, D.: Bone cancer detection using feature extraction based machine learning model. Comput. Math. Methods Med. **2021**(1), 7433186 (2021) https://doi.org/10.1155/2021/7433186
13. Lv, J.J., et al.: Contour extraction of medical images using an attention-based network. Biomed. Signal Process. Control **84**, 104828 (2023). https://doi.org/10.1016/j.bspc.2023.104828
14. Reyes, A.A., Paheding, S., Deo, M., Audette, M.: Gabor filter-embedded U-net with transformer-based encoding for biomedical image segmentation. In: Li, X., Lv, J., Huo, Y., Dong, B., Leahy, R.M., Li, Q. (eds.) Multiscale Multimodal Medical Imaging, pp. 76–88. Springer, Cham (2022). https://doi.org/10.1007/978-3-031-18814-5_8
15. Srivastava, G., Chaudhary, V., Balodi, S., Khari, G., García-Díaz, A., M.: Vicente: an enhanced texture-based feature extraction approach for classification of biomedical images of CT-scan of lungs. Int. J. Interact. Multimed. Artif. Intell. (IJIMAI) **6**(7), 18–25 (2021). https://doi.org/10.9781/ijimai.2020.11.003
16. Yang, C., Guo, X., Chen, Z., Yuan, Y.: Source free domain adaptation for medical image segmentation with Fourier style mining. Med. Image Anal. **79**, 102457 (2022). https://doi.org/10.1016/j.media.2022.102457
17. Nawaz, S.A., Li, J., Shoukat, M.U., Bhatti, U.A., Raza, M.A.: Hybrid medical image zero watermarking via discrete wavelet transform-resnet101 and discrete cosine transform. Comput. Electr. Eng. **112**, 108985 (2023). https://doi.org/10.1016/j.compeleceng.2023.108985
18. Heriawati, S.O., Harsono, T., Bachtiar, M.M., Hernaningsih, Y.: Blood cells classification for identification of acute lymphoblastic leukemia on microscopic images using image processing. In: 2021 International Electronics Symposium (IES), pp. 215–220 (2021). https://doi.org/10.1109/IES53407.2021.9593939
19. Rácz, A., Bajusz, D., Héberger, K.: Effect of dataset size and train/test split ratios in QSAR/QSPR multiclass classification. Molecules **26**(4) (2021). https://doi.org/10.3390/molecules26041111
20. Gong, M.: A novel performance measure for machine learning classification. Int. J. Manag. Inf. Technol. (IJMIT) **13**(1) (2021). https://doi.org/10.12865/CHSJ.48.02.09
21. Sucuy Zhingre, R.: Trypanosoma Cruzi detection pipeline (2025). https://github.com/RichardSucuy/trypanosoma-cruzi-pipeline. Accessed 29 Jan 2025

Moral Rebels and Self-affirmation: Exploring Attitudinal Effects in a Blocked Experimental Design

Jimmy Ronald Riojas Rivera[1(✉)], Cesar Ángel Durand Gonzales[2], Constantino Miguel Nieves Barreto[2], and Enrique Pinchi Ugarte[2]

[1] Universidad César Vallejo, Trujillo, Spain
jrriojas@upn.pe
[2] Universidad Nacional del Callao, Bellavista, Peru

Abstract. This study replicated the "moral rebels" experiment using a sample of 75 participants randomly assigned to three conditions (Obedient, Rebel Control, and Rebel Affirmed with a self-affirmation task). The primary dependent variable—average attitude toward the rebel—was computed by averaging scores for willingness to work with, befriend, and respect the rebel (Other.Work, Other.Friend, Other.Respect). Full-block ANOVA models (gender, ethnicity, and age group) and a multivariate MANOVA using Pillai's trace were conducted; in all analyses, neither the experimental condition nor the demographic blocks showed significant effects ($p > 0.05$). Only Other.Work revealed a marginal trend between the Obedient and Rebel Affirmed conditions ($p \approx 0.08$). These results suggest that self-affirmation did not alter perceptions of the moral rebel in this context, and highlight the need for larger samples and experimental designs capable of assessing the manipulation's differentiated effects on each component of the overall attitude toward the rebel.

Keywords: Moral Rebels · Self-Affirmation · Social Perception · Experimental Psychology

1 Introduction

Observing morally exemplary actions typically elicits positive emotions—such as admiration and respect—toward the individual performing them. When such actions are particularly costly or uncommon, they may even evoke moral elevation, an emotion characterized by a desire to connect with the individual, emulate their behavior, and strive to become a better person. However, this is not always the case. At times, the actions of the moral rebel—a person who refuses to remain silent or conform to the behavior of others when it contradicts their own principles and values—can provoke rejection and resentment [1].

The tendency to be a moral rebel reflects individuals' propensity to resist, oppose, and confront morally problematic situations when failing to do so would compromise their values [2]. Self-affirmation, understood as the process through which individuals

J. L. Ballesteros et al. (Eds.): CITIS 2025, LNNS 1780, pp. 364–371, 2027.
https://doi.org/10.1007/978-3-032-14287-0_36

reinforce their self-esteem and sense of personal integrity, has been the subject of study across various areas of social psychology [3]. This concept gains particular relevance when examined in the context of how individuals perceive and accept those who challenge established ethical norms within a group. Moral rebels—those who question and oppose practices deemed unjust or immoral within their social environment—often face resistance and rejection from their peers [3]. However, self-affirmation may play a crucial role in modulating these responses, facilitating greater acceptance and understanding of such individuals [4].

At the societal level, individual actions taken in defense of moral principles may be essential for sustaining cooperative relationships and a peaceful world[5]. In light of this, the present study aims to determine whether a self-affirmation trait increases acceptance attitudes toward a "moral rebel"—that is, the willingness to work with, befriend, and respect such a person (Other.Work, Other.Friend, and Other.Respect)—in comparison to two control groups (Conditions: obedient, rebel control, and rebel affirmed), while also controlling for potential sources of demographic variability (gender, ethnicity, and age group).

The observation of morally exemplary actions typically elicits positive emotions such as admiration or respect toward the individual who performs them. However, when such actions are particularly costly or unusual, they may even evoke moral elevation—an emotion that fosters the desire to connect with the actor, emulate their behavior, and strive to become a better person. Nevertheless, as noted by [6], this is not always the case. Under certain circumstances, the actions of a moral rebel—someone who refuses to remain silent or conform to group behavior when it contradicts their core principles and values—may provoke rejection and resentment.

In this context, [2] conceptualize the tendency toward moral rebellion as "the propensity of individuals to resist, oppose, and confront morally problematic situations when, otherwise, failing to do so would compromise their personal values". This disposition reflects a deep commitment to personal integrity which, paradoxically, may be both admired and rejected by one's peers. Indeed, [7] have empirically documented that "people who witness acts of moral rebellion often experience ambivalent reactions: on one hand, admiration for the moral courage displayed; on the other, resentment due to the implicit threat to their own moral behavior".

In parallel, self-affirmation—understood as the process through which individuals reinforce their self-esteem and sense of personal integrity—has been extensively studied across various areas of social psychology. [8] suggest that "individuals are motivated to maintain a positive self-image as moral, adaptive, and capable people. When this image is threatened, they tend to seek compensatory mechanisms that allow them to restore their personal integrity". This concept becomes particularly relevant when examined in the context of how individuals perceive and accept those who challenge the ethical norms established within a group.

Delving into the underlying mechanisms, Sherman and Cohen (2006) argue that "self-affirmation exercises have been shown to reduce psychological defensiveness and increase openness to information that may be threatening to one's self-concept" (p. 187), which may help explain its influence on the perception of those who challenge group norms. Complementarily, [9] explain that "self-affirmation operates through a self-expansion mechanism, allowing individuals to contextualize specific threats within a broader and more complex self-concept, thereby reducing their negative impact".

Regarding the social perception of moral dissent, [10] have identified that "the perception of a moral rebel is influenced by contextual factors such as the clarity of the ethical situation, the personal cost of dissent, and the presence of explicit or implicit group norms" (p. 305). Similarly, [11] argue that "judgments about moral rebels are influenced by attributional biases, whereby individuals tend to attribute more positive motives to their own moral actions than to those of others—particularly when the latter involve social criticism".

Despite theoretical advancements, it is important to acknowledge, as [12] point out, that "studies on moral perception and ethical rebellion face unique methodological challenges, particularly concerning the ecological validity of experimental manipulations and the measurement of complex moral constructs" For this reason, it is essential to develop robust experimental designs that allow for the precise assessment of the factors that influence the perception of moral rebels.

In light of the foregoing, this study aims to determine whether a self-affirmation trait increases acceptance attitudes toward a "moral rebel"—that is, the willingness to work with, befriend, and respect such an individual (Other.Work, Other.Friend, and Other.Respect)—in comparison to two control groups (Conditions: obedient, rebel control, and rebel affirmed), while also controlling for potential sources of demographic variability (gender, ethnicity, and age group). The evidence derived from this research will not only contribute to the theoretical understanding of the interaction between self-affirmation and moral perception, but may also have practical implications for the design of interventions aimed at promoting the acceptance of morally exemplary behavior in diverse social contexts.

2 Methods

The present study employed an experimental design to investigate the perception and acceptance of moral rebels. University students were randomly assigned to one of three experimental conditions: obedience, rebel control, or rebel affirmed (self-affirmation). Participants were asked to imagine that a robbery had just occurred and were presented with three photographs of possible suspects, along with descriptive profiles. Based on these materials, they were instructed to select the individual they believed was most likely to be the thief and to complete a form justifying their decision. Two of the photographs depicted white male individuals, while the third depicted an African American male whose description indicated a criminal record and lack of an alibi; the descriptions were constructed to make him appear the most likely suspect.

After completing the form, participants were shown a second form that was allegedly submitted by another participant in the study (though it was, in fact, fabricated by the experimenters). Participants in the **obedience** condition were shown a form that identified the African American man as the thief. In contrast, participants in both **rebel conditions** were shown a form from someone who had written: "I refuse to choose," explaining that "it is offensive to make a Black man the obvious suspect." Following this, participants answered questions about the individual whose form they had just read.

The crucial distinction between the **rebel control** and **rebel affirmed** conditions lay in the self-affirmation intervention. Participants in the **rebel affirmed** group, between completing their own form and viewing the fabricated one, were asked to write a brief essay about "a recent experience in which they demonstrated a quality they deeply value and which made them feel good about themselves." In contrast, participants in the other two groups were given a placebo task: they were asked to describe what they had eaten over the past two days.

The sample consisted of 75 participants: 20 in the obedience condition, 28 in the rebel control condition, and 27 in the rebel affirmed condition. Participants were randomly assigned to conditions. Data were analyzed using the R statistical software. One-way analysis of variance (ANOVA) was conducted to compare mean levels of moral rebel perception and acceptance across the control and affirmation conditions.

The study employed a **Randomized Complete Block Design (RCBD)** in three versions, using a one-factor ANOVA model with blocks.

Block by Gender

$$Yi\,jk = \mu + \tau i + Gj + \varepsilon i\,jk \tag{1}$$

Σi.
τi = 0, Σj.
Gj = 0.
Where:
Yi jk = average attitude of subject k in condition i and gender j.
μ = general mean.
τi = effect of the experimental condition (i = 1,2,3).
Gj = effect of gender (j = 1 Male, j = 2 Female).
εi jk ~ N(0,σ2).

Block by Ethnicity

$$Yikl = \mu + \tau i + Ek + \varepsilon ikl \tag{2}$$

Σi.
τi = 0, Σk.
Ek = 0.
Where:
Yikl = average attitude of subject l in condition i and ethnicity k.
• Ek = effect of ethnicity (k = 1,..., 6).

Block by Age Group

$$Yimn = \mu + \tau i + Am + \varepsilon imn \tag{3}$$

Σi.
τi = 0, Σm.
Am = 0.
Where:
Yimn = average attitude of subject n in condition i and age group m.
Am = effect of the age block (m = 1,..., 5, based on ranges: 29, 30 – 39,...).

3 Results

Figure 1. Distribution of attitude towards the rebel by condition and age blocks, ethnicity

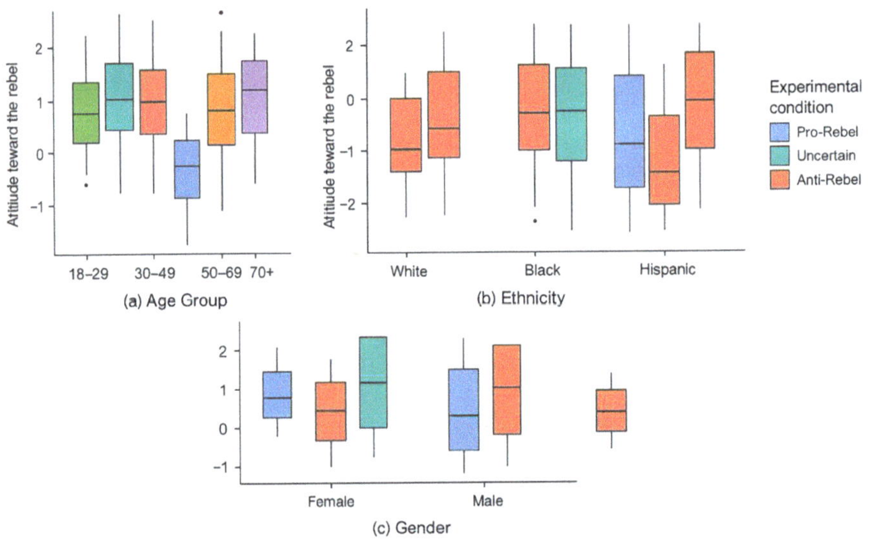

Fig. 1. Distribution of attitude toward the rebel according to condition and blocks of age, ethnicity, and age group (Distribution of attitude toward the rebel according to condition and blocks)

Given that the composite variable could potentially obscure differential effects in each dimension, a multivariate analysis of variance (MANOVA) was performed on Other.Work, Other.Friend, and Other.Respect simultaneously, including the three blocking factors in the model. The Pillai test revealed no significant effects for Condition ($p = 0.228$), Gender ($p = 0.790$), Ethnicity ($p = 0.402$), or Age Group ($p = 0.702$). Post-MANOVA univariate ANOVAs confirmed that only Other.Work showed a marginal effect of Condition ($p = 0.080$), while Other.Friend and Other.Respect did not show significant differences (Fig. 1).

To further explore Other.Work, post-hoc comparisons were conducted with Bonferroni correction. The largest mean difference was observed between Rebel Affirmed and Obedient ($p = 0.085$), followed by Rebel Control versus Obedient ($p = 0.146$). No comparison reached $p < 0.05$, reinforcing the absence of statistically robust effects. However, the comparison between Obedient and Rebel Affirmed suggests a potential trend whereby self-affirmation slightly reduces the willingness to work with the rebel.

Overall, the data indicate that the experimental condition did not produce significant differences in attitude toward the rebel, neither globally (MANOVA) nor in each dimension separately (univariate ANOVA and post-hoc tests), and that the demographic blocks of gender, ethnicity, and age did not alter this effect. A weak signal was only detected in Other.Work for the Obedient vs. Rebel Affirmed comparison, which could be explored in future studies with a larger sample size or directed contrasts.

The results suggest that the experimental condition (Obedient, Rebel Control, Rebel Affirmed) did not generate statistically significant differences in average attitude toward the rebel, even when controlling for gender, ethnicity, or age group. This finding indicates that the simple act of self-affirmation may not be sufficient to clearly alter participants' predisposition toward a "moral rebel." The marginal signal observed in the Other.Work dimension ($p \approx 0.08$) points to a possible trend—less willingness to work with the rebel after self-affirmation—but the effect lacks strength and could be due to chance or uncontrolled factors.

The absence of significant effects for gender, ethnicity, and age as blocks suggests that these demographic characteristics do not interact meaningfully with the experimental manipulation to influence perceptions of the rebel. This is consistent with the hypothesis that equity norms and self-affirmation operate relatively homogeneously across different population subgroups. However, it is possible that the limited sample size reduced the effects of these variables.

A critical point is the choice of the composite variable as the sole primary measure. Although the multivariate analysis (MANOVA) did not reveal global effects, the univariate analysis by dimension showed that Other.Work may behave differently from Other.Friend and Other.Respect. This suggests that future work could benefit from more refined approaches, such as mixed-model analysis or including trait covariates (e.g., perceived morality of the participant) to better explain residual variability. Additionally, replicating this experiment with larger sample sizes and in in-person settings would help validate the reliability and validity of the findings.

In conclusion, the manipulation of self-affirmation, as implemented in the Rebel Affirmed condition, did not produce a statistically significant change in the average attitude toward the moral rebel, compared to the Obedient and Rebel Control groups. Although multivariate analysis and univariate ANOVAs suggested a slight trend in the Other.Work dimension, this difference did not reach the conventional significance threshold ($p < 0.05$), and thus should be interpreted as an exploratory indication rather than a consolidated effect.

Moreover, the demographic characteristics of the participants—gender, ethnicity, and age group—did not act as moderating factors for the experimental impact. In all three block models analyzed, none of these variables demonstrated significant additional explanatory power in shaping perceptions of the rebel, suggesting that the manipulation of self-affirmation produces a homogeneous effect across diverse demographic subgroups, at least within the context and sample size of this study.

The marginal finding in the Other.Work dimension suggests that different attitude dimensions (e.g., willingness to collaborate versus sympathy or respect) may respond differentially to self-affirmation. Therefore, future studies should explore these dimensions more thoroughly by separating them, using mixed designs, or including covariates that capture individual traits (such as the participant's perceived morality) to reduce residual variability and improve the precision of estimates.

Finally, it is recommended to expand the sample size and diversify the studied populations, as well as to combine multivariate analyses with more complex modeling techniques (e.g., mixed models or structural equations) to strengthen the validity of the results. This will help determine with greater certainty whether self-affirmation influences perceptions of "moral rebels" and under which contextual and demographic conditions this influence is most pronounced.

4 Results Discussion

The results show that the experimental condition (Obedient, Rebel Control, Rebel Affirmed) did not generate statistically significant differences in the average attitude toward the moral rebel, even when controlling for gender, ethnicity, or age group. These findings contrast with previous literature on self-affirmation and social perception. For instance, [13] found that self-affirmation exercises significantly increased openness to dissenting opinions in politically polarized contexts, whereas our results suggest that this effect may not necessarily extend to the domain of moral rebellion.

The marginal signal observed in the *Other.Work* dimension ($p \approx 0.08$) warrants particular attention, as it points to a potential trend—less willingness to work with the rebel after self-affirmation. This partial finding could be related to what [14] refer to as the "moral contrast effect," where self-affirmation, by reinforcing the individual's moral self-image, may paradoxically increase the perceived distance from those who exhibit morally challenging behaviors. The authors argue that "self-affirmation exercises centered on values can temporarily raise the moral standards we apply to others, resulting in stricter evaluations of behaviors perceived as morally ambiguous".

The absence of significant effects for gender, ethnicity, and age as blocks suggests that these demographic characteristics do not interact meaningfully with the experimental manipulation. This result aligns with recent research by [15], who found that "equity norms and self-affirmation processes operate relatively homogeneously across different population subgroups, suggesting the universality of certain fundamental psychological mechanisms" (p. 203). However, as [16] note, perceptions of moral rebels may be modulated by specific cultural factors not adequately captured by our study, such as individualistic versus collectivist orientations.

A critical point in our research is the choice of the composite variable as the sole primary measure. The multivariate analysis revealed that *Other.Work* might behave differently from *Other.Friend* and *Other.Respect*, suggesting differentiated dimensions of attitude toward the moral rebel. This is in line with [17] propose, arguing that "the perception of moral dissenters operates through multiple evaluative channels that may differentially respond to psychological interventions such as self-affirmation". Similarly, [18] identified that "willingness to professionally associate with ethical dissenters follows distinct patterns compared to the willingness to form friendship ties, suggesting the activation of different evaluative frameworks depending on the type of relationship considered".

In conclusion, although our study did not find significant effects of self-affirmation on the global perception of the moral rebel, the marginal trend observed in the *Other.Work* dimension and recent literature suggest complex psychological mechanisms that warrant further exploration. As [19] point out, "the intersection of self-affirmation, moral identity, and social perception represents a fertile area for theoretical advancements that could shed light on how societies respond to individuals who advocate for ethical principles in the face of conformist pressures" Future studies with more robust designs, larger samples, and more refined measures hold the potential to deepen our understanding of these phenomena, which are fundamental to the moral functioning of contemporary societies.

References

1. Conejero, S., Vasco, P., Herriko, E.: Cuando la acción moral no provoca admiración sino resentimiento: el rechazo del rebelde moral, pp. 79–93 (2022)
2. Sonnentag, T.L., Wadian, T.W.: Moral rebels: Measuring individuals' tendency for principled resistance to morally compromising situations. Pers. Individ. Dif. **184**, 111210 (2022). https://doi.org/10.1016/j.paid.2021.111210
3. Aarts, A.A., et al.: Estimating the reproducibility of psychological science. Science (80-.) **349**(6251) (2015). https://doi.org/10.1126/science.aac4716
4. Cohen, G., Sherman, D.: The psychology of change: self-affirmation and social psychological intervention. Annu. Rev. Psychol. **65**, 333–371 (2014). https://doi.org/10.1146/annurev-psych-010213-115137
5. Fehr, E., Gächter, S.: Altruistic Punishment in Humans. Nature **415**, 137–140 (2002). https://doi.org/10.1038/415137a
6. Pascual, A., Conejero, S., Etxebarria, I.: Cuando la acción moral no provoca admiración sino resentimiento: El rechazo del rebelde moral. Rev. Ciencias Soc. **93**, 79–93 (2022)
7. Monin, B., Sawyer, P., Marquez, M.: The rejection of moral rebels: resenting those who do the right thing. J. Pers. Soc. Psychol. **95**, 76–93 (2008). https://doi.org/10.1037/0022-3514.95.1.76
8. Ding, M., Liu, Y., Qing, H.: The interpersonal impact of social comparison. Psychology **09**(04), 797–808 (2018). https://doi.org/10.4236/psych.2018.94051
9. Critcher, C.R., Dunning, D.: Self-affirmations provide a broader perspective on self-threat. Personal. Soc. Psychol. Bull. **41**(1), 3–18 (2015). https://doi.org/10.1177/0146167214554956
10. Hashimoto, K., Nishina, K., Miura, A.: Excessive altruism and its underlying motivation: effects on likability in Japan and the US. Lett. Evol. Behav. Sci. **14**(2), 48–52 (2023). https://doi.org/10.5178/lebs.2023.109
11. Dong, M., van Prooijen, J.W., Lange, P.: Self-enhancement in moral hypocrisy: Moral superiority and moral identity are about better appearances. PLoS ONE **14**, e0219382 (2019). https://doi.org/10.1371/journal.pone.0219382
12. Aarts, A., et al.: Estimating the reproducibility of psychological science. Science (80-.) **349** (2015). https://doi.org/10.1126/science.aac4716
13. B. Lyons et al.: Self-Affirmation and Identity-Driven Political Behavior (2020). https://doi.org/10.1017/XPS.2020.46
14. Chan, F., Gibbs, C.: Integrated Theories of White-Collar and Corporate Crime, pp. 191–207 (2019)
15. Smith, P.B., Peterson, M.F., Schwartz, S.H.: Cultural values, sources of guidance, and their relevance to managerial behavior: A 47-nation study. J. Cross-Cult. Psychol. **33**(2), pp.188–208 (2002)
16. Darwish, A.F., Huber, G.: Individualism vs. collectivism in different cultures: a cross-cultural study. Intercult. Educ. **14**, 47–56 (2003). https://doi.org/10.1080/1467598032000044647
17. Leach, C., Ellemers, N., Barreto, M.: Group virtue: the importance of morality (vs. competence and sociability) in the positive evaluation of in-groups. J. Pers. Soc. Psychol. **93**, 234–249 (2007). https://doi.org/10.1037/0022-3514.93.2.234
18. Frawley, J.: 2020_Book_CulturalCompetenceAndTheHigher.pdf. (2020)
19. Stets, J.: The Social Psychology of the Moral Identity, pp. 385–409 (2010)

Antibiotic Resistance in Probiotics and Its Potential Impact on the Spread of Resistance Genes

Joe Dylan Cáceres Taya[1], Samantha Natasha Marcillo Ontaneda[1], Inés Catalina Echeverría Llumipanta[1], and Marco Orlando Fuel Herrera[1,2](✉)

[1] Bioquímica y Farmacia, Universidad Central del Ecuador, Quito, Ecuador
mfuel@ups.edu.ec
[2] Enfermería, Universidad Politécnica Salesiana, Quito, Ecuador

Abstract. Probiotics are essential for maintaining a healthy balance of human microbiota. However, their misuse could be associated with the spread of antimicrobial resistance genes, constituting a danger to the health field. The aim of this systematic review was to analyze antibiotic resistance in probiotic strains of *Lactobacillus, Enterococcus, Bifidobacterium*, etc. and the impact on the dissemination of resistance genes in the gut microbiota. A search was carried out in *PubMed* ® and *Scopus* ®, using the keywords: "*Lactobacillus*", "*Probiotics*", "*Antibiotic resistance*", "*Resistance genes*", "*Horizontal gene transfer*". We included studies with probiotic strains isolated from food, supplements, or commercial products, reporting phenotypic and/or genotypic resistance. Research without microbiological or genetic analysis was excluded. Titles/abstracts were reviewed and, finally, out of 171, fifteen articles that met the established criteria were included. The strains analyzed (mainly *Lactobacillus spp.)* exhibit genes such as *tet(M), tet(W), erm(B), blaTEM*, *strA/strB* and *aadA*, associated with resistance to tetracyclines, macrolides, β-lactams, and aminoglycosides. Several studies indicated the presence of mobile genetic elements (*MGEs*) that facilitate the horizontal transfer of these genes. In addition, some conjugation assays showed transmission of resistance genes to *E. coli, S. aureus,* or *E. faecalis* when incubated with antibiotic-adapted probiotic strains with recipient bacteria. Probiotics can harbor and transfer resistance genes, posing a potential risk to public health if they spread in the gut microbiota, so it is recommended to evaluate antibiotic resistance in probiotics prior to marketing and to carry out gene transfer tests to assess their safety.

Keywords: Lactobacillus · probiotics · antibiotic resistance · gene transfer

1 Introduction

In recent years gut microbiota has been the subject of increasing interest due to its impact on human health, contributing to host defense against pathogens, regulating energy metabolism, supporting immune system function, and maintaining mucosal integrity. Dysbiosis, or microbial imbalance, has been linked to obesity, metabolic syndrome, diabetes, inflammatory bowel disease, neurological disorders and various allergic or autoimmune conditions [1].

J. L. Ballesteros et al. (Eds.): CITIS 2025, LNNS 1780, pp. 372–380, 2027.
https://doi.org/10.1007/978-3-032-14287-0_37

For these reasons, Food products companies, supplements, and others. In recent years, they have decided to implement different probiotic strains based on the global guidelines of the World Gastroenterology Organization, which define probiotics as "live microorganisms that, when administered in adequate amounts, are beneficial to the health of the host." Despite the benefits attributed to probiotics. The growing demand in the market as in 2023 the leadership by Asia-Pacific, which reaches a size of US$ 3.7 billion and exhibits the fastest growth in the period 2019–2023 (CAGR 15.8%). The Americas maintain the second position with US$ 3.0 billion; their advance was much more moderate (5.8%) and is expected to drop to 4.4%, reflecting an already relatively mature market, led by the U.S. Europe adds US$ 2.0 billion and, finally, the Middle East and Africa start from a small base (US$ 0.1 billion). This massification and growth of this industry raises concerns about their role as reservoirs and vectors of antibiotic resistance genes [2].

This is because probiotic supplements contain large amounts of bacteria that could facilitate horizontal gene transfer through mechanisms such as conjugation, transformation, or transduction. This could facilitate the spread of antimicrobial resistance in the intestinal ecosystem to commensal or pathogenic bacteria with significant clinical implications [3].

Microorganisms can become competent when they possess the ability to incorporate external DNA and undergo genetic transformation. This ability is genetically regulated and varies among bacterial species. In most bacteria that can acquire DNA from the environment, competence is controlled by specific mechanisms, which limit their efficiency. However, natural transformation with high efficiency is a rare phenomenon in the bacterial world [4].

Therefore, the aim of the study was to analyze the evidence on antibiotic resistance in strains of probiotic bacteria, considering both phenotypic and genotypic characteristics, and their impact on the dissemination of genes associated with resistance, to understand the associated risks and their relevance.

2 Method

This review was conducted in accordance with the guidelines established by the PRISMA-ScR model for systematic reviews and meta-analyses, taking into consideration the following eligibility criteria:

- Studies evaluating antibiotic resistance in probiotic strains isolated from food, supplements, or commercial products.
- Research to identify antibiotic resistance genes with horizontal transfer potential.
- Studies included genotypic and phenotypic analyses of antibiotic-resistant probiotics.

Exclusion criteria: Research with qualitative approaches, without microbiological or genetic analysis of antibiotic resistance and studies focused on non-probiotic bacterial strains.

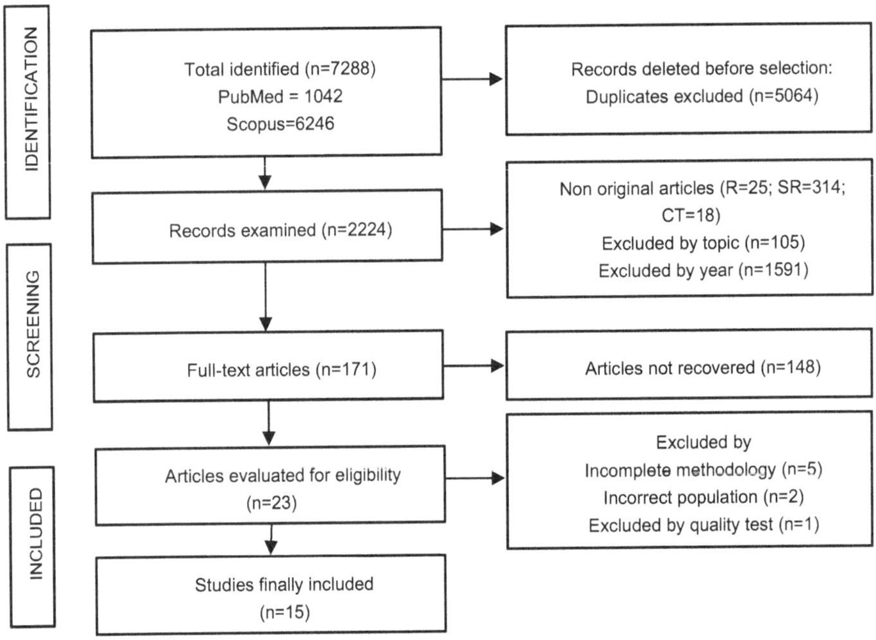

Fig. 1. Flow diagram of the eligible studies included in this systematic review.

2.1 Sources of Information

The information search was carried out in two databases (PubMed and Scopus) from 01 January 2017 to 10 April 2025. We retrieved full articles from all relevant studies without language restriction and hand searched.

2.2 Search

The bibliographic search strategy in the databases was carried out using the following key terms: "*Lactobacillus*", "Probiotics", "Antibiotic resistance", "Resistance genes, "Horizontal gene transfer", "antibiotic resistance in Lactobacillus". A boolean operator (NOT, AND, OR) was used for a more advanced search.

2.3 Data Extraction

During the selection process of the research articles, the authors independently reviewed and extracted the selected data. According to Cohen's Kappa statistical test, a value of 0.85 was obtained, which indicates that there was good agreement between both researchers. Any discrepancies were resolved by consensus or with the correspondence author. A quality test was applied to each article focused on methodology, results, discussions, and conclusions. Where a value ≤ 6 indicated low quality 7–14 good quality: 15–20 excellent quality. The extracted data is summarized in tables. Figure 1 presents the general outline of the bibliographic search.

3 Results

From the search, 171 articles were identified. Subsequently, an initial review of titles, abstracts and conclusions was conducted, and 23 studies were selected for further analysis. Finally, 15 articles that met the inclusion criteria were included, covering research conducted in 9 countries: Slovenia, India, Nigeria, Mexico, China, Russia, Italy, Korea and Bangladesh. The studies analyzed examined probiotics from different sources, including commercial products, dietary supplements, fermented feeds, and probiotics for veterinary use in poultry. Most of the investigations identified strains of Lactobacillus spp., with special emphasis on *L. acidophilus*, *L. casei*, *L. rhamnosus* and *L. paracasei*. In addition, other lactic acid bacteria such as *Bifidobacterium spp.*, *Enterococcus spp.* And *Streptococcus spp.* Were detected (Table 1). The most frequent resistance genes were *tet(M/W/K/O/L)*, *erm*(B/C), *blaTEM/Z*, *vanE/X*, *strA/strB*, *aadA* and *aac* (6′) *aph* (2″) (Table 2). Nine studies identified MGEs (plasmids and Tn916/Tn1545 transposons), and four demonstrated *in vitro* transfer to pathogenic bacteria such as *E. coli*, *S. aureus*, and *E. faecalis tet* (Tetracycline resistance genes); *erm* (Erythromycin ribosomal methylase); *bla* (β-lactamase genes); *str* (Streptomycin phosphotransferase); *aad* (Aminoglycoside adenylyl transferase); *aac* (Aminoglycoside acetyltransferase; *aph* (Aminophoshotransferase); **van** (Vancomycin resistance operons); *qnr* (Quinolone-resistance protein); *parC/gyrA* (Topoisomerase IV (C) and gyrase (A) subunits).

Table 1. Characteristics of the included studies

Author (year)	Country	Sample Type	Strains identified
Qiwen Tian [5] (2025)	China/USA	Commercial probiotic supplements	*L. plantarum, L. rhamnosus, L. casei, L. acidophilus, L. reuteri*
Lobna Badr [6] (2024)	Saudi Arabia	Human breast milk	*Enterococcus faecalis*
Ilenia Campedelli [7] (2024)	Italy	International *Lactobacillus* collections	*Lactobacillus spp.*
Yiming Li [8] (2024)	China	Fermented milk products (yogurts, dairy drinks)	*Lactobacillus spp., Streptococcus spp.*
Rachael T. Duche [9] (2023)	Nigeria	Fermented foods and human samples	*Lactobacillus spp.*
Md. Rokon-Uz-Zaman [10] (2023)	Bangladesh	Veterinary probiotics (Poultry)	*Lactobacillus spp.*

(*continued*)

Table 1. (*continued*)

Author (year)	Country	Sample Type	Strains identified
Eunju Shin [11] (2023)	Korea	Commercial probiotic products	*Lactobacillus spp., Lactococcus spp., Bifidobacterium spp.*
Elizaveta Anisimova [12] (2022)	Russia	Probiotic and dietary supplements	*Lactobacillus spp.*
Anup Kumar [13] (2022)	India	Traditional fermented milk (curd/yogurt)	*Lactobacillus fermentum*
Ting Li [14] (2020)	China	Food and medicine probiotic strains	*Lactobacillus spp., Bifidobacterium spp., Enterococcus spp., Streptococcus spp.*
Vita Rozman [15] (2020)	Slovenia	Commercial strains	*Bifidobacterium spp., Lactobacillus spp.*
Joseph Selvin [16] (2020)	India	Commercial probiotic supplements	*L. acidophilus, L. sporogenes, L. rhamnosus, B. longum, S. faecalis, B. mesentericus, S. boulardii*
Ana May [17] (2020)	Mexico	Commercial products	*L. acidophilus, L. casei, Leuconostoc mesenteroides, B. bifidum*
Devika J. Das [18] (2020)	India	Probiotic supplements	*Lactobacillus spp.*
Huiling Guo [19] (2017)	China	Traditional fermented products (yak, cow, mare milk)	*L. brevis, L. casei, L. plantarum, L. delbrueckii*

Table 2. Resistance genes and evidence of transfer

Author (year)	Genes detected (major families)	MGEs	Proven transfer
Qiwen Tian [5] (2025)	*tet, erm, msr, str, aad, bla, van*	Plasmids, TN916/TN1545	Yes, to *E. faecalis, S. aureus, E. coli*

(*continued*)

Table 2. *(continued)*

Author (year)	Genes detected (major families)	MGEs	Proven transfer
Lobna Badr [6] (2024)	*lsaA*, *tetM*, *ermB*, *gyrA*/*parC*	Plasmids, transposons	Potential risk (not demonstrated)
Ilenia Campedelli [7] (2024)	*ermC*, *tetM/K*, *aac*(6′)-*aph*(2″), *aph*(3′)-III, *gyrA*	Plasmids	Potential risk
Yiming Li [8] (2024)	*tetM*, *strA/B*, *aac*(6′)-*aph*(2″), *aph*(3′)-III	Plasmids	Potential risk
Rachael T. Duche [9] (2023)	*parC*, *aac*(6′)-Ii, *ermB*/*C*, *tetM*, *vanE*	TN916/TN1545	Potential risk
Md. Rokon-Uz-Zaman [10] (2023)	*tetA*, *qnrS*	Plasmids	Not reported
Eunju Shin [11] (2023)	lnu(A), *tetW*	Plasmids	Potential risk
Elizaveta Anisimova [12] (2022)	*vanX*, *parC*, *blaTEM*, *tetK*	Plasmids	Transfer risk
Anup Kumar [13] (2022)	*ermB*, *tetM/K*	Plasmids	Potential risk
Ting Li [14] (2020)	*blaZ*, *tetM*, *ermB*	Plasmids	Potential risk
Vita Rozman [15] (2020)	tetW	MGEs	Potential risk
Joseph Selvin [16] (2020)	β-lactamase	ESBL	ESBL activity (dissemination)
Ana May [17] (2020)	*blaTEM*, *tetW*	Not evaluated	Not evaluated
Devika J. Das [18] (2020)	*tetM/O/W/L/K*, *ermB/C/T*	Plasmids, transposons	Potential risk
Huiling Guo [19] (2017)	*vanX/E*, *gyrA*, *tetW*, *ermB*, *aph*(3″)-III, *aadB*, *dfrD*, *rpoB*	Plasmids, transposons	No (negative assay)

4 Discussion

The studies reviewed show that commercial probiotics may contain antibiotic resistance genes (ARGs), some of which have the potential to be transferred horizontally. Among the most common genes identified are *tetW*, *tetM*, *ermB*, and b*laTEM*. The *tetW* and *tetM* genes act through ribosomal protection, preventing antibiotics from binding to the ribosomes of the bacterium generating resistance to tetracyclines. The *ermB* gene confers resistance by methylating ribosomal RNA, blocking the action of antibiotics such as macrolides, lincosamides and streptogramins B (MLSB phenotype). The *blaTEM* gene encodes a narrow-spectrum β-lactamase, allowing the hydrolysis of β-lactam antibiotics such as penicillin and some cephalosporins.

Previous research has indicated that the *tet(W)* gene, associated with tetracycline resistance, is one of the most prevalent in antibiotic-resistant bacteria, being detected mainly in *Bifidobacterium* and *Lactobacillus* and it was observed that most of the resistance genes identified were linked to mobile genetic elements (MGEs), which suggests a high potential for dissemination within the gut microbiota to other bacteria. Commensal and pathogenic bacteria [7].

These findings contrast with previous studies that argue for a minimal risk of horizontal dissemination in *Lactobacillus*, arguing that, under normal conditions, these microorganisms do not have efficient machinery for plasmid conjugation. [20]. However, although the risk of horizontal spread is low, acquired, and transferable resistance through mobile genetic elements (MGEs) is a cause for concern. Therefore, it is recommended to perform antibiotic susceptibility tests on probiotic strains before approval for human consumption.

This systematic review included studies from various geographical regions, which allowed us to obtain a global perspective on antibiotic resistance in probiotic strains. In addition, both phenotypic and genotypic tests were evaluated, which allowed a more complete analysis on antibiotic resistance in probiotics. However, one of the main limitations found is that some studies did not carry out *in vitro* or *in vivo* gene transfer tests, which prevents the actual risk of dissemination of resistance genes within the gut microbiota from being accurately determined.

5 Conclusions

The fifteen studies analyzed demonstrate the recurrent presence of ARGs, mainly *tet, erm, bla* and *van*, in probiotic strains for human and veterinary use. Nine studies detected plasmids or transposons compatible with horizontal transfer and four demonstrated effective conjugation to opportunistic bacteria *in vitro*. Although the actual frequency of *in vivo* transfer remains low, the circulation of probiotic strains with ARGs is an additional vector in the global spread of antimicrobial resistance. It is recommended to encourage regulations requiring antimicrobial susceptibility profiling to authorize new probiotic products on the market and to verify the absence or minimal presence of MGEs with transfer potential.

References

1. Valdovinos-Díaz, M.: Intestinal microbiota in digestive disorders. Probiotics, prebiotics and symbiotics. Rev. Gastroenterol. Mex. **78**, 25–27 (2013). https://doi.org/10.1016/j.rgmx.2013.06.008
2. Kabir, S.M.L., Islam, S.S., Tuhin-Al-Ferdous, Akhter, A.H.M.T.: Production, cost analysis, and marketing of probiotics. In: Amaresan, N., Dharumadurai, D., Babalola, O.O., (eds.) Food Microbiology Based Entrepreneurship, pp. 305–326. Springer, Cham (2023). https://doi.org/10.1007/978-981-19-5041-4_16

3. Zavišić, G., Popović, M., Stojkov, S., Medić, D., Gusman, V., Jovanović Lješković, N., et al.: Antibiotic resistance and probiotics: knowledge gaps, market overview and preliminary screening. Antibiotics. **12**(8), 1281 (2023). https://doi.org/10.3390/antibiotics12081281
4. Carvalho, G., Fouchet, D., Danesh, G., Godeux, A.S., Laaberki, M.H., Pontier, D., et al.: Bacterial transformation buffers environmental fluctuations through the reversible integration of mobile genetic elements. In: Koonin, E.V., Zhulin, I.B. (eds.). mBio. **11**(2), e02443-19 (2020). https://doi.org/10.1128/mbio.02443-19
5. Tian, Q., Ye, H., Zhou, X., Wang, J., Zhang, L., Sun, W., et al.: Evaluating the health risk of probiotic supplements from the perspective of antimicrobial resistance. In: Parveen, S. (ed.) Microbiol. Spectr. **13**(1), e00019-24 (2025). https://doi.org/10.1128/spectrum.00019-24
6. Badr, L., Yasir, M., Alkhaldy, A.A., Soliman, S.A., Ganash, M., Turkistani, S.A., et al.: Genomic evaluation of the probiotic and pathogenic features of Enterococcus faecalis from human breast milk and comparison with the isolates from animal milk and clinical specimens. PeerJ **12**, e18392 (2024). https://doi.org/10.7717/peerj.18392
7. Campedelli, I., Mathur, H., Salvetti, E., Clarke, S., Rea, M.C., Torriani, S., et al.: Genus-wide assessment of antibiotic resistance in lactobacillus spp. In: Björkroth, J. (ed.) Appl. Environ. Microbiol. **85**(1), e01738–18 (2019). https://doi.org/10.1128/AEM.01738-18
8. Li, Y., Li, L., Kromann, S., Chen, M., Shi, L., Meng, H.: Antibiotic resistance of lactobacillus spp. and streptococcus thermophilus isolated from Chinese fermented milk products. Foodborne Pathog. Dis. **16**(3), 221–228 (2019). https://doi.org/10.1089/fpd.2018.2516
9. Duche, R.T., Singh, A., Wandhare, A.G., et al.: Antibiotic resistance in potential probiotic lactic acid bacteria of fermented foods and human origin from Nigeria. BMC Microbiol. **23**, 142 (2023). https://doi.org/10.1186/s12866-023-02883-0
10. Rokon-Uz-Zaman, M., Bushra, A., Pospo, T.A., Runa, M.A., Tasnuva, S., Parvin, M.S., et al.: Detection of antimicrobial resistance genes in Lactobacillus spp. from poultry probiotic products and their horizontal transfer among Escherichia coli. Vet. Anim. Sci. **20**, 100292 (2023). https://doi.org/10.1016/j.vas.2023.100292
11. Shin, E., Paek, J.J., Lee, Y.: Antimicrobial resistance of seventy lactic acid bacteria isolated from commercial probiotics in Korea. J. Microbiol. Biotechnol. **33**(4), 500–510 (2023). https://doi.org/10.4014/jmb.2210.10041
12. Anisimova, E., Gorokhova, I., Karimullina, G., Yarullina, D.: Alarming antibiotic resistance of lactobacilli isolated from probiotic preparations and dietary supplements. Antibiotics. **11**(11), 1557 (2022). https://doi.org/10.3390/antibiotics11111557
13. Ojha, A.K., Shah, N.P., Mishra, V.: Characterization and transferability of erm and tet antibiotic resistance genes in lactobacillus spp. isolated from traditional fermented milk. Curr. Microbiol. **79**, 339 (2022). https://doi.org/10.1007/s00284-022-02980-9
14. Li, T., Teng, D., Mao, R., Hao, Y., Wang, X., Wang, J.: A critical review of antibiotic resistance in probiotic bacteria. Food Res. Int. **136**, 109571 (2020). https://doi.org/10.1016/j.foodres.2020.109571
15. Rozman, V., Mohar Lorbeg, P., Accetto, T., Bogovič Matijašić, B.: Characterization of antimicrobial resistance in lactobacilli and bifidobacteria used as probiotics or starter cultures based on integration of phenotypic and in silico data. Int. J. Food Microbiol. **314**, 108388 (2020). https://doi.org/10.1016/j.ijfoodmicro.2019.108388
16. Selvin, J., Maity, D., Sajayan, A., Kiran, G.S.: Revealing antibiotic resistance in therapeutic and dietary probiotic supplements. J. Glob. Antimicrob. Resist. **22**, 202–205 (2020). https://doi.org/10.1016/j.jgar.2020.02.007
17. May-Torruco, A.L., Corona-Cruz, A.I., Jiménez, A.L.L., Cortés, N.G., Vera, R.J.: Sensibilidad y Resistencia a Antibióticos de Cepas Probióticas Empleadas en Productos Comerciales. Eur. Sci. J. **16**(18), 43 (2020). https://doi.org/10.19044/esj.2020.v16n18p43

18. Das, D.J., Shankar, A., Johnson, J.B., Thomas, S.: Critical insights into antibiotic resistance transferability in probiotic Lactobacillus. Nutrition **69**, 110567 (2020). https://doi.org/10.1016/j.jgar.2020.02.007
19. Guo, H., Pan, L., Li, L., Lu, J., Kwok, L., Menghe, B., et al.: Characterization of antibiotic resistance genes from lactobacillus isolated from traditional dairy products. J. Food Sci. **82**(3), 724–730 (2017). https://doi.org/10.1111/1750-3841.13645
20. Van Reenen, C.A., Dicks, L.M.T.: Horizontal gene transfer amongst probiotic lactic acid bacteria and other intestinal microbiota: what are the possibilities? A review. Arch. Microbiol. **193**(3), 157–168 (2011). https://doi.org/10.1007/s00203-010-0668-3

Renewable Energies and Industry

Influence of Exclusive Anaerobic Fermentation of Cocoa Beans on Beans, Paste and Chocolate Aromatics Compounds

Mero Jaritza, Aguirre Pamela, Cabrera Daniel, and Sánchez Manoella(✉)

Facultad de Ciencia E Ingeniería en Alimentos y Biotecnología, Universidad Técnica de Ambato, Ambato, Ecuador
ma.sanchezg@uta.edu.ec

Abstract. Cocoa beans are the main raw material for chocolate production; however, they must undergo biochemical changes that allow the development of a sensory profile suitable for the industry, which is achieved through fermentation, which occurs in two stages: anaerobic and aerobic; however, during the second stage, undesirable microorganisms tend to spread; therefore, this study evaluated the influence of anaerobic fermentation of CCN-51 cocoa beans on the concentration of compounds responsible for the organoleptic profile of chocolate. Six treatments were designed combining fermentation times of 15, 30 and 45 days and temperatures of 25 degrees Celsius and 35 degrees Celsius. Parameters such as pH, acidity and microbial activity were analyzed to characterize the biochemical transformation of the beans; subsequently, the beans were processed into paste and chocolate to analyze the concentration of phenols and alkaloids (theobromine and caffeine) by spectrophotometry, observing a progressive decrease of these compounds, with caffeine being the most stable. The results of the sensory analysis indicated that the 30-day treatment at 35 degrees Celsius was the most appropriate to balance the conservation of bioactive compounds with an optimal sensory profile, since the chocolate obtained from cocoa beans under these fermentation conditions presented the best organoleptic characteristics and the highest sensory acceptance. These findings underline the importance of optimizing fermentation conditions to improve both the functional and sensory quality of chocolate.

Keywords: Anaerobic · Fermentation · Cocoa · Phenols · Theobromine · Caffeine

1 Introduction

Cocoa (*Theobroma cacao*) is the base plant for the entire chocolate industry (Ho et al., 2015). Each year, more than 4.4 million tons of beans are produced worldwide, reaching a current market value of close to $10000 per ton (ICCO, 2024a, 2024b). Its fruits are not only appreciated for the delicious pulp, but also for the valuable beans it possesses, which are the essence of a wide variety of products with high nutritional value (Afoakwa, 2016).

J. L. Ballesteros et al. (Eds.): CITIS 2025, LNNS 1780, pp. 383–395, 2027.
https://doi.org/10.1007/978-3-032-14287-0_38

Cocoa bean quality is evaluated through standards covering physicochemical and sensory aspects, as well as its nutritional value (ECA et al., 2015). These characteristics are not only influenced by the genetics of the plant and the geographic region in which it is produced, but also by the practices performed post-harvest such as fermentation and drying (Aprotosoaie et al., 2016). Even though Ecuador is a leader in cocoa exports in Latin America and ranks fourth worldwide, many farmers do not follow proper fermentation techniques or omit them, due to the use of additional resources required or even ignorance of the importance of this process, which significantly reduces the commercial value of this product (Alcivar et al., 2022; MAGAP, 2024).

The fermentation process plays a crucial role in optimizing the sensory profile of cocoa beans by inducing a series of biochemical reactions that produce fundamental precursors for the development of aroma and flavor compounds characteristic of chocolate (De Vuyst & Weckx, 2016). However, traditional fermentation presents certain risks for producers, since during the aerobic phase of the process there is a high probability of contamination of the raw material, due to the proliferation of fungi and other undesirable microorganisms that not only generate visual contamination, but can also produce mycotoxins hazardous to human health, which can result in significant losses (Subroto et al., 2023).

Research on improved fermentation practices has led to the production of high-quality grains, which in turn has allowed farmers to obtain more competitive and favorable prices in the market, this double positive impact underlines the importance of continuing to invest in research and development in the field of fermentation, not only to innovate in the food industry, but also to improve the economic conditions of agricultural producers (Calvo et al., 2021).

2 Materials and Methods

2.1 Obtaining Cocoa Beans

The cocoa beans were obtained from a manual collection of healthy cobs without the presence of fungi at the "Carmita" farm located in "Quinindé-Esmeraldas" (0.4139347, -79.5632176). They were transported in polypropylene sacks to Ambato - Tungurahua on the same day of collection. The cobs were opened for the extraction of the grains and immediately placed in the respective fermenters for the beginning of the fermentation process at a constant temperature of 25 ± 2 °C and 35 ± 2 °C inside incubators.

2.2 Fermentation Process

The fermentation process of the cocoa beans was carried out in borosilicate containers with hermetic lids with dimensions of 21.2 cm x 18.8 cm x 5.2 cm. Each fermenter was uniformly filled with 500 g of cocoa pulp and seeds, using a digital balance to ensure accuracy. A total of 6 fermenters were obtained according to the number of treatments and replicates required. Subsequently, the fermenters were placed in incubators at 25 ± 2 °C and 35 ± 2 °C during the respective fermentation process of each treatment.

2.3 Titratable Acidity and pH Analysis During Fermentation

For pH Analysis, the methodology of García González et al. (2018) and Nazaruddin et al. (2006) was used. A solution was prepared by manually crushing 10 g of fermented cocoa seeds and mixing them with 90 mL of distilled water. The resulting mixture was sieved to remove solid residues, and from this 10 mL aliquots were taken to measure the pH using a previously calibrated potentiometer. To determine the titratable acidity, the same solution was prepared again, adding 100 mL of distilled water after sieving to clarify the sample. From this solution, 25 g were transferred to a flask, adding 3 drops of phenolphthalein as indicator. The titration was then performed by slowly adding 0.1 M NaOH from a burette until a slight color change was observed, indicating the end point.

2.4 Microbial Counts of Bacteria and Yeasts During Fermentation

For the counting of microorganisms after cocoa fermentation, the parameters described by García González et al. (2018) and Salazar et al. (2017) were followed. First, serial 1:10 dilutions of the cocoa samples were performed in peptone water as diluent, once the fermentation periods of 15, 30 and 45 days were completed. The diluted samples were inoculated on selective media to quantify two specific microbial groups: yeasts and LAB. Rose Bengal agar was used for yeast counts, and MRS agar was used for LAB. 100 μL of the 10–6 and 10–5 dilution of each sample was taken and deposited in the center of a Petri dish with the respective agars, distributing the liquid uniformly using the spread seeding technique. The inoculated plates were incubated at 37 °C for 48 h in the case of MRS agar, and at 28 °C for the same period for the Rose Bengal agar plates. After incubation, the number of colonies forming units (CFU) per gram of cocoa was recorded using a colony counter.

2.5 Roasted Drying and Grinding of Fermented Cocoa Beans

For the drying of the cocoa beans, the methodology of Guehi et al. (2010) was used as a reference, with some modifications. The fermented cocoa beans were distributed in a uniform layer approximately 2 cm thick on an aluminum tray, ensuring that each bean occupied a single space inside the container, which was placed in an oven at 45 °C for 72 h. The grains were stirred twice a day to ensure homogeneous drying. At the end of each drying cycle, the moisture percentage was measured to confirm that the values were in the desired range of 7–8%.

Then in the roasting process, the methodology described by Oracz & Nebesny., (2019)) was used, with certain modifications. The previously dried beans were subjected to heat treatment in an oven at 100 °C for 5 h, with the objective of reducing their moisture content to 2%. After roasting, the beans were manually shelled, removing the shells to obtain only the edible portion of the bean necessary for the preparation of cocoa paste, following the procedure described by (Matavaca et al., 2022).

Subsequently, the beans were ground, applying the methodology of (Matavaca et al., 2022). This process consisted of grinding the beans in an automatic disc mill in a single pass, obtaining cocoa paste as the final product.

2.6 Production of Dark Chocolate

Using the methodology of Samanta et al., (2022), the amounts of ingredients for a 250 g tablet were established. For each sample, 175 g of pure cocoa paste, 50 g of sugar and 25 g of cocoa butter were used. The cocoa paste was subjected to a manual conching process, during which it was aerated and stirred at a temperature of 45–50 °C for approximately 4 h. In this process, the cocoa butter and sugar were progressively integrated, ensuring a uniform mixture.

Subsequently, tempering was carried out. For this, three quarters of the mixture was spread on a metal surface until a temperature of 27–28 °C was reached. Once this temperature was reached, this part was reincorporated into the remaining mixture and homogenized until a final temperature of 30–32 °C was reached, at which time the chocolate was molded at room temperature.

2.7 Extraction and Quantification of Phenols and Alkaloids (Theobromine and Caffeine)

To obtain the phenolic extract, 0.5 g of previously defatted sample was weighed and placed in a 15 ml Falcon tube, to which 1.25 ml of methanol was added. The mixture was subjected to an ultrasonic bath (Branson 2800 sonicator) for 30 min to facilitate extraction. Subsequently, it was centrifuged at 3600 g for 15 min, collecting the supernatant. The process was repeated with the residual pellet, and both supernatants obtained were combined. Finally, the volume of the extract was adjusted to 5 ml with methanol.

The total content of phenolic compounds was determined by the Folin-Ciocalteu spectrophotometric method. For this purpose, 0.13 ml of the extract was added to a 20 ml volumetric flask, to which 0.7 ml of Folin-Ciocalteu's reagent, 8 ml of distilled water and 2 ml of a 20% Na_2CO_3 solution were added. The final volume was adjusted with distilled water to 15 ml. The mixture was allowed to stand for 2 h at room temperature in the dark and subsequently the absorbance of the solution was measured at 765 nm against a blank (Laličić-Petronijević et al., 2016).

On the other hand, for the quantification of theobromine and caffeine, 0.25 g of sample was taken and added to 20 mL of distilled water previously heated to 80 °C, the mixture was kept at this temperature with constant stirring for 5 min. Subsequently, 1.25 mL of Carrez I solution was added, stirring for 1 min, and then allowed to stand until room temperature was reached. Then, 0.2 g of $NaHCO_3$ was added until precipitation ceased. The mixture was filtered, and the filtrate was diluted with distilled water to a total volume of 25 mL.

With the extracts obtained, 10 mL were transferred to 15 mL Falcon tubes and approximately 1 mL of 1 M NaOH was added to adjust the pH between 12.5 and 12.7. Subsequently 1. mL of chloroform was added, then centrifuged at 2500 rpm x 15 min. The extract phases were separated to quantify TB (aqueous phase) and CF (organic phase) at a wavelength of 275 nm and that of CF at 273 nm respectively (González-Nuñez & Cañizares-Macías, 2011).

2.8 Statistical Analysis

A factorial AxB experimental design was employed, with temperature (25 °C and 35 °C) and fermentation time (15, 30, and 45 days) as the independent variables. The response variables included total phenolic content, alkaloid levels, microbial counts, and pH. Prior to analysis, assumptions of normality, homogeneity of variances, and independence were assessed. Data were analyzed using analysis of variance (ANOVA), followed by Tukey's post hoc test to identify statistically significant differences among treatment groups.

3 Results and Discussion

3.1 Anaerobic Fermentation of Cocoa (*Theobroma Cacao*)

Table 1. Microorganisms in fermented cocoa beans.

Treatment	Yeast CFU/g	BAL CFU/g
0	< 9	< 9
15–25	$5.1 \text{x} 10^7 \pm 2.16$	$8.8 \text{x} 10^7 \pm 1.47$
15–35	$5.0 \text{x} 10^7 \pm 2.94$	$7.0 \text{x} 10^7 \pm 2.09$
30–25	$1.37 \text{x} 10^8 \pm 1.03$	$1.1 \text{x} 10^8 \pm 2.89$
30–35	$5.6 \text{ x} 10^7 \pm 1.83$	$9.0 \text{x} 10^7 \pm 1.41$
45–25	$3.8 \text{ x} 10^7 \pm 2.37$	$5.8 \text{x} 10^7 \pm 2.13$
45–35	$1.0 \text{ x} 10^6 \pm 0.82$	$1.3 \text{x} 10^7 \pm 1.03$

Note. The concentration of microorganisms is expressed in colony forming units per gram (CFU/g). The control represents the initial sample without fermentation. Concentration values correspond to the average of six replicates for each treatment

During the anaerobic fermentation of cocoa beans, the pH of the cotyledons undergoes significant time- and temperature-dependent changes in the process, influencing its microbial population (Calvo et al., 2021). The control treatment obtained a relatively low pH, Obinze et al., (2022), suggests that this is mainly due to the presence of citric acid in the pulp, which creates an acidic environment. However, the pH increased as a function of the time elapsed, this increase can be attributed to the microbial activity of cocoa, such as yeasts, which use sugars and citric acid as a carbon source.

Also, a pH below 4.75 is considered as over fermented or very acidic, significantly reducing the final quality of the beans, while a pH between 4.75 and 5.19 indicates a good fermentation process Obinze et al., (2022) and Afoakwa et al., 2008 indicates in their study that a better generation of aromatic and sensory profiles in cocoa is obtained when the pH is maintained between 5 and 5.2, with a temperature below 40 °C. Therefore, the 30-day treatments at 25 °C and 35 °C show better results, with a pH more in line with what is established in the literature (Tables 1 and 2).

Table 2. pH and titratable acidity of fermented cocoa beans.

Treatment	pH	Titratable acidity meq NaoH/100 g
0 (control)	4.65 ± 0.03	0.44 ± 0.02
15–25	5.48 ± 0.05	0.2 ± 0.02
15–35	5.60 ± 0.04	0.11 ± 0.02
30–25	5.18 ± 0.02	0.32 ± 0.02
30–35	5.08 ± 0.07	0.28 ± 0.02
45–25	5.75 ± 0.04	0.07 ± 0.02
45–35	6.06 ± 0.05	0.04 ± 0.02

Note. The pH was measured at the end of each fermentation treatment and the values presented correspond to the average of six repetitions for each treatment.

On the other hand, total titratable acidity is indirectly proportional to pH, so that as pH increases, it decreases. Thus, throughout the fermentation process, variations in total titratable acidity (TTA) are observed, which are closely related to the production of organic acids, such as lactic acid, acetic acid and other metabolic compounds (Chagas et al., 2021). These acids can be consumed or transformed through specific chemical or metabolic reactions, influenced by factors such as substrate availability, microbial composition, process duration and temperature (do Carmo Brito et al., 2017).

3.2 Processed Cocoa Beans

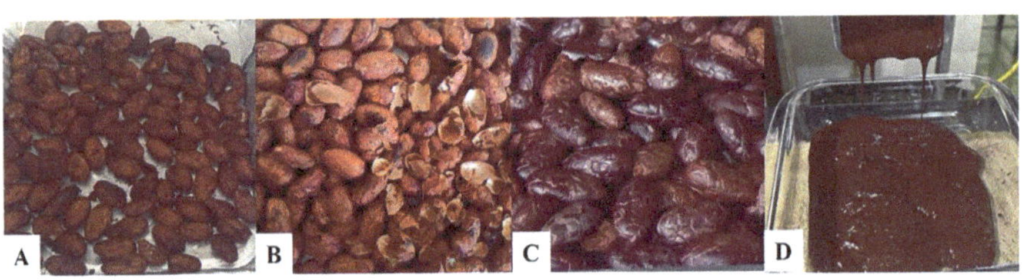

Fig. 1. Stages of cocoa bean processing.

Note. A) Dried beans; B) Roasted beans; C) Shelled beans; and D) Ground beans (cocoa paste) (Fig. 1).

Studies such as that of Mujaffar et al., (2018), propose that the drying of cocoa beans is mainly aimed at reducing moisture from an initial level of approximately 45%–60% to values close to 6%-8%, considered optimal to inhibit microbial activity. During drying, water evaporation also causes the removal of undesirable volatile compounds, such as free fatty acids, which can negatively influence the sensory profile of cocoa (Giacon et al.,

(2021)). However, Hii et al., (2009), highlights that a high temperature drying method can increase the concentration of fatty acids, which are responsible for higher acidity and off-flavors in beans. The roasting process is influenced by the residual moisture content in the kernels, during this process moisture is directly related to the Maillard reaction and caramelization of sugars (Barišić et al., 2019). These processes influence the formation of volatile compounds responsible for aroma and flavor, such as aldehydes, ketones and pyrazines, which contribute characteristic notes of chocolate, nutty and caramel, additionally, high moisture content during the roasting process hinders uniform heat transfer, which can lead to inconsistency in the texture and flavor of the final product Chen et al., (2021).

Cocoa paste was obtained by grinding previously roasted and dehulled beans, where the friction and heat generated facilitated the release of cocoa butter, which acted as a dispersing medium for the solid particles, allowing the formation of a homogeneous and fluid matrix Chen et al., (2021). Simultaneously, the breakdown of cellular structures promoted the availability of bioactive compounds such as polyphenols, caffeine and theobromine, which contribute to the functional properties of the product Pokharel, (2023). However, these compounds are susceptible to oxidation by exposure to heat and oxygen, which decreases the astringency and bitterness characteristic of cocoa (Grassia et al., 2019).

3.3 Chocolate Processing

A B

Fig. 2. 70% dark chocolate

Note. A. Chocolates obtained from fermentation treatments. B. Cross-section of a chocolate sample (Fig. 2).

To produce dark chocolate, cocoa butter is a fundamental part in the manufacture of chocolate, mainly because, according to research by Chen et al., (2021), it helps with the formation of the fat phase, which helps to regulate certain characteristic properties of chocolate such as its brightness, texture and melting. Being composed of lipids, especially triglycerides, which are responsible for the formation of crystals for the formation of a firm texture, optimal shine, fracture sensation and their tendency to melt slightly above body temperature, enhancing the sensory experience (Barišić et al., 2019).

The tempering process is crucial to ensure the correct crystallization of cocoa butter. Samanta et al., (2022) and (Oracz & Nebesny, 2019), propose that this process takes place in several stages, where, first, the chocolate is heated to high temperatures to melt any unstable glass; It is then cooled in a controlled manner to promote the formation of stable crystal structures; and finally, a slight reheating is carried out to remove unwanted crystals and stabilize the fat structure. This procedure guarantees a chocolate with the ideal properties of firmness, resistance to fatty "bloom" or appearance of white spots in the final product and a pleasant texture on the palate (Zhao & James, 2019), obtaining a final appearance like the reported result.

3.4 Quantification of Total Phenols, Caffeine and Theobromine

Table 3. Phenol concentrations (mg/g) in cocoa beans and their derivatives

Treatment	Beans	Paste	Chocolate
0	36.454 ± 0.11	27.205 ± 0.17	19.459 ± 0.15
15 – 25	27.205 ± 0.09**A**	18.778 ± 0.17**A**	11.067 ± 0.17**A**
15 – 35	18.004 ± 0.23**B**	10.327 ± 0.26**B**	6.865 ± 0.21**B**
30 – 25	9.424 ± 0.18**C**	6.771 ± 0.20**C**	5.421 ± 0.21**C**
30 – 35	6.771 ± 0.16**D**	5.292 ± 0.20**D**	3.004 ± 0.12**D**
45 – 25	5.421 ± 0.12**E**	2.534 ± 0.16**E**	1.220 ± 0.16**E**
45 – 35	3.250 ± 0.19**F**	1.607 ± 0.14**F**	0.680 ± 0.18**F**

Note. Analysis using the Tukey method reveals significant differences in phenols, content between treatments in cocoa beans, pastas and chocolates grouped into different levels of average with a 95% confidence interval

Table 4. Alkaloids concentrations (mg/g) in cocoa beans and their derivatives

Theobromine mg/g			Caffeine mg/g		
Beans	Paste	Chocolate	Beans	Paste	Chocolate
13.518 ± 0.05	11.170 ± 0.05	10.615 ± 0.02	4.538 ± 0.05	3.735 ± 0.04	3.146 ± 0.08
12.624 ± 0.03**A**	10.455 ± 0.0**A**	8.003 ± 0.04**A**	3.742 ± 0.04**A**	2.975 ± 0.04 **A**	2.611 ± 0.04**A**
11.369 ± 0.04**B**	9.673 ± 0.04**B**	7.142 ± 0.04**B**	3.615 ± 0.04**A**	2.745 ± 0.05 **B**	2.340 ± 0.05**B**
10.736 ± 0.04**C**	8.597 ± 0.03**C**	7.015 ± 0.04**C**	3.341 ± 0.03 **AB**	2.574 ± 0.06 **C**	1.998 ± 0.07**C**
9.887 ± 0.05**D**	7.100 ± 0.03**D**	5.903 ± 0.02**D**	3.106 ± 0.05 **BC**	2.330 ± 0.05 **D**	1.536 ± 0.04**D**
8.924 ± 0.04**E**	6.588 ± 0.03**E**	5.470 ± 0.03**E**	2.972 ± 0.03 **CD**	1.975 ± 0.05 **E**	1.182 ± 0.03**E**
7.615 ± 0.03**F**	5.688 ± 0.02 **F**	4.758 ± 0.03**F**	2.718 ± 0.03**D**	1.322 ± 0.05**F**	0.747 ± 0.05**F**

Note. Analysis using the Tukey method reveals significant differences in theobromine and caffeine, content between treatments in cocoa beans, pastas and chocolates grouped into different levels of average with a 95% confidence interval

The data obtained on the concentration of phenolic compounds in seeds, paste and chocolate show a progressive decrease as cocoa processing progresses. In seeds, particularly those that were not subjected to fermentation treatments, a high initial concentration of phenols was observed. This trend is consistent with the report by De Vuyst & Leroy, (2020), who indicate that unfermented cocoa beans may contain between 12 and 18% of polyphenols, with concentrations ranging from 21,7 and 38.7 mg/g. The reduction of these compounds during processing is mainly associated with the combined effect of time and temperature, where more intensive and prolonged heat treatments accelerate their degradation (Deus et al., 2018) (Tables 3 and 4).

This behavior was also replicated in the analysis of pasta and chocolate, with a more pronounced decrease under conditions of higher temperature and duration. The loss of phenolic compounds at each stage is the result of a series of key biochemical transformations. Firstly, during fermentation, a significant reduction of polyphenols occurs due to the activity of the enzyme polyphenol oxidase, which catalyzes the oxidation of epicatechins and catechins. This process results in the formation of quinones, which can polymerize and give rise to high molecular weight tannins, responsible for the brown or violet color of fermented grain and the reduction of astringency and cocoa bitterness (Rojas et al., 2021; Camu et al., (2008)).

The drying process, whose main purpose is to reduce moisture content to prevent microbial growth, also contributes to the loss of bioactive compounds. During this stage, some residual enzyme activities that negatively affect polyphenol levels may persist (Deus et al., 2018). Subsequently, the roasting, grinding and conching stages subject cocoa to high temperatures that intensify oxidative reactions, promoting a greater thermal degradation of these compounds. Jimenez et al., 2018, highlight the critical role of temperature in the degradation of bioactive compounds, noting that its effect is more severe when combined with extended exposure times.

As for alkaloids, the results obtained indicate that fermentation has a considerable impact on theobromine and caffeine concentrations in seeds, pastas and chocolates. Theobromine shows a more marked decrease in prolonged fermentations (up to 45 days) and at elevated temperatures (35 °C), while caffeine, although more chemically stable, also tends to decrease under these conditions. This is consistent with previous studies indicating that theobromine is more susceptible to chemical and microbial degradation, being able to transform into secondary metabolites during extensive fermentations (Adler et al., 2014; Chagas Junior et al., 2021). On the other hand, the more stable structure of caffeine allows it to retain higher concentrations, although it also shows moderate losses with prolonged treatments (Ozturk & Young, 2017).

Comparing the different products (seeds, pasta and chocolate), a general trend of decreasing levels of theobromine and caffeine was identified as cocoa advances in processing. Seeds had the highest concentrations, followed by paste and finally chocolate (Grassia et al., 2019). This cumulative loss is due not only to the effects of fermentation, but also to subsequent thermal processes such as roasting and grinding, which enhance the degradation of these bioactive compounds (Rusinek et al., 2023). (Amorim et al., 2017)reinforce this observation, noting that the sensitivity of alkaloids to heat increases with exposure time during post-fermentation processing.

Lemarcq et al., (2020) also highlights the crucial role of temperature in reducing caffeine and theobromine, especially at stages such as drying and roasting. Although caffeine is relatively stable during fermentation, its degradation intensifies during heat treatments (López Cerino & Chávez García, 2018). However, it is important to note that thermal processes alone do not guarantee the development of an optimal flavor and aroma profile in the final chocolate, this underlines the need for an integrated process management that balances the conservation of functional compounds with the desired sensory properties (Hermund et al., 2024).

Finally, the results of this research suggest that anaerobic fermentation constitutes a viable alternative to the traditional process, allowing a significant reduction of phenols, caffeine and theobromine in the absence of oxygen.

4 Conclusion

Anaerobic fermentation of CCN-51 cocoa beans significantly influences their physical quality and chemical composition, especially when the temperature and duration of the process are properly regulated. At 35 °C, fermentation is accelerated, while 25 °C favors the growth of beneficial microorganisms that improve internal attributes of the bean. Treatments of up to 30 days make it possible to balance parameters such as pH and acidity, fundamental aspects for obtaining high quality grains. It has also been shown that the fermentation, roasting and milling stages directly affect the concentration of bioactive compounds such as phenols, caffeine and theobromine. Although alkaloids are more stable, all these compounds tend to degrade under prolonged treatment or high temperatures, with phenols being the most susceptible.

References

Adler, P., Frey, L.J., Berger, A., Bolten, C.J., Hansen, C.E., Wittmann, C.: The key to acetate: metabolic fluxes of acetic acid bacteria under cocoa pulp fermentation-simulating conditions. Appl. Environ. Microbiol. **80**(15), 4702–4716 (2014). https://doi.org/10.1128/AEM.01048-14

Afoakwa, E.: Ciencia y tecnología del chocolate. In: Ciencia y tecnología del chocolate (2nd ed., p. 49). John Wiley & Sons (2016)

Afoakwa, E.O., Paterson, A., Fowler, M., Ryan, A.: Flavor formation and character in cocoa and chocolate: a critical review. Crit. Rev. Food Sci. Nutr. **48**(9), 840–857 (2008). https://doi.org/10.1080/10408390701719272

Alcivar, L., Cobos, F., Alvarado, A., Obando, M.: Índices de calidad en la comercialización del cacao (theobroma cacao L.) En ECUADOR. J. Sci. Res. **7**, 230–248 (2022)

Amorim, G.M., Oliveira, A.C., Gutarra, M.L.E., Godoy, M.G., Freire, D.M.G.: Solid-state fermentation as a tool for methylxanthine reduction and simultaneous xylanase production in cocoa meal. Biocatal. Agric. Biotechnol. **11**, 34–41 (2017). https://doi.org/10.1016/j.bcab.2017.05.009

Aprotosoaie, A.C., Luca, S.V., Miron, A.: Flavor chemistry of cocoa and cocoa products—an overview. Compr. Rev.Food Sci. Food Safety **15**(1), 73–91 (2016). https://doi.org/10.1111/1541-4337.12180

Barišić, V., et al.: The chemistry behind chocolate production. Molecules **24**(17), 3163 (2019). https://doi.org/10.3390/molecules24173163

Calvo, A.M., Botina, B.L., García, M.C., Cardona, W.A., Montenegro, A.C., Criollo, J.: Dynamics of cocoa fermentation and its effect on quality. Sci. Rep. **11**(1), 16746 (2021). https://doi.org/10.1038/s41598-021-95703-2

Camu, N., De Winter, T., Addo, S.K., Takrama, J.S., Bernaert, H., De Vuyst, L.: Fermentation of cocoa beans: influence of microbial activities and polyphenol concentrations on the flavour of chocolate. J. Sci. Food Agric. **88**(13), 2288–2297 (2008). https://doi.org/10.1002/jsfa.3349

Chagas, G.C.A., Ferreira, N.R., Gloria, M.B.A., Martins, L.H.S., Lopes, A.S.: Chemical implications and time reduction of on-farm cocoa fermentation by Saccharomyces cerevisiae and Pichia kudriavzevii. Food Chem. **338**, 127834 (2021). https://doi.org/10.1016/j.foodchem.2020.127834

Chagas Junior, G.C.A., Ferreira, N.R., Lopes, A.S.: The microbiota diversity identified during the cocoa fermentation and the benefits of the starter cultures use: an overview. Int. J. Food Sci. Technol. **56**(2), 544–552 (2021). https://doi.org/10.1111/ijfs.14740

Chen, J., Ghazani, S.M., Stobbs, J.A., Marangoni, A.G.: Tempering of cocoa butter and chocolate using minor lipidic components. Nat. Commun. **12**(1), 5018 (2021). https://doi.org/10.1038/s41467-021-25206-1

De Vuyst, L., Weckx, S.: The cocoa bean fermentation process: from ecosystem analysis to starter culture development. J. Appl. Microbiol. **121**(1), 5–17 (2016). https://doi.org/10.1111/jam.13045

Deus, V. L., et al.: Influence of drying methods on cocoa (Theobroma cacao L.): antioxidant activity and presence of ochratoxin A. Food Sci. Technol. **38**(suppl 1), 278–285 (2018). https://doi.org/10.1590/fst.09917

De Vuyst, L., Leroy, F.: Functional role of yeasts, lactic acid bacteria and acetic acid bacteria in cocoa fermentation processes. FEMS Microbiol. Rev. **44**(4), 432–453 (2020). https://doi.org/10.1093/femsre/fuaa014

do Carmo Brito, B.N., Campos Chisté, R., da Silva Pena, R., Abreu Gloria, M.B., Santos Lopes, A.: Bioactive amines and phenolic compounds in cocoa beans are affected by fermentation. Food Chem. **228**, 484–490 (2017). https://doi.org/10.1016/j.foodchem.2017.02.004

ECA, CAOBISCO, FCC.: Cacao en Grano: Requisitos de Calidad de la Industria del Chocolate y del Cacao (2015)

Garcia Gonzalez, E., et al.: Study of the spontaneous cocoa fermentation (Theobroma Cacao l.) and evaluation of bean quality in a productive unit in small-scale, Vol. 5, Issue 2 (2018)

Giacon, T.G., de Gois e Cunha, G.C., Eliodório, K.P., de Souza Oliveira, R.P., Basso, T.O.: Homo-and heterofermentative lactobacilli are differently affected by lignocellulosic inhibitory compounds (2021). https://doi.org/10.1101/2021.01.18.427060

González-Nuñez, L.N., Cañizares-Macías, M.P.: Focused microwaves-assisted extraction of theobromine and caffeine from cacao. Food Chem. **129**(4), 1819–1824 (2011). https://doi.org/10.1016/j.foodchem.2011.05.118

Grassia, M., Salvatori, G., Roberti, M., Planeta, D., Cinquanta, L.: Polyphenols, methylxanthines, fatty acids and minerals in cocoa beans and cocoa products. J. Food Measur. Charact. **13**(3), 1721–1728 (2019). https://doi.org/10.1007/s11694-019-00089-5

Guehi, T.S., Zahouli, I.B., Ban-Koffi, L., Fae, M.A., Nemlin, J.G.: Performance of different drying methods and their effects on the chemical quality attributes of raw cocoa material. Int. J. Food Sci. Technol. **45**(8), 1564–1571 (2010). https://doi.org/10.1111/j.1365-2621.2010.02302.x

Hermund, D.B., et al.: Fate of flavonoids and theobromine in cocoa beans during roasting: Effect of time and temperature. J. Am. Oil Chem. Soc. (2024). https://doi.org/10.1002/aocs.12853

Hii, C.L., Law, C.L., Cloke, M., Suzannah, S.: Thin layer drying kinetics of cocoa and dried product quality. Biosys. Eng. **102**(2), 153–161 (2009). https://doi.org/10.1016/j.biosystemseng.2008.10.007

Ho, V.T.T., Zhao, J., Fleet, G.: The effect of lactic acid bacteria on cocoa bean fermentation. Int. J. Food Microbiol. **205**, 54–67 (2015). https://doi.org/10.1016/j.ijfoodmicro.2015.03.031

ICCO. Production Of Cocoa Beans (thousand tonnes). International Cocoa Organization (2024a)

ICCO.: Cocoa Market Review. International Cocoa Organization (2024b)

Jimenez, J.C., et al.: Differentiation of Ecuadorian National and CCN-51 cocoa beans and their mixtures by computer vision. J. Sci. Food Agric. **98**(7), 2824–2829 (2018). https://doi.org/10.1002/jsfa.8790

Laličić-Petronijević, J., et al.: Effect of chocolate storage on total phenolics, Flavan-3-Ols, proanthocyanidins, oxidative stability and antioxidant capacity. Food Technol. Biotechnol. **54**(1) (2016). https://doi.org/10.17113/ftb.54.01.16.4014

López Cerino, I., Chávez García, E.: Eficacia de secador solar tipo túnel con cacao (Theobroma Cacao L.) en Tabasco. Revista Mexicana de Ciencias Agrícolas **21**, 4395–4405 (2018). https://doi.org/10.29312/remexca.v0i21.1528

MAGAP. Ecuador es el primer exportador de cacao en grano de América. Ministerio de Agricultura y Ganadería (2024)

Matavaca, B., et al.: APplied Biotechnology in Cocoa Paste Production. Neuro Quantolog 1303–5150 (2022). https://doi.org/10.14704/nq.2022.20.9.NQ440077

Mujaffar, S., Sukha, D., Ramroop, A.: Comparison of the drying behavior of fermented cocoa (Theobroma cacao L.) beans dried in a cocoa house, greenhouse and mechanical oven. International Symposium on Cocoa Research (ISCR) (2018)

Nazaruddin, R., Seng, L.K., Hassan, O., Said, M.: Effect of pulp preconditioning on the content of polyphenols in cocoa beans (Theobroma Cacao) during fermentation. Ind. Crops Prod. **24**(1), 87–94 (2006). https://doi.org/10.1016/j.indcrop.2006.03.013

Obinze, S., Ojimelukwe, P.C., Eke, B.A.: Box fermentation and solar drying improve the nutrient composition and organoleptic quality of chocolate from cocoa beans. Front. Sustain. Food Syst. **6** (2022).https://doi.org/10.3389/fsufs.2022.1023123

Oracz, J., Nebesny, E.: Effect of roasting parameters on the physicochemical characteristics of high-molecular-weight Maillard reaction products isolated from cocoa beans of different Theobroma cacao L. groups. Eur. Food Res. Technol. **245**(1), 111–128 (2019). https://doi.org/10.1007/s00217-018-3144-y

Ozturk, G., Young, G.M.: Food evolution: the impact of society and science on the fermentation of cocoa beans. Compr. Rev. Food Sci. Food Safety **16**(3), 431–455 (2017). https://doi.org/10.1111/1541-4337.12264

Pokharel, B.: Cocoa Bean Fermentation: Impact on Chocolate Flavor and Quality. Int. J. Sci. Res. (IJSR) **12**(6), 1668–1674 (2023). https://doi.org/10.21275/SR23614230652

Rojas, K., Hernández, C., Mencía, A.: Biochemical transformations of cocoa (theobroma cacao l.) during a controlled fermentation process. Agron. Costarricense (2021). https://www.scielo.sa.cr/scielo.php?pid=S0377-94242021000100053&script=sci_arttext&utm_source=chatgpt.com

Rusinek, R., et al.: Effect of the roasting level on the content of bioactive and aromatic compounds in Arabica coffee beans. Int. Agrophys. **38**(1), 31–42 (2023). https://doi.org/10.31545/intagr/176300

Salazar, L., Rojas, R., Hurtdo, J.: Aislamiento e identificación de microorganismos presentes durante el proceso de fermentación de Theobroma cacao L., variedad "Chuncho" del Cuzco. International Symposium on Cocoa Research (2017). https://www.icco.org/wp-content/uploads/T5.102.AISLAMIENTO-E-IDENTIFICACION-DE-MICROORGANISMOS-PRESENTES-DURANTE-EL-PROCESO-DE-FERMENTACION-DE-THEOBROMA.pdf

Samanta, S., et al.: Dark chocolate: an overview of its biological activity, processing, and fortification approaches. Current Res. Food Sci. **5**, 1916–1943 (2022). https://doi.org/10.1016/j.crfs.2022.10.017

Subroto, E., Djali, M., Indiarto, R., Lembong, E., Baiti, N.: Microbiological activity affects post-harvest quality of cocoa (Theobroma cacao L.) Beans. Horticulturae, **9**(7), 805–830 (2023). https://doi.org/10.3390/horticulturae9070805

Zhao, H., James, B.J.: Fat bloom formation on model chocolate stored under steady and cycling temperatures. J. Food Eng. **249**, 9–14 (2019). https://doi.org/10.1016/j.jfoodeng.2018.12.008

Development and Evaluation of a Parameterized 3D Model for the Manufacturing of Rearview Mirrors Using Infusion and Additive Manufacturing Processes

Sandino Torres[1], Cristian Redroban[2], Edisson Calderón[1](✉), and Roberto Ortega[3]

[1] Departamento de Materiales, Escuela Politécnica Nacional, Quito CP170525, Ecuador
edisson.calderon@epn.edu.ec
[2] Escuela Superior Politécnica de Chimborazo, Riobamba CP060106, Ecuador
[3] Instituto Superior Universitario Central Técnico, Quito CP170502, Ecuador

Abstract. Access to advanced technology in resin infusion processes is limited in the country due to economic restrictions, lack of access to equipment, supplies, specialized material and lack of knowledge of the process. This study develops a parameterized 3D model in CATIA V5 for the manufacture of motorcycle mirrors. The mold is printed on a Creality CR-10S Pro V2 and optimized for high strength and low roughness. A release agent is applied to facilitate removal. Manufacturing uses veil fiber, mat fiber and orthophthalic polyester resin, which is introduced into the hermetically sealed mold using a vacuum pump. The resin infusion process impregnates the fibers in 4 min after applying vacuum for 20 min, achieving the final product. The main objective of the project is to develop a parameterized 3D model of motorcycle mirrors using CATIA V5, evaluating the use of 3D printing technologies in mold production and its impact on product quality, costs and manufacturing times. To this end, applied research and technological development are carried out, focusing on the creation and improvement through an innovative process for the manufacture of prototypes. In addition, the surface quality of the mould, the precision of the final product and the costs associated with both production methods were measured. The results suggest that additive manufacturing presents significant advantages in personalisation and time reduction, without compromising the quality of the final product.

Keywords: Resins · Automotive Parts · Prototype · Fused Deposition Modeling · Molds

1 Introduction

Composite manufacturing processes use different types of fibers or fabrics impregnated or not in resin to achieve the desired design and final fabrication of the composite part [1]. The selection of any type of process is made taking into account economic, logistical, engineering, and other factors. There are various types of composite manufacturing, such

J. L. Ballesteros et al. (Eds.): CITIS 2025, LNNS 1780, pp. 396–405, 2027.
https://doi.org/10.1007/978-3-032-14287-0_39

as manual lamination lay-up of prepregs and infusion. In most processes, it is necessary to obtain a mold, which can be machined or obtained from a 'master' model, in order to reproduce the parts. These processes can require many hours of work and can be complex in the manufacturing process [2].

The resin infusion process has the ability to produce high quality components [3], which is why many industries have adopted it. Infusion is performed by vacuum-assisted resin transfer, several layers of fiber are stacked in the mold and a vacuum bag is placed in the mold, which is sealed to the mold with adhesive tape to form a sealed chamber [4]. Vacuum inlets and outlets are connected to this sealed chamber [5], from which the final product is obtained [6]. A schematic of the resin infusion process is shown, in which the mold that is part of the process can be observed (See Fig. 1).

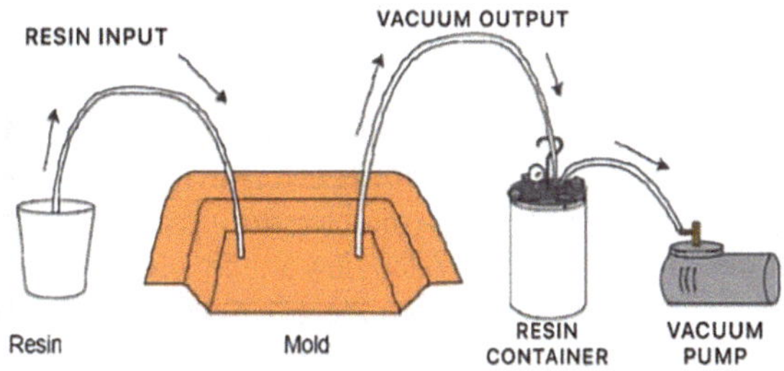

Fig. 1. Resin infusion process (Source: Fibre Glast Developments Corporation, "Vacuum Infusion Complete Guide").

This is why attempts have been made to replace the process of obtaining the mold, which is traditionally performed with conventional technologies, with new techniques that are adapted or combined with other processes [7]. Additive manufacturing technology is applied both in the manufacture of final products and in the production of molds, tooling and fixtures. Molds produced using additive manufacturing methods can improve the manufacturing process and product characteristics [8].

On the other hand, manufacturing products with composite materials using conventional processes incurs high time and costs. Nowadays, the concept of additive manufacturing and fused deposition modeling is increasingly applied in product development [9], covering various industrial areas such as automotive, aerospace, medical, mechanical and prototyping. Similarly, the manufacturing processes of parts with composite materials require a lot of manual labor, which generates additional costs [10]. The combination or integration of additive manufacturing (AF) and composite manufacturing (CM) is very limited or even non-existent, indicating the difficulties in bringing these two technologies together.

Consequently, the use of 3D printing for tooling and rapid manufacturing has promised to produce components with complex geometries according to computer designs, as well as its potential applications in the fields of biomedical, electronics, and aerospace engineering. The formation methodology and performance of polymer composites reinforced with particles, fibers, and nanomaterials are emphasized [11].

Also, additive manufacturing technologies and composite materials have been integrated, especially in customized processes [12] where they seek to reduce manufacturing times and improve performance through rapid 3D printed molds. A study was conducted in which they manufactured polymer matrix composites using vacuum-assisted resin infusion molding [13]. The variables involved in the process were monitored, allowing better control and optimization of the process. Tooling, especially for prototypes or small series, can be very expensive. In addition, prototyping of fiber-reinforced thermoplastic sheath structures can rely on time-consuming manual efforts.

The cost and time of current mold building technologies often preclude going through these verification stages. Using the 3D mold printing technology developed by the company iGestek, the time and price of molds used for prototype construction is significantly reduced. The prices of 3D printed molds can be up to 10 times cheaper compared to those manufactured by conventional methods [14]. In addition, it is important to note that molds with customized dimensions can be created for both prototypes and small-scale productions. Additive manufacturing offers great advantages in product customization, which is increasingly motivating manufacturers of customized products to replace conventional manufacturing methods. In the case of auto parts, it is often necessary to offer innovative ideas [15]. In the specific case of motorcycle mirrors, some drivers are looking for customized designs that give them a better view and functionality of the product [16]. In this way, it seeks to provide a solution to the manufacture of customized products, integrating additive manufacturing methods and resin infusion. The study focuses on the user's design to generate a mold using FDM, which will allow the manufacture of a final product made of fiberglass and polyester resin.

2 Methodology

2.1 Research Approach and Experimental Design

The methodological approach is framed within an applied research and technological development; therefore, an infusion mold with geometric measurements established by the additive manufacturing process is presented and compared with the results obtained once the infusion process is finished, in that sense, the relevant parameters of the process such as the length of the geometric body obtained by the processing of composite materials are highlighted.

2.2 Model Design in 3D

Product customization begins with the customer's or user's idea, from which a parameterized 3D model is created using computer-aided design software. This software allows the geometry of the product to be modified if necessary. For this design, the CATIA V5 program is used due to the features it offers in the generation of complex surfaces. This allows us to model the part according to the needs and functionality required for a motorcycle rear-view mirror. Starting from the general measurements of the product, a surface is created based on the technical parameters that guarantee the functionality of

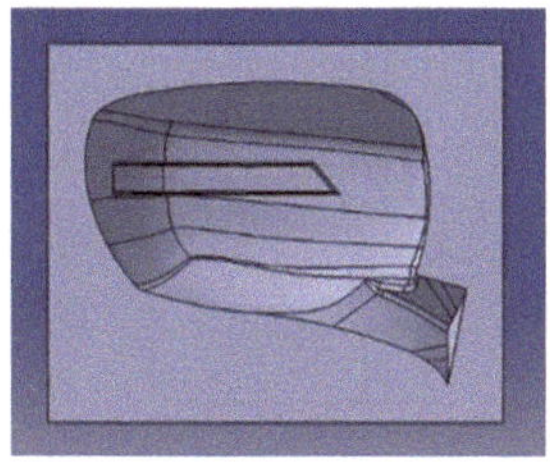

Fig. 2. Rearview mirror mold

the product. Finally, the solid of the part is obtained, which represents the virtual prototype of the product to be manufactured. When the 3D model of the part is approved, the 3D model of the mold is generated (See Fig. 2).

The mold is manufactured using additive manufacturing technology. The dimensions of the mold are established, taking into account the restrictions imposed by the size of the printing table. In this project, the Creality CR-10S Pro V2 printer is used, which has a printing table of 300 × 300 × 400 mm. White polylactic acid (PLA) filament is used as the printing material.

2.3 Mold Fabrication

The mold must be able to withstand the mechanical loads related to the shrinkage experienced by the laminate during the curing process. In addition, it must withstand the pressure during curing, which in vacuum resin infusion processes is typically around 0.1 MPa, along with a resin that cures at room temperature. Also, the mold must maintain a certain stiffness and strength due to the curing temperature, which can reach around 180 °C. It is important that the mold surface has low surface roughness. Since the material used for the preparation of the mold is porous, a polyester filler is used to provide greater mechanical resistance to the mold and guarantee a seal on its surface. Once this filler has been applied, it must be ensured that the surface has a uniform thickness that does not exceed 750–1000 μm, in order to faithfully reproduce the details of the final product. This is achieved through a sanding and polishing process. Finally, a release agent is applied to facilitate the removal of the part.

2.4 Production Process

As for the manufacturing process, the infusion process is used. Although this process can be somewhat complex, it is chosen as a demonstration in this project for obtaining a motorcycle rear-view mirror. However, the methodology shown can be applied in other processes, such as hand lay-up that are simpler and more economical. The steps performed in the infusion process are shown in the infusion are shown in the schematic. (See Fig. 3).

Once the mold is prepared and cleaned, with a uniform layer of wax, a release fluid is applied to its surface. Then, layers of fiber are manually cut to be used as reinforcement. These layers are placed on the surface of the mold following the position and orientation of the design of the part to be manufactured.

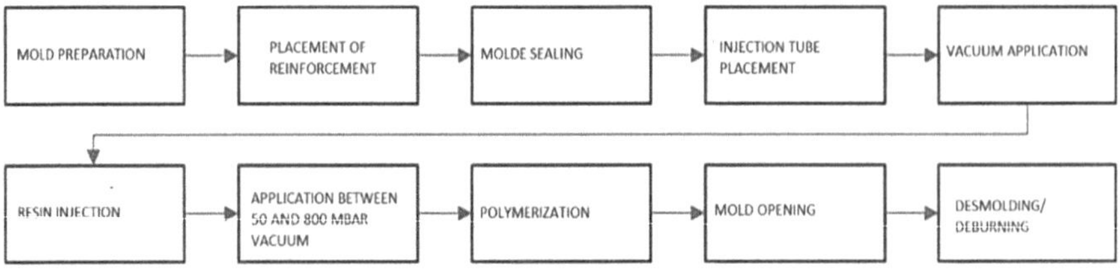

Fig. 3. Procedure for resin infusion

Excess fiber is left behind and then cut off. As a first layer, a fiber called "surface veil" with a surface weight of 30 g/m^2 is placed, followed by a layer of "mat" fiber of 450 g/m^2 and finally another layer of surface veil. On this set of fibers is placed the release cloth or "Peel Ply", which facilitates the detachment, additionally a flow mesh is placed that allows a faster and more fluid advance of the resin. On the edges or perimeter of the mold, a sealing tape or "Tacky Tape" is placed in order to achieve a uniform sealing on the entire surface. Once this is completed, sealing is done using a plastic bag and the sealing tape, which creates an airtight environment.

Next, the vacuum system is assembled, with the vacuum inlets and outlets, the vacuum bag is punctured to connect the vacuum valve to the corresponding tube, which allows the resin to enter. These connections are made perpendicular to the mold, placing the inlet connection at the lower limit of the mold and the outlet connection at the upper limit of the mold, as shown in Fig. 4a, the assembly of the infusion system together with a description of each of the elements that are part of this assembly. In this setup, a hose is connected to the vacuum pump. For this process, a ½ HP, 5.5 CFM vacuum pump is used.

2.5 Testing and Evaluation

Initially, tests are performed to ensure that there are no leaks in the system. Once it is verified that the system is completely airtight, resin infusion is performed.

This project uses unsaturated polyester orthophthalic resin, to cure or perform the polymerization reaction, an organic peroxide catalyst is used to generate the hardening process of the resin and obtain the thermoplastic.

The resin has a Brookfield viscosity of between 2000 and 2200 cP, an acid number of between 21 and 27 mg KOH/g resin, and a gel time at 25 °C of between 12 and 14 min. Once infusion takes place, the air in the mold is evacuated and the pressure difference between the fiber bundle at the specified vacuum level and the circulating air at atmospheric pressure allows this bundle to compress.

When the desired vacuum level is reached which remains constant during infusion, the inlet valve is opened to allow the resin to enter. At that point, the resin is sucked from the container and permeates the fiber bundle due to the pressure difference. Figure 4b shows how the resin infusion process is carried out to obtain the mirror. The time taken to impregnate the fibers was 4 min and 50 s and vacuum is applied in the system for 20 min until the resin polymerization process is completed.

Then, demolding is carried out, which consists of removing the vacuum bag and all the elements that are part of the infusion process, in order to extract the final product, i.e., the finished mirror. Finally, with the product obtained, a comparative analysis of the costs associated with the manufacture of the product obtained using the described processes, additive manufacturing versus conventional methods, is performed.

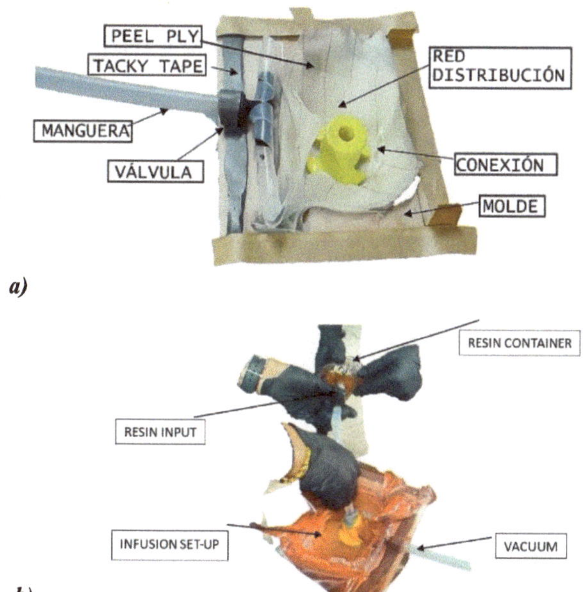

Fig. 4. (a) Elements in the rear-view mirror infusion process, (b) Process set-up.

3 Results

In general, the appropriate choice of composite mirror manufacturing process depends on factors such as design complexity, production volumes, costs and required mechanical properties. The 3D parameterized model was developed using CATIA V5, (See Fig. 5(a)); allowing accurate visualization of the mirror prior to fabrication. This digital visualization provided the basis for the creation of the mold using additive manufacturing technologies. Figure 5(b) shows the demolding process of the fabricated product, evidencing the successful transition from digital design to physical product. A detailed comparison is made between the digital model and the final product to ensure that the intended dimensions and features were maintained. Accordingly, general measurements are performed using the MITUTOYO Crysta Plus M coordinate measuring machine to check the differences between the digital model and the final product, (See Fig. 5(c)).

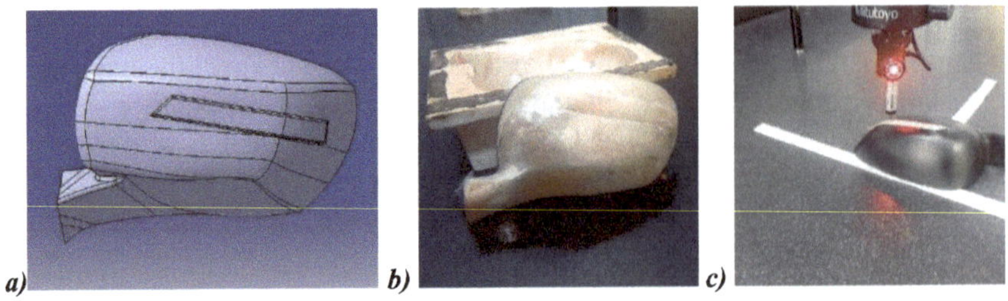

Fig. 5. (a) Parameterized 3D model, (b) Demolding of the product, (c) Measurement of the rearview mirror.

Main dimensions are established and compared with the digital model and the measurements of the final product. Figure 6 shows the plan with the reference measurements used to verify the dimensions of the final product. The analysis shows that the difference between the digital model and the prototype is almost null, with an average of approximately 0.5%, thus showing that the 3D model parameterized in CATIA V5 was translated accurately, meeting the expected dimensional requirements. In addition, a measurement of the thickness of the part is carried out, which remained constant throughout the contour with a value of 3 mm.

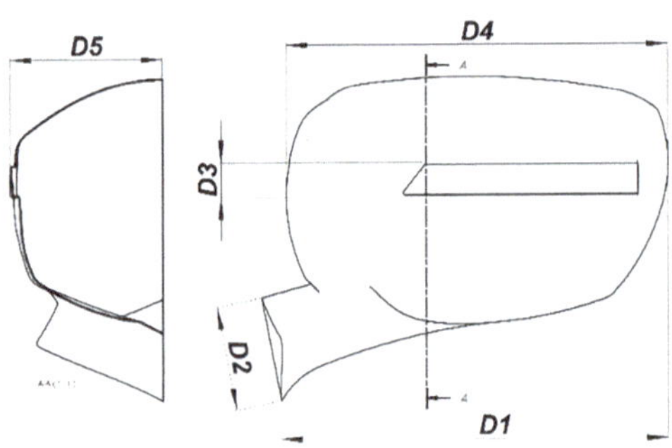

Fig. 6. Rearview mirror plan

Table 1. Percentage of similarity.

Cota	Digital Model	Part manufactured by infusion	% Difference
D1	146.2	145.4	99.45
D2	38.30	38.128	99.55
D3	11.30	11.238	99.45
D4	144.3	144.027	99.81
D5	57.60	57.256	99.4

Table 1 shows the percentages of similarity between the dimensions of the digital model and the manufactured product. These data reinforce the accuracy of the method used, indicating that the mold printing methodology and subsequent resin infusion guarantee a faithful reproduction of the dimensions of the original design. In this process, a rigid mold printed in PLA was obtained, with a roughness of Ra = 1.1 μm. This allows the mold surface to guarantee a faithful reproduction of the part. The final weight of the mold is 0.293 kg [13, 17]. The study compares the costs between molds machined in aluminum and molds made by additive manufacturing for the manufacture of medical devices, demonstrating a considerable reduction in costs. Figure 7 shows a manufacturing simulation performed in CATIA V5 software.

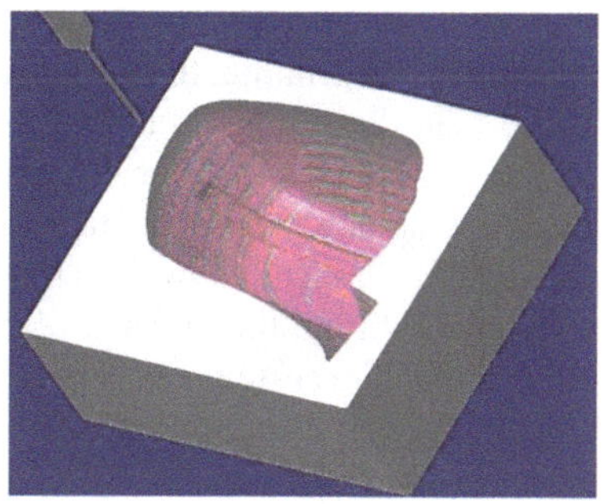

Fig. 7. Computerized numerical control machining simulation

Table 2. Manufacturing methods.

Type of mold	Manufacturing time (hours)	Cost	Weight (Kg)
Machined in aluminum	5	$ 700	1.7
Printed in polylactic acid (PLA)	12	$ 70	0.293

While the proposed method demonstrates cost and time advantages for prototypes and small series, scaling to mass production introduces challenges such as mold durability under high-volume demands, material supply chain logistics, and automation requirements. Further research is needed to evaluate process repeatability and the adaptation of additive manufacturing infrastructure to industrial standards, ensuring compatibility with automated resin infusion systems.

4 Discussion

It is important to note that the printer used is a low-cost desktop printer, which makes it accessible even for home use. In addition, in the case of developing a "master" model to later obtain a composite mold, the cost is very high, so the proposed method is innovative. The proposed method is innovative. Table 2 shows the cost comparison of the manufacturing methods.

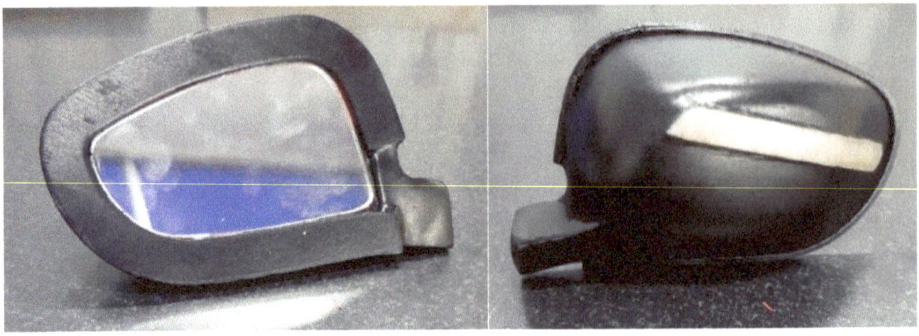

Fig. 8. Results obtained

On the other hand, in the case of 3D printing, the overall manufacturing cost of the mold is reduced. It is observed that, during the infusion process, both in casting and polymerization of the resin, the mold does not suffer deformations, which guarantees the quality of the final product, to maintain the proper shape and dimensions.

The results show that the mold manufactured by 3D printing techniques in PLA is resistant and presents good behavior during the infusion of composite materials. This approach makes it possible to manufacture customized parts at low cost compared to traditional methods. The motorcycle rearview mirror turns out to be practical and functional, Fig. 8 shows the result obtained in this project.

5 Conclusions

This research presents the methodology for the development of prototypes using printed molds and resin infusion, and demonstrates the advantages of applying additive manufacturing to obtain parts with composite materials. The project develops a fully functional customized motorcycle rear-view mirror. The use of CATIA V5 for the development of the parameterized model and its subsequent manufacture by resin infusion have proven to be viable and effective methods, meeting the established dimensional and functional requirements, which validates the ability of this approach for the design and manufacture of complex components.

The approach used in this project has great potential to be applied to the manufacture of other complex components. Mold 3D printing technology and resin infusion have proven to be effective and economical methods for the production of custom parts. The resin infusion process used in the production was successful, showing a correct impregnation of the fibers with the resin, meeting the established curing times. The research shows that it not only reduces the production time, but also improves the precision and quality of the final product, demonstrating the feasibility for the production of customized components. The infusion process guarantees a constant thickness of the part, and the measurements obtained are remarkably close to the digital model.

The comparative cost analysis performed in this study shows that additive manufacturing can offer a significant economic advantage over traditional methods, especially in the production of prototypes and short series. The reduction in manufacturing costs, combined with the reduction in production times, positions this methodology as a competitive and cost-effective option. In summary, the study demonstrates that the combination of printed molds and resin infusion is a viable strategy for the production of high-quality components in the automotive industry. This methodology not only meets technical and economic objectives, but also offers a path to innovation and continuous improvement in manufacturing processes.

References

1. Fleischer, J., Teti, R., Lanza, G., Mativenga, P., Möhring, H.-C., Caggiano, A.: Composite materials parts manufacturing. CIRP Ann. **67**(2), 603–626 (2018)
2. Iernutan, R.A., Babota, F., Istoan, R.: Carbon fibre reinforced aluminium mesh composite materials. Procedia Manuf. **32**, 901–907 (2019)
3. Hindersmann, A.: Confusion about infusion: An overview of infusion processes. Compos. Part A Appl. Sci. Manuf. **126**, 105583 (2019)
4. Ma, W., Elkin, R.: Laminating processes of thermoset sandwich composites, in Sandwich Structural Composites, CRC Press, pp. 125–157 (2021)
5. Ouezgan, A., Mallil, E.H., Echaabi, J.: Manufacturing routes of vacuum assisted resin infusion: numerical investigation. J. Compos. Mater. **56**(21), 3221–3236 (2022)
6. Van Oosterom, S., Allen, T., Battley, M., Bickerton, S.: An objective comparison of common vacuum assisted resin infusion processes. Compos. Part A Appl. Sci. Manuf. **125**, 105528 (2019). https://doi.org/10.1016/j.compositesa.2019.105528
7. Khosravani, M.R., Nasiri, S.: Injection molding manufacturing process: review of case-based reasoning applications. J. Intell. Manuf. **31**, 847–864 (2020)
8. Pereira, T., Kennedy, J.V., Potgieter, J.: A comparison of traditional manufacturing vs additive manufacturing, the best method for the job. Procedia Manuf. **30**, 11–18 (2019)
9. Daminabo, S.C., Goel, S., Grammatikos, S.A., Nezhad, H.Y., Thakur, V.K.: Fused deposition modeling-based additive manufacturing (3D printing): techniques for polymer material systems. Mater. Today Chem. **16**, 100248 (2020)
10. Hamidi, F., Aslani, F.: Additive manufacturing of cementitious composites: materials, methods, potentials, and challenges. Constr. Build. Mater. **218**, 582–609 (2019)
11. Wang, X., Jiang, M., Zhou, Z., Gou, J., Hui, D.: 3D printing of polymer matrix composites: a review and prospective. Compos. B Eng. **110**, 442–458 (2017). https://doi.org/10.1016/j.compositesb.2016.11.034
12. Pal, A.K., Mohanty, A.K., Misra, M.: Additive manufacturing technology of polymeric materials for customized products: recent developments and future prospective. RSC Adv. **11**(58), 36398–36438 (2021)
13. Loos, A.C., Rattazzi, D., Batra, R.C.: A Three-Dimensional Model of the Resin Film Infusion Process. Compos. Mater. **36**(10), 1255–1273 (2022). https://doi.org/10.1106/002199802023168
14. Wang, P.H., Kim, G., Sterkenburg, R.: Investigating the effectiveness of a 3D printed composite mold. Int. J. Aeros. Mech. Eng. **13**(11), 705–709 (2019)
15. Candi, M., Beltagui, A.: Effective use of 3D printing in the innovation process. Technovation **80**, 63–73 (2019)
16. Bunsell, A.R., Joannès, S., Thionnet, A.: Fundamentals of fibre reinforced composite materials. CRC Press (2021)
17. Munoz-Guijosa, J.M., Martínez, R., Cendrero, A.M., Lantada, A.D.: Rapid prototyping of personalized articular orthoses by lamination of composite fibers upon 3D-printed molds. Materials **13**(4) (2020). https://doi.org/10.3390/ma13040939

Methodology for Smoothing the Altitude Profile of Urban Bus Routes in Mountainous Cities

Sebastián Quizhpi-Salamea(✉), Jairo Castillo-Calderón, Diego Díaz Sinche, and Elmer Arias-Montaño

eX-MoVeT, Research Group, Facultad de la Energía, las Industrias y los Recursos Naturales no Renovables, Universidad Nacional de Loja, Avenida Pio Jaramillo Alvarado y Reinaldo Espinosa, 110103 Loja, Ecuador
{kleber.quizhpi,jdcastilloc,diego.v.diaz,elmer.arias}@unl.edu.ec

Abstract. Public ground transportation studies are essential for designing more sustainable, efficient, safe, and accessible transportation systems, contributing to the quality of life in cities and the development of more sustainable urban environments. Currently, mobile devices, such as smartphones, are economical and efficient means of collecting data associated with bus routes using GPS. In this regard, the quality of the collected data is imperative, as it can be affected by external factors and the accuracy of the monitoring equipment. Therefore, it is necessary to process the data and ensure that they are consistent with the actual driving experience before using them for subsequent analyses, such as calculating energy or fuel consumption of the transport fleet. This study presents a methodology for smoothing altitude data from urban bus routes in Loja, a mountainous city in the Ecuadorian highlands. The Gaussian kernel probability density function is applied to smooth GPS data. Additionally, the correct selection of the bandwidth that allows homogenizing the measured data and reducing the loss of information associated with the smoothing process is evaluated.

Keywords: kernel gaussiano · regla de Silverman · ancho de banda · GPS

1 Introduction

In the context of any scientific research, the data collection procedure is of utmost importance, since various methodologies will be applied to the acquired data to validate the proposed hypotheses. However, due to the experimental nature of data collection where external factors such as environmental conditions, telecommunication service quality, or the effectiveness of measuring equipment can affect data quality the measured data may lose representativeness [1], directly influencing the research outcomes [2].

Supported by organization x.

J. L. Ballesteros et al. (Eds.): CITIS 2025, LNNS 1780, pp. 406–417, 2027.
https://doi.org/10.1007/978-3-032-14287-0_40

Given this, scientific research should ideally rely on raw data, that is, data without any manipulation to faithfully observe the events that occurred during the experiment. Nevertheless, in some cases, it becomes necessary to process the data in order to filter out anomalous values caused by the aforementioned external factors such as values outside the expected measurement range, or simply to remove noise caused by measurement variations. This aims to achieve uniform quality in the measured data, thereby improving the research results [3,4].

This process of data homogenization is known as **smoothing.**

In Fig. 1, the GPS based altitude measurements of a bus route are shown, along with different techniques used to smooth the measured data.

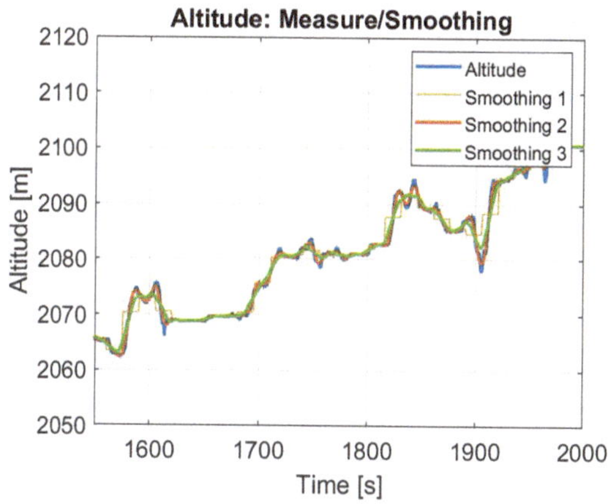

Fig. 1. Smoothing data comparison.

Smoothing method 1 was performed by averaging the measured values over a fixed data range. Smoothing method 2 was implemented using the moving average technique, which calculates the average of measured values over a defined range with a variable mean. Smoothing method 3 employed the probability density function using a Gaussian kernel, which homogenizes the data through the construction of a probability function.

Figure 1 illustrates that some smoothing techniques are more effective than others. More importantly, it highlights how the smoothed data deviate from the measured values, either aligning more closely or becoming distorted in relation to them—this deviation can be interpreted as a loss of information.

The objective of this study is to explore the smoothing technique based on Gaussian kernel probability density distribution, and to analyze the effect of the parameter known as bandwidth on this technique. The goal is to define an optimal dataset that minimizes the mean squared error (MSE) between the original measured data vector and the smoothed data vector, thereby minimizing information loss.

The methodology section of this article outlines the data collection process, the sampling strategy, and the mathematical foundation that enables the construction of the probability function using the Gaussian kernel. In the results section, a comparison of the bandwidth effect is presented to demonstrate the effectiveness of the proposed methodology through the optimization of the Mean Squared Error (MSE).

2 Metodology

2.1 Data Collection

As part of a research project aimed at determining the energy consumption of the urban bus fleet in the city of Loja, Ecuador [5] data collection was carried out for both outbound and return trips on three bus lines operating within the city: Line 2, Line 8, and Line 11. These routes are illustrated in Fig. 2. In total, 40 outbound and 40 return trips were recorded for each bus line. For the purpose of this study, it is not necessary to use all recorded trips; a representative sample from each is sufficient.[6].

2.2 Sampling Strategy

The sampling strategy consisted of collecting trip data using a mobile device equipped with GPS, [7,8] through the Torque Pro application [9]. The application provides data on time, latitude, longitude, altitude, and speed, recorded at a frequency of 1 Hz. For this study, only time and altitude data are used specifically, the altitude vector to which the smoothing process is applied.

2.3 Data Processing/noise Filtering

Among the various techniques available for signal smoothing, this research focuses on smoothing through the probability density distribution using a Gaussian kernel, and evaluates its effectiveness by varying the parameter known as bandwidth. [10].

Probability Distribution Smoothing: Kernel Construction
The probability distribution of a vector of random variables representing the measured altitude values is determined through a function known as a density function [11], represented by the Eq. (1):

$$P\left(a \leq X \leq b\right) = \int_a^b f(x)dx \tag{1}$$

where X is a continuous random variable composed of the elements $x_1, x_2, x_3, \cdots, x_n$.

A kernel is a real-valued function K, where its elements belong to the set of real numbers. In this context, the values considered for K are non-negative.

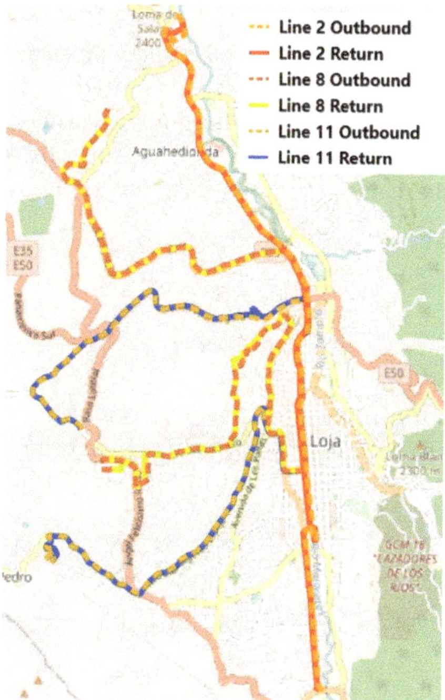

Fig. 2. Outbound - Return urban busses trajectories: 2, 8 y 11.

Kernel Parameterization

The kernel function K of the random variable X is defined in the Eq. (2) [11]:

$$K_h(X) = \frac{1}{h} K\left(\frac{X}{h}\right) \\ x \in [-h, h] \tag{2}$$

Here, h esents the bandwidth, which, in simple terms, is the numerical parameter that determines the boundaries of the kernel function K.

Kernel Translation

Since the values of the random variable X are grouped into a numerical vector of non-negative elements on the real line, and because the size of the kernel function explicitly depends on the bandwidth, it is necessary to translate the kernel across the elements of the random variable X. The translation of the kernel is defined by the Eq. (3):

$$K_h(X) = \frac{1}{h} K\left(\frac{x - x_i}{h}\right) \\ x \in [x_i - h, x_i + h] \\ x_i \in \mathbf{R} \tag{3}$$

Kernel Models: Gaussian Kernel
Probability density functions constructed using the kernel model can be developed from various mathematical functions, as cited in the literature [11]. For smoothing GPS-measured data, the Epanechnikov kernel model may be used, as cited in [10]. In this study, however, we use the Gaussian kernel model K_G [12], named for its use of the Gaussian probability function, given in the Eq. (4):

$$K_G(x) = \frac{1}{\sqrt{2\pi}} \exp^{-\frac{x^2}{2}} \qquad x \in (-\infty, \infty) \tag{4}$$

Construction of Probability Density Functions
To obtain the probability density function $\widehat{f}(x)$, in other words, the smoothed function, we apply the formula specified in (5) [11,13,14]. :

$$\begin{aligned} \widehat{f}(X) &= \frac{1}{n} \sum_{i=1}^{n} K_h(x) \\ &= \frac{1}{n} \sum_{i=1}^{n} \frac{1}{h} K\left(\frac{x - x_i}{h}\right) \\ x \in [x_i - h, x_i + h] &\quad \text{in each kernel i} \\ \widehat{f}(x) &: [x_1 - h, x_i + h] \\ x_1 \leqslant x_2 &\leqslant x_3 \leqslant \cdots \leqslant x_n \end{aligned} \tag{5}$$

where n is the number of elements in the random variable X.

Optimal Bandwidth Calculation
As previously mentioned, the bandwidth is a critical factor when smoothing data using a probability density function, as it directly influences the amount of information lost during the analysis process.

There is no formal method to determine this value; in most cases, it depends on the analyst's experience and the specific goals of the study.

The effect of data smoothing can be categorized in two ways: aggressive smoothing and moderate smoothing [15,16]. Figure 3 illustrates the smoothing effect on the measured data values (in blue) using the Gaussian kernel model. The curve in orange, corresponding to a small bandwidth, closely follows the original signal.

The yellow curve shows how certain peaks in the original data vector are smoothed out. In this case, the smoothing is considered moderate, with only a slight loss of information. In the case of smoothing 3 (purple curve), the effect of aggressive smoothing is evident, with a significant loss of information [7].

Bandwidth Estimation Based on Empirical Rules
The bandwidth value is independent of the selected smoothing method and can be used with different kernel models or with other smoothing techniques such as moving mean.

To compare the efficiency of different bandwidths, we use the estimator proposed in the Eq. (6), [17,18], where the bandwidth is calculated based on the

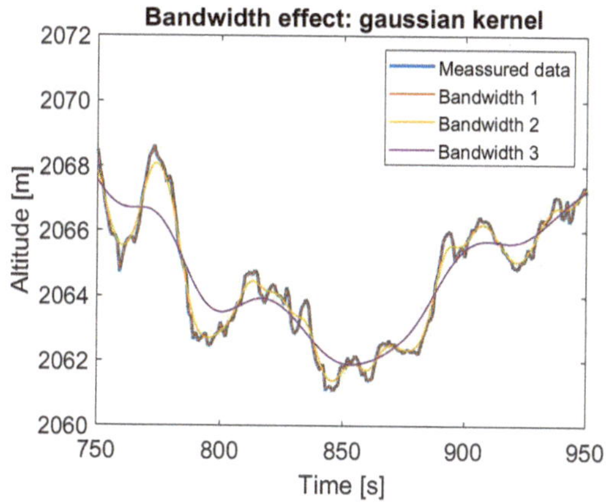

Fig. 3. Altitude profile smoothing.

variance σ of the random variable X and the number n of elements. This bandwidth h_1 is denoted as $Bw1$:

$$h_1 = \left(\frac{4\sigma^5}{3n}\right)^{1/5} \tag{6}$$

The second bandwidth value is derived from B.W. Silverman's rule [9,18], [19], defined in (7). To calculate this value, the variance and the minimum value of the interquartile range (IQR). This bandwidth h_2 is denoted as $Bw2$:

$$h_2 = 0.9 \min\left(\sigma, \frac{IQR}{1.34}\right) n^{-1/5} \tag{7}$$

Efficiency of the Estimator $\widehat{f}(x)$ with Respect to $f(x)$
The efficiency of the kernel model's probability density function, which determines the amount of information lost during the smoothing process, depends exclusively on the bandwidth value. It is calculated using the Mean Squared Error (MSE) between the values of the measured data and the smoothed data, using the following equation (8) [20]:

$$MSE = \frac{1}{n}\sum_{i=1}^{n}\left(\widehat{f}(x) - f(x)\right)^2 \tag{8}$$

3 Results

Figure 4 shows the measured altitude profiles for the outbound and return trips of urban bus lines 2, 8, and 11, where the data processing will be applied:

To smooth the altitude measurements, data analysis was performed using MATLAB programming software. First, null values `NaN` , in the measured data

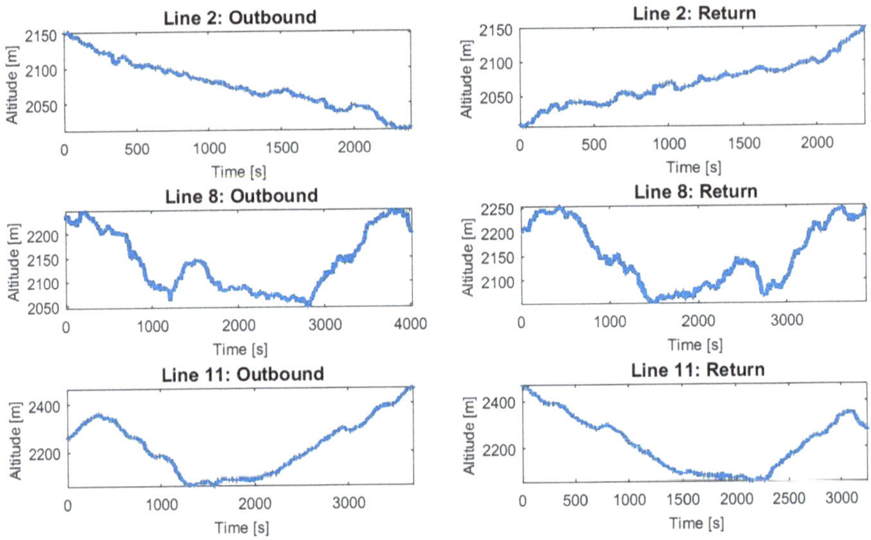

Fig. 4. Altitude profiles of autobuses lines: 2, 8 y 11.

vector were removed and replaced with the mean of the measured values. To apply the Gaussian kernel model to the data vector, the command \`smoothdata` was implemented with the gaussian configuration `gaussian` .

```
ysmooth = smoothdata(alt1, 'gaussian', bw);
```

Here, the parameter `Bw` represents the bandwidth value. The number of data points to analyze varies depending on the route; therefore, the mean, variance, and interquartile range values also differ between trips, making the bandwidth value unique for each case. The calculated bandwidth values are specified in Table 1.

Table 1. Bandwidth values

Trajectories	Bandwidth 1 Bw1	Bandwidth 2 Bw2
L2 Outbound	7.9228	6.7269
L2 Return	7.0819	6.0130
L8 Outbound	12.8082	10.8749
L8 Return	13.2585	11.2572
L11 Outbound	23.6573	20.0863
L11 Return	24.7977	21.0547

For visualization purposes, a representative section of one hundred seconds was selected from each route to observe the effect of the bandwidth on smoothing.

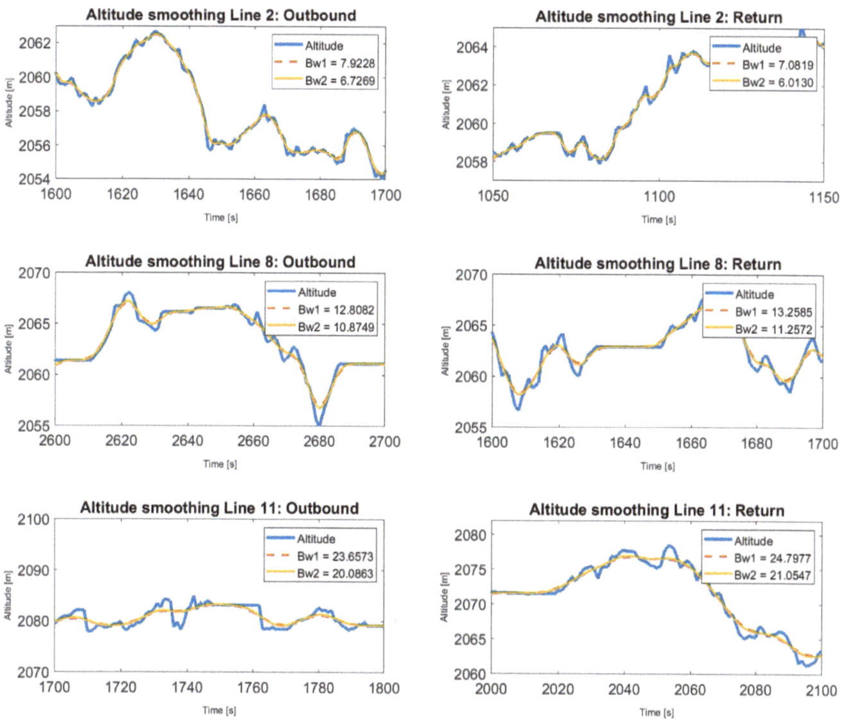

Fig. 5. Altitude smoothing of outbound-return trajectories.

In Fig. 5, the blue line represents the altitude measured by the Torque Pro application, the orange dashed line shows the smoothed data using the $Bw1$ bandwidth from Eq. (5) and the yellow line shows the smoothed data using the $Ab2$ bandwidth from Eq. (6). At a glance, it can be observed that both the orange and yellow lines eliminate certain spike-shaped noise from the route, producing a more uniform data vector. In neither case is there evidence of under-smoothing or over-smoothing. Although the yellow and orange lines follow nearly the same path, there are slight variations between them.

To statistically assess the efficiency of the method, the MSE was calculated for the smoothed values corresponding to each bandwidth, compared to the original data vector. Figure 6 shows the behavior of the MSE in the data vector: the blue line represents the MSE values calculated with bandwidth $Bw1$ (Eq. (5)) while the orange line represents the MSE values calculated with bandwidth $Ab2$ (Eq. (6)). Table 2 shows the MSE maximun values.

The MSE calculated with bandwidth 1 consistently shows higher values than the MSE calculated with bandwidth 2 across all routes. Thus, Silverman's rule from Eq. (6) minimizes the MSE value.

Table 2. MSE maximun values

Trajectory	Bandwidth 1	Bandwidth 2
	MSE	MSE
L2 Outbound	0.0019	0.0063
L2 Return	0.1685	0.1483
L8 Outbound	0.0194	0.0514
L8 Return	0.0019	0.0055
L11 Outbound	0.0116	0.0073
L11 Return	0.0157	0.0480

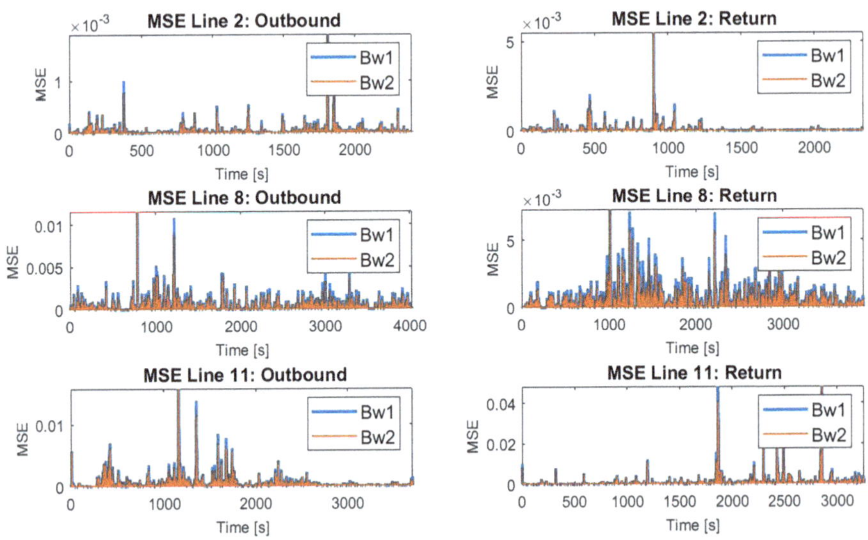

Fig. 6. Autobuses lines MSE: 2, 8 and 11.

4 Discussion

After applying the probability density smoothing using the Gaussian kernel model with two bandwidth values on the outbound and return trips of bus lines 2, 8, and 11, two smoothed data vectors were obtained, one for each bandwidth.

Among the various smoothing techniques, such as moving average or polynomial fitting, the probability density method was chosen due to its effectiveness compared to the others.

The probability density method allows for the use of different mathematical functions for the kernel function, such as the rectangular kernel, triangular kernel, or Epanechnikov kernel [11]. However, the Gaussian kernel function proved to be more efficient, showing a lower Mean Squared Error MSE.

Bandwidth selection is often done empirically, manually adjusting the data without a supporting calculation. Determining a bandwidth based on dispersion measures of the data vector allows for optimization of the selection, thereby reducing information loss.

Figure 5 graphically shows the smoothed data, where the first noticeable effect is the elimination of peak-shaped noise from the original data vector.

Each of the two smoothed data vectors fulfills its function, and at first glance, there is no significant difference between them.

Table 1 shows that the calculated bandwidth values using Equations (5) and (6), are not identical, yet they both result in similar smoothing.

Figure 6 illustrates the MSE behavior, where the general trend shows lower MSE values for bandwidth 2 compared to bandwidth 1. This demonstrates that the optimal bandwidth value is that defined by Silverman's rule $Bw2$ in Equation (6).

Table 2 shows the maximum MSE values per run, where the MSE corresponding to bandwidth $Ab2$ is lower compared to the bandwidth values calculated in $Ab1$ across the six case studies addressed, confirming the previously stated findings.

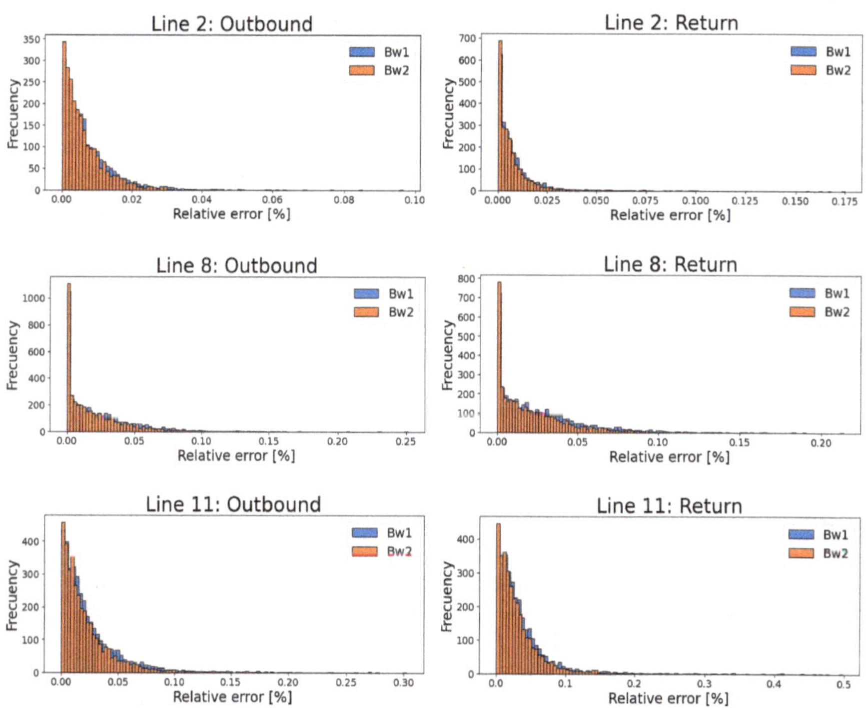

Fig. 7. Outbound-return trajectories relative errors.

To corroborate this, Fig. 7 shows the relative errors of the smoothed vectors using the respective bandwidth values in relation to the original data vector. In both cases, it can be observed that the highest frequency is concentrated around the value 0, and that the bandwidth value $Bw2$ is lower than the bandwidth $Bw1$.

By applying this smoothing technique with the chosen bandwidth value, noise is effectively filtered out, resulting in a more homogeneous data vector.

5 Conclusions

The quality of monitored variables under real driving conditions in motor vehicles is subject to the precision of monitoring equipment and external factors, including environmental conditions, road type, and driving style. Accurately defining the altitude profile of bus routes is crucial prior to conducting analyses related to energy consumption, fuel consumption, emission factor determination, route selection, etc., for a given location or road segment.

The selection of the smoothing method and the optimal bandwidth value positively influences the methodology for determining the energy consumption of the bus routes in the city of Loja, by removing the implicit noise in the measurements, homogenizing the data, and reducing information loss.

Based on the results obtained, this study concludes that applying the Gaussian kernel probability density function with the bandwidth defined by Silverman's rule is an appropriate method for smoothing altitude profile data of bus routes. This is supported by the negligible difference in the MSE when compared to the original data vector. The proposed methodology gains further reliability as it was evaluated across three different bus routes in the city, which exhibit notable variations in altitude and slope profiles for both outbound and return trips.

This smoothing procedure must be applied to each of the 40 outbound and 40 return trips for each of the three routes, as a preliminary step prior to the subsequent energy analysis. This means that the optimal bandwidth value will vary for each trip and, in every case, will be optimized based on the mean squared error MSE.

References

1. Martino, L., Read, J.: A joint introduction to Gaussian processes and relevance vector machines with connections to Kalman filtering and other kernel smoothers. Inf. Fusion **74**, 17–38 (2021)
2. Keusch, F., Struminskaya, B., Antoun, C., Couper, M.P., Kreuter, F.: Willingness to participate in passive mobile data collection. Public Opin. Q. **83**(S1), 210–235 (2019)
3. Tsouros, I., et al.: From raw data to informed decisions: the development of an online data repository and visualization dashboard for transportation data. Procedia Comput. Sci. **231**, 161–167 (2024)

4. Bhowmick, D., et al.: Collecting population-representative bike-riding GPS data to understand bike-riding activity and patterns using smartphones and Bluetooth beacons. Travel Behav. Soc. **38**, 100919 (2025)
5. Castillo-Calderón, J., Jaura, R.C., Sinche, D.D., Panchana, B.:. Estimation of traction energy consumption of urban service buses in an intermediate Andean City. **2**(5) (2022)
6. Kaya, S., Kilic, N., Kocak, T., Gungor, C.: A battery-friendly data acquisition model for vehicular speed estimation. Comput. Electr. Eng. **50**, 79–90 (2016)
7. Korpilo, I., Virtanen, T., Lehvävirta, S.: Smartphone GPS tracking-inexpensive and efficient data collection on recreational movement. Landsc. Urban Plan. **157**, 608–617 (2017)
8. Do, T.M.T., Dousse, O., Miettinen, M., Gatica-Perez, D.: A probabilistic kernel method for human mobility prediction with smartphones. Pervasive Mob. Comput. **20**, 13–28 (2015)
9. Jun, J., Guensler, R., Ogle, J.: Smoothing methods designed to minimize the impact of GPS random error on travel distance, speed, and acceleration profile estimates. Transp. Res. Rec. (2005)
10. Brady, J., O'Mahony, M.: Development of a driving cycle to evaluate the energy economy of electric vehicles in urban areas. Appl. Energy **177**, 165–178 (2016)
11. Ojeda, L.R.: Construcción de Kernels y Funciones de densidad de probabilidad. **2**(5) (2014)
12. Rasmussen, C.E., Williams, C.K.I.: Gaussian processes for machine learning. **2**(5) (2006)
13. Hellinga, B., Chan, T.N.: Issues related to quantifying the environmental impacts of transportation strategies using gps data. **2**(5) (2022)
14. Nadaraya, E.A.: On estimating regression. **2**(5) (1964)
15. Fried, R., Einbeck, J., Gather, U.: Weighted repeated median smoothing and filtering. J. Am. Stat. Assoc. **102**(480), 1300–1308 (2007)
16. Campulova, M.: Comparison of methods for smoothing environmental data with an application to particulate matter PM_{10}. **2**(5) (2018)
17. Silverman, B.W.: Estimation for statistics and data analysis. **2**(5), (1998)
18. Flores-Garnica, J.G., Macías-Muro, A.: Selección de ancho de banda para la estimación de densidad kernel de incendios forestales. **2**(5) (2018)
19. Silverman, B.W.: Density estimation for statictic analysis. **2**(5) (1996)
20. Ghimire, S., Deo, R.C., Pourmousavi, S.A., Casillas-Pérez, D., Salcedo-Sanz, S.: Point-based and probabilistic electricity demand prediction with a Neural Facebook Prophet and Kernel Density Estimation model. Eng. Appl. Artif. Intell. **135**, 108702 (2024)

Predictive Current Control with Delay Compensation for 3×3 Matrix Converters

Alex Moreta-Masaquiza[1], Julio Pérez Naranjo[2(✉)], Kevin Reyna-Guarderas[3], and Marcelo Pozo-Palma[3]

[1] Departamento de Automatización y Control Industrial, Escuela Politécnica Nacional, Quito, Ecuador
alex.moreta@epn.edu.ec

[2] Departamento de Energía Eléctrica, Escuela Politécnica Nacional, Quito, Ecuador
julio.perez@epn.edu.ec

[3] Departamento de Automatización y Control Industrial, Escuela Politécnica Nacional, Quito, Ecuador
{kevin.reyna,marcelo.pozo}@epn.edu.ec

Abstract. Matrix converters (MC) enable efficient AC-AC conversion, offering a compact, transformerless solution for power conversion. Unlike conventional converters, MCs do not require a direct current (DC) bus, which reduces losses associated with energy storage in capacitors and improves the system's energy efficiency.

This study proposes a predictive current control strategy with delay compensation for a 3 × 3 MC, aiming to improve tracking accuracy and reduce current ripple. The developed model incorporates first-order (ODE1) and fourth-order (ODE4) numerical methods to optimize the system response. Simulations were carried out under four different scenarios to assess the effectiveness of the strategy: predictive control with and without delay compensation using ODE1, and predictive control with and without delay compensation using ODE4.

The results showed that the technique that best follows the current reference is ODE4 with delay compensation. These findings validate the proposed approach and highlight its potential to enhance the performance of matrix converters in practical applications. By eliminating the DC bus, the system becomes more energy-efficient while also benefiting from reductions in both size and cost, positioning it as a suitable choice for energy-sensitive applications.

Keywords: Current control · Delay compensation · 3x3 Matrix converter · Predictive control · ODE1 · ODE4

1 Introduction

The 3 × 3 AC/AC matrix converter (MC) is a single-stage converter that enables direct energy transfer between multiphase sources and loads. It offers advantages

J. L. Ballesteros et al. (Eds.): CITIS 2025, LNNS 1780, pp. 418–428, 2027.
https://doi.org/10.1007/978-3-032-14287-0_41

such as frequency-independent operation, near-unity power factor, and bidirectional power flow, making it suitable for applications like motor drives, precision machinery, electric vehicles, and renewable energy systems [5].

This work presents the implementation of a Model Predictive Control (MPC) strategy using ODE1 and ODE4 integration methods applied to the MC. Unlike studies focused on simplified converter models or voltage-source inverters, this paper includes detailed time-domain simulations with realistic commutation and delay compensation. The comparative analysis of ODE1 and ODE4 provides practical insights into the trade-offs between current quality, switching performance, and computational complexity in MPC schemes. [2,4,5].

2 Modeling of the 3 × 3 Matrix Converter

The filter shown in Fig. 1 serves two main purposes: it suppresses over-voltages caused by the source's short-circuit impedance during fast switching of the phase currents (i_{eA}, i_{eB}, i_{eC}), and it reduces the presence of high-frequency harmonics in the input current waveform (i_A, i_B, i_C). Each switch is denoted by S_{xy}, where $x \in \{A, B, C\}$ and $y \in \{a, b, c\}$. The control signal determines the conduction state: [8].

- $S_{xy} = 1$: switch is closed (conducting)
- $S_{xy} = 0$: switch is open

To avoid damaging over-voltages due to inductive loads, the load current must remain continuous. Additionally, short circuits at the input must be avoided. Therefore, the switching signals must satisfy the constraint: [8].

$$S_{Ay} + S_{By} + S_{Cy} = 1 \quad \forall y \in \{a, b, c\} \tag{1}$$

The input voltage to the matrix converter $(\mathbf{V_i})$ and output load voltage $(\mathbf{V_o})$ are expressed as the following vectors: [8].

$$\mathbf{V_i} = \begin{bmatrix} v_{eA}(t) \\ v_{eB}(t) \\ v_{eC}(t) \end{bmatrix}, \quad \mathbf{V_o} = \begin{bmatrix} v_a(t) \\ v_b(t) \\ v_c(t) \end{bmatrix}, \tag{2}$$

The connection between input and output voltages can be represented as: [8].

$$\begin{bmatrix} v_a(t) \\ v_b(t) \\ v_c(t) \end{bmatrix} = \begin{bmatrix} S_{Aa} & S_{Ba} & S_{Ca} \\ S_{Ab} & S_{Bb} & S_{Cb} \\ S_{Ac} & S_{Bc} & S_{Cc} \end{bmatrix} \cdot \begin{bmatrix} v_{eA}(t) \\ v_{eB}(t) \\ v_{eC}(t) \end{bmatrix} \tag{3}$$

The matrices that describe the behavior of the input current to the matrix converter $(\mathbf{i_i})$ and output load current $(\mathbf{i_o})$ through the switches (SW) are defined as: [8].

$$\mathbf{i_i} = \begin{bmatrix} i_{eA}(t) \\ i_{eB}(t) \\ i_{eC}(t) \end{bmatrix}, \quad \mathbf{i_o} = \begin{bmatrix} i_a(t) \\ i_b(t) \\ i_c(t) \end{bmatrix} \tag{4}$$

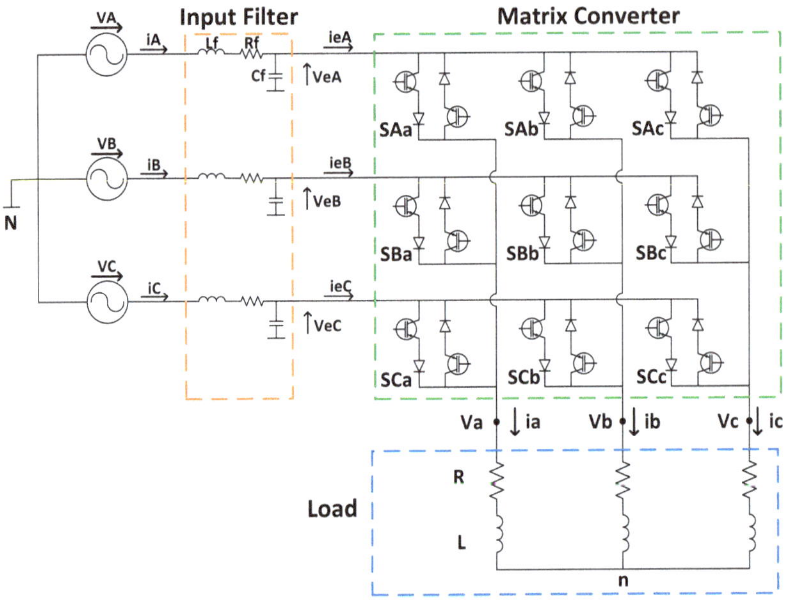

Fig. 1. 3 × 3 Matrix Converter Circuit.

3 Model Predictive Current Control

Model Predictive Control (MPC) is a modern control technique that employs dynamic models to anticipate system behavior and compute the best control actions over a defined time horizon. Its goal is to minimize tracking errors and ensure compliance with operational constraints [8].

3.1 Euler Method (ODE1) and Delay Compensation

The Euler technique serves as a numerical method to approximate the solutions of first-order ordinary differential equations (ODEs). It operates under the assumption that the function's slope at any given point can be estimated through the tangent line at that point [6]. Although it is simple and easy to implement, the Euler method is a first-order method and may introduce significant errors if the time step is too large or if the differential equation is highly complex. In predictive control, Eqs. (5) and (6) are employed to establish the relationship between the instantaneous input and output currents and voltages. Furthermore, 27 distinct switching states are taken into account when predicting the load current. For the implementation of predictive control, a formula is required to predict the load current at the next sampling point, considering all 27 possible switching configurations of the converter. A general equation representing the load, assuming an inductive nature (Fig. 1), is given by: [1,8].

$$L \cdot \frac{d\mathbf{i_o}(t)}{dt} = \mathbf{V_o}(t) + R \cdot \mathbf{i_o}(t) \tag{5}$$

where R and L represent the resistance and inductance of the load, respectively, T_s is the sampling time, and using the Euler approximation for the derivative of the output current, we obtain: [8].

$$\frac{d\mathbf{i_o}(t)}{dt} \approx \frac{\mathbf{i_o}(k+1) - \mathbf{i_o}(k)}{T_s} \tag{6}$$

Substituting Eqs. (5) in (6) leads to the predictive equation (7):

$$\mathbf{i_o}(k+1) = \left(1 - \frac{RT_s}{L}\right) \cdot \mathbf{i_o}(k) + \frac{T_s}{L} \cdot \mathbf{V_o}(k) \tag{7}$$

In this context, $\mathbf{i_o}(k+1)$ denotes the predicted current for the sampling period $k+1$, corresponding to each of the 27 switching state voltages $\mathbf{v_o}(k)$. In predictive control systems, delay compensation is used to mitigate the effects of delays inherent in the models and the communication between the controller and the system. This is achieved by adjusting the system's future predictions to account for these delays, improving the tracking of the current reference [1,8]. The general equation to incorporate delay compensation into predictive control is: [8].

$$\mathbf{i_o}(k+2) = \left(1 - \frac{RT_s}{L}\right) \cdot \mathbf{i_o}(k+1) + \frac{T_s}{L} \cdot \mathbf{V_o}(k+1) \tag{8}$$

3.2 Fourth-Order Runge-Kutta Method (ODE4) and Delay Compensation

The ODE 4 is employed to solve the differential equation that characterizes the current dynamics in an RL circuit. The current $\mathbf{i_o}$ is updated at each iteration using the values k_1, k_2, k_3, and k_4, which are combined to estimate the current at the next time step. The update rule for $\mathbf{i_o}$ is given by: [7].

$$\mathbf{i_o}(k+1) = \mathbf{i_o}(k) + \frac{T_s}{6}(k_1 + 2k_2 + 2k_3 + k_4) \tag{9}$$

where T_s is the time step and the slope coefficients are defined as: [7].

$$k_1 = \frac{T_s}{L}(\mathbf{V_o} - R \cdot \mathbf{i_o}) \tag{10}$$

$$k_2 = \frac{T_s}{L}\left[\mathbf{V_o} - R \cdot \left(\mathbf{i_o} + \frac{T_s}{2}k_1\right)\right] \tag{11}$$

$$k_3 = \frac{T_s}{L}\left[\mathbf{V_o} - R \cdot \left(\mathbf{i_o} + \frac{T_s}{2}k_2\right)\right] \tag{12}$$

$$k_4 = \frac{T_s}{L}\left[\mathbf{V_o} - R \cdot (\mathbf{i_o} + T_s k_3)\right] \tag{13}$$

Thus, Eqs. (9) ensures that the change in current $\mathbf{i_o}$ is accurately estimated by considering variations at different points within the time interval T_s. This improves precision in simulating dynamic systems. The delay compensation is introduced by adjusting the predictions of current based on the delayed state. The modified update rule is as follows:

$$k_{11} = \frac{T_s}{L}\left(\mathbf{V_o}(k+1) - R \cdot \mathbf{i_o}(k+1)\right) \tag{14}$$

$$k_{21} = \frac{T_s}{L}\left(\mathbf{V_o}(k+1) - R \cdot \left(\mathbf{i_o}(k+1) + \frac{T_s}{2} \cdot k_{11}\right)\right) \tag{15}$$

$$k_{31} = \frac{T_s}{L}\left(\mathbf{V_o}(k+1) - R \cdot \left(\mathbf{i_o}(k+1) + \frac{T_s}{2} \cdot k_{21}\right)\right) \tag{16}$$

$$k_{41} = \frac{T_s}{L}\left(\mathbf{V_o}(k+1) - R \cdot \left(\mathbf{i_o}(k+1) + T_s \cdot k_{31}\right)\right) \tag{17}$$

This adjustment accounts for the impact of delay on current evolution, enhancing the simulation accuracy and allowing for more precise tracking of the reference current. As a result, the predicted current with delay compensation more accurately reflects the future behavior of the system under real operating conditions, as shown below.

$$\mathbf{i_o}(k+2) = \mathbf{i_o}(k+1) + \frac{T_s}{6}(k_{11} + 2k_{21} + 2k_{31} + k_{41}) \tag{18}$$

4 Cost Function and Predictive Current Control Strategy

The current predictive control structure is depicted in Fig. 2, where an RL load configuration is assumed. This model is used to generate future predictions of the output current. Using these predictions, a cost function is evaluated to select the optimal switching state, which is subsequently applied to the converter [8]. The cost function G serves as the evaluation criterion for the control algorithm, helping to identify the optimal switching state to be applied during the next sampling interval. The mathematical expression for the cost function is as follows: [3,8].

$$G = \left|i^*_{o\alpha}(k+1) - i_{po\alpha}(k+1)\right| + \left|i^*_{o\beta}(k+1) - i_{po\beta}(k+1)\right| \tag{19}$$

where:

- $i^*_{o\alpha}$: Real component of the reference output current vector.
- $i^*_{o\beta}$: Imaginary component of the reference output current vector.
- $i_{po\alpha}$: Real component of the predicted output current vector.
- $i_{po\beta}$: Imaginary component of the predicted output current vector.

This cost function quantifies the absolute deviation between the reference and predicted currents in both real and imaginary components. The objective is to minimize these deviations. A lower value of G indicates that the predicted current is closer to the reference, which is essential for precise tracking

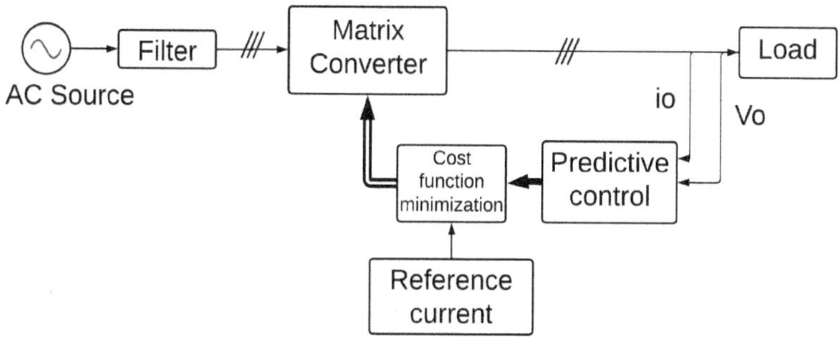

Fig. 2. Diagram of the model predictive current control system

of the output current [3,8]. The following flowcharts (Fig. 3a, Fig. 3b) show the predictive control algorithm, which evaluates all switching states to select the optimal one, using ODE1 or ODE4 with or without delay compensation.

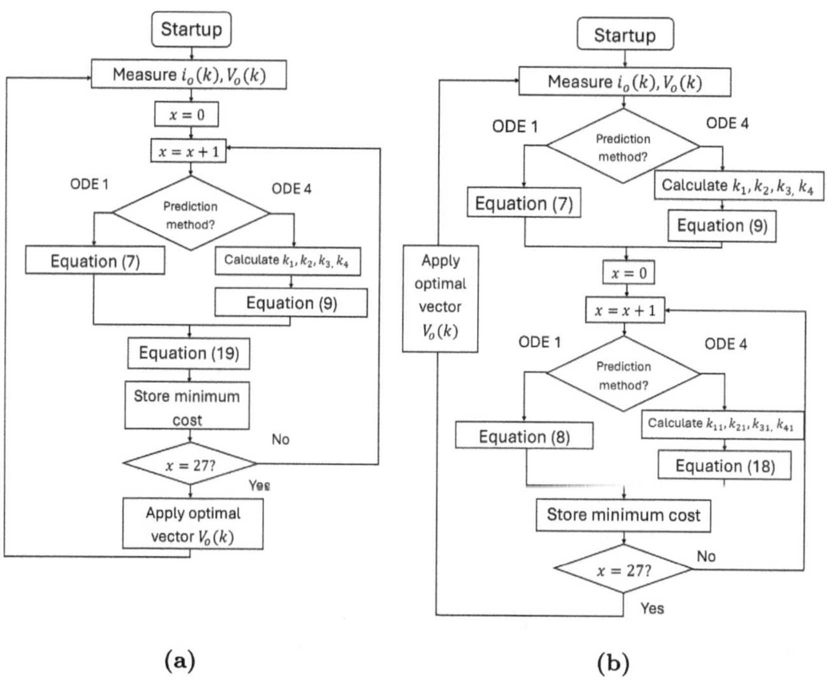

Fig. 3. (a) Flowchart for ODE 1 and ODE 4 without delay compensation and (b) Flowchart for ODE 1 and ODE 4 with delay compensation.

5 Test and Results

The simulations were carried out in MATLAB(R2024b)/Simulink, using a 3×3 matrix converter model with space vector modulation. The predictive control algorithm was implemented with discrete blocks to simulate real-time operation. Table 1 presents a summary of the parameters employed in the system simulation.

Table 1. Simulation Parameters

Parameter	Name	Value	Unit
f	Input grid frequency	50	Hz
V_s	Input voltage	180	V
L_f	Input filter inductance	400	μH
R_f	Input filter resistance	0.5	Ω
C_f	Input filter capacitance	21	nF
L	Load inductance	0.10	H
R	Load resistance	10	Ω
T_s	Sampling time	0.1	μs
I_{ref}	Reference current	0.5	A

The real model was evaluated under four simulation scenarios over a period of 0.025 [s], as presented in Table 2.

Table 2. Proposed Case Studies

No.	The predictive control method employed
1	ODE 1 (Euler)
2	ODE 1 (Euler) with delay compensation
3	ODE 4 (Runge–Kutta)
4	ODE 4 (Runge–Kutta) with delay compensation

Results of case study 1, method using euler (ODE1) without delay compensation: The purpose of applying this method is to evaluate the system's performance regarding current tracking and waveform quality, using a basic predictive approach without delay compensation. The next figure illustrates the output current behavior achieved with this method, compared to the reference current. As seen in Fig. 4a, the maximum positive peak of the output current reached 0.72 [A], while the maximum negative peak was −0.78 [A]. These peaks demonstrate excessive fluctuations, especially when compared to the reference current value of 0.5 [A] at the maximum point of the waveform.

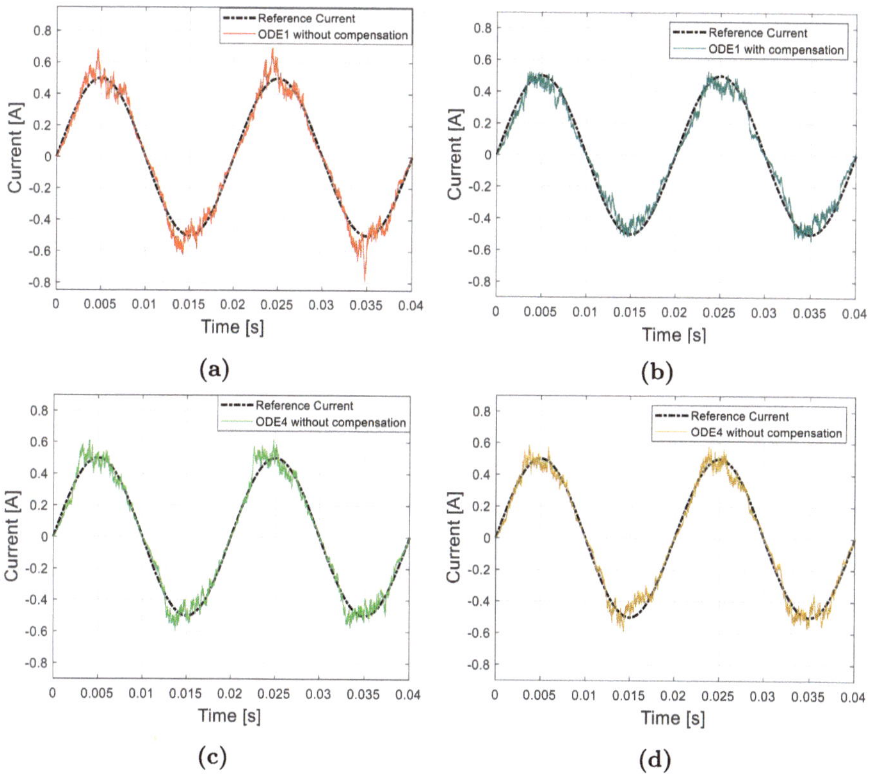

Fig. 4. Output and reference currents of phase A: (a) ODE1 without delay compensation, (b) ODE1 with delay compensation, (c) ODE4 without delay compensation, (d) ODE4 with delay compensation.

Results of Case Study 2, using Euler (ODE1) with delay compensation: Fig. 4b shows the output current behavior, where the maximum positive and negative peaks were reduced to 0.56 [A] and −0.58 [A], respectively. However, the phase shift between the reference and output current increased.

Results of case study 3, method using runge-kutta (ODE4) without delay compensation: The Fig. 4c presents the system's response obtained using the ODE4 without applying delay compensation, where the maximum positive and negative peaks of the output current were reduced to 0.61 [A] and −0.63 [A], respectively. These values are closer to the ideal reference waveform, which reaches ±0.5 [A] at its maximum amplitude, in addition to reducing the phase shift between the reference current and the output current.

Results of case study 4, method using runge-kutta (ODE4) with delay compensation: As shown in Fig. 4d, the maximum positive and negative peaks of the output current are set at 0.54 [A] and −0.54 [A], respectively. These values are close to the ideal reference waveform compared to the previous three case studies, and the phase shift is 4.74×10^{-4} degrees, indicating that this method

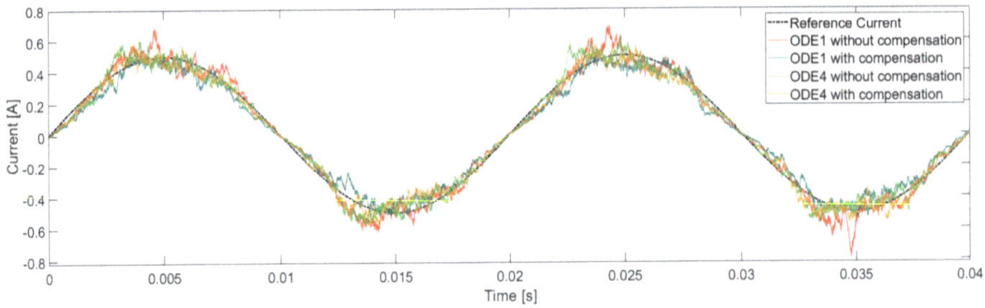

Fig. 5. Reference waveforms and output currents corresponding to the four solution methods.

Table 3. Peak output current values for different numerical methods.

No.	Method	Positive Peak	Negative Peak	Phase Angle (Degrees)
1	ODE 1 (Euler)	0.72	−0.78	1.54×10^{-3}
2	ODE 1 (Euler) with delay compensation	0.56	−0.58	1.97×10^{-3}
3	ODE 4 (Runge-Kutta)	0.61	−0.63	8.54×10^{-4}
4	ODE 4 (Runge-Kutta) with delay compensation	0.54	−0.54	4.74×10^{-4}

achieves better performance. In Fig. 5 the four current waveforms corresponding to the four solution methods are presented, compared to the reference current. The Table 3 provides a consolidated overview of the outcomes obtained from the case studies analyzed.

6 Discussion and Conclusions

The analysis of different numerical integration methods used in current control for 3×3 matrix AC/AC converters reveals important insights into each technique's effectiveness, particularly regarding waveform behavior, phase angle, and accuracy in tracking the reference current, as shown in Table 3 and Fig. 5.

In the case of ODE1 without delay compensation, a significantly distorted output current waveform was observed, with peaks reaching 0.72 [A] and −0.78 [A], considerably deviating from the reference ±0.5 [A]. The pronounced oscillations indicate poor tracking performance and ineffective control. This underscores the importance of considering delay effects, as the system without compensation fails to mitigate undesirable fluctuations. However, when ODE1 with delay compensation was applied, system performance improved significantly: current peaks reduced to 0.56 [A] and −0.58 [A], and the response was more stable.

This confirms delay compensation as crucial for improving tracking and reducing oscillations, thus enabling more precise and efficient control.

On the other hand, the ODE4 method without delay compensation, known for higher numerical accuracy, significantly improved waveform quality. Output peaks dropped to 0.61 [A] and −0.63 [A], producing a smoother, more stable waveform with minimal distortion compared to ODE1. Although no delay compensation was used here, tracking performance was notably better, demonstrating ODE4's superiority in accuracy. Combining ODE4 with delay compensation yielded the best results: peaks at 0.54 [A] and −0.54 [A], excellent tracking, and the most stable waveform, minimizing overshoots and residual fluctuations. This shows that combining a high-order numerical method with delay compensation provides the most robust, precise, stable, and efficient current control solution.

Higher-order methods like ODE4 significantly enhance current control in AC/AC matrix converters even without delay compensation, as presented in Table 3 and Fig. 5. The improved waveform quality and fluctuation reduction show that ODE4 outperforms ODE1. Applying delay compensation to both methods further enhances tracking by reducing fluctuations, phase angle error, and increasing precision. This is critical for optimizing system performance, especially in high-demand applications where stability and efficiency are key factors.

The 3×3 MC stands out for structural efficiency by not requiring a DC link, eliminating intermediate conversion stages. This reduces energy losses related to rectification and storage and decreases system volume, weight, and complexity. Such configuration suits applications demanding dynamic response and energy efficiency. For current control in $3{\times}3$ matrix AC/AC converters, combining advanced numerical methods like ODE4 with delay compensation offers a robust, efficient solution that minimizes fluctuations and ensures precise reference tracking, significantly improving system performance.

References

1. Abdullah, A.T., Idrus, S.M., Ayob, S.M.: Model predictive control using Euler method for switched-battery boost-multilevel inverter. In: 2023 International Journal of Power Electronics and Drive Systems, pp. 1497–1508 (2023). https://doi.org/10.11591/ijpeds.v14.i3
2. Apap, M., Clare, J., Wheeler, P., Bradley, K.: Analysis and comparison of ac-ac matrix converter control strategies. In: 2003 IEEE 34th Annual Conference on Power Electronics Specialist, PESC 2003, vol. 3, pp. 1287–1292. IEEE (2003). https://doi.org/10.1109/PESC.2003.1216774
3. Arahal, M.R., Kowal, A., Barrerro, F., Castilla, M.D.M.: Optimizaci6n de funciones de coste para control predictivo de máquinas de inducci6n multifásicas. Rev. Iberoamericana Autom. Inform. Industr. **16**(1), 48–55 (2019). https://doi.org/10.4995/riai.2018.9771
4. Friedli, T., Kolar, J.W., Rodriguez, J., Wheeler, P.W.: Comparative evaluation of three-phase ac-ac matrix converter and voltage DC-link back-to-back converter systems. IEEE Trans. Industr. Electron. **59**(12), 4487–4510 (2011). https://doi.org/10.1109/TIE.2011.2179278

5. von Jouanne, A., Agamloh, E., Yokochi, A.: A review of matrix converters in motor drive applications. Energies **18**(1), 164 (2025). https://doi.org/10.3390/en18010164
6. Karamanakos, P., Liegmann, E., Geyer, T., Kennel, R.: Model predictive control of power electronic systems: methods, results, and challenges. IEEE Open J. Ind. Appl. **1**, 95–114 (2020). https://doi.org/10.1109/OJIA.2020.3020184
7. Malarvizhi, M., Karunanithi, S., Gajalakshmi, N.: Numerical analysis using RK-4 in transient analysis of RLC circuit. Adv. Math. Sci. J **9**, 6115–6124 (2020). https://doi.org/10.37418/AMSJ.9.8.79
8. Rodriguez, J., Cortes, P.: Predictive Control of Power Converters and Electrical Drives. Wiley (2012). https://doi.org/10.1002/9781119941446

Prospective Energy Substitution Modeling: Assessing the Technical, Environmental, and Financial Feasibility of Geothermal Power in Ecuador

Cristian Cuji(✉)

Universidad Politécnica Salesiana, Quito, Ecuador
ccuji@ups.edu.ec

Abstract. This study aims to assess the technical, environmental, and economic feasibility of progressively replacing fossil-fuel-based thermal power generation with geothermal energy in Ecuador. Based on an analysis of national installed capacity and available geothermal potential, energy transition scenarios are modeled for the 2025–2030 period. The analysis integrates projections of the Levelized Cost of Electricity (LCOE), CO_2 emission reductions, and financial returns. An ARIMA (1,1,1) statistical model was applied to forecast the evolution of geothermal LCOE through 2027, using a 95% confidence interval. Results indicate a sustained downward trend, projecting an LCOE of 0.045 USD/kWh by 2027, positioning geothermal energy as a competitive and reliable source within the national energy matrix [1].

At the national level, over 1,400 MW of geothermal potential were identified across 19 sites, with Chalupas, Chachimbiro, and Tufiño highlighted as priority zones. Under a scenario replacing 50% of fossil thermal capacity (~1,700 MW), up to 28 million tons of CO_2 emissions could be avoided annually. The required investment for this scenario is estimated at USD 6 billion, with annual revenues from electricity sales and carbon credits exceeding USD 2.5 billion, and a cumulative Net Present Value (NPV) of over USD 20 billion within 25 years.

As a methodological contribution, this work proposes an innovative integration of statistical forecasting, multivariate scenario simulation, and climate impact analysis, generating a useful tool for the formulation of public policies and energy transition plans in Andean countries.

Keywords: Geothermal Energy · Decarbonization · Energy Transition · ARIMA Models · Ecuador

1 Introduction

For decades, hydropower has been one of the primary sources of electricity generation in Ecuador and several Latin American countries. However, its efficiency is highly dependent on climatic conditions. The year 2024 was recorded as the warmest year in Ecuador's history, with prolonged droughts leading to power outages and energy

J. L. Ballesteros et al. (Eds.): CITIS 2025, LNNS 1780, pp. 429–438, 2027.
https://doi.org/10.1007/978-3-032-14287-0_42

rationing throughout the year causing blackouts lasting up to 14 h per day. This situation resulted in significant economic losses in the industrial sector and disrupted the continuity of essential public services. Consequently, it is imperative to consider alternative sources of energy generation that are more resilient to climate change [2]. Geothermal energy represents a strategic alternative for Ecuador, as it can provide a continuous and stable supply of electricity without being inherently dependent on weather conditions. Nevertheless, its share in the global energy matrix remains marginal, often limited by high initial investment costs, limited exploration data, and institutional challenges in energy planning [3]. According to [2], geothermal energy accounts for only 0.9% of renewable energy sources worldwide—a figure significantly lower compared to technologies such as solar photovoltaics (41.8%) or renewable hydropower (27.5%). According to [4], Ecuador possesses favorable geographic conditions for the development of geothermal energy. Nineteen suitable geothermal sites have been identified across the country. The areas with the highest estimated generation capacity are: Tufiño–Chiles–Cerro Negro (330 MW), Chalupas (283 MW), and Chachimbiro (178 MW).

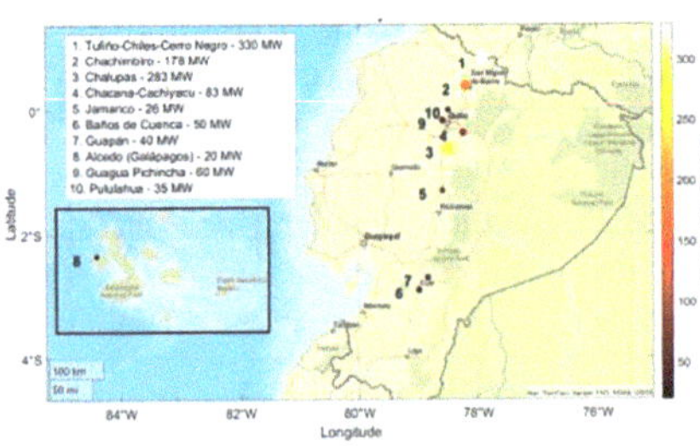

Fig. 1 Geothermal potential in key sites in Ecuador

Based on this context, the present study aims to contribute to the identification of geothermal energy as a viable alternative to strengthen Ecuador's power generation matrix. Additionally, it proposes a methodology that integrates economic forecasting, energy substitution scenario modeling, and multivariable financial analysis, in order to assess the impact of progressively replacing fossil thermal generation with geothermal energy in terms of investment requirements, avoided emissions, and projected profitability by the year 2030.

2 Proposed Methodology for the Energy Transition

The proposed methodology begins with a structural analysis of Ecuador's energy matrix, focusing particularly on the evolution of installed capacity between 2010 and 2020. As shown in Fig. 1, hydropower capacity has grown steadily, reaching 6,000 MW, while fossil-based thermal generation has remained relatively constant at around 2,300 MW. The contribution of non-hydropower renewable sources has been marginal, with a peak of only 450 MW in 2015.

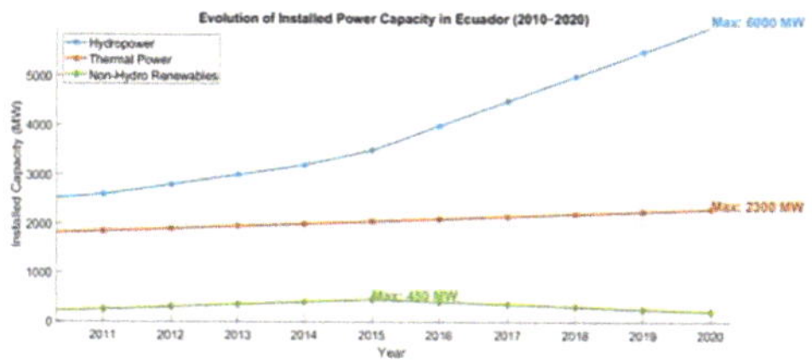

Fig. 2 Evolution of Ecuador's installed power generation capacity (2010–2020)

Figure 2 confirms that geothermal energy maintains a capacity factor close to 90%, significantly outperforming hydropower (~50%) and solar photovoltaics (~14%). Moreover, its Levelized Cost of Electricity (LCOE) has remained competitive over recent years.

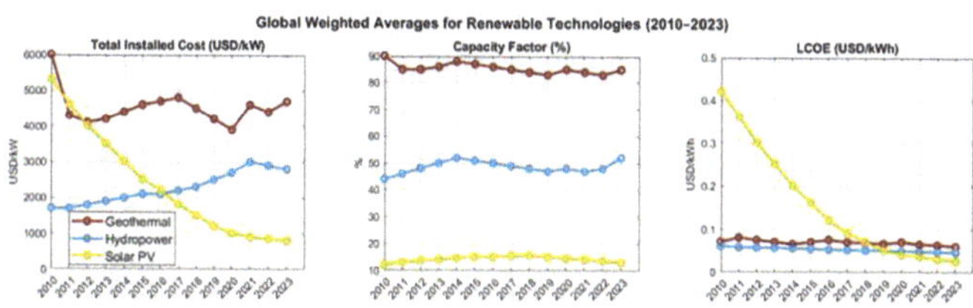

Fig. 3 Global weighted averages of renewable technologies (2010–2023)

Based on this background, this research aims to develop an energy substitution model that evaluates the progressive replacement of fossil thermal generation with national geothermal capacity. To achieve this, an energy balance function is employed, which relates the required installed geothermal capacity (C_g) to the capacity factor (CF) and the annual demand to be covered (D).

$$D = C_g.CF.H_a$$

Where:

- ***D*****:** represents the annual electricity demand to be covered by geothermal energy $\left(\frac{\text{MWh}}{\text{año}}\right)$.
- $\boldsymbol{C_g}$**:** Annual electricity demand to be covered by geothermal energy (MW).
- ***CF*****:** Capacity factor of geothermal energy (0.88)
- $\boldsymbol{H_a}$**:** Total number of hours in a year (8.760 h).

Based on this relationship, the required capacity to cover different percentages of the electricity currently generated by fossil thermal power plants [5], is calculated. Additionally, the avoided emissions are estimated using the following standard emission factor:

$$\text{CO}_2 = \text{E}_{\text{th}} * \text{EF}$$

Where:

- CO_2: Amount of carbon dioxide avoided ($\frac{tCO_2}{\text{año}}$)
- E_{th}: Annual energy generated by fossil thermal power plants ($\frac{\text{MWh}}{\text{año}}$)
- ***EF:*** Emission factor of fossil thermal generation ($\frac{0.8\, tCO_2}{\text{MWh}}$).

The model also incorporates global projections of installed geothermal capacity (Fig. 3), which show sustained growth reaching 16.4 GW by 2024, reflecting the technological viability and increasing market maturity of the sector. Methodologically, replacement scenarios ranging from 10% to 50% of the national fossil thermal capacity are established. For each scenario, the total investment cost (CAPEX), projected revenues from electricity sales and carbon credits, and the Net Present Value (NPV) over a 25-year project lifetime are evaluated.

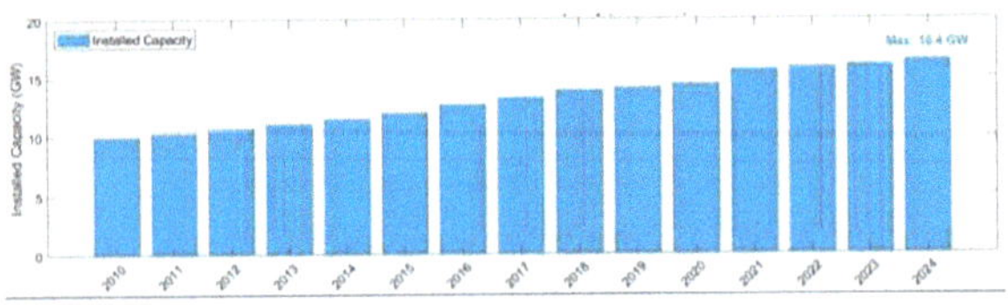

Fig. 4 Installed Geothermal Capacity Worldwide (2010–2024)

This approach enables the identification of the most efficient technical and economic thresholds to initiate the transition toward a resilient, low-carbon energy matrix, centered on the progressive utilization of the country's geothermal potential.

3 Results Analysis

3.1 Geothermal LCOE Projection Using ARIMA (2010–2027)

Figure 4 shows the historical trend and forecast of the Levelized Cost of Electricity (LCOE) for geothermal energy using an ARIMA model [6]. The LCOE has progressively decreased from 0.09 USD/kWh in 2010 to 0.056 USD/kWh in 2022, demonstrating a strong trend component that supports the application of a time series model for projection purposes. The model applied was an ARIMA (1,1,1) [6], whose general formula is:

$$Y_t = \alpha + \phi Y_{t-1} + \theta \varepsilon_{t-1} + \varepsilon_t$$

Where:

- Y_t: Projected value of the LCOE in year t,
- ϕ: Autoregressive coefficient
- θ: Moving average coefficient
- ε_t: White noise error term

This model was fitted using observed values from 2010 to 2022, generating a forecast for the 2023–2027 period with a 95% confidence interval. The projected trend suggests that the geothermal LCOE will continue to decline, reaching values close to 0.045

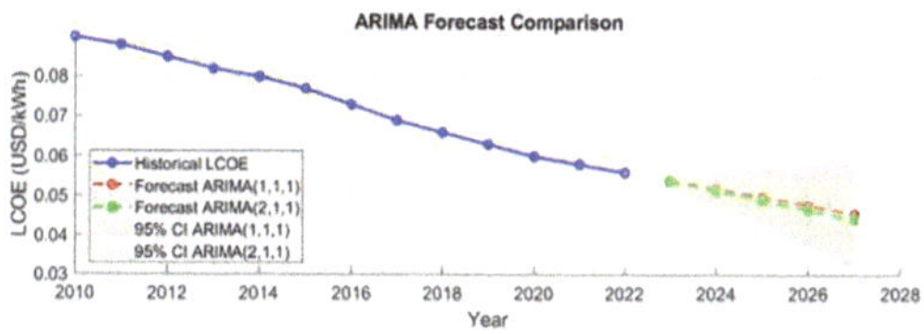

Fig. 5 Forecast of geothermal LCOE using ARIMA (2010–2027)

USD/kWh by 2027, positioning it as one of the most competitive sources in the energy market. The sustained decrease in LCOE is particularly relevant, as it directly affects project profitability and enhances competitiveness compared to fossil fuels [7] and other renewable technologies with lower capacity factors [8].

3.2 Geothermal LCOE Projection Using ARIMA (2010–2027)

Figure 5 illustrates the economic and environmental impact of different scenarios for replacing fossil thermal generation with geothermal energy, based on the LCOE projection derived from the ARIMA model [9], and incorporating a ± 5% uncertainty margin. The left axis represents the total installation cost (CAPEX), calculated as:

$$CAPEX = C_g * 1000 * Costo_{USD/kW}$$

Where:

- C_g Installed geothermal capacity (MW)
- $Cost_{USD/kW}$: Adjusted based on the projected LCOE and economies of scale.

The right axis shows the annual CO_2 emissions avoided, in millions of tons, estimated using the following formula:

$$CO_2 = E * EF$$

Where:

- *E:* Energy generated by the substituted capacity (in MWh),
- *EF:* Emission factor of fossil thermal generation ($\frac{0.8\ tCO_2}{\text{MWh}}$)

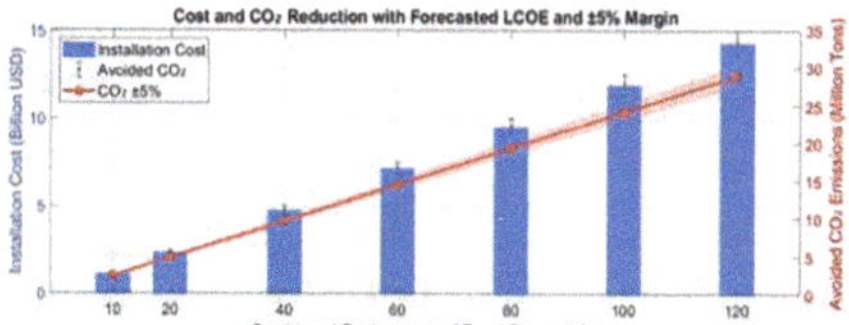

Fig. 6 Cost and CO_2 reduction with projected LCOE and ± 5% margin

The analysis reveals that a 60% replacement of fossil thermal capacity (approximately 1,380 MW) would require an estimated investment of USD 8.5 billion. This scenario would avoid nearly 20 million tons of CO_2 emissions annually. Under a full (100%) replacement scenario, the investment would exceed USD 14 billion, while more than 30 $MtCO_2$ would be avoided—an amount equivalent to nearly the entire emissions of Ecuador's power sector during years of high demand. The shaded area around the emissions curve reflects the model's sensitivity to variations in the projected LCOE, which is crucial for informed decision-making under uncertainty.

3.3 Progressivity of Thermal Generation Displacement (2025–2030)

Figure 6 presents a staggered energy transition model, projecting a progressive expansion of geothermal capacity in Ecuador over the 2025–2030 period, alongside a gradual reduction in fossil thermal generation.

This model considers a baseline thermal capacity of 3,200 MW (reference value from the 2023 National Energy Balance), and projects that this capacity will be progressively replaced through the annual addition of approximately 300 MW of geothermal capacity [10]. The transition is described as follows:

$$C_{g,t} = C_{g,t-1} + \Delta C_g y C_{th,t} = C_{th,t-1} + \Delta C_g$$

Where:

- $C_{g,t}$: Geothermal capacity in year t,
- $C_{th,t}$:Remaining thermal capacity
- ΔC_g: Annual geothermal capacity increment (approximately 300 MW).

This type of orderly transition mitigates the risk of energy shortages, allows for gradual planning of public and private investment, and provides clear signals to the energy market.

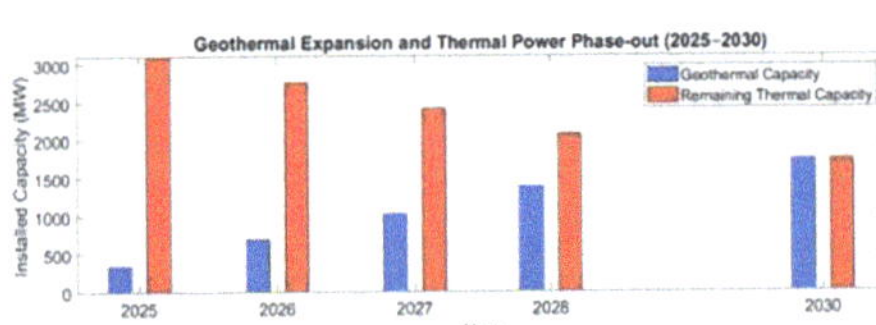

Fig. 7 Geothermal expansion and progressive phase-out of thermal energy (2025–2030)

The model confirms that by 2030, Ecuador could achieve geothermal coverage equivalent to 50% of its current thermal capacity without compromising the stability of the national power system.

3.4 Comparative Evolution of Renewable Vs. Non-renewable Capacity

Figure 7 shows how the share of renewable energies hydropower, solar, wind, and geothermal increases its dominance over the total installed capacity [11], increasing from 4,800 MW in 2025 to 6,300 MW in 2030.

The relationship between renewable capacity (CR) and non-renewable capacity (CNR) can be expressed as:

$$Renewable\ Share = \frac{C_R}{C_R + C_{NR}} \times 100\%$$

Applying this formula, it is observed that the share of renewables in total installed capacity would increase from 57% in 2025 to over 74% by 2030.

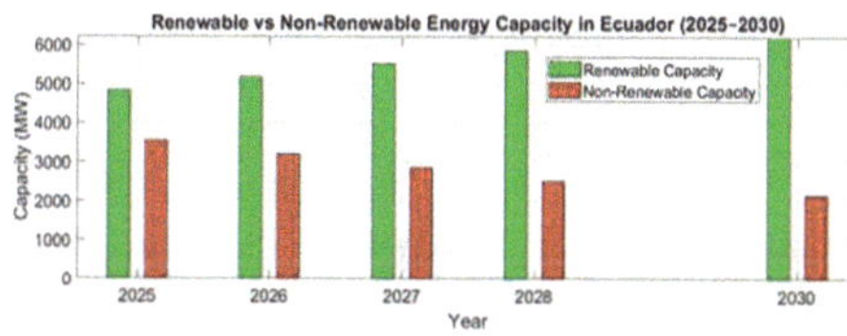

Fig. 8 Renewable and non renewable energy capacity in Ecuador (2025–2030)

The phase-out of fossil technologies from 3,500 MW to 2,200 MW is driven by three key factors: the sustained reduction of geothermal LCOE, the high operational stability of geothermal energy (capacity factor > 85%), and the reduction of greenhouse gas emissions, which strengthens the country's compliance with its climate commitments (NDCs) [1].

3.5 Unit Cost and Strategic Value of Geothermal Energy

Figure 8 presents the analysis of cumulative costs per megawatt (MW) installed and per kilowatt-hour (kWh) generated during the 2025–2030 period. While total investment (CAPEX) increases linearly with the expansion of geothermal capacity, the unit cost per kWh remains stable at approximately 0.397 USD/kWh. This reflects a steady and predictable economy of scale, which is ideal for long-term financial planning.

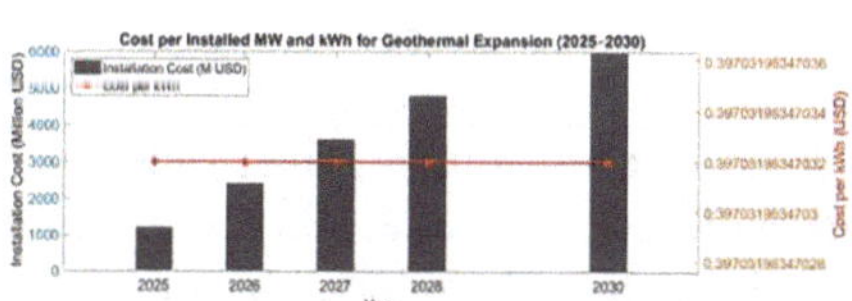

Fig. 9 Cost per MW and kWh installed for geothermal expansion (2025–2030)

As shown in Fig. 9, the total CAPEX and the cumulative operation and maintenance (O&M) costs over a 25-year period reveal that values increase proportionally with installed capacity, without exponential rises that would compromise scalability (Figs. 10, 11, 12).

In terms of revenue, the projects generate dual benefits: electricity sales and carbon credits, reaching a combined annual return of approximately USD 2.8 billion under the 100% substitution scenario.

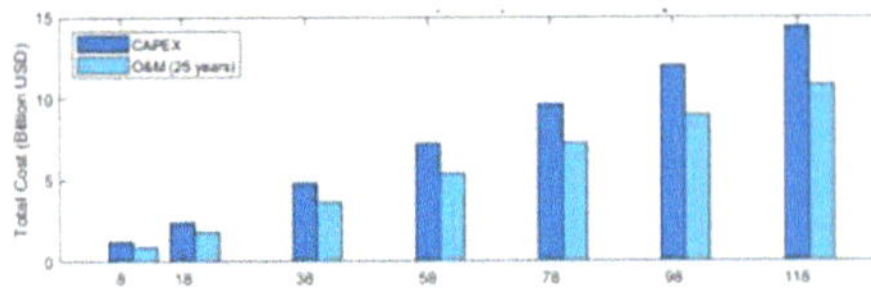

Fig. 10 Geothermal investment and operation & maintenance over 25 Years

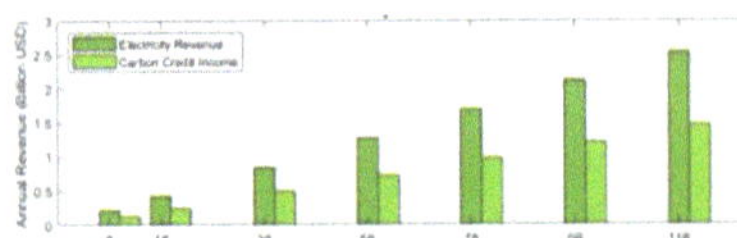

Fig. 11 Annual revenues: Electricity sales vs. Carbon credit income

On the environmental axis, a progressive reduction in CO_2 emissions is observed, exceeding 28 million tons per year in the most ambitious scenario. Finally, the Net Present Value (NPV) increases steadily across all scenarios, reaching USD 29 billion—validating geothermal energy as an alternative with high economic profitability and a significant positive climate impact.

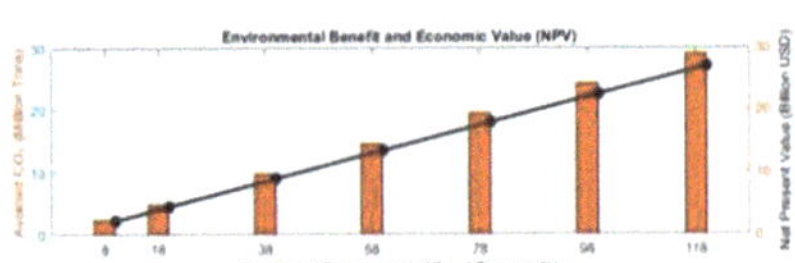

Fig. 12 Environmental benefit and economic value (NPV)

4 Discussion and Future Research Directions

This study demonstrates that geothermal energy represents a strategic option for the decarbonization of Ecuador's electrical system. With a capacity factor of 88%, near-zero operational emissions, and a lifespan exceeding 30 years, the progressive substitution of fossil thermal plants could avoid up to 28 million tons of CO_2 annually by 2030—representing a significant reduction compared to the current emission levels of the national power sector. Furthermore, the projected Levelized Cost of Electricity (LCOE) for 2027 (0.045 USD/kWh) positions geothermal as a competitive alternative to other renewable sources, even when accounting for high initial investment costs.

As future lines of research, it is recommended to deepen the analysis of hybrid electrical networks incorporating geothermal and energy storage, geospatial prioritization of high-potential geothermal sites, and financial mechanisms such as carbon credits and green bonds. These findings support the formulation of public policies aimed at accelerating a just and sustainable energy transition in Ecuador.

5 Conclusions

a. An integrated methodology was proposed, combining energy, financial, and environmental analysis. It incorporates ARIMA-based projections, multivariable scenario simulations, and climate impact assessment. Its main contribution lies in empirically demonstrating that the progressive substitution of fossil thermal energy with geothermal power is both technically and economically viable in Ecuador.
b. Geothermal energy represents a viable and strategic source for diversifying Ecuador's energy matrix. With an estimated potential of over 1,400 MW distributed across more than 15 identified sites nationwide, and a capacity factor of 88%, this source could replace up to 50% of the current fossil thermal generation. This would avoid more than 28 million tons of CO_2 emissions annually.
c. The main obstacle to geothermal expansion is its high initial investment cost, estimated at over USD 3 billion for a 1,200 MW expansion over the next five years. This is due to the technical challenges of deep exploration, geological risks, and the lack of a robust regulatory framework that incentivizes this technology in Ecuador.

References

1. Qu, W., et al.: A comprehensive multicriteria decision analysis framework for shallow geothermal energy suitability evaluation in Yinchuan area, China. Sci. Rep. **15**(1), 6420 (2025). https://doi.org/10.1038/s41598-025-90559-2
2. Penman, D., Muhammad-Sukki, F., Abu-Bakar, S.H., Bani, N.A., Mas'ud, A.A.: Use of deep geothermal energy to supply heat to a village in Scotland. Interactions. **246**(1), 39 (2025). https://doi.org/10.1007/s10751-025-02265-0
3. Luo, J., et al.: Investigation of shallow geothermal potentials for different types of ground source heat pump systems (GSHP) of Wuhan city in China. Renew. Energy. **118**, 230–244 (2018). https://doi.org/10.1016/j.renene.2017.11.017
4. Miladinović, A., Grujić, A.S., Kijevčanin, M., Stijepović, V., Alnouri, S.Y., Stijepović, M.: Towards zero carbon emissions: electrification and decarbonization of an ethylene plant's utility system. Comput. Chem. Eng. **198**, 109117 (2025). https://doi.org/10.1016/j.compchemeng.2025.109117
5. Barman, P., et al.: Renewable energy integration with electric vehicle technology: a review of the existing smart charging approaches. Renew. Sust. Energ. Rev. **183**, 113518 (2023). https://doi.org/10.1016/j.rser.2023.113518
6. Cai, M., et al.: Coupling CO_2 reduction and energy storage by electrolytic zinc. Energy Storage Mater. **77**, 104165 (2025). https://doi.org/10.1016/j.ensm.2025.104165
7. De la Cruz, R.D., Tipán, L.F., Cuji, C.C.: Brief analysis of the location and determination of maximum capacity of distributed generation in electrical systems considering demand scenarios in Ecuador. Energies. **17**(10), 2308 (2024). https://doi.org/10.3390/en17102308
8. Ministerio de Energía, Minas del Ecuador: Balance Energético Nacional 2023 – Capítulo 1: Energía, Sociedad y Ambiente, Quito, Ecuador (2024). [En línea]. Disponible en: https://www.recursosyenergia.gob.ec/wp-content/uploads/2024/08/BEN_2023-cap_1.pdf. Accedido el: 23 de abril de 2025.
9. International Renewable Energy Agency (IRENA): Data and statistics. IRENA, Abu Dhabi (2025). [En línea]. Disponible en: https://www.irena.org/Data. Accedido el: 23 de abril de 2025.

10. Agaal, A., Farkash, H.M., Essgaer, M., Ahessin, A.: Towards efficient electricity management in Benghazi: forecasting demand and load shedding with ARIMA models. Solar Energy Sustain. Dev. **14**, 110–136 (2025). https://doi.org/10.51646/jsesd.v14iFICTS-2024.446. FICTS-2024
11. Pierre, A.A., Akim, S.A., Semenyo, A.K., Babiga, B.: Peak electrical energy consumption prediction by ARIMA, LSTM, GRU, ARIMA-LSTM and ARIMA-GRU approaches. Energies. **16**(12), 4739 (2023). https://doi.org/10.3390/en16124739
12. Xiang, W., Zhou, L.: Planning research on electrically coupled integrated energy system based on ARIMA-LSTM model. Appl. Math. Nonlinear Sci. **9**(1), 20230643 (2024). https://doi.org/10.2478/amns.2023.2.00643
13. Alirahmi, S.M., Rahmani Dabbagh, S., Ahmadi, P., Wongwises, S.: Multi-objective design optimization of a multi-generation energy system based on geothermal and solar energy. Energy Convers. Manag. **205**, 112426 (2020). https://doi.org/10.1016/j.enconman.2019.112426

Comparative Analysis of FC and PINN Architectures in Solar Panel Power Estimation

Francis-Nicole Benitez-Montalvo(✉), Pablo-Andrés Buestán-Andrade, and Nathalia-Michelle Peralta-Vasconez

Unidad Académica de Informática, Ciencias de La Computación e Innovación Tecnológica, Universidad Católica de Cuenca, Cuenca, Ecuador
francis.benitez@est.ucacue.edu.ec

Abstract. The integration of physical principles into machine learning models is an innovative strategy to improve accuracy in energy systems. This study proposes a physics-informed neural network (PINN) methodology to predict solar panel power, comparing it to fully connected networks (FC) using traditional loss functions. Correlation analysis identified solar irradiance, ambient temperature, and module temperature as key inputs. The FC-PINN model incorporated a physical approximation in its loss function, combining historical data with thermodynamic constraints. This hybrid model reduced the mean absolute error (MAE) by 3.17% and increased the coefficient of determination (R2) by 0.47% compared to the mean squared error (MSE) approach. It also showed greater adaptability to sharp fluctuations, validating robustness under variable conditions. These results highlight the importance of integrating physical knowledge into AI models to optimize photovoltaic systems and reduce operational costs.

Keywords: Solar Panel · FC · PINN · Time Series Forecasting · Power Prediction

1 Introduction

Over the years, the world has faced serious environmental pollution issues. According to the World Health Organization (WHO), about 6.7 million deaths linked to air pollution occur annually [1]. Despite standards for pollutants like PM10, PM2.5, SO2, NO2, O3, and CO [2], only 17% of cities meet these limits, driving a global search for effective solutions [3].

In this context, transitioning to clean energy is not only a mitigation strategy but also a fundamental solution to climate change. Solar energy, in particular, has grown at an average annual rate of 26% since 2014, with projections exceeding 1500 GW installed by 2025 [4], fueled by environmental concerns, technological advances, and policy support [5]. However, photovoltaic systems still face challenges such as radiation variability, weather conditions, and panel maintenance, which affect energy generation efficiency [6].

One effective alternative to optimize performance, reduce operation and maintenance (O&M) costs, and increase efficiency is to predict generated power based on

J. L. Ballesteros et al. (Eds.): CITIS 2025, LNNS 1780, pp. 439–448, 2027.
https://doi.org/10.1007/978-3-032-14287-0_43

climatic and operational variables [7]. Artificial intelligence (AI) has proven effective in analyzing large datasets, anticipating photovoltaic system behavior, and supporting real-time decisions [8]. Among the most innovative techniques are physics-informed neural networks (PINNs), which integrate physical principles into learning, producing more robust, accurate, and sustainable models over time [9].

This study proposes a PINN-based methodology to predict solar panel power, with main contributions:

- Optimal input variable selection via correlation analysis, sliding window, and Random Search.
- Development of a mathematical approximation for power calculation based on atmospheric variables.
- Comparison between traditional loss function models and PINN-based models using common error metrics.

The article is structured as follows: Sect. 2 reviews techniques and the case study; Sect. 3 details training, validation, and model comparison; Sect. 4 presents conclusions.

2 Materials, Methods and Case Study

Machine learning (ML) models capture complex relationships between variables in SCADA systems [10]. Fully connected neural networks (FC) minimize prediction errors using traditional loss functions like mean squared error (MSE), without explicit physical constraints [11]. In contrast, PINNs embed differential equations into the loss function, enforcing physical principles during training. This improves accuracy under variable conditions and reduces dependence on large datasets [12]. The following sections describe both methods and their application to solar power prediction.

2.1 Fully Connected Neural Network (FC)

FCs are high-dimensional models where neurons from one layer are fully connect-ed to the next through weighted synapses. During training, the network adjusts these weights to minimize prediction error [13]. An example of FC is shown in Fig. 1.

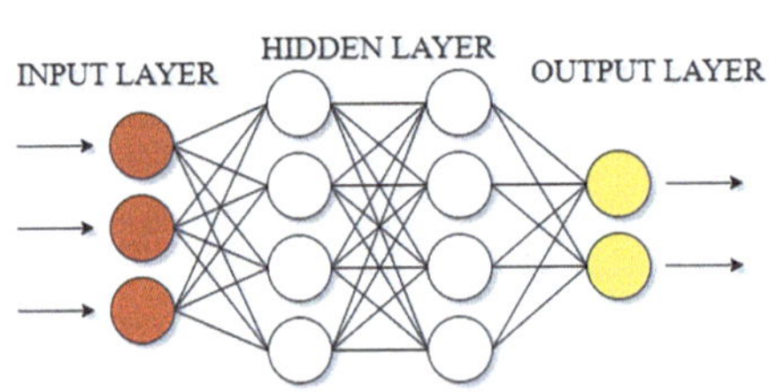

Fig. 1. FC Architecture.

2.2 Physics-Informed Neural Networks (PINN

In recent years, PINNs (Fig. 2) have gained prominence over traditional methods for solving partial differential equations (PDEs). Unlike techniques like computational fluid dynamics, which require mesh construction to represent and simulate the physical space, PINNs solve these equations directly [9].

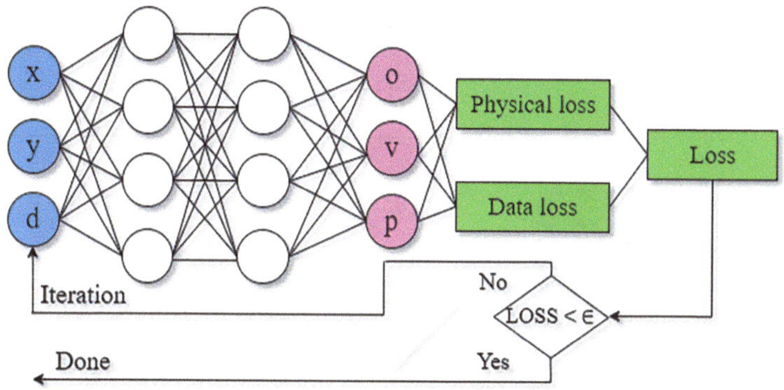

Fig. 2. PINN Architecture.

Unlike FC networks, PINNs incorporate the relevant physical laws directly into the loss function by integrating differential equations that describe the system's behavior [14]. This enables the model to learn not only from historical data but also to follow fundamental physical principles, leading to more robust and generalizable predictions [15]. In predicting solar panel power, PINNs use inputs such as solar radiation, ambient temperature, and other atmospheric variables during training.

Application in Solar Energy Prediction

The power generated by a photovoltaic system is modeled using a physical equation that incorporates solar irradiance (G_i), module temperature (T_m), and construction parameters. Based on the formulation of PVWatts (NREL) [16], which relates the nominal power (P_{dc0}) to a temperature correction:

$$P_{dc} = \frac{G_i}{1000} P_{dc0}\left(1 + \gamma\left(T_{cell} - T_{ref}\right)\right) \quad (1)$$

where P_{dc0} is the power under standard conditions (STC: 1000 W/m2, 25 °C). By substituting $P_{dc0} = \eta_{ref} \bullet G_{ref} \bullet A$ and considering G_i as the actual irradiance, we obtain:

$$P_{DC} = G_i \cdot A \bullet \eta_{ref} \cdot \left(1 - \gamma\left(T_m - T_{ref}\right)\right) \quad (2)$$

- G_i: Actual irradiance (W/m2) on the plane of the panel, determining the available energy.
- A: Area of the module (m2), which defines the surface area that receives radiation.
- η_{ref}: Efficiency under STC (approximately 15–22%), the fraction of solar energy converted into electricity.
- γ: Temperature coefficient (%/°C), quantifies the loss of efficiency due to an increase in Tm (typical: -0.3%/°C).

- T_m: Temperature measured on the back of the panel, influenced by irradiance under environmental conditions.
- T_{ref}: Reference temperature (25 °C), the basis for calculations under STC.

Figure 3 illustrates the training of the architecture using Eq. 2, referred to in this study as FC-PINN.

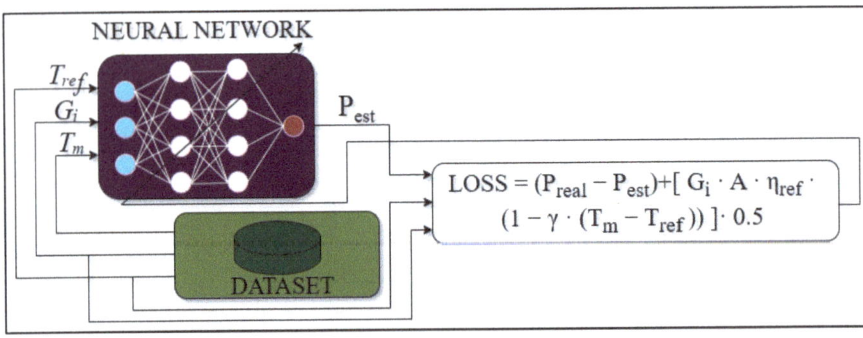

Fig. 3. Training of the FC-PINN architecture.

Figure 3 shows the PINN architecture, a fully connected network with three inputs: Irradiance (G_i), Module Temperature (T_m), and Ambient Temperature (T_ref). The output is predicted power (P_est). During training, data loss—the gap between real values and predictions—and physical loss—the deviation from the equation—are computed simultaneously. In each iteration, internal parameters and constants are adjusted until the combined loss falls below a predefined threshold; otherwise, the optimizer continues until convergence.

Study Case

This research used two public databases from a photovoltaic plant in India [17], covering 34 days with 15-min intervals, totaling 3,158 samples. The data includes energy generation (recorded at the inverter level) and environmental measurements (recorded at the plant level). Table 1 shows the structure and content of both datasets.

Table 1. Structure of the generation and atmospheric data.

Column	Description
DC_POWER	Direct current output power of the inverter (W)
AMBIENT_TEMPERATURE	Ambient temperature measured at the site (°C)
MODULE_TEMPERATURE	Surface temperature of the photovoltaic modules (°C)
IRRADIATION	Solar irradiance measured on the modules (W/m^2)

3 Results

3.1 Data Preprocessing

The databases were initially read using the Pandas library in Python and converted into DataFrames. The "SOURCE_KEY" columns were removed, and the datasets were merged using an inner join on "DATE_TIME," keeping only records with exact time matches. Figure 4 illustrates the database histogram over time.

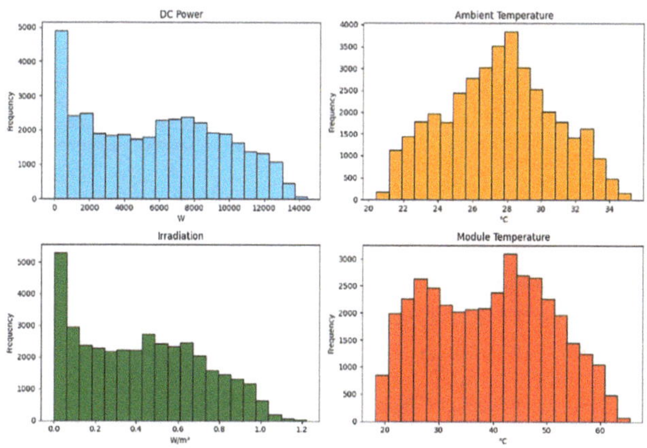

Fig. 4. Histogram of the time series.

From the resulting dataframe, key variables for analysis AMBIENT_TEMP, IRRADIATION, and MODULE_TEMP were selected. All rows with null values were removed, and only records from daylight hours (6:00 AM to 6:00 PM) were kept. Variable selection was based on Pearson's correlation analysis (Fig. 5).

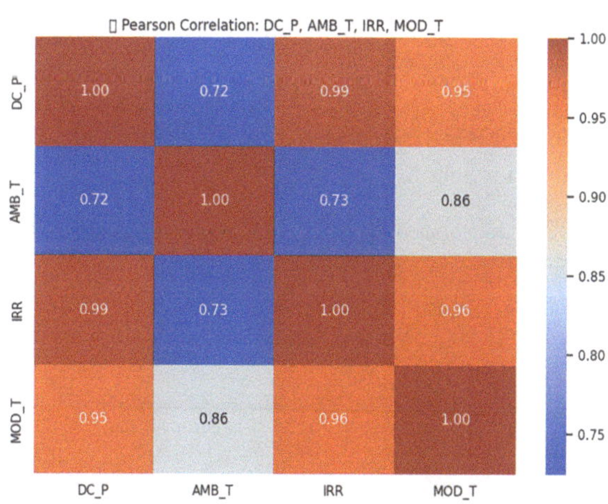

Fig. 5. Correlation of the variables in the time series.

Consequently, the dataset was divided into a training subset (80%) and a testing subset (20%).

Finally, a standardization methodology was applied. Standardization involves adjusting the characteristics of the data to achieve a mean of 0 and a standard deviation of 1. Equation (3) serves as the analytical method used for this calculation in the context of this research.

$$X_s = \frac{x - \overline{x}}{\sigma_x} \tag{3}$$

- x: value that requires normalization.
- $\overline{x}$: arithmetic mean of the distribution.
- σ_x: standard deviation of the distribution.

3.2 Training

Table 2 summarizes the hyperparameters used in each model. These values were selected after conducting a systematic hyperparameter optimization (combining random search with 5-fold cross-validation to explore learning rates, batch sizes, dropout rates, and network depth) in order to identify the configuration that provided the best validation performance.

Table 2. Hyperparameters for predicting solar panel power output.

Hyperparameter	FC-MSE	FC-PINN
Architecture	2 hidden layers de 64 neurons + Dropout 0.2	
Activation (hidden)	ReLU	
Activation (outputs)	Lineal	
Optimizer	Adam	
Learning rate	0.001	
Loss function	MSE	MSE + PINN
Physics weight	-	0.5
Epochs	50	
Batch size	32	
Metrics	MAE	

3.3 Comparison and Performance

Evaluation Metric

The Mean Absolute Error (MAE) measures the mean magnitude of the differences between predicted values and actual values, without considering their direction (Eq. 4).

$$MAE = \frac{1}{n}\sum_{i=1}^{n}\left|y_i - \hat{y}_i\right| \tag{4}$$

- n: number of observations.
- y_i: observed values.
- $\hat{y}_i$: predicted values.

The coefficient of determination (R^2) indicates the variance in the dependent variable explained by the regression model and its predictors. It ranges from 0 to 1, where values near 1 suggest the model accurately fits the observed data (Eq. 5).

$$R^2 = 1 - \frac{\sum_{i=1}^{n}(y_i - \hat{y}_i)^2}{\sum_{i=1}^{n}(y_i - \overline{y})^2} \tag{5}$$

- $\overline{y}$: Average of all y_i.
- $\sum_{i=1}^{n}(y_i - \hat{y}_i)^2$: Squared difference between predictions and observations.
- $\sum_{i=1}^{n}(y_i - \overline{y})^2$: Sum of the squared deviations of the observed values from their mean.

The Mean Absolute Percentage Error (MAPE) shows, on average, what percentage the predictions are off compared to the actual values; it can be sensitive when the actual values are very small (Eq. 6).

$$MAPE = \frac{100\%}{n}\sum_{i=1}^{n}\left|\frac{y_i - \hat{y}_i}{y_i}\right| \tag{6}$$

The Root Mean Square Error (RMSE) determines the square root of the mean of the squared discrepancies between the predicted and actual values. It expresses the error in the same units as the original data (Eq. 7).

$$RMSE = \sqrt{\frac{1}{n}\sum_{i=1}^{n}(y_i - \hat{y}_i)^2} \tag{7}$$

Interpretation of Results

The results obtained (Table 3) show that the FC-PINN model consistently outperforms the FC-MSE model across all evaluated metrics. The MAE decreases from 0.0609 to 0.0589, representing a 3.17% improvement in average accuracy. Similarly, the RMSE drops from 0.0791 to 0.0771, indicating a reduction in large errors and greater overall accuracy. Although the MAPE also decreases slightly, its values remain very high (on the order of 1013) due to the nature of the metric: dividing absolute error by actual values causes instability when those values approach zero. This issue, reported in recent literature (e.g., Renewable Energy World, 2023), is attributed in our case to irradiance, which can reach minimal values under extreme cloudiness, directly impacting power generation. These observations, while problematic for MAPE, are critical to assess model robustness under drastic solar resource variations. Therefore, it's essential to complement MAPE with metrics like MAE and RMSE, which are independent of the reference value. The coefficient of determination (R2) also improves with the FC-PINN model, increasing from 0.9138 to 0.9181, reflecting a 0.47% gain in explanatory capability.

Table 3. Results obtained from the two architectures.

FC Model	MAE	RMSE	MAPE	R2
FC MSE	0.060857	0.079104	$1.115713X10^{13}$	0.913789
FC PINN	**0.058930**	**0.077112**	**$1.042794X10^{13}$**	**0.918078**

On the other hand, Fig. 6 compares the actual generated power values (solid blue line) with the predictions of the FC-PINN model (dashed orange line) and the FC-MSE model (dotted green line) for the first 100 test examples. The FC-PINN predictions more closely follow the actual behavior, adapting better to the peaks and valleys of the time series, while the FC-MSE model shows larger deviations, particularly at extremes and abrupt changes. This demonstrates that incorporating physical information into the FC-PINN training enables the model to capture real power fluctuations more accurately, reducing systematic errors and enhancing robustness under sudden variations. Therefore, the visualization confirms that integrating physical constraints not only improves data fitting but also yields more consistent predictions in complex or poorly represented scenarios, supporting the quantitative results in Table 3.

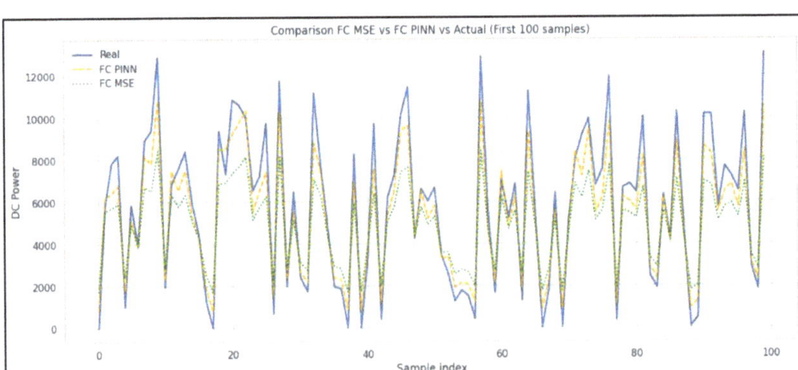

Fig. 6. Comparison of the predictions of the FC-MSE model vs. FC-PINN and the actual values.

4 Conclusions and Future Work

The results obtained in this study demonstrate that integrating physical information through PINNs significantly improves the accuracy of power prediction in solar panels compared to traditional data-driven models.

Compared to [18], where PINNs are applied to multi-VP problems related to partial differential equations, this study demonstrates the effectiveness of the approach in real energy applications, particularly for photovoltaic power prediction. The results also surpass those presented at the European Symposium on Artificial Neural Networks 2024, where, despite incorporating physical knowledge, the adaptability under high variability conditions is less notable than in the proposed model [19]. In addition, as highlighted in Frontiers in Energy Research [20], which emphasizes the potential of PINNs to enhance predictions in energy systems, this work confirms that incorporating explicit physical constraints strengthens robustness against noise and limited data.

The FC-PINN model achieved a lower MAE and higher R2, reflecting a better capacity to generalize and capture the real dynamics of the photovoltaic system. Visual comparisons further support that the PINN model closely follows actual values, especially under variable or abrupt conditions. These findings reinforce the relevance of including physical constraints during deep learning model training for energy applications, enabling more robust and reliable predictions even in challenging scenarios.

As future work, we propose exploring more complex architectures, such as recurrent or attention-based models, to capture longer-term temporal dependencies in solar generation. It is also important to evaluate the generalization of the PINN approach across different geographic locations and climate conditions, and to incorporate additional variables (e.g., cloud cover or wind speed) to enrich the physical model.

References

1. WHO. Ambient (outdoor) air pollution. In: World Health Organization (2024). https://www.who.int/news-room/fact-sheets/detail/ambient-(outdoor)-air-quality-and-health. Accessed 23 Feb 2025
2. WHO. Air Quality Standards database. In: World Health Organization (2025). https://www.who.int/tools/air-quality-standards. Accessed 19 Mar 2025
3. Stanway, D.: Only seven countries met WHO air quality standards in 2024, data shows. In: Reuters (2025). https://www.reuters.com/business/healthcare-pharmaceuticals/only-seven-countries-met-who-air-quality-standards-2024-data-shows-2025-03-11/. Accessed 31 Mar 2025
4. Garside, K.: 2025: the year of the solar? In: Integrity Energy (2025). https://www.integrityenergy.com/blog/2025-the-year-of-the-solar/. Accessed 9 Apr 2025
5. United Nations Five ways to jump-start the renewable energy transition now (2024). https://www.un.org/en/climatechange/raising-ambition/renewable-energy-transition. Accessed 9 Jan 2025
6. Bamisile, O., Acen, C., Cai, D., et al.: The environmental factors affecting solar photovoltaic output. Renew. Sustain. Energy Rev. **208**, 115073 (2025)
7. Palomo, W., Quinatoa, C., Mullo, M., Castillo, J.: Predicción de radiación solar en sistemas fotovoltaicos utilizando técnicas de aprendizaje automático. Conectividad **6**, 338–355 (2025)

8. Burgos, J.G.C., Vargas, F.J.F., Revilla, A.R.D., et al.: Una revisión del desarrollo de los modelos matemáticos para la estimación de la potencia eléctrica generada en un sistema fotovoltaico. Ciencia Latina Rev. Científica Multidiscip. **7**, 8002–8013 (2023)
9. Bhargava, S., Chamakuri, N.: Enhancing arterial blood flow simulations through physics-informed neural networks. arXiv preprint arXiv:2404.16347 (2024)
10. Cooley, M., Shankar, V., Kirby, R.M., Zhe, S.: Fourier PINNs: from strong boundary conditions to adaptive fourier bases. arXiv preprint arXiv:2410.03496 (2024)
11. Wang, H., Zhou, H., Cheng, S.: Dynamical system prediction from sparse observations using deep neural networks with Voronoi tessellation and physics constraint. Comput. Methods Appl. Mech. Eng. **432**, 117339 (2024)
12. Cuomo, S., Di Cola, V.S., Giampaolo, F., et al.: Scientific machine learning through physics-informed neural networks: where we are and what's next. J. Sci. Comput. **92**, 88 (2022). https://doi.org/10.1007/s10915-022-01939-z
13. Scabini, L.F.S., Bruno, O.M.: Structure and performance of fully connected neural networks: emerging complex network properties. Phys. A **615**, 128585 (2023)
14. Monaco, S., Apiletti, D.: Training physics-informed neural networks: one learning to rule them all? Results Eng. **18**, 101023 (2023)
15. Wu, Y., Sicard, B., Gadsden, S.A.: Physics-informed machine learning: a comprehensive review on applications in anomaly detection and condition monitoring. Expert Syst. Appl. 124678 (2024)
16. Dobos, A.P.: PVWatts version 5 manual. National Renewable Energy Lab. (NREL), Golden, CO, USA (2014)
17. Kannal, A.: Solar power generation data. In: Kaggle (2020). https://www.kaggle.com/datasets/anikannal/solar-power-generation-data. Accessed 14 Jan 2025
18. Biromiro, B.: PINN-MultiVP: physics-informed neural networks for multi-value problems. GitHub repository (2022). https://github.com/biromiro/pinn-multivp
19. Tomašević, N., Pešić, A., Milinković, D. Physics-based learning for solar power forecasting. In: Proceedings of the European Symposium on Artificial Neural Networks (ESANN 2024), Bruges, Belgium (2024)
20. Weng, L., Xu, J., Zhao, Y., Zhang, X.: Physics-informed neural networks for energy systems: a review. Front Energy Res **10**, 902722 (2022). https://doi.org/10.3389/fenrg.2022.902722

Factorial Analysis and Optimization of Temperature and Sodium Bicarbonate Effects on the Hydration of Red and White Beans for Canning

Juan Gabriel Mollocana-Lara[1(✉)], Anderson Steve Caranqui Chanalata[2], and Katiuska Jajaira Obando-Proaño[3]

[1] Grupo de Investigación Ambiental en el Desarrollo Sustentable GIADES, Carrera de Ingeniería Industrial, Universidad Politécnica Salesiana, Quito 170525, Ecuador
jmollocana@ups.edu.ec
[2] Carrera de Ingeniería Industrial, Universidad Politécnica Salesiana, Quito 170525, Ecuador
[3] Dirección Técnica de Ecosistemas de Emprendimiento e Innovación, Universidad Politécnica Salesiana, Quito 170525, Ecuador

Abstract. This study investigates the combined effects of temperature and bicarbonate concentration on the final moisture content of red and white beans in a controlled experimental setup. The objective was to optimize the hydration process, a common time-consuming step in the canning industry, where hydration can last up to 24 h without the addition of any accelerating agent. A 4 × 3 factorial design was used, varying temperature (25 °C, 35 °C, 45 °C, and 55 °C) and bicarbonate concentration (0%, 1.5%, and 2%). Statistical analysis, including the validation of parametric assumptions, ANOVA, and Tukey post-hoc test, revealed that both factors and their interaction had significant impacts on moisture retention, with temperature exhibiting a particularly strong effect at higher levels. The interaction between temperature and bicarbonate was more pronounced in red beans compared to white beans, suggesting variety-specific responses. The study demonstrates that increasing temperature and bicarbonate concentration can significantly improve processing efficiency and reduce cycle times in canning production, while ensuring consistent product quality.

Keywords: Hydration process · Experimental Design · Moisture Content · ANOVA

1 Introduction

The hydration of beans is key in the production of canned goods, as it affects texture, cooking time, and the quality of the product. Increasing the temperature accelerates water absorption, improving efficiency (Mba et al., 2019), while ultrasound facilitates water penetration through cavitation (Yadav et al., 2021). Additionally, sodium bicarbonate accelerates grain softening by altering its cellular structure (Nkhata et al., 2022).

J. L. Ballesteros et al. (Eds.): CITIS 2025, LNNS 1780, pp. 449–457, 2027.
https://doi.org/10.1007/978-3-032-14287-0_44

The bean canning process includes stages like reception, selection, cleaning, hydration, cooking, packaging, and sterilization, all with strict quality controls. Hydration ensures uniform water absorption, influenced by factors like water type, soaking time, and pre-treatment (Schoeninger et al., 2017). The increase in soaking water temperature has been shown to enhance water absorption in legumes by increasing the kinetic energy of the water molecules, which accelerates diffusion into the grains (Miano et al., 2019). Cooking focuses on starch gelatinization and protein denaturation for better digestibility and palatability, while aseptic packaging ensures microbiological stability (Bekele & Emire, 2023).

The main challenge in industrial bean hydration is the extended time needed for proper softening, often exceeding 24 h, negatively impacting production efficiency. Prolonged soaking can also encourage microbial growth if water temperature is not properly controlled (Guajardo-Flores et al., 2013). To accelerate the process, alternatives such as alkaline solutions and non-thermal technologies have been proposed, preserving the grain's sensory properties (Didinger et al., 2023). Studies on soaking show that increasing water temperature enhances absorption, but excessive temperatures can damage texture and reduce nutritional value (Miano & Augusto, 2016). The addition of sodium bicarbonate improves hydration by breaking cell wall bonds, but high concentrations may cause discoloration and flavor changes (Ghafoor et al., 2014). Therefore, determining the optimal bicarbonate concentration is essential to balance efficiency and quality (Fernandes et al., 2010).

Several studies have used a Design of Experiments (DOE) methodology to optimize the hydration process of beans by considering key physicochemical variables. For instance, Cichy et al. (2023) applied a completely randomized factorial design to assess the effect of prior soaking on biofortified bean varieties, while Abud-Archilla et al. (2008) optimized osmotic treatment in yam using an L9 design, and Kim et al. (2024) showed that ultrasound intensity and soaking time significantly affect hydration in legumes. Additionally, Makdoud & Rosentrater (2017) employed experimental designs to enhance gluten-free pasta properties using amaranth and quinoa. Gabriele et al. (2023) explored ultrasound-assisted fortification of beans with phenolic compounds from grape skins to improve the nutritional value of the product.

A DOE can be applied to analyze the effect of variables such as temperature, bicarbonate concentration, and soaking time in red and white bean hydration (Andrade, 2024; White et al., 2024). High temperatures accelerate water absorption, but excessive heat can affect texture and nutritional value (Kon, 1979). Bicarbonate improves softening, although high concentrations can alter flavor and increase the percentage of broken beans (Munthali et al., 2022). DOE, using methods like response surface methodology, optimizes these factors to improve cooking efficiency (Fernandez Castaneda et al., 2024).

The aim of this research is to optimize the hydration process of red and white beans for canning production by analyzing the effects of temperature and sodium bicarbonate concentration. A factorial experimental design will be used to identify optimal conditions that maximize water absorption within a set time, comparing the behavior of both varieties. This optimization aims to improve industrial process efficiency by reducing cycle times in the canning process.

2 Methodology

2.1 Design of Experiments

The 4 × 3 factorial design combines two factors, temperature with four levels (25 °C, 35 °C, 45 °C, 55 °C) and sodium bicarbonate concentration with three (0%, 1.5%, 2%), generating 12 unique treatments to evaluate the main effects and their interactions on the moisture uptake. Each experimental combination was tested with three repetitions per bean type (red and white), increasing the accuracy of the results. A total of 72 trials were conducted, allowing for the evaluation of the treatment conditions and performance of the hydration process.

The selection of sodium bicarbonate ($NaHCO_3$) concentrations for bean hydration is based on studies that evaluate its impact on cooking time and the nutritional and sensory quality of the grain. Research has shown that $NaHCO_3$ concentrations above 2% m/m do not significantly reduce cooking time and may alter flavor and texture (Huma et al., 2008). In contrast, concentrations around 1% m/m are effective in softening the beans and improving protein digestibility without notable adverse effects on sensory properties (El-Adawy et al., 2000).

2.2 Experiment Description

The experiments carried out aimed to optimize the hydration of red and white beans under a 4x3 factorial experimental design. For each treatment, 25 g of beans were placed in 600 mL beakers with 500 mL of potable water at four temperatures: 25 °C, 35 °C, 45 °C, and 55 °C. Sodium bicarbonate was added at three concentrations: 0%, 1.5%, and 2% (m/m), using a volumetric flask to ensure proper dissolution. Samples were soaked in a preheated water bath for 3 h. Afterwards, excess surface moisture was removed, and samples were transferred to pre-dried Petri dishes. Moisture content was determined through oven drying at 105 °C, following the ISO 712:2009 standard. Sample weights were recorded every 30 min using an analytical balance until they stabilized. To prevent reabsorption of moisture, samples were cooled in a desiccator before each weighing. Moisture percentage was calculated based on weight loss, allowing for the assessment of the effects of temperature and bicarbonate concentration on final bean hydration.

2.3 Statistical Analysis

The statistical analysis was based on validating the assumptions required for analysis of variance (ANOVA). Hypothesis tests were used to assess normality, homoscedasticity, and independence of the data. The Shapiro-Wilk test was applied to check the normality of residuals, while Levene's test assessed homoscedasticity for temperature, bicarbonate concentration, and their interaction. The Durbin-Watson test was used to verify data independence, ensuring the validity of the ANOVA model and the reliability of the results. Finally, a Tukey post-hoc analysis was used to determinate difference between groups.

3 Results

Table 1, Fig. 1 and Fig. 2 show the average moisture content and standard deviation for three repetitions for red and white beans based on bicarbonate concentration and temperature. As temperature and bicarbonate concentration increase, moisture content rises, particularly at 2% m/m bicarbonate and 55 °C, with final moisture reaching 58.9% for red beans and 69.7% for white beans.

Table 1. Average Moisture and Standard Deviation for Red and White Beans.

Treatment	Red beans		White beans	
	Average Moisture (%)	Standard deviation (%)	Average Moisture	Standard deviation (%)
0% m/m - 25 °C	38.6	0.45	39.3	0,97
0% m/m - 35 °C	42.0	0,56	45.8	0,59
0% m/m - 45 °C	45.6	0,56	52.7	0.39
0% m/m - 55 °C	49.7	0,72	58.6	0,5
1,5% m/m - 25 °C	40.0	1.25	47.9	1.71
1,5% m/m - 35 °C	42.8	1.76	51.8	2.57
1,5% m/m - 45 °C	48.0	0.69	58.7	1.24
1,5% m/m - 55 °C	56.3	1.21	62.2	3.05
2% m/m - 25 °C	45.0	0,74	51.2	0.90
2% m/m - 35 °C	49.6	0,59	56.0	1.17
2% m/m - 45 °C	52.2	0,53	62.3	2.72
2% m/m - 55 °C	58.9	0.65	69.7	0.90

3.1 Statistical Analysis

The ANOVA results show that both temperature and bicarbonate significantly impact the final moisture content of red and white beans (Table 2), with p-values < 0.05 for both factors and their interaction in red beans.

The analysis of normality, independence and homoscedasticity assumptions for red and white beans data meet the necessary assumptions for ANOVA results acceptance, with appropriate p-values in the Shapiro-Wilk and Levene tests, and a value close to 2 for DW statistics (Table 3).

The Tukey test performed a total of 66 pairwise comparisons between all treatment combinations included in the ANOVA. To simplify the analysis, Fig. 3 and Fig. 4 present

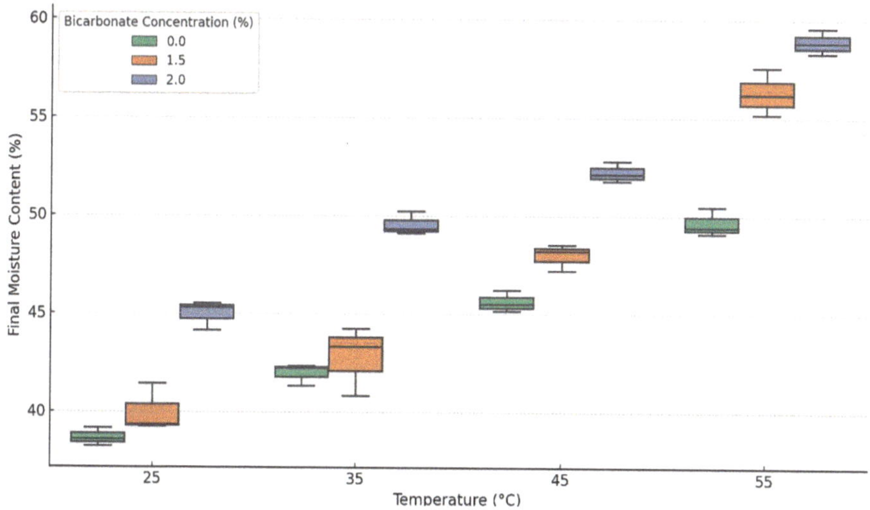

Fig. 1. Effect of temperature and bicarbonate concentration on final moisture in for red beans.

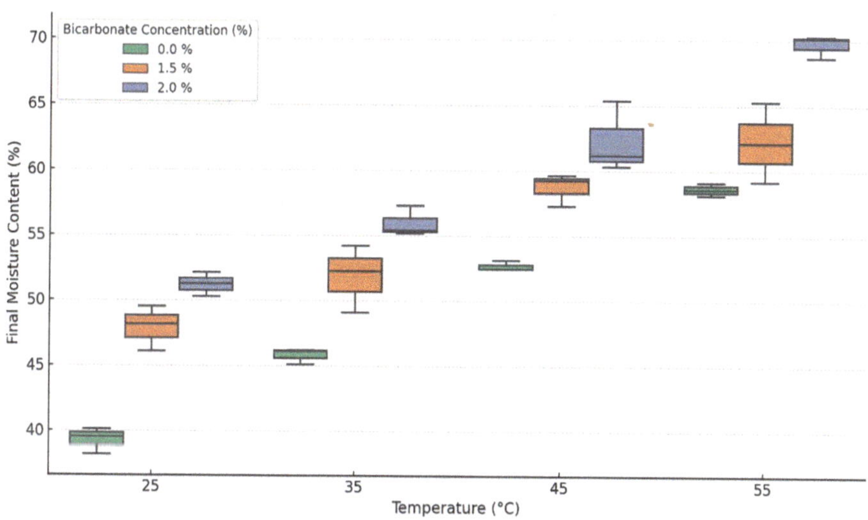

Fig. 2. Effect of temperature and bicarbonate concentration on final moisture in for white beans.

the four treatment pairs with the largest differences in mean moisture content for both red and white beans. In both cases, it can be observed that increasing the sodium bicarbonate concentration at 45 °C yields similar results to those obtained by raising the temperature to 55 °C with lower bicarbonate concentration. This suggests that, for achieving similar final moisture levels, either increasing temperature or increasing bicarbonate concentration are valid strategies. From a practical standpoint, it is necessary to determine whether the operational costs associated with using more bicarbonate are preferable to those of raising the process temperature.

Table 2. ANOVA Results for Red and White Beans.

Factor	Red beans		White beans	
	F	p-value	F	p-value
Temperature	393.180	1.58E−20	192.101	6.59E−17
Bicarbonate	213.535	5.15E−16	127.635	1.62E−13
Interaction	7.368	1.50E−4	1.686	0.168

Table 3. Assumptions analysis for ANOVA red and white beans (RB and WB).

Assumption	Hypothesis Test	RB p-value	WB p-value
Normality	Shapiro-Wilk	0.5331	0.4074
Homoscedasticity	Levene (for temperature)	0.7798	0.8991
Homoscedasticity	Levene (for bicarbonate)	0.3571	0.6296
Homoscedasticity	Levene (for interaction)	0.9347	0.6058
Independence	Durbin-Watson	DW = 2.7691	DW = 2.3586

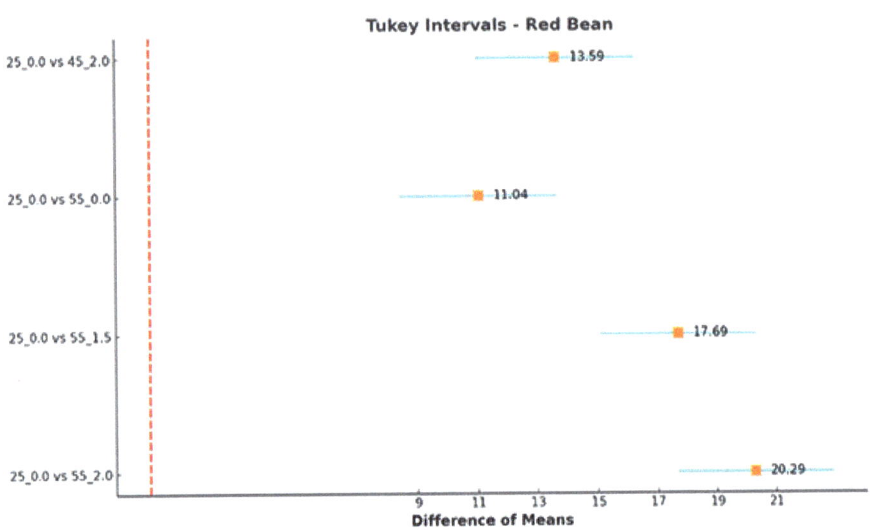

Fig. 3. Tukey intervals for red beans major mean differences.

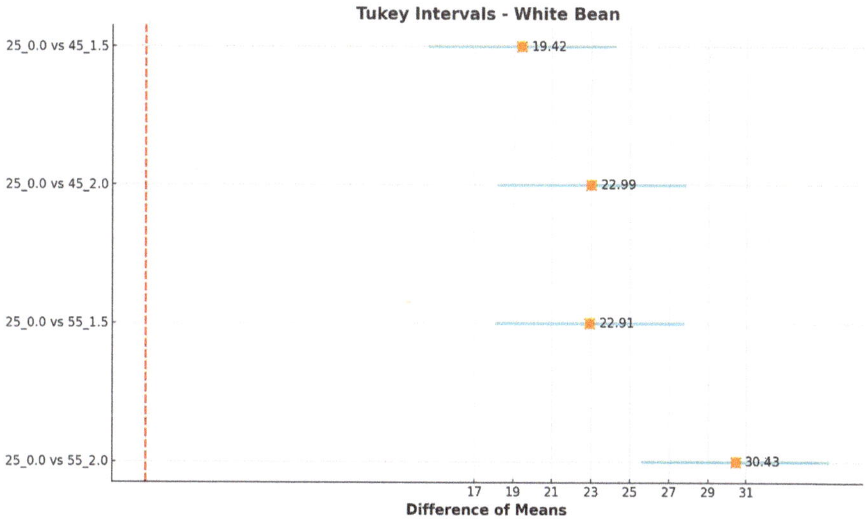

Fig. 4. Tukey intervals for white beans major mean differences.

4 Discussion

The results of this study confirm that both temperature and sodium bicarbonate concentration significantly influence the final moisture content of red and white beans during the hydration process. The factorial ANOVA showed that these two factors, as well as their interaction in red beans, had statistically significant effects, validating the relevance of jointly analyzing them to optimize canning hydration.

The interaction effect was particularly marked in red beans, indicating that this variety is more sensitive to variations in temperature and bicarbonate concentration. This may be due to differences in seed coat composition, cell wall structure, or permeability compared to white beans. For red beans, increasing the sodium bicarbonate concentration at 45 °C achieved moisture levels similar to those obtained by raising the temperature to 55 °C with lower bicarbonate levels. This implies that comparable hydration outcomes can be reached either by adjusting the thermal conditions or by adding chemical agents that modify water absorption dynamics.

In the other hand, white beans also showed increased final moisture content with rising temperature and bicarbonate concentration, but the magnitude of change was less pronounced, suggesting that this variety is less responsive to the combined effect of both factors. This reinforces the need for variety-specific process adjustments in industrial settings.

The results from the Tukey post hoc analysis further support the identification of specific treatment pairs with significantly different hydration outcomes. These differences are crucial for industrial decision-making, as they offer flexibility in process optimization. From a practical standpoint, the choice between increasing temperature or adding sodium bicarbonate should consider not only hydration efficiency but also operational costs, energy consumption, and potential sensory impacts on the final product.

Additionally, the statistical validation of ANOVA assumptions, including normality, homoscedasticity, and independence, confirms the robustness of the analysis and reinforces the reliability of the observed trends. The consistent increase in final moisture with temperature and bicarbonate supports the application of factorial experimental designs to model and optimize bean hydration processes.

Overall, the findings demonstrate the potential of using factorial analysis to define efficient and adaptable hydration protocols that can be tailored to different bean varieties, ultimately improving productivity and quality in the canning industry.

5 Conclusions

Factorial analysis confirmed that both temperature and sodium bicarbonate concentration significantly influence bean hydration, with variety-specific responses. A strong interaction effect was observed in red beans, highlighting their higher sensitivity to processing conditions and reinforcing the need for tailored protocols. Statistical validation supports the robustness of these findings and their application in optimizing industrial processes.

As for future work, it is important to determine whether increasing temperature or bicarbonate concentration is more efficient, since both yielded comparable hydration outcomes. Additionally, the impact of these treatments on the final product after cooking should be assessed to ensure quality and consumer acceptability.

References

1. Abud-Archila, M., et al.: Optimization of osmotic dehydration of yam bean (Pachyrhizus erosus) using an orthogonal experimental design. J. Food Eng. **84**(3), 413–419 (2008). https://doi.org/10.1016/j.jfoodeng.2007.06.003
2. Andrade, C.: Understanding factorial designs, main effects, and interaction effects: Simply explained with a worked example. Indian J. Psychol. Med. **46**(2) (2024). https://doi.org/10.1177/02537176241237066
3. Bekele, D.W., Emire, S.A.: Effects of pre-drying treatment and particle sizes on physicochemical and structural properties of pumpkin flour. Heliyon **9**(11) (2023). https://doi.org/10.1016/j.heliyon.2023.e21609
4. Cichy, K.A., Wiesinger, J.A., Mendoza, F.A.: Cooking properties and nutrient retention of biofortified common bean varieties in Burundi. Front. Sustain. Food Syst. **7** (2023). https://doi.org/10.3389/fsufs.2023.1062978
5. Didinger, C., O'Leary, A., Mitchell, D.C.: The effects of elevation and soaking conditions on dry bean cooking time. Legume Sci. **5**(1) (2023). https://doi.org/10.1002/leg3.207
6. El-Adawy, T.A., Rahma, H., Rahma, H., Sobihah, T.Y.: Effect of soaking process on nutritional quality and protein solubility of some legume seeds. Food/Nahrung **44**(5) (2000). https://doi.org/10.1002/1521-3803(20001001)44:5<339::AID-FOOD339>3.0.CO;2-T
7. Fernandes, A.C., Nishida, W., Uchimura, N.S.: Influence of soaking on the nutritional quality of common beans (Phaseolus vulgaris L.) cooked with or without the soaking water. Int. J. Food Sci. Technol. **45**(11) (2010). https://doi.org/10.1111/j.1365-2621.2010.02395.x
8. Fernandez Castaneda, L.A., et al.: Optimizing soaking and boiling time in the development of tempeh-like products from faba bean (Vicia faba L.). Fermentation **10**(8) (2024). https://doi.org/10.3390/fermentation10080407

9. Gabriele, M., Pucci, L., Tiano, L., Longo, V.: Hydration and fortification of common bean (Phaseolus vulgaris L.) with grape skin phenolics: effects of ultrasound application and heating. Antioxidants **12**(5) (2023). https://doi.org/10.3390/antiox13050615
10. Ghafoor, M., Misra, N.N., Mahadevan, K., Tiwari, B.K.: Ultrasound assisted hydration of navy beans (Phaseolus vulgaris). Ultrasonics Sonochem. **21**(1) (2014). https://doi.org/10.1016/j.ultsonch.2013.05.016
11. Guajardo-Flores, D., Serna-Saldívar, S.O., Gutiérrez-Uribe, J.A.: Evaluation of the antioxidant and antiproliferative activities of extracted saponins and flavonoids from germinated black beans (Phaseolus vulgaris L.). Food Chem. **141**(2) (2014). https://doi.org/10.1016/j.foodchem.2013.04.010
12. Huma, N., Anjum, F.M., Sehar, S., Khan, M.I., Hussain, S.: Effect of soaking and cooking on nutritional quality and safety of legumes. Nutr. Food Sci. **38**(6) (2008). https://doi.org/10.1108/00346650810920187
13. Kim, H., et al.: Optimization of ultrasound-assisted pretreatment for accelerating rehydration of Adzuki bean (Vigna angularis). J. Microbiol. Biotechnol. **34**(4) (2024). https://doi.org/10.4014/jmb.2401.01004
14. Kon, S.: Effect of soaking temperature on cooking and nutritional quality of bean. J. Food Sci. **44**(5) (1979). https://doi.org/10.1111/j.1365-2621.1979.tb06432.x
15. Makdoud, S., Rosentrater, K.A.: Development and testing of gluten-free pasta based on rice, quinoa and amaranth flours. J. Food Res. **6**(4) (2017). https://doi.org/10.5539/jfr.v6n4p91
16. Mba, O.I., Kwofie, E.M., Ngadi, M.: Kinetic modelling of polyphenol degradation during common beans soaking and cooking. Heliyon **5**(5) (2019). https://doi.org/10.1016/j.heliyon.2019.e01613
17. Miano, A.C., Augusto, P.E.D.: Enhancing mung bean hydration using the ultrasound technology: description of mechanisms and impact on its germination and main components. Sci. Rep. **6** (2016). https://doi.org/10.1038/srep38996
18. Miano, A.C., Sabadoti, V.D., Augusto, P.E.D.: Combining ionizing irradiation and ultrasound technologies: effect on beans hydration and germination. J. Food Sci. **84**(11) (2019). https://doi.org/10.1111/1750-3841.14819
19. Munthali, J., Nkhata, S.G., Masamba, K., Mguntha, T., Fungo, R., Chirwa, R.: Soaking beans for 12 h reduces split percent and cooking time regardless of type of water used for cooking. Heliyon **8**(9) (2022). https://doi.org/10.1016/j.heliyon.2022.e10561
20. Nkhata, S.G., Munthali, J., Masamba, K.G., Fungo, R.: Soaking beans for 12 h reduces split percent and cooking time regardless of type of water used for cooking. Heliyon **8**(9) (2022). https://doi.org/10.1016/j.heliyon.2022.e10561
21. Schoeninger, V., Coelho, S.R. M., Bassinello, P.Z.: Industrial processing of canned beans. Ciência Rural **47**(5) (2017). https://doi.org/10.1590/0103-8478cr20160672
22. White, I.R., Szubert, A.J., Choodari-Oskooei, B., Walker, A.S., Parmar, M.K.B.: When should factorial designs be used for late-phase randomised controlled trials? Clin. Trials **21**(1) (2024). https://doi.org/10.1177/17407745231206261
23. Yadav, S., Mishra, S., Pradhan, R.C.: Ultrasound-assisted hydration of finger millet (Eleusine coracana) and its effects on starch isolates and antinutrients. Ultrasonics Sonochem. **73** (2021). https://doi.org/10.1016/j.ultsonch.2021.105542

Taxonomic Characterization of Yeasts from Joyapa (*Macleania Rupestris*) with Fermentative Potential

Myriam Ximena Mancheno Cárdenas(✉) and Guissella de los Ángeles Rivera Yunga

Universidad Politécnica Salesiana, Quito, Ecuador
mmancheno@ups.edu.ec, griveray@est.ups.edu.ec

Abstract. This study aims to identify and characterize a yeast species present in thc fruit of joyapa (*Macleania rupestris*) with potential for alcoholic fermentation. Ripe fruits were collected and cultured in YPD agar at 36 °C for 72 h. Fermentation and ethanol resistance tests were performed using ethanol concentrations of 0%, 3%, 6%, and 9%. Fermentation capacity was evaluated using Durham tube assays, and ethanol resistance was assessed based on gas production and visible growth. Three yeast isolates showing the highest ethanol tolerance were selected and identified through NGS ITS86F-ITS4 sequencing. The results revealed that the dominant species was Saccharomyces cerevisiae, a yeast widely used in the fermentation industry due to its robust fermentative capacity. These findings suggest that native S. cerevisiae strains from M. rupestris could be valuable for industrial applications, reducing dependence on commercial yeasts. Additionally, the adaptation of these yeasts to local environmental conditions may offer advantages in terms of fermentation efficiency and product complexity. This study highlights the potential of Ecuadorian native yeasts in the development of sustainable and innovative fermentation processes.

Keywords: Fermentation · Alcohol Tolerance · *Saccharomyces cerevisiae*

1 Introduction

Alcoholic fermentation is a cornerstone of traditional biotechnology, as it utilizes the metabolic capabilities of microorganisms to convert carbohydrates into ethanol under anaerobic conditions [1]. Among the many microorganisms employed in this process, Saccharomyces cerevisiae stands out as a fundamental biocatalyst [2]. This yeast species is recognized for its ability to convert sugars into ethanol, making it indispensable in industries such as winemaking, brewing, and bioethanol production [3].

Saccharomyces cerevisiae is not only cultivated in controlled environments but is also naturally present in various fruits, where it plays an essential role in spontaneous fermentation. However, spontaneous fermentation is not the result of a single yeast strain but rather a consortium of multiple yeast strains working in synergy. This diversity

J. L. Ballesteros et al. (Eds.): CITIS 2025, LNNS 1780, pp. 458–466, 2027.
https://doi.org/10.1007/978-3-032-14287-0_45

of microbial populations contributes to the complexity and uniqueness of fermented products [2].

By 2018, the WHO reported that in Ecuador, the total alcohol consumption per person was 7.2 L, most of which was in the form of beer, as it is consumed during various occasions such as family gatherings, football matches, and traditional festivities [4]. By 2021, approximately 250 craft beer brands were produced in Ecuador, representing a 20% increase compared to the number of craft beer brands before the pandemic [5].

However, in Ecuador, the fermented beverage industry still depends on the importation of yeasts. This dependency arises from a combination of factors, including the nature of traditional fermentation processes, the transition to industrial production, and the specific requirements of commercial yeast strains. While Ecuadorian traditional fermentation processes use a wide range of microorganisms, local microbial diversity may not fully meet the specific demands of industrial production due to the limited study of these microorganisms [6, 7].

As the craft beer market has grown continuously, consumer interest and expectations have increased, leading to the production of distinctive and innovative beers. Thus, non-conventional yeasts play a leading role in the development of new products [8]. Therefore, it is crucial to identify native yeasts with good alcohol resistance to reduce production costs and promote innovation in the sector.

Studies on other fruits and beverages have shown that native yeasts can adapt to specific environmental conditions, improving fermentation efficiency and flavor complexity [9, 10].

Macleania rupestris, known as joyapa, is an Andean fruit with high sugar and organic acid content. Andean communities consume it for its sweet flavor, both fresh and processed into jams, juices, liquors, etc., making it an ideal substrate for yeast growth [11, 12]. The high sugar content of joyapa is crucial for yeast fermentation, as sugars serve as the primary energy source for yeast metabolism [13, 14]. The organic acids present in the fruit can influence yeast growth and the production of fermentation byproducts, such as fusel alcohols [15]. This study aims to determine the presence of native yeasts in the fruit of joyapa and evaluate their potential for alcoholic fermentation.

2 Materials and Methods

2.1 Sample Collection

Ripe Macleania rupestris fruits were collected from a 1.5-hectare farm located in the Punzara Grande sector, in the province of Loja, Ecuador (coordinates: −4.054230, −79.212800). The experimental phase was carried out in the Microbiology and Biology I laboratories of Universidad Politécnica Salesiana, Cuenca campus. The fruits were transported to the laboratory and placed in flasks with YPD agar medium, incubating at 36 °C for 72 h.

Due to the limited population of Macleania rupestris plants in the selected area, samples were taken from all twelve plants present, which represented the twelve experimental units used in the study. Each sample was placed in an individual flask labeled with a letter from A to L, corresponding to the twelve collected plants.

2.2 Yeast Isolation and Identification

Through streak plating on Petri dishes, yeast colonies were isolated and evaluated microscopically. To eliminate bacteria, 10% gentamicin was used.[16].

Colony characteristics such as color and shape were assessed. A smear of each pure culture obtained was prepared with methylene blue staining. The samples were examined under a microscope using a 100X objective.

2.3 Fermentation and Alcohol Resistance Tests

Fermentation tests were conducted in test tubes with Durham tubes, and alcohol resistance was evaluated in media with 0%, 3%, 6%, and 9% ethanol. The pH was adjusted to 4.2. Colony growth was monitored at 48, 72, and 96 h. For viability analysis, the number of live and dead cells in a 1:10 diluted sample was counted using a Neubauer chamber.[17].

2.4 Genetic Sequencing

The yeast strains with the highest ethanol tolerance were preserved in 30% glycerol at −30 °C and sent to BioSequence for NGS ITS86F-ITS4 sequencing. It is important to highlight that the ITS region is considered a universal barcode for the study of fungi and yeast taxonomy [18].

3 Results and Discussion

3.1 Isolation and Microscopic Characterization

Yeast growth was observed in four out of the twelve flasks. Although colony development was detected in all samples, only flasks corresponding to samples B, C, I, and J exhibited colonies with macroscopic characteristics consistent with typical yeast morphology. Table 1 presents only the samples that showed yeast growth.

It should be emphasized that yeast isolation was only achieved from four of the twelve initial flasks, despite the fact that diverse colony growth was evidenced in all the flasks, however, not all of them presented colony growth with yeast characteristics.

$$\text{Percentage} = \frac{4}{12}(100) = 33.33\%$$

A 33.33% of flasks with yeast mycelia growth was obtained.

3.2 Fermentation and Alcohol Tolerance Tests

In the fermentation tests, the five isolated strains were evaluated, and the results are presented in Table 2.

At 24 h, samples B2, B4, and I2 demonstrated fermentative capacity, as CO2 production was observed. However, by 48 and 72 h, all samples had undergone fermentation,

Table 1. Appearance of cells as seen under the microscope.

Code Petri Dishes	Yeast growth	Bacterial growth
B1	–	+
B2	+	–
B3	–	–
B4	+	–
C1	–	+
C2	+	–
I1	–	–
I2	+	–
J1	–	–
J2	+	–

Note: The appearance of the cells is evaluated, with (+) being consistent with the characteristics and (–) not visualized

Table 2. Fermentation test results.

Yeast code	24 h	48 h	72 h
B2	+	+	+
B4	+	+	+
C2	–	+	+
I2	+	+	+
J2	–	+	+

Note: Fermentative capacity is evaluated, with (+) indicating gas production in the Durham tube and (–) indicating no gas production

confirming their ability to ferment glucose. Nevertheless, B2, B4, and I2 exhibited a faster metabolism in sugar conversion.

Viability was assessed at 48, 72, and 96 h. The results are presented in Table 3.

The viability percentages at 48 h with 0% ethanol were quite high, ranging from 93.33% to 100%. According to theoretical recommendations, yeast cell viability should be at least > 95% in brewery yeast management, as viability is the primary analysis criterion for assessing yeast quality and condition [19]. Additionally, viability is expected to decrease over time. Experiments have shown that external ethanol affects cell viability, determining that after 3 days of incubation at 30 °C, no viable cells could be detected in the presence of 12% ethanol [20].

The Shapiro-Wilk test was applied to evaluate data normality, confirming a normal distribution. Therefore, parametric tests such as ANOVA were used.

Table 3. Viability percentage results.

Yeast code	0% Ethanol			3% Ethanol			6% Ethanol			9% Ethanol		
	48 h	72 h	96 h	48 h	72 h	96 h	48 h	72 h	96 h	48 h	72 h	96 h
B2	97,44	88,89	49,21	97,67	83,05	46,77	82,50	72,22	38,89	63,58	67,36	61,11
B2	95,35	91,30	34,04	95,12	79,31	50,77	81,25	77,19	37,29	71,47	69,81	33,33
B2	100,00	94,00	55,10	100,00	74,51	40,43	86,54	83,93	43,24	62,22	64,51	31,03
B4	100,00	93,75	44,90	87,50	78,18	47,83	90,70	80,36	51,79	63,00	58,50	33,33
B4	98,56	97,92	48,00	92,11	83,48	43,90	86,00	83,72	36,96	75,47	73,95	33,33
B4	95,83	100,00	46,67	93,75	82,98	35,19	81,25	79,17	27,27	68,93	69,92	45,16
C2	93,33	92,50	36,00	95,00	76,47	38,64	65,91	74,42	30,77	0,00	0,00	9,00
C2	96,69	91,49	49,02	94,12	79,63	42,59	63,16	72,22	44,19	0,00	0,00	12,00
C2	98,65	95,45	38,89	89,13	74,00	33,93	63,89	69,49	34,88	0,00	0,00	0,00
I2	99,16	87,76	43,40	93,18	75,00	48,33	90,70	68,97	34,69	65,75	74,51	27,27
I2	98,79	93,48	50,00	95,00	80,00	42,86	81,82	80,00	32,65	70,73	68,60	33,33
I2	100,00	90,70	32,50	95,45	76,00	28,89	87,80	75,00	50,00	75,93	55,35	42,62
J2	98,77	93,33	34,15	88,89	75,00	32,69	72,09	71,43	43,90	35,93	21,89	0,00
J2	96,56	90,57	41,82	95,45	83,72	41,67	66,00	71,74	27,08	31,97	19,97	0,00
J2	94,15	93,75	53,45	88,37	83,67	31,11	73,47	82,22	32,50	32,00	19,68	0,00

After applying ANOVA, the population means for 0%, 3%, and 6% ethanol at 96 h were found to be equal, indicating no significant difference. However, when evaluating the population exposed to 9% ethanol at 96 h, the following ANOVA result was obtained (Table 4):

The Tukey test was applied at 96 h to obtain the subsets (Table 5).

Subset 3 consists of the strains that exhibited the highest viability at 96 h in a medium with 9% ethanol: I2, B4, and B2. It is noteworthy that these three samples also showed the highest viability at 48 and 72 h in the 9% ethanol medium.

3.3 Molecular Identification

NGS sequencing revealed that the selected strains belong to *Saccharomyces cerevisiae* (Table 6).

Analyzing the results, the samples predominantly relate to the Saccharomyces genus, as all three samples showed a percentage higher than 99%. Additionally, other genera, such as Kazachstania, were present in smaller proportions. Regarding the most representative species-level taxa, between 47.15% and 49.42% of the three samples corresponded to unclassified species. The next most representative classification was Saccharomyces sp., accounting for 46.324% in B2, 43.75% in B4, and 45.10% in I2.

Table 4. ANOVA for viability with 9% ethanol.

		Sum of squares	df	Mean squares	F	Sig
48 h * Yeast code	Between groups (combined)	11405,546	4	2851,386	150,221	,000
	Within groups	189,813	10	18,981		
	Total	11595,359	14			
72 h * Yeast code	Between groups (combined)	12202,488	4	3050,622	90,275	,000
	Within groups	337,925	10	33,793		
	Total	12540,413	14			
96 h * Yeast code	Between groups (combined)	4401,896	4	1100,474	12,926	,001
	Within groups	851,374	10	85,137		
	Total	5253,269	14			

Table 5. Tukey Test Subsets at 96 h and 9% Ethanol.

Yeast code	Subset		
	N	1	2
J2	3	,000000000000000	
C2	3	7,000000000000000	
I2	3		34,409670475244200
B4	3		37,275985663082400
B2	3		41,825197956577234
Sig		,879	,856

Note: The significance level is compared with α = 0.05, where lower values indicate a significant difference

Table 6. Most Representative Taxa at the Species Level.

Sample		Classification	Reads	%
B2	1	Unclassified at Species level	86,926	47,15%
B2	3	Saccharomyces Saccharomyces_sp	85,236	46,24%
B2	4	Saccharomyces Saccharomyces_eubayanus	9,191	4,99%
B2	5	Saccharomyces Saccharomyces_mikatae	1,153	0,63%
B2	2	Saccharomyces Saccharomyces_kudriavzevii	499	0,27%
B2	6	Ascochyta Ascochyta_rabiei	162	0.09%
B2	7	Talaromyces Talaromyces_ruber	78	0.04%
B2	8	Lachancea Lachancea_lanzarotensis	73	0.04%
B4	1	Unclassified at Species level	117,222	49,42%
B4	2	Saccharomyces Saccharomyces_sp	103,768	43,75%
B4	3	Saccharomyces Saccharomyces_eubayanus	12,312	5,19%
B4	4	Saccharomyces Saccharomyces_mikatae	1,380	0,58%
B4	5	Saccharomyces Saccharomyces_kudriavzevii	632	0,27%
B4	6	Ascochyta Ascochyta_rabiei	264	0.11%
B4	7	Talaromyces Talaromyces_ruber	115	0.05%
B4	8	Lachancea Lachancea_lanzarotensis	96	0.04%
I2	1	Unclassified at Species level	118,369	48,11%
I2	2	Saccharomyces Saccharomyces_sp	110,967	45,10%
I2	3	Saccharomyces Saccharomyces_eubayanus	12,578	5,11%
I2	4	Saccharomyces Saccharomyces_mikatae	1,474	0,60%
I2	5	Saccharomyces Saccharomyces_kudriavzevii	720	0.29%
I2	6	Ascochyta Ascochyta_rabiei	249	0.10%
I2	7	Talaromyces Talaromyces_ruber	120	0.05%
I2	8	Penicillium Penicillium_corylophilum	91	0.04%

Note: A total of 191 taxonomic categories were identified at the species level. This table presents the top 8 classifications out of 191, obtained from BioSequence

4 Conclusions

From the ripe joyapa (Macleania rupestris) fruits sampled in the city of Loja, Loja province, it was possible to isolate pure cultures with yeast colony characteristics, achieving a yield of 33.33% from the obtained samples.

The B2, B4, and I2 yeast cultures exhibited a more efficient fermentative capacity for sugars such as glucose compared to C2 and J2, which only showed fermentation starting at 48 h.

Yeasts capable of fermenting sugars and exhibiting alcohol resistance were successfully isolated from the B2, B4, and I2 cultures, as they were able to grow in a medium with 9% ethanol under conditions of 36 °C and pH 4.2. Through NGS sequencing (Amp ITSA1, 20000R, ITS86F-ITS4) for fungal and yeast phylogenetics, the three isolated samples were taxonomically classified as Saccharomyces sp. This finding opens up possibilities for utilizing native yeasts in the fermentation industry.

It is necessary to practically apply the yeast obtained in this experimental study to determine its capacity and efficiency in ethanol production, as well as to evaluate its potential for industrial applications.

References

1. Machado, A.R., Machado, M.I.R., Viana, L.F.: Use of biotechnology to produce alcoholic and acetic fermentation: a review. Res. Soc. Dev. **12**(12), e102121223288 (2023). https://doi.org/10.33448/rsd-v12i12.23288
2. Praveen, M., Brogi, S.: Microbial fermentation in food and beverage industries: innovations, challenges, and opportunities. Foods **14**(1), 114 (2025). https://doi.org/10.3390/foods14010114
3. Elhalis, H.: Expanding the hrizons of saccharomyces cerevisiae: nutrition, oenology, and bioethanol production. Sustainability **16**(24), 11151 (2024). https://doi.org/10.3390/su16241151
4. Cajas, P., Cauja, L., León, F.: Plan de negocio para la creación de una empresa dedicada a la producción y comercialización de la cerveza artesanal en la ciudad de Guayaquil, p. 6 (2021). https://repositorio.uide.edu.ec/handle/37000/4631
5. Coba, G.: El número de marcas de cerveza artesanal crece un 20% en Ecuador (2021). https://www.primicias.ec/noticias/economia/marcas-cerveza-artesanal-crecimiento-ecuador/. Accessed 22 Apr 2022
6. Mayorga, G.A., Chacón Palma, G.B., Arias Sandoval Cañas, G.J., Ordoñez Araque, R.H.: Ancestral fermented indigenous beverages from South America made from cassava (Manihot esculenta). Food Sci. Technol. **41**(suppl 1), 360–367 (2021). https://doi.org/10.1590/fst.15220
7. Jimenez, M.E., O'Donovan, C.M., de Ullivarri, M.F., Cotter, P.D.: Microorganisms present in artisanal fermented food from South America. Front. Microbiol. **13** (2022). https://doi.org/10.3389/fmicb.2022.941866
8. Burini, J.A., Eizaguirre, J.I., Loviso, C., Libkind, D.: Non-conventional yeasts as tools for innovation and differentiation in brewing. Rev. Argent. Microbiol. **53**(4), 359–377 (2021). https://doi.org/10.1016/j.ram.2021.01.003
9. Castrillo, D., Blanco, P.: Preliminary study on yeasts associated with the production of 'Tostado'—a traditional sweet wine from Galicia (NW Spain). Front. Biosci.-Elite **16**(1) (2024). https://doi.org/10.31083/j.fbe1601010
10. Franco, W., Benavides, S., Valencia, P., Ramírez, C., Urtubia, A.: Native yeasts and lactic acid bacteria isolated from spontaneous fermentation of seven grape cultivars from the Maule Region (Chile). Foods **10**(8), 1737 (2021). https://doi.org/10.3390/foods10081737
11. Caiza Caizalitin, K.G.: Elaboración de una barra energética a base de Macleania rupestris (Kunth) A.C. Sm. como suplemento alimenticio. (8.5.2017), 2003–2005 (2022)
12. Siccha Sánchez, R.E.: Aislamiento de levaduras Saccharomyces spp a partir de bayas de uva para su aplicación en procesos de fermentación de cervez. Universidad Nacional Mayor de San Marcos (2021)
13. Ortiz Rodríguez, K.: Evaluación de métodos de propagación de Macleania rupestris (Joyapa) mediante la germinación de semillas y micropropagación de meristemas (2019)
14. Amaya, N., Rincón, H.: Contribución al estudio de la uva camarona (Macleania rupestris HBK) y su posible utilización en la industria alimenticia (1989)

15. Whiting, G.C.: Organic acid metabolism of yeasts during fermentation of alcoholic beverages-a review. J. Inst. Brew. **82**(2), 84–92 (1976). https://doi.org/10.1002/j.2050-0416.1976.tb03731.x
16. Heydariyan, Z., Soofivand, F., Dawi, E.A., Al-Kahdum, S.A.A., Hameed, N.M., alavati-Niasari, M.: A comprehensive review: different approaches for encountering of bacterial infection of dental implants and improving their properties (2023). https://doi.org/10.1016/j.jddst.2023.104401
17. Heredia Peñafiel, K.E., Kwok Tacan, E.C.: Aislamiento e identificación de las taxa de levaduras presentes en el fruto de taxo (Passiflora mollissima), con capacidad fermentativa y resistencia alcohólicas. https://dspace.ups.edu.ec/handle/123456789/9052
18. Esteve-Zarzoso, B., Belloch, C., Uruburu, F., Querol, A.: Identification of yeasts by RFLP analysis of the 5.8S rRNA gene and the two ribosomal internal transcribed spacers. Int. J. Syst. Evol. Microbiol. **49**(1), 329–337 (1999). https://doi.org/10.1099/00207713-49-1-329
19. Hill, A.E.: Brewing Microbiology: Managing Microbes, Ensuring Quality and Valorising Waste. Elsevier (2015)
20. Ghareib, M., Youssef, K.A., Khalil, A.A.: Ethanol tolerance ofSaccharomyces cerevisiae and its relationship to lipid content and composition. Folia Microbiol. **33**(6), 447–452 (1988). https://doi.org/10.1007/BF02925769

Performance Assessment of Mother Wavelets for Compression and Reconstruction of Transient Electrical Signals

Milton Ruiz(✉), Jonathan Vasco, Diego Carrión, and Marcelo García

Department of Electrical Engineering, GIREI Research Group, Universidad Politécnica Salesiana, Quito 170525, Ecuador
mruizm@ups.edu.ec

Abstract. The present research addresses improving the storage and processing of electrical signals within smart grids by developing and using an advanced compression and restoration algorithm based on mother wavelets. The main focus of the study is to determine the optimal balance between compression efficiency and restoration accuracy by evaluating various wavelets and different compression levels. Specific metrics are employed to assess the algorithm's performance, such as the Normalized Mean Squared Error (NMSE), the Relative Time Error (RTE), and the Correlation (COR). The research process includes selecting and experimenting with at least five different types of mother wavelets and testing ten different compression levels. An iterative method is applied for compressing and subsequently restoring transient electrical signals, including events such as voltage sags (SAG), voltage swells (SWELL), voltage impulses, and stable signals. Simulations carried out in MATLAB have identified Daubechies 4 (db4) and Symlet 3 (sym3) as the most effective wavelets. These wavelets have demonstrated excellent performance, with RTE values of 0.499 and 0.5, NMSE values of 0.1716 and 0.1715, and correlations of 0.7 and 0.707, respectively. The results suggest that the use of these wavelets, because of their high efficiency and compression capability, could be especially beneficial for optimizing signal management in smart electrical grids. This finding opens new avenues for future research focused on integrating these algorithms into real operating systems to improve response capability and energy efficiency in modern network infrastructures.

Keywords: Compression Algorithm · Wavelet Transform · Data Processing

1 Introduction

The analysis and processing of electrical signals is an essential component to ensure the reliability and quality of today's electrical systems. Data compression and restoration, based on frequency analysis techniques, are presented as vital tools to optimize both the transmission and storage of electrical signals, ensuring their full integrity and facilitating subsequent analysis [1].

J. L. Ballesteros et al. (Eds.): CITIS 2025, LNNS 1780, pp. 467–480, 2027.
https://doi.org/10.1007/978-3-032-14287-0_46

Electrical signal compression aims to reduce the amount of data required to accurately represent a signal while retaining the essential and relevant information. Compressed sensing, a prominent technique in this field, facilitates the acquisition and reconstruction of sparse signals from a reduced number of samples. This technique has proven particularly effective for detecting harmonics in electrical signals, which significantly contributes to power quality improvement. In technical terms, compressed sensing employs advanced algorithms to identify the significant components of a signal, thereby allowing the original signal to be reconstructed with astonishing fidelity using fewer data than traditionally required [15].

Frequency analysis plays a crucial role in decomposing a signal into its frequency components. It provides a clear representation in the frequency domain, which is fundamental for identifying and separating features such as noise and harmonics that can be critical in many practical applications, from electrical engineering to acoustics and medicine. The Fourier transform, for example, is a key tool in this analysis since it converts signals from the time domain to the frequency domain, enabling a more detailed and specific analysis [4].

Besides the Fourier transform, other mathematical models and transforms, such as the Laplace transform and the Hilbert-Huang transform, are also frequently used to analyze non-stationary signals or signals with significant transient components. These mathematical tools are essential for engineers and scientists who work with complex and varied data that requires in-depth analysis and understanding [5].

The combination of compression techniques based on frequency analysis not only optimizes data management and storage but also improves key performance indicators such as the Normalized Mean Squared Error (NMSE), Relative Time Error (RTE), and Correlation (COR) of electrical signals. For example, the appropriate selection of the mother wavelet is fundamental for the frequency analysis of transient signals, as it directly influences both the compression precision and the quality of data restoration [12].

The application of the wavelet transform in data compression has extensive and varied practical use, benefiting multiple sectors. In telecommunications, for instance, it enables high-speed data transmission with low bandwidth consumption, which is essential for real-time applications and emerging technologies such as the Internet of Things (IoT). In data science and big data analysis, wavelets help efficiently manage and analyze large volumes of data, facilitating the detection of patterns and anomalies [6].

The practical applications of these algorithms range from data transmission in networks to cloud storage and archiving, the reconstruction and analysis of scientific data, biomedical signal processing, and even the compression of images and videos. In each of these fields, the ability to efficiently compress and reconstruct data not only improves storage economy and transmission speed but also ensures that the information's integrity and quality remain intact.

For example, in network data transmission, the use of wavelet-based compression algorithms can significantly reduce the bandwidth required, a crucial aspect for environments with limited connectivity, such as smart grids and mobile networks. In cloud storage, efficient compression allows the economical storage of enormous volumes of data, including historical records of faults and measurements, while providing quick access to high-quality information.

In conclusion, the development of algorithms that integrate compression and restoration techniques based on frequency analysis represents a promising strategy to improve the efficiency and precision of electrical signal processing. This not only contributes to the advancement of more robust and efficient electrical systems but also plays a crucial role in adapting these technologies to the increasingly demanding and dynamic needs of the modern world.

This comprehensive and multifaceted approach to the analysis and management of electrical signals highlights the importance of continuous innovation and research in electrical and electronic engineering, ensuring that the systems and technologies relying on these principles can continue to evolve and meet future challenges.

2 Related Work

The authors [1] propose the compression of electrical signals applied to smart grids, owing to the large number of devices required for real-time monitoring and observation in distribution systems. Key compression techniques are reviewed, such as lossless and lossy methods, including Fourier, Wavelet, and hybrid techniques that balance the compression-distortion relationship. The authors identify challenges, such as the creation of standardized databases and specific evaluation metrics, and propose developing cognitive and cooperative techniques to improve efficiency in smart grids.

The research [2] presents a real-time compression method applied to electrical signals. It implements the 2D wavelet lifting transform and the Deflate algorithm. This method transforms 1D periodic signals into 2D, optimizing the reduction of redundancies within cycles and between them. Then, the Deflate algorithm is applied to achieve lossless compression. Tests conducted with real data from dynamic monitoring devices achieved a compression ratio close to 4, validating the effectiveness and applicability of the approach for managing large volumes of data in electrical grids. This process significantly improves the storage and transmission of information without compromising reliability.

The authors [3] have developed an efficient data compression algorithm based on Singular Value Decomposition (SVD) for smart distribution systems, aimed at addressing challenges related to the storage and transmission of large volumes of information. Through simulations carried out on the IEEE 14-bus system under various signal conditions (voltage, harmonics, and faults), it was shown that the algorithm can significantly reduce data size while improving the signal-to-noise ratio (SNR) and reducing the mean squared error (MSE), outperforming techniques such as the Discrete Wavelet Transform (DWT). This confirms its effectiveness in optimizing communication and ensuring data quality in smart electrical systems.

The article [4] proposes an efficient data compression algorithm for smart distribution systems using Singular Value Decomposition (SVD). Its main objective is to address the challenge of massive storage and transmission of data generated by smart devices in electrical grids. The algorithm uses SVD to reduce the dimensionality of data matrices while preserving essential characteristics for reconstruction. The results obtained with simulated signals from the IEEE 14-node system demonstrate that the algorithm achieves a significant reduction in data size, improves the signal-to-noise ratio (SNR), and reduces the mean squared error (MSE), surpassing traditional techniques such as the discrete wavelet transform. This approach shows an optimal balance between the compression ratio (CR) and precision, allowing its implementation in smart grids to improve the quality and efficiency of data communication.

The authors [7] propose a novel data compression technique for electrical signals, combining wavelet transforms and compressed sensing. The methodology includes three algorithms: the first identifies key signal features, such as zero-crossings and amplitude; the second applies a biorthogonal wavelet transform, adjusting shifts and amplitudes with error correction and filtering to minimize noise; and the third implements Compressive Sampling Matching Pursuit, creating a dictionary with orthogonal bases to represent the signals as sparse vectors. The results show a compression ratio of 1020:1 (99.90% compression), with outstanding quality metrics (RTE = 0.9938, NMSE = 0.0098, COR = 0.99), surpassing previous high-impact studies. The authors [8] present an algorithm based on the Haar wavelet transform for the detection and classification of faults in transmission lines. Voltage and current signals, which encompass all information of the power system, experience characteristic disturbances during faults, such as changes in amplitude, phase, and harmonics. The property of wavelets to respond to sudden signal changes allows faults to be identified in the time domain. Experimental results demonstrate that at frequencies above 100 kHz, faults can be detected with 100% accuracy using only 4 samples of the signal processed with the Haar wavelet. Detection times range between 0.31 ms and 1.15 ms, while classification times vary between 0.94 s and 1.31 s.

The research [9] analyzes the use of the Hilbert-Huang Transform (HHT) and the Discrete Wavelet Transform (DWT) to monitor deviations in electrical magnitudes in real time in power systems with distributed generation (DG). The HHT decomposes signals using Empirical Mode Decomposition (EMD) into Intrinsic Mode Functions (IMF) ordered by frequency, whereas the DWT separates signals into detail and trend components. A comparative analysis reveals that the DWT offers greater precision and temporal resolution for identifying Power Quality (PQ) events, such as voltage sags, swells, and harmonics. The real-time implementation of the DWT, processing signals sequentially through a buffer, allows the extraction of key attributes for classifying disturbances. Simulations in MATLAB/Simulink and experimental validation confirm that the DWT detects disturbances with high precision and minimal delay, establishing it as an effective tool for real-time PQ analysis.

The article [10] proposes a novel methodology for data compression of power quality signals using the Matching Pursuit Compressed Sensing approach. This methodology integrates wavelet techniques and compressed sensing to improve the efficiency in handling power quality data. This research advances the management of large volumes of data generated by modern power systems, aiming to enhance real-time monitoring and management of power quality through advanced data compression and reconstruction techniques. This could have broad implications for the energy sector, particularly in improving the efficiency of distribution and power quality monitoring systems.

In the article [13], the authors propose a technique focused on improving the efficiency of electrical signal compression by using novelty detection and frequency estimation methods. Its objective is to allow continuous and effective recording of waveforms in power quality systems, thus optimizing bandwidth usage and data storage in electrical network monitoring systems.

The research [14] on data compression using the Fourier Transform (FT) and the Fast Fourier Transform (FFT) has shown significant advances in various fields. These techniques are particularly useful in compressing images and signals, optimizing both storage efficiency and data transmission. For example, the FFT, being a computationally efficient version of the FT, allows rapid and effective spectrum analysis, ideal for digital signal and image processing. It is used to reduce the number of arithmetic operations required, significantly accelerating the data compression and decompression process. In the context [15], in image compression, the combination of FFT with the Discrete Cosine Transform (DCT) has allowed achieving higher compression ratios, especially in wireless sensor networks where transmission efficiency is crucial due to energy and bandwidth limitations. Furthermore, [16] explored the application of FT and FFT in audio and video compression, where data blocks are manipulated to minimize discontinuities between frames, thus reducing file size while maintaining acceptable quality. This technique has been used in popular algorithms such as JPEG for images and MP3 for audio.

In summary, [17], FT and FFT are fundamental in the development of data compression techniques that seek a balance between size reduction and quality preservation. These methods continue to evolve, offering new possibilities for the efficient management of large volumes of data in various technological applications.

He worked on this topic [11], presenting advanced methods for compressing electrical power signals, which is crucial for efficient monitoring and power quality management in modern electrical systems.

This is an example of a quote.

3 Problem and Methodology

In today's era, smart electrical grids are undergoing a significant transformation with the extensive incorporation of a wide range of sensors distributed across all layers of the power system (SEP). This phenomenon has triggered an exponential growth in the amount of generated data, which in turn poses challenges and opportunities in processing and managing this information. Mother wavelets, known for their effectiveness in compressing and restoring electrical signals, have emerged as a promising solution. These tools allow for an effective representation of complex signals, ensuring the retention of their most crucial characteristics, which is essential for maintaining the integrity and utility of the processed information.

The variety of available mother wavelets offers a range of distinct properties, each suitable for different types of signals and applications. This presents a significant challenge: the need for a meticulous evaluation to select the optimal mother wavelet that best suits the specific needs of each application. This selection process is vital to optimize both the compression and restoration of signals within the context of smart electrical grids.

Additionally, assessing the quality of compression and restoration of electrical signals requires robust analytical tools. Quantitative metrics such as Relative Time Error (RTE), the correlation coefficient (COR), and the Normalized Mean Squared Error (NMSE) are indispensable. These metrics provide a clear measure of the performance of the algorithms used by evaluating their accuracy, response time, and the degree of correlation with the original signals. Such evaluations are crucial to ensure that the compression and restoration are not only efficient but also maintain a quality that does not compromise the functionality of the dependent applications.

Consequently, the central challenge of this study is to conceptualize and develop a versatile algorithm capable of using various mother wavelets and their different compression levels to compress and restore a variety of electrical signals. This algorithm must seek an ideal balance between compression, precision, and efficiency—elements necessary for high-quality data processing in smart grid environments. Such a balance is fundamental to support advanced and efficient data processing, which is essential in modern energy systems management, where the speed and accuracy of the processed information can have significant implications on the operation and safety of the electrical system.

3.1 Case Studies

Electrical systems are frequently subjected to transient disturbances that degrade performance and shorten the lifespan of critical components such as sensors, meters, and other connected devices. These disturbances—collectively referred to as *transient signals*—comprise a range of phenomena with distinct characteristics that impact the grid in various ways.

Figure 1 illustrates the main types of transient events commonly observed in power systems:

- **Voltage Sag (SAG):** A temporary drop in voltage, typically ranging from 10% to 90% of the nominal value. Common causes include short circuits, motor startups, and excessive load demand. Sags can lead to equipment malfunction, automation process failures, and increased operational costs. Mitigation strategies include infrastructure improvements, voltage regulation systems, and dynamic load balancing.
- **Voltage Swell (SWELL):** An overvoltage condition where the voltage exceeds 110% of nominal, often caused by poor transformer regulation or sudden load disconnections. Swells may result in insulation breakdown, overheating, and equipment damage. Countermeasures include the use of Transient Voltage Surge Suppression (TVSS) devices and automatic voltage stabilizers.
- **Voltage Impulse:** Brief but intense voltage spikes, typically lasting microseconds to milliseconds, caused by lightning, switching events, or transformer failures. These impulses can damage sensitive electronic components and communication systems. Protection involves surge protection devices (SPD) and effective grounding systems.
- **Stable Signal:** Represents a normal operating condition where voltage and current remain within nominal values. Stability ensures optimal system performance, energy efficiency, and longevity of equipment. Maintaining this condition requires real-time monitoring, reactive compensation, and compliance with standards like IEEE 519 and IEC 61000.

3.2 Solution Methods

In the study presented in this article, an algorithm will be implemented for the analysis of mother wavelets at different compression and restoration levels. The first stage consists of data preparation and selection—in this case, transient electrical signals such as voltage sags (SAG), voltage swells (SWELL), voltage impulses, and a stable signal. Next, a selection and testing will be carried out with the following mother wavelets:

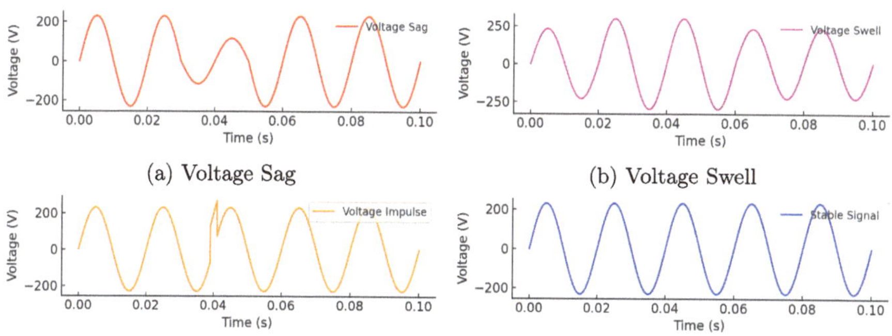

(a) Voltage Sag (b) Voltage Swell

(c) Voltage Impulse (d) Stable Signal

Fig. 1. Typical transient signals observed in power systems.

- Daubechies (db);
- Symlets (sym);
- Coiflets (coif);
- Haar;
- Biorthogonal (bior);
- Mexhaar;
- Gauss;

An iterative module is implemented to test each wavelet and assign the compression level (ranging from level one to ten). The objective of this stage is to implement the transforms of the mother wavelets for the theoretical and experimental evaluation of their compression properties. Subsequently, signal reconstruction is performed from the compressed coefficients for each mother wavelet, and the NMSE, RTE, and COR indices are calculated for each selected wavelet. Tests conducted under multiple scenarios and conditions ensure the robustness of the algorithm through systematic comparison of performance for each signal type—both transient and stable—with optimization and parameter adjustment based on the obtained results.

The figures illustrate the behavior of average Normalized Mean Square Error (NMSE) and Relative Time Error (RTE) across different decomposition levels using various mother wavelets. It is observed that as the decomposition level increases, the NMSE rises significantly, indicating a notable deterioration in performance at higher decomposition levels. Conversely, the RTE decreases progressively with increasing compression levels, approaching values close to zero, demonstrating improved temporal accuracy at higher compression levels (Fig. 2).

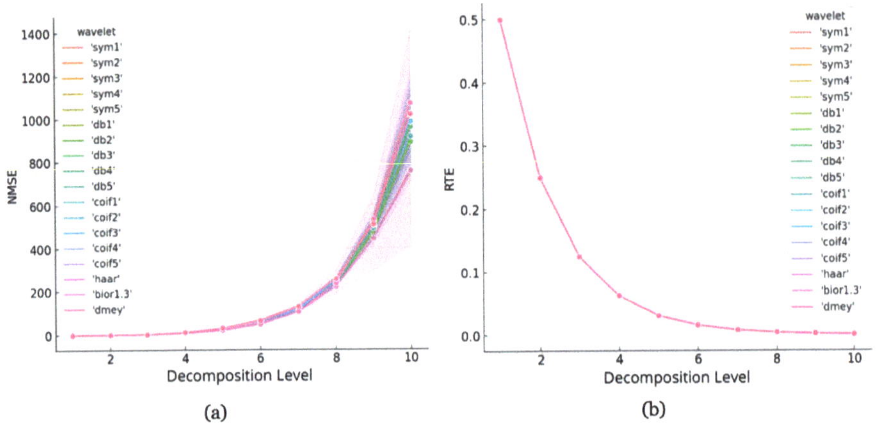

Fig. 2. Performance analysis of NMSE and RTE for different decomposition levels.

Comparison of correlation values across decomposition levels for various mother wavelets when applied to an impulse signal. The graph illustrates a general decreasing trend in correlation as the decomposition level increases. This behavior suggests that higher levels of decomposition may result in a loss of structural fidelity during signal reconstruction. Among all evaluated wavelets, coif1 maintains one of the highest correlation values at the initial decomposition levels, indicating superior performance in preserving the characteristics of the original impulse signal (Fig. 3).

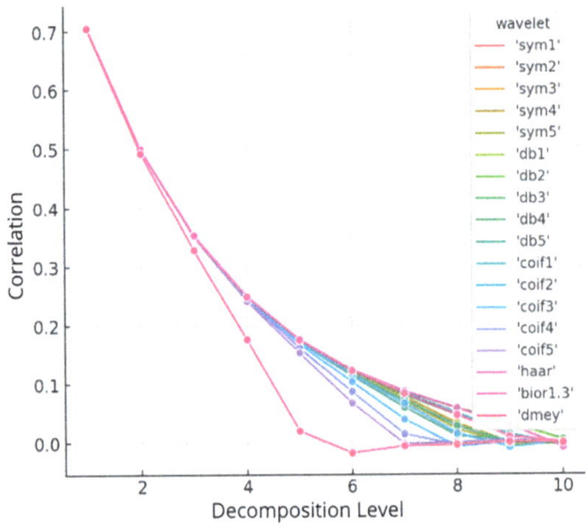

Fig. 3. Performance analysis of correlation for different decomposition levels.

The results demonstrate an effective compression-reconstruction relationship, characterized by high-quality performance indicators such as a compression time of 0.002698 seconds and a restoration time of 0.0017876 seconds. Additionally, the method achieved outstanding evaluation metrics (RTE = 0.9998, NMSE = 0.00998, COR = 0.99991), surpassing those reported in previous high-impact studies. These findings enabled the identification of the most suitable mother wavelet for the compression and reconstruction of electrical signals.

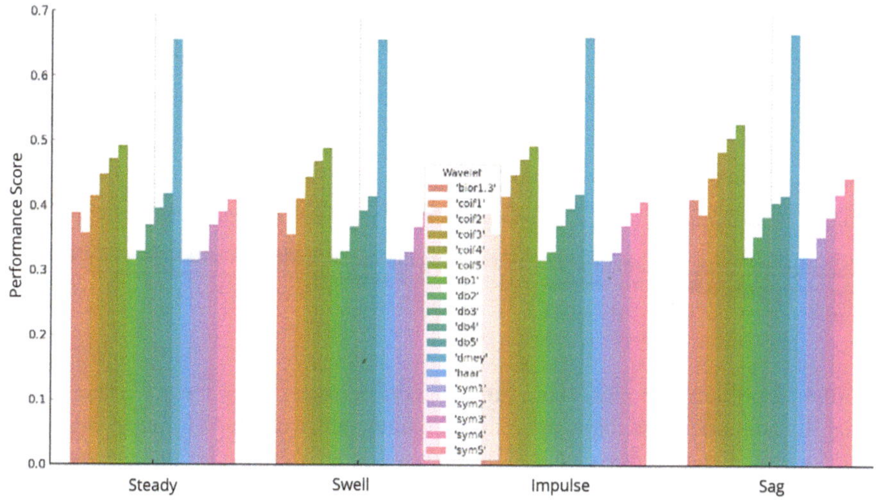

Fig. 4. Performance Metrics of the Proposed Compression and Reconstruction Method for Electrical Signals.

As shown in Table 1, a summary of all performance indices is presented, providing a combined evaluation of the mother wavelets across all tested signals. The table highlights representative values obtained through the iterative algorithm applied to each signal included in the study.

Table 1. Combined Performance of Mother Wavelets.

Wavelet	Compression Time (s)	Restoration Time (s)	RTE	v	Correlation
bior1.3	0.27592	0.00307	0.09988	203.10813	0.22505
coif1	0.21846	0.00308	0.09992	183.26275	0.22555
coif2	0.21688	0.00316	0.09991	194.98896	0.21768
coif3	0.22145	0.00325	0.09992	198.33720	0.21163
coif4	0.22984	0.00313	0.09992	199.28620	0.20678
coif5	0.23026	0.00302	0.09993	199.63327	0.20265
db1	0.20772	0.00300	0.09997	161.84359	0.22791
db2	0.20466	0.00311	0.09992	178.17021	0.22960
db3	0.20906	0.00303	0.09991	188.41594	0.22483

(*continued*)

Table 1. *(continued)*

Wavelet	Compression Time (s)	Restoration Time (s)	RTE	v	Correlation
db4	0.21184	0.00295	0.09991	192.74337	0.22080
db5	0.21216	0.00298	0.09991	195.66129	0.21720
dmey	0.24445	0.00335	0.10002	201.74549	0.17050
haar	0.23721	0.00308	0.09997	161.84359	0.22791
sym1	0.19034	0.00343	0.09997	161.84359	0.22791
sym2	0.20780	0.00312	0.09992	178.17021	0.22960
sym3	0.20451	0.00317	0.09991	188.41594	0.22483
sym4	0.20225	0.00319	0.09991	192.90004	0.22167
sym5	0.20498	0.00327	0.09991	195.98515	0.21912

Table 2 presents the most representative performance values for the voltage swell signal. It summarizes the maximum and minimum values obtained for each evaluation metric across all mother wavelets and decomposition levels. The data include indicators such as Relative Time Error (RTE), Normalized Mean Squared Error (NMSE), correlation, and the respective compression and restoration times. This comparison highlights the wavelets that yielded the best and worst performance under different decomposition configurations for this specific transient signal.

Table 2. Representative Values for the Voltage Swell Signal.

Voltage Swell Signal	Wavelet	Level	RTE	NMSE	Correlation	Compression Time (s)	Restoration Time (s)
Max RTE	'dmey'	1	0.5007	0.1748	0.7066	0.0045	0.0037
Min RTE	'dmey'	10	0.0007	1439.4139	0.0004	0.0081	0.0021
Max NMSE	'dmey'	10	0.0007	1439.4139	0.0004	0.0081	0.0021
Min NMSE	'coif2'	1	0.5001	0.1715	0.7071	0.0029	0.0038
Max Correlation	'coif4'	1	0.5002	0.1716	0.7072	0.0030	0.0025
Min Correlation	'dmey'	6	0.0169	61.6699	−0.0141	0.0091	0.0029
Max Compression Time	'sym1'	1	0.5000	0.1716	0.7071	0.0523	0.0146
Min Compression Time	'sym2'	1	0.5000	0.1716	0.7071	0.0027	0.0035

(continued)

Table 2. (*continued*)

Voltage Swell Signal	Wavelet	Level	RTE	NMSE	Correlation	Compression Time (s)	Restoration Time (s)
Max Restoration Time	'sym1'	1	0.5000	0.1716	0.7071	0.0523	0.0146
Min Restoration Time	'db5'	10	0.0010	991.6124	-0.0024	0.0063	0.0018

4 Analysis of Results

The performance analysis of the frequency analysis-based data compression and restoration algorithm, aimed at improving the NMSE, RTE, and COR indices of both stable and transient electrical signals, was carried out on the MATLAB platform. Regarding the mother wavelets, Mexhaar and Gauss, the results of the analyzed indices were out of range and did not meet the study parameters. Therefore, they were discarded from this analysis since, during compression and restoration, they exhibited significant differences in amplitude and shape, suggesting that important coefficients may be lost or omitted during compression. Both stable and transient signal scenarios were considered to validate the algorithm's efficiency, which are detailed below.

4.1 Performance of Mother Wavelets for Stable Signal

Different simulations were conducted to evaluate the algorithm's effectiveness concerning the NMSE indices, where the 'coif1' wavelet shows the lowest value, indicating its best performance. In contrast, the 'bior1.3' and 'coif4' wavelets exhibit the highest values, indicating greater reconstruction error. The RTE index shows similar values across the tested wavelets, so it is not a decisive factor in selecting a particular wavelet, whereas in terms of correlation, the 'coif1' wavelet has the highest average value, suggesting the best match between the original and reconstructed signal. As seen in Figure 4, an average analysis of the NMSE for all mother wavelets with their level of decomposition shows that as the level increases, the normalized mean squared error increases significantly.

4.2 Performance of Mother Wavelets for Voltage Swell Signal

Different simulations were carried out to evaluate the algorithm's effectiveness in terms of NMSE regarding the performance of the wavelets for the voltage swell signal. The 'coif1' wavelet shows the lowest value, indicating better performance in minimizing the mean squared error, whereas the 'bior1.3' and 'coif4' wavelets present higher values, suggesting poorer error performance. In terms of the correlation index, the 'coif1' wavelet exhibits the highest value, demonstrating its superior ability to maintain the relationship between the reconstructed and original signal. Considering the RTE, any of the wavelets would be acceptable due to the uniformity of the results.

4.3 Performance of Mother Wavelets for Impulse Signal

Similarly, when analyzing the performance of the wavelets for the impulse signal, the NMSE index indicates that the 'coif1' wavelet has the lowest value (0.17), making it the most precise in minimizing error. In contrast, the 'bior1.3' and 'coif4' wavelets exhibit much higher values (1118.82), indicating lower precision. For the impulse signal, the 'coif1' wavelet provides the best correlation, while the 'coif3' and 'coif4' wavelets show the lowest correlations, which may indicate lower fidelity in the reconstruction.

4.4 Performance of Mother Wavelets for Voltage Sag Signal

Finally, regarding the NMSE index for voltage sag signals, the 'coif1' wavelet exhibits a lower value, indicating superior precision in error minimization, while the 'coif4' wavelet shows the highest value, suggesting poorer performance. For the RTE index, the 'coif4' wavelet shows a slightly lower RTE compared to the others. Meanwhile, the wavelet with the best correlation is 'bior1.3', closely followed by 'coif1', whereas 'coif4' presents the lowest correlation value, indicating lower reconstruction fidelity.

4.5 General Analysis of Wavelets and Electrical Signals

The 'coif1' wavelet has consistently shown the best performance regarding NMSE and correlation across nearly all signal types, while the 'bior1.3' wavelet is competitive in correlation but exhibits a higher NMSE, which presents a comparative analysis of all mother wavelets based on performance level. In comparison with previous studies, we were able to define the ideal mother wavelets for compressing both stable and transient electrical signals. The results demonstrate a compression ratio with quality indicators such as a compression time of 0.002698 seconds, a restoration time of 0.0017876 seconds, and outstanding performance indices (RTE = 0.9998, NMSE = 0.00998, COR = 0.99991), surpassing previous high-impact studies, thereby determining the ideal wavelet for the compression and restoration of electrical signals.

5 Conclusions

In this recent study, an advanced technique for compressing and restoring electrical signals was developed through an exhaustive evaluation of different wavelets and their compression levels. It was found that mother wavelets differ significantly in their effectiveness, especially when comparing metrics such as Normalized Mean Squared Error (NMSE) and the correlation between the different wavelets.

Specifically, wavelets such as Daubechies 4 (db4) and Symlet 3 (sym3) stood out due to their superior performance, presenting low NMSE values and high correlations. This makes them particularly stable and reliable for the analysis of electrical signals. It was observed that wavelets with lower decomposition levels tend to be more accurate in signals with lower complexity and lower noise levels, making these wavelets especially useful in environments where signal simplicity and clarity are critical.

In transient situations, such as voltage sags (SAG), voltage swells (SWELL), and voltage impulses wavelets with lower decomposition levels exhibit better performance by adapting quickly to abrupt fluctuations in the signal. Conversely, wavelets with higher decomposition levels are more effective at capturing and faithfully representing detailed signal characteristics in more stable and constant conditions, which benefits a deeper and more thorough analysis.

Daubechies 4 (db4) and Symlet 3 (sym3) wavelets prove particularly suitable across a variety of conditions, offering robust and reliable performance over a wide range of scenarios. These findings suggest that the choice of the correct wavelet is critically dependent on the type of signal and its specific application context, underscoring the importance of careful selection to optimize both the compression and restoration of electrical signals in various smart grid environments.

References

1. Anikiev, D., et al.: Machine learning in microseismic monitoring. Earth-Sci. Rev. **239**(August 2022), 104371 (2023). https://doi.org/10.1016/j.earscirev.2023.104371
2. Chen, F., Jin, W., Shen, D., Chen, T.: Wavelet transform-based overvoltage feature extraction for high-speed moving trains. In: 2022 IEEE International Conference on Power Systems and Electrical Technology, PSET 2022, no. 1, pp. 8–12 (2022). https://doi.org/10.1109/PSET56192.2022.10100406
3. Ding, H., Liu, H., Yang, F.: A real-time compression method of power system waveform data based on 2-D lifting wavelet transform and deflate algorithm. In: 2021 3rd Asia Energy and Electrical Engineering Symposium, AEEES 2021, pp. 626–630 (2021). https://doi.org/10.1109/AEEES51875.2021.9403107
4. Dwork, N., et al.: Utilizing the wavelet transform's structure in compressed sensing. Signal Image Video Process. **15**(7), 1407–1414 (2021). https://doi.org/10.1007/s11760-021-01872-y
5. Guido, R.C.: Wavelets behind the scenes: practical aspects, insights, and perspectives. Phys. Rep. **985**, 1–23 (2022). https://doi.org/10.1016/j.physrep.2022.08.001
6. Hassan, A.M.A., Alzaidi, M.S., Ghoneim, S.S., Nahal, W.E.: Efficient data compression of ECG signal based on modified discrete cosine transform. Comput. Mater. Continua **71**(2), 4391–4408 (2022). https://doi.org/10.32604/cmc.2022.024044
7. Huang, J., Liu, J., Gong, H., Deng, X.: A comprehensive review of loosening detection methods for threaded fasteners. Mech. Syst. Signal Process. **168**(December 2021), 108652 (2022). https://doi.org/10.1016/j.ymssp.2021.108652
8. Karthika, S., Rathika, P.: An efficient data compression algorithm for smart distribution systems using singular value decomposition. In: IEEE International Conference on Intelligent Techniques in Control, Optimization and Signal Processing, INCOS 2019, pp. 1–7 (2019). https://doi.org/10.1109/INCOS45849.2019.8951340
9. Konovalov, A.B.: Compressed-sensing-inspired reconstruction algorithms in low-dose computed tomography: a review. Phys. Med. **124**(July), 104491 (2024). https://doi.org/10.1016/j.ejmp.2024.104491
10. Lin, L., Huang, N., Huang, W.: Review of power quality signal compression based on wavelet theory. In: Proceedings of the International Symposium on Test and Measurement, vol. 1, pp. 235–238 (2009). https://doi.org/10.1109/ICTM.2009.5412951
11. Ruiz, M., Jaramillo, M., Aguila, A., Ortiz, L., Varela, S.: A novel data compression methodology focused on power quality signals using compressive sampling matching pursuit. Energies **15**(24) (2022). https://doi.org/10.3390/en15249345
12. Song, B., Wei, P., Wu, S., Lin, Y., Zhou, W.: A survey on deep-learning-based image steganography. Expert Syst. Appl. **254**(April), 124390 (2024). https://doi.org/10.1016/j.eswa.2024.124390

13. Soto-Varela, R., Boumadan, M., Ortega-Rodríguez, P.J., Poyatos-Dorado, C.: The Inclusion of ICT-based innovative educational in primary schools and their impact on the academic performance of students. Rev. Electron. Interuniversitaria Formacion Profesorado **26**(1), 41–53 (2023). https://doi.org/10.6018/reifop.545011
14. Tcheou, M.P., et al.: The compression of electric signal waveforms for smart grids: state of the art and future trends. IEEE Trans. Smart Grid **5**(1), 291–302 (2014). https://doi.org/10.1109/TSG.2013.2293957
15. Tian, Y., Gao, J., Wang, D.: The multisynchrosqueezing optimal basic wavelet transform and applications to sedimentary cycle division. IEEE Geosci. Remote Sens. Lett. **19**, 1–5 (2022). https://doi.org/10.1109/LGRS.2021.3108911
16. Wang, J., Han, Y., Cao, Z., Xu, X., Zhang, J., Xiao, F.: Applications of optical fiber sensor in pavement engineering: a review. Construct. Build. Mater. **400**(July), 132713 (2023). https://doi.org/10.1016/j.conbuildmat.2023.132713
17. Yan, F.J., Li, B.Z.: Multi-dimensional grap fractional Fourier transform and its application to data compression. In: 2022 IEEE International Conference on Power Systems and Electrical Technology, PSET 2022, p. 9 (2022)

Short-Term Wind Energy Prediction Based on Machine Learning Techniques

Bryan David Llerena Ortiz(✉), Jessica Nataly Castillo Fiallos, Luigi Orlando Freire Martinez, and Gonzalo Xavier Vizuete Fiallos

Facultad de Ciencias de La Ingeniería y Aplicadas, Universidad Técnica de Cotopaxi, Latacunga, Ecuador
bryan.llerena1490@utc.edu.ec

Abstract. The following research work presents a short-term prediction model of the electric potential elaborated with Machine Learning techniques. A historical data base was used between 2022 and 2024 from the hydro-meteorological network of Tungurahua, specifically in the Chiquihurcu reservoir at 3875 m above sea level, which contains the variables speed, wind direction, temperature, relative humidity and generated electric power. By statistical analysis, a variable power generation was found throughout the period (standard deviation of 72.79 W) and a moderately low wind (mean of 2.46 m/s); whose values were validated by the Weibull distribution with a shape parameter (k) = 2.38 and scale (c) = 2.77. Using Python, three machine learning models were trained with the objective of predicting the electrical power in the short term (6 months), the models used were Random Forest, Decision Tree and XGBoost, as a result Random Forest showed the best performance when adapting to the variability of the data, generalizing them correctly and handling the errors very well, this was confirmed with the following validation metrics MSE = 0.108; MAE = 0.174 and R2 = 99.97%. Despite the relatively low wind speeds, machine learning techniques can provide an accurate prediction of electrical power, opening the door to autonomous energy systems and creating the basis for future smart microgrids in isolated communities.

Keywords: Wind Energy · Machine Learning · Prediction · Power · Wind · Speed · Renewable Energies

1 Introduction

Worldwide the shift towards the use of renewable energies has gained considerable importance in recent years due to the urgency of reducing dependence on fossil fuels and reducing the harmful effects on the environment [1]. Within this framework, wind energy has been established as a feasible option due to its accessibility and low environmental impact. However, the natural variability of wind poses an obstacle to the effective incorporation of this source into electricity grids, especially in areas where wind conditions are moderate [2].

J. L. Ballesteros et al. (Eds.): CITIS 2025, LNNS 1780, pp. 481–490, 2027.
https://doi.org/10.1007/978-3-032-14287-0_47

To combat this situation and to be able to optimize operational planning while ensuring grid stability, the prediction of short-term wind power generation has become a critical element [3]. Throughout history, statistical approaches and methods have been used with the objective of predicting generation, however, they present several complications in the ease of capturing the complexity of weather patterns [4].

Thus, machine learning has emerged as an effective solution for predicting energy production with greater accuracy and adaptability, being highly effective in the use of time series, even when working with highly variable data [5]. Python has an extensive list of libraries such as Scikit-learn, XGBoost and LightGBM, which makes it a powerful tool when working with machine learning, which is why it has been selected as the reference tool for the development of the predictive model [6].

Statistical techniques were fundamental in the preparation of the data; in this research, the interpolation method was effective at the time of autocompletion of missing data and fundamental in the training of each of the prediction models [7]. Decision trees, Random Forest and XGBoost are only a few machine learning techniques to make predictions but effective when using them, for example, in [8] used a decision tree model to make the prediction of photovoltaic generation in Tabacundo- Ecuador, the machine learning model was reliable in the short term, obtaining a precision of 0.976 and an accuracy of 0.974 demonstrating the effectiveness of the model and the importance of including photovoltaic generation in off-grid or grid-connected systems.

On the other hand, in [9] they present a computational tool developed in Python for data processing and short-term prediction of electricity demand in the national interconnected system in which they compare the ARIMA statistical algorithm with machine learning models, thus obtaining that Random Forest obtained the best projection performance, so it was concluded that machine learning models are much more effective in predicting adapting to data variability such as climate.

Unlike most of the research conducted in the south of the country, this study is carried out in the Chiquihurcu reservoir in the central part of the country, thus contributing to territorial diversification and inspecting wind power generation in different sectors, thus being able to implement microgrids in the future, including hybrid generation systems that could help vulnerable rural sectors to improve their quality of life and support the growth and local development of the population.

2 Methodology

2.1 Location of Study

Figure 1 shows the selected study site located in Ecuador, province of Tungurahua, parish of San Fernando. The site is located at an altitude of 3875 masl, its geographical coordinates correspond to latitude −1.2108677 and longitude −78.809198.

Fig. 1. Chiquihurcu Reservoir.

2.2 Data Collection and Preparation

Historical data of meteorological information was collected from January 2022 to December 2024, the dataset contains 26305 rows with an interval of 1 h, measured and recorded by the weather station including the variables Temperature [°C]; Relative Humidity [%]; Wind Speed [m/s]; Wind Direction [°]; Precipitation [mm]; Dew Point (dl) [°C]; Wind Speed [m/s]; Wind Direction [°]; Precipitation [mm]; Dew Point (dl) [°C].

2.3 Statistical Analysis of Meteorological Variables

A statistical analysis was performed using Python for each of the meteorological variables, and the results of the analysis can be found in Table 1.

Table 1. Statistical analysis of climatological data.

	Power [W]	Wind Speed (m/s)	Wind Direction (°)	Pressure (hPa)	Temperature (°C)	Relative Humidity (%)
count	25596	25596	25596	25596	25596	25596
mean	68,289	2,459	109,046	718,699	6,852	85,008
std	72,797	1,109	49,680	1,109	2,655	11,332
min	0,010	0,150	35,370	715,100	−2,430	31,290
25%	11,128	1,570	69,580	717,900	5,170	78,630
50%	42,237	2,460	96,750	718,700	6,480	88,275
75%	102,415	3,300	137,740	719,500	8,500	93,600
max	343,399	4,920	312,210	722,500	16,140	99,900

The average wind speed is 2.459 m/s with a standard deviation of 1.109 m/s, which could indicate a moderate dispersion around the mean; there is a minimum wind speed of 0.150 m/s. Twenty-five percent of the time the wind speed was equal to or less than 1.57 m/s, while 75% of the time it was equal to or less than 3.30 m/s, the maximum recorded wind speed value is 4.92 m/s.

With the correct data treatment, we conclude that for each meteorological variable we have 25596 observed data, 80% of which we use for training and 20% for testing.

2.4 Prediction of Generation Power

For generation prediction, models were developed using Machine Learning algorithms, specifically 3; Decision Tree Regressor, Random Forest Regressor and XGBoost (Extreme Gradient Boosting), validation metrics such as mean square error (MSE), mean absolute error (MAE) and coefficient of determination (R2) were used to evaluate the performance of each regression model, this was achieved by adjusting hyperparameters such as max_depth, n_estimators and learning_rate in the case of XGBoost.

Decision Tree Regressor.

A Decision Tree is formed by a node called root, which lacks input or incident arcs, and by a group of nodes that have an input arc. The nodes that have output arcs are called internal nodes or test nodes, while those that do not have output arcs are identified as leaves, terminal nodes or decision nodes [10].

The decision tree performs (binary) partitions of the feature space and at each decision node a feature $j \in \{1, \ldots, d\}$ and a threshold value $\mathcal{S} \in \mathbb{R}$, so that the set is divided into two left subsets $\{x|x_j \leq \mathcal{S}\}$ and right $\{x|x_j > \mathcal{S}\}$, , this with the objective of minimizing an error measure that for regression we use the Mean Squared Error (MSE). The optimal partition criterion (j^*, s^*) is:

$$\left(j^*, s^*\right) = \text{argmin}_{j,s}\left[\frac{N_{left}}{N}\right].Var\left(D_{left}\right) + \frac{N_{right}}{N}.Var\left(D_{right}\right) \tag{1}$$

where:

Var(D) means the variance of the values and in the set D

N_{left} y N_{right} are the sizes of the left and right sets, respectively.

Once the decision tree is obtained and to make a prediction on a data x, it is enough to follow the path of the tree according to the values of x until culminating in a leaf, which has a prediction value that most of the time is of this form:

$$\hat{y} = \frac{1}{N_{leaf}} \sum\nolimits_{i \in leaf} yi \tag{2}$$

Random Forest Regressor.

Having a set N of different observations, a random sample N with replacement will be chosen, this technique is known as bootstrapping and is a model used in different machine learning techniques, it also introduces randomness because each tree is formed in a slightly different way [11].

Finally, the Random Forest prediction is the average of the individual predictions of the decision trees:

$$\hat{y} = \frac{1}{M} \sum\nolimits_{m=1}^{M} h_m(x) \tag{3}$$

XGBoost (Extreme Gradient Boosting).

The principle of XGBoost is based on creating an ensemble with another machine learning model called ensemble, which is basically a combination of simple individual models that together create a much more powerful one, this starts by initially fitting a model that in our study is the decision tree to the data, then creates a second model that aims to predict the observations that the first model predicted wrong, this process is repeated several times and each next model aims to correct the errors of the past model in Fig. 6 a clear example can be seen [12].

Objectic Function.

XGBoost minimizes the following function:

$$\mathcal{L}(\emptyset) = \sum_{i=1}^{N} l(y_i, \hat{y}_i) + \sum_{k=1}^{K} \Omega(f_k) \quad (4)$$

where:

$l(y_i, \hat{y}_i)$ is a loss function (e.g., quadratic error for regression)

$\Omega(f_k)$ is a regularization term that penalizes the complexity of decision trees to avoid overfitting.

Figure 4 shows the flow diagram corresponding to the electric potential prediction process (Fig. 2).

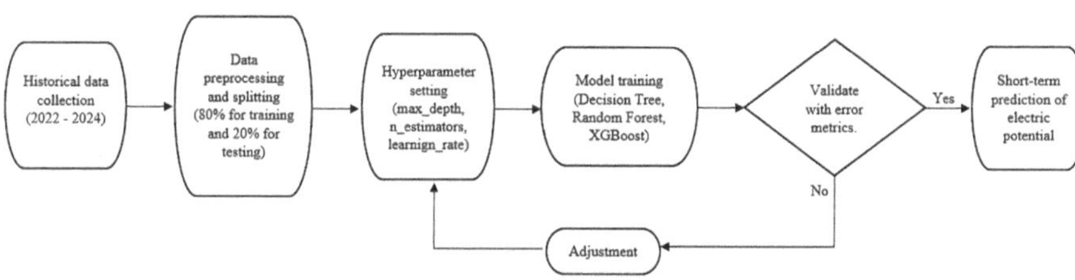

Fig. 2. Flow chart of the prediction model.

3 Results and Discussion

Three short-term prediction models (6 months) were implemented using the Python programming language. Figures 3, 4 and 5 show the comparison between real values, predicted values and the 6-month future prediction for each type of automatic learning, showing power peaks in the future prediction as a consequence of the high variability of the wind and its impact on energy generation.

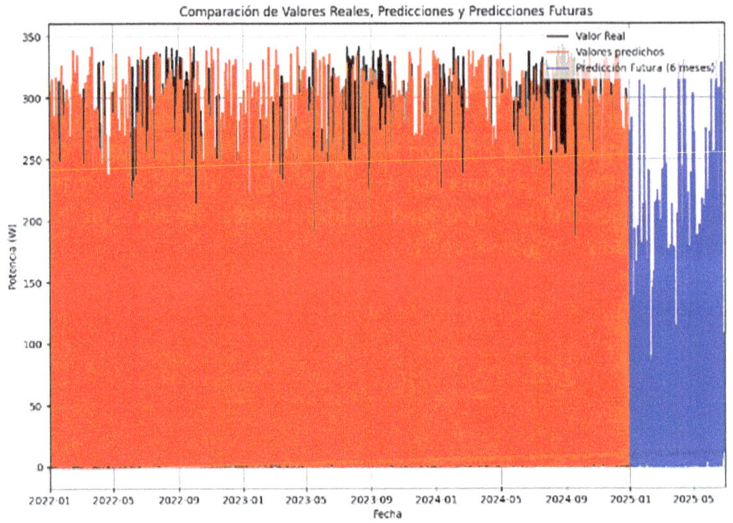

Fig. 3. Comparison of actual values, predictions and future predictions using Decision Tree Regressor.

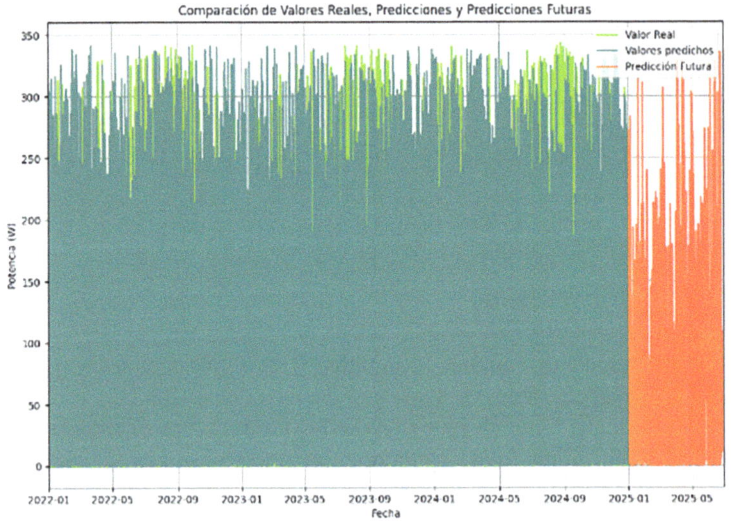

Fig. 4. Comparison of actual values, predictions and future predictions using XGB Regressor.

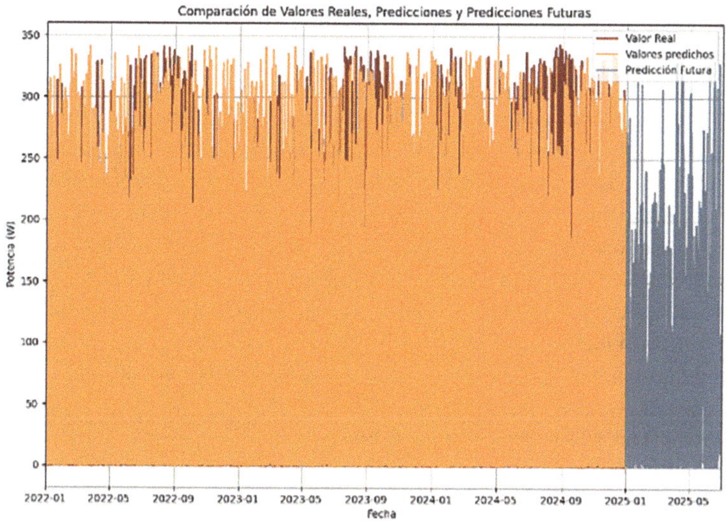

Fig. 5. Comparison of actual values, predictions and future predictions using Random Forest.

Validation Metrics.

The results of each of the machine learning techniques were evaluated through validation metrics; Table 2 shows the values for each of the metrics:

Table 2. Validation metric values for each machine learning technique

Metrics	Decision Tree Regressor	XGB Regressor	Random Forest Regressor
Mean Squared Error (MSE)	0,212	0,626	0,109
Mean Absolute Error (MAE)	0,198	0,376	0,175
(R^2)	0,960	0,966	0,979
Explained Variance Score [%]	99,995	99,988	99,997
MASE	0,053	0,173	0,075
MSE Training	4,091E-14	1,321E-06	4,176E-07
MSE Test	1,798E-6	5,307E-06	9,226E-07

The Random Forest model obtained the best results, consolidating itself as the most effective model, with the lowest Mean Squared Error (MSE) in the test (9.226E-7), the lowest Mean Squared Error (MSE) in the prediction (0.109) and the highest coefficient of determination R2 (0.979), which indicates that the model adjusts correctly to the training data, generalizes the new data in a better way and avoids overfitting.

On the other hand, the Decision Tree Regressor model shows a quite low training MSE (4.091E-14) which ends up being really excellent, but the test MSE (1.798E-6) is much higher than that of Random Forest, so it can be understood that the model memorizes but does not generalize, Similarly, the MSE of the prediction obtained a value of 0.212 higher than that of Random Forest, however, the median absolute error of the prediction was the best (0.053), indicating that, in a different scenario in which there are no severe outliers and high variability, a single tree can be quite effective.

XGBoost Regressor is an advanced and powerful model, but fails to outperform regression tree based models, having a test MSE of 5.307E-6 and a MAE of the prediction set of 0.376 the performance was declining, this because the model is highly sensitive to the correct configuration of hyperparameters such as learning_rate, max_depth, subsample and lambda, with an environment in variable weather conditions and low wind speeds XGBoost was not effective (see Table 3).

Statistical Analysis of Prediction Values.

Table 3. Statistical analysis of power prediction [W]

	Power prediction (W)
count	4320
mean	53,745
std	63,964
min	0,053
25%	6,144
50%	27,995
75%	78,112
max	330,685

Regarding the measures of distribution and central tendency we have a mean of 53.745 W and the median of 27.995 W which indicates a positive asymmetry, this means that there are high values that push the mean upwards, which corroborates the values of the quartiles of precisely 75% that tell us that they are below 78.112 W and some values even reach 330.685 W.

Figure 8 shows the boxplot of the prediction values of the machine learning model, the diagram indicates that most of the power prediction is below the mean (orange line) with the exception of some high values that increase the value of the mean, 50% of the central values are in the range of the first quartile Q1 = 6.144 and the third quartile Q3 = 78.112, which indicates that there is enough dispersion in the mean values, on the other hand, the whiskers extend approximately up to 190 W which represents a large part of the data without taking into account the extreme values. There are outliers that are above 190 W reaching even up to the maximum value of 330 W, these outliers do not represent an error in the generation values but it is a feature that is taken into account in the wind time series, this because there will be intense wind gusts which cause significant moments of power generation.

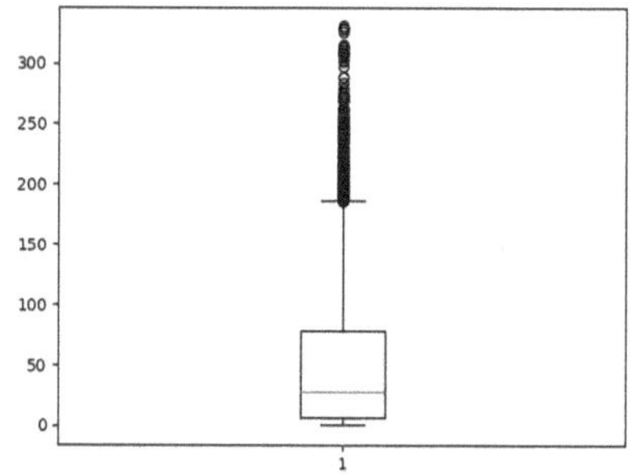

Fig. 6. Boxplot diagram of generation prediction.

Residue Analysis.

Table 4 shows the residual analysis of the real power with the power prediction, it should be taken into account that the statistical analysis of the real power is for 3 years (2022 - 2024) and the statistical analysis of the power prediction is for 6 months, the lowest residual value corresponding to the minimum values is highlighted, obtaining 0.043 W, the average value is $14,453$, this positive value indicates that the model underestimates the real power values, which means that the prediction tends to be lower than the real power, this may be caused perhaps because the model does not use all the relevant input variables in the power, this could cause possible deficits in generation, or that the components are not supporting the real load.

Table 4. Residual analysis of actual and predicted values.

	Real Power (W)	Power prediction (W)	Residues
mean	68,288	53,745	14,453
std	72,797	63,964	8,833
min	0,0095	0,053	0,043
25%	11,127	6,144	4,983
50%	42,236	27,995	14,241
75%	102,415	78,112	24,303
max	343,399	330,685	12,714

4 Conclusions

The prediction model (Random Forest Regressor) was the one that obtained the best results in the validation metrics, this demonstrates its ability to generalize and its ease of adaptation to low wind speed scenarios. Wind energy is very unpredictable and it is a complicated task to know the energy capacity in the future, in addition, it depends on the weather and environmental conditions, this means that it is difficult to adapt any forecast model, the use of ensemble algorithms such as Random Forest developed in

Python programming language are born as a practical and efficient strategy for a correct prediction of wind power in the short term in rural areas and low wind speed.

The correct use of hyperparameters is fundamental to increase the efficiency of the models and avoid underestimation or overestimation of prediction values, using more complex methods such as neural networks could improve and get even closer to the real values of wind potential.

References

1. Shabbir, N., et al.: Short-term wind energy forecasting using deep learning-based. ResearchGate **71**(1), 1017–1033 (2022)
2. Gómez Limia, C.: Aplicación de técnicas de aprendizaje automático a la predicción de la producción de parques eólicos. Universidad Pontificia Comillas (2024)
3. Baer, D.M.: Modelo de predicción a corto plazo de la generación eléctrica en parques eólicos. Universidad de las Islas Balcares (2020)
4. Ouanani Allachi, M.: Predicción de energía eólica utilizando técnicas de aprendizaje automático. Universidad Carlos III de Madrid (2015)
5. Ibargüengoytia González, P.H., Reyes Ballesteros2, A., Borunda Pacheco y U, M., García López, A.: Predicción de potencia eólica utilizando técnicas modernas de Inteligencia Artificial. Ingeniería Investigación y Tecnología **19**(4) (2018)
6. Gil Vera y, V.D., Seguro Gallego, C: Machine Learning aplicado al análisis del rendimiento de desarrollos de software. Revista Politécnica **18**(35), 128–139 (2022)
7. Mezgebu Ayene y, S., Mohammed Yibre, A.: Wind power prediction based on deep learning models: the case of Adama wind farm. ScienceDirect **10**(21) (2024)
8. Lojano Navas, C.S.: Diseño de una herramienta de predicción mediante machine learning para la generación fotovoltaica en usuarios urbanos y rurales. Universidad Técnica de Cotopaxi, Latacunga (2024)
9. Gallo, A., Perez y, F., Salinas, D.: Minería de Datos y Proyección a Corto Plazo de la Demanda de Potencia en el Sistema Eléctrico Ecuatoriano. Revista Técnica energía **18**(1), 72–85 (2021)
10. Arcega, A.F.: Arboles de decisión para grandes conjuntos de datos. Instituto Nacional de Astrofísica, Óptica y Electrónica, Puebla (2010)
11. Ruiz de León, M.G.: Analisis de sensibilidad mediante Random Forest. Escuela Técnica Superior Ingenieros Industriales (UPM), Madrid (2018)
12. Chen y, T., Guestrin, C.: XGBoost: a scalable tree boosting system. In: Proceedings of the 22nd ACM SIGKDD International Conference on Knowledge Discovery and Data Mining (KDD '16), pp. 785–794 (2016)

Paris Agreement, Energy Intensity and Climate Change Mitigation: Evidence for Latin American Countries

Segundo F. Vilema-Escudero[1(✉)], Cinthya Game-Varas[2], Juan Carlos Lata-García[3], and Mirian Solis-Ramón[4]

[1] Universidad ECOTEC, Km 13 1/2 Vía Samborondón, Samborondón, Guayas, Ecuador
svilema@ecotec.edu.ec
[2] Universidad Espíritu Santo, Samborondón, Ecuador
cigame@uees.edu.ec
[3] Universidad Politécnica Salesiana, Guayas, Ecuador
jlatag@ups.edu.ec
[4] Universidad Católica de Santiago de Guayaquil, Guayaquil, Ecuador
mirian.solis@cu.ucsg.edu.ec

Abstract. This study re-examines whether gains in energy efficiency translate into lower CO_2 emissions in Latin America and under what conditions. Using a balanced panel of 13 countries (2000–2021) and country-fixed-effects regressions, we test three hypotheses: (H1) lower energy intensity reduces emissions; (H2) this effect strengthens after the 2015 Paris Agreement; and (H3) it is amplified by environmental governance, proxied by ISO 14001 certification density. The baseline model finds no significant decarbonising effect of energy-intensity improvements. The post-Paris interaction likewise remains insignificant, suggesting treaty commitments have yet to alter regional mitigation trajectories. By contrast, the intensity–governance interaction is negative and statistically significant ($p < 0.05$), indicating that strong institutional frameworks convert efficiency gains into measurable emission cuts. Robustness checks confirm a consistent, negative impact of renewable-energy penetration and a positive association between GDP growth and emissions, while industrialisation shows no independent effect. The results underscore that technological efficiency alone is insufficient; without credible environmental governance, Latin-American economies remain locked into carbon-intensive growth. Strengthening institutional capacity should therefore accompany efficiency and renewable-deployment policies if the region is to meet its Nationally Determined Contributions.

Keywords: Energy Intensity · Environmental Governance · CO_2 Emissions

1 Introduction

Anthropogenic CO_2 emissions remain the principal driver of global warming, threatening ecosystems, economic stability and public health [1]. The Intergovernmental Panel on Climate Change (IPCC) underscores that unabated fossil-fuel combustion will push

J. L. Ballesteros et al. (Eds.): CITIS 2025, LNNS 1780, pp. 491–500, 2027.
https://doi.org/10.1007/978-3-032-14287-0_48

the global mean temperature well beyond the Paris Agreement targets of "well below 2 °C"—preferably 1.5 °C—by the end of this century [2, 3]. Because mitigation pledges under the Agreement must now translate into verifiable domestic action, the twin levers of energy efficiency and institutional quality have moved to the centre of climate-policy debates [4].

Energy intensity (EI) captures the combined effect of technological upgrading, structural change and behavioural adaptation [5, 6]. Declining EI is therefore widely interpreted as evidence of absolute decoupling, where economic growth proceeds without a commensurate rise in emissions [7, 8]. Yet the empirical record is mixed and region-specific pathways remain poorly understood, particularly outside the OECD [9–11].

Latin America combines rapid, demand-led growth in final energy use with persistent fossil-fuel dependence, despite world-class renewable endowments [12]. The region's average EI is still above the global mean and shows only a modest downward trend [13]. At the same time, institutional capacity and environmental regulation vary markedly among countries, shaping the effectiveness of efficiency programmes and the stringency of emissions controls [14, 15]. These contrasts provide a natural laboratory for testing whether similar efficiency gains translate into different emission outcomes under divergent governance regimes.

Existing Latin-American evidence is fragmentary: most studies analyse single countries, short horizons or omit governance factors altogether [16]. Three gaps persist: a) Governance–efficiency nexus: ISO 14001 adoption has been used as a micro-level proxy for environmental management elsewhere, yet its moderating role on the EI–CO_2 link has not been quantified for Latin America [17], b) Paris Agreement effectiveness: Quasi-experimental evaluations of the post-2015 period remain scarce for the region, leaving it unclear whether the Agreement has altered the EI–emissions elasticity [18], c) Regional heterogeneity: No study to date tests whether the combined influence of EI, governance and the Paris framework differs systematically across Latin-American economies.

The remainder proceeds as follows. Section 2 details data sources, variable construction and econometric strategy. Section 3 presents the core results and a suite of robustness checks. Section 4 discusses policy implications for achieving Nationally Determined Contributions (NDCs) in heterogeneous institutional settings. Section 5 concludes with recommendations for research and practice.

2 Methods

Reductions in energy intensity indicate more efficient energy use, often linked to technological upgrades and structural economic shifts, which lower carbon output per unit of GDP [19]. In Latin America, improvements in energy efficiency have significantly contributed to emission mitigation, especially in industrial and transportation sectors [20]. These patterns suggest that decreasing energy intensity is a viable strategy for carbon reduction [21]. Moreover, empirical models confirm that energy productivity gains correlate with lower CO_2 output in developing economies [22]. Therefore, Hypothesis **H1** is proposed: Lower energy intensity is associated with lower CO_2 emissions in Latin America.

After the Paris Agreement, policy alignment toward emission reduction has intensified globally [23]. Enhanced commitments have promoted energy efficiency strategies, reinforcing the negative link between energy intensity and CO_2 emissions [24]. In Latin America, post-2015 reforms prioritized low-carbon technologies and regulatory frameworks targeting efficiency improvements [25]. Empirical assessments reveal that post-Paris, reductions in energy intensity have produced stronger emission declines [26]. Therefore, Hypothesis **H2** is proposed: The relationship between energy intensity and CO_2 emissions becomes more negative after the Paris Agreement.

Environmental governance enhances energy policy implementation and monitoring [27]. Strong institutions ensure compliance, enabling sustained efficiency gains [9]. Improved rule of law and regulatory quality increase the effectiveness of emission-reduction strategies [28]. Empirical evidence confirms that governance quality moderates the efficiency-emissions link in developing regions [29]. In Latin America, transparency and institutional strength have increased energy policy impacts on emissions [13]. Therefore, Hypothesis **H3** is proposed: Stronger environmental governance strengthens the impact of energy efficiency on emissions.

Data from 13 Latin American countries were used, covering the period from 2000 to 2021. A balanced panel was constructed to analyze the effect of energy intensity on CO_2 emissions in Latin America, considering the Paris Agreement and environmental governance. The countries selected in the sample include Chile, Costa Rica, Colombia, Mexico, Argentina, Brazil, Peru, Panama, Venezuela, Ecuador, Bolivia, Paraguay, and Uruguay. The data were obtained from the CEPALSTAT statistical platform, developed by the Economic Commission for Latin America and the Carib-bean (ECLAC) of the United Nations (UN) [30], which compiles information from the main official sources in Latin American countries. Table 1 shows a description of the variables used in the study.

Table 1. Variable description

Variable	Code	Description	Unit
Dependent	emissions	Carbon dioxide (CO2) emissions, total excluding land use change and deforestation	Millions of tons of carbon dioxide equivalent (MtCO2e)
Independent	intensity	Energy Intensity of the Gross Domestic Product (GDP) (Final energy consumption/GDP at constant prices in 2018 dollars)	Primary energy supply (in thousands of barrels of oil equivalent) per million dollars of GDP (at constant 2018 prices)
Moderator	paris	Post-Paris Agreement	1 = > 2016; 0 = < 2016
	environmental	Environmental governance measured by the number of companies certified with ISO14001 in relation to the gross domestic product (GDP) in constant 2010 dollars	Number of companies per $1 billion of GDP (at constant 2010 prices)

(continued)

Table 1. *(continued)*

Variable	Code	Description	Unit
Control	energy	Energy transition measured by the proportion of renewable energy in total final energy consumption	Percentage
	industrial	Industrialization measured by the growth rate of the gross domestic product of the industrial sector at constant prices	Percentage
	growth	Economic growth measured by the Gross Domestic Product per inhabitant generated during a year by a country at market prices for each year	Dollars per inhabitant at current prices

To validate the study hypotheses, a fixed-effects panel data model was estimated. To better stabilize the variance and mitigate heteroskedasticity in the data, variables were transformed into natural logarithms and normalized [31]. This facilitates clearer interpretations and proportional comparisons between economic variables, improving the robustness of the econometric model [32]. The reference panel data models for testing the hypotheses are presented below:

$$emissions_{it} = \beta_0 + \beta_1 intensity_{it} + \beta_j \sum\nolimits_n^{j=2} X_{it} + \mu_i + \alpha_t + \varepsilon_{it} \tag{1}$$

$$emissions_{it} = \beta_0 + \beta_1 intensity_{it} + \beta_2 paris_t + \beta_3 (intensity * paris)_{it} + \beta_j \sum\nolimits_n^{j=4} X_{it} + \mu_i + \alpha_t + \varepsilon_i \tag{2}$$

$$emissions_{it} = \beta_0 + \beta_1 intensity_{it} + \beta_2 environmental_{it} + \beta_3 (intensity * environmental)_{it} + \beta_j \sum\nolimits_n^{j=4} X_{it} + \mu_i + \alpha_t + \varepsilon_i \tag{3}$$

where, $emissions_{it}$ is the dependent variable representing Carbon dioxide (CO2) emissions from country i at time t, $intensity_{it}$ is the independent variable that represents energy intensity. The moderator variables $paris_t$ is a dummy variable representing the implementation of the Paris Agreement in 2016, $environmental_{it}$ represents environmental governance. $\sum_n^{j=2} X_i$ represents the information from the control variables that describe energy transition, industrialization and economic growth. Finally, $\beta_0, \beta_1, ..\beta_6$ are the regression coefficients represent the relationship between the variables and ε_{it} is the error term, which captures the variation not explained by the independent variables and time.

3 Results

Table 2 presents descriptive statistics for 286 observations across six key variables. Average CO_2 emissions reached 107.18 $MtCO_2e$, with a maximum of 511.62 and a minimum of 3.52. Mean energy intensity was 0.835, ranging from 0.378 to 1.872. Environmental governance averaged 1.672 companies per $1 billion GDP, with a peak of 9.812. Renewable energy usage averaged 29.51%, with values between 7.65% and 70.41%. Industrial growth averaged 1.36%, reaching up to 21.44% and falling as low as −54%. GDP per capita averaged $7,425.23, with a high of $19,100.10 and a low of $888.21.

Table 2. Descriptive statistics

Variable	N	Mean	SD	Max	Min
emissions	286	107.177	143.041	511.620	3.520
intensity	286	0.835	0.293	1.872	0.378
environmental	286	1.672	1.684	9.812	0.019
energy	286	29.514	16.246	70.410	7.650
industrial	286	1.363	8.650	21.440	−54.000
growth	286	7425.230	4.324.588	19100.100	888.212

Table 3 reports fixed-effects panel regressions testing the three hypotheses using 204 observations. Model 1 tested the direct effect of energy intensity on emissions. The coefficient for intensity was positive (0.239) and not statistically significant ($p > 0.10$), indicating no robust direct relationship. Environmental governance showed a positive effect (0.0291) but remained insignificant. Renewable energy displayed a significant negative coefficient (-0.564, $p < 0.01$), confirming its role in emission reduction. Economic growth was positively associated with emissions (0.293, $p < 0.01$), while industrialization showed no significant effect (0.00225, $p > 0.10$).

Model 2 tested the moderating effect of the Paris Agreement. The interaction term intensity × paris yielded a positive coefficient (0.207) without statistical significance, suggesting no differential effect post-2015. The paris dummy was also insignificant (−0.0994, $p > 0.10$). The governance variable became marginally significant (0.0263, $p < 0.10$). Energy and growth effects remained consistent in magnitude and significance. Industrial growth stayed non-significant.

Model 3 evaluated the moderating role of environmental governance. The interaction intensity × environmental produced a positive and statistically significant coefficient (0.0292, $p < 0.05$), supporting Hypothesis 3. This result suggests that governance improves the effect of energy efficiency on emissions. The main effect of intensity became significant (0.263, $p < 0.10$) only when governance moderation was included. Environmental governance alone lost significance (0.00813, $p > 0.10$), indicating its influence depends on interaction. Renewable energy remained negatively significant (−0.574, $p < 0.01$). Economic growth maintained a consistent positive effect (0.290, $p < 0.01$), while industrialization remained insignificant (0.00479, $p > 0.10$). Model fit improved slightly from Model 2 (R2 = 0.306) to Model 3 (R2 = 0.334), suggesting better explanatory power with the governance interaction.

The right side of Fig. 1shows the differential effect of energy intensity on emissions before and after the Paris Agreement. The fitted lines show a steeper positive slope post-2015, indicating a weaker mitigation effect. Before 2016, lower energy intensity corresponded to modest reductions in CO_2 emissions. After 2015, reductions in energy intensity were associated with smaller decreases or even increases in emissions.

The left side is shown the moderating role of environmental governance in the energy intensity–emissions relationship. Two fitted lines are shown: one for low and one for high governance levels. Under stronger governance, the slope between energy intensity

Table 3. Panel data regression models with fixed effects

Variables	Model 1	Model 2	Model 3
intensity	0.239 (0.142)	0.168 (0.123)	0.263* (0.124)
paris		−0.0994 (0.0984)	
intensity*paris		0.207 (0.143)	
environmental	0.0291 (0.0172)	0.0263* (0.0137)	0.00813 (0.0174)
intensity*environmental			0.0292** (0.0121)
energy	−0.564*** (0.0660)	−0.517*** (0.0667)	−0.574*** (0.0649)
industrial	0.00225 (0.00431)	0.00102 (0.00388)	0.00479 (0.00507)
growth	0.293*** (0.0243)	0.273*** (0.0265)	0.290*** (0.0215)
Constant	2.925*** (0.325)	2.919*** (0.322)	2.948*** (0.325)
N	204	204	204
R- R-squared	0.324	0.306	0.334

Standard errors in parentheses
$^{*}p < 0.10$, $^{**}p < 0.05$, $^{***}p < 0.01$

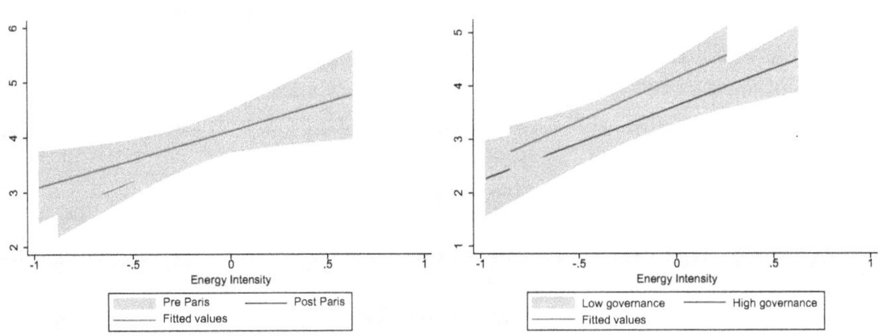

Fig. 1. Differential effect of energy intensity versus Energy intensity by governance

and emissions is negative and steeper. This suggests that energy intensity reductions lead to larger emission declines when governance improves.

4 Discussion

The analysis yields three main findings. First, energy-intensity reductions are not, by themselves, a reliable lever for decarbonisation in Latin America. In Model 1 the coefficient on energy intensity is positive and insignificant, and only attains significance once institutional quality is introduced in Model 3. This pattern implies that efficiency gains translate into lower emissions only when embedded in robust governance frameworks, thus partially confirming Hypothesis 1.

Second, the Paris Agreement has not yet altered the regional energy-emissions elasticity. The interaction between energy intensity and the post-2015 dummy in Model 2 remains insignificant and—even more telling—Fig. 1 depicts a flatter slope in the post-Paris subsample. Hypothesis 2 is therefore rejected. The most plausible explanations lie in heterogeneous Nationally Determined Contributions (NDC) ambition, uneven enforcement capacity and the time lag between treaty ratification and on-the-ground implementation.

Third, environmental governance emerges as a decisive moderator. The positive and significant interaction term in Model 3 ($p < 0.05$) indicates that higher ISO 14001 penetration amplifies the decarbonising effect of lower energy intensity, fully supporting Hypothesis 3. Figure 2 visualises this governance premium: under high certification density, a one-unit drop in energy intensity produces nearly double the marginal CO_2 reduction observed in low-governance settings.

These results reinforce the view that institutional capacity is a prerequisite for turning technical efficiency into realised mitigation [9, 28]. They also echo evidence from Asia and Africa showing negligible post-Paris effects where enforcement mechanisms are weak [26]. Our contribution lies in demonstrating—for the first time at the regional level—that ISO-based governance metrics systematically condition the EI–CO_2 link across 13 Latin-American economies over two decades.

5 Conclusions

This study shows that energy-intensity gains translate into lower CO_2 emissions only when embedded in strong environmental-governance frameworks. In baseline models the intensity coefficient is indistinguishable from zero; its mitigating effect emerges—and doubles in magnitude—once interacted with ISO 14001 density, underscoring institutional quality as a necessary condition for decarbonisation.

By contrast, the Paris Agreement has not yet produced a measurable post-2015 shift in the regional energy-emissions elasticity. The insignificant interaction suggests that, in Latin America, treaty commitments remain too recent, too heterogeneous or too weakly enforced to alter national mitigation trajectories.

The analysis further confirms three ancillary patterns: a) Governance premium – Countries with higher certification coverage reap significantly larger emission cuts per unit of efficiency improvement, validating governance as a critical moderator, b) Renewable leverage – A robust, negative renewable-energy coefficient highlights the complementary role of clean-power expansion in meeting Nationally Determined Contributions, c) Growth lock-in – The positive GDP effect signals continued carbon dependence of prevailing development models; industrialisation, however, shows no independent influence, implying sectoral heterogeneity or limited recent structural change.

For policy implications, efficiency measures must be paired with credible oversight. Regulators should strengthen monitoring, reporting and verification mechanisms and incentivise ISO 14001 uptake to unlock the full mitigation dividend of energy efficiency. International accords need domestic teeth. Harmonising legal transposition of NDCs and accelerating implementation timetables would convert Paris pledges into tangible emission reductions. Renewables remain the surest near-term lever. Scaling wind, solar and hydro capacity can offset growth-driven emissions while institutional reforms mature.

For Limitations and research avenues, residual endogeneity between energy intensity and economic structure warrants instrumental-variable or dynamic-panel approaches. Sub-national analyses could capture within-country variation in governance capacity. Future work should triangulate ISO-based metrics with broader rule-of-law and climate-policy indices and replace the single post-2015 dummy with policy-specific implementation dates.

References

1. Nunes, L.J.R.: The rising threat of atmospheric CO2: a review on the causes, impacts, and mitigation strategies. Environments **10**, 66 (2023). https://doi.org/10.3390/environments10040066
2. Matthews, H.D., Wynes, S.: Current global efforts are insufficient to limit warming to 1.5°C. Science **376**, 1404–1409 (2022). https://doi.org/10.1126/science.abo3378
3. O'Neill, B.C., et al.: IPCC reasons for concern regarding climate change risks. Nat. Clim. Change **1**(7), 28–37 (2017). https://doi.org/10.1038/nclimate3179
4. Bakhsh, S., Zhang, W., Ali, K., Oláh, J.: Strategy towards sustainable energy transition: the effect of environmental governance, economic complexity and geopolitics. Energ. Strat. Rev. **52**, 101330 (2024). https://doi.org/10.1016/j.esr.2024.101330
5. Chen, W., Alharthi, M., Zhang, J., Khan, I.: The need for energy efficiency and economic prosperity in a sustainable environment. Gondwana Res. **127**, 22–35 (2024). https://doi.org/10.1016/j.gr.2023.03.025
6. Namahoro, J.P., Wu, Q., Zhou, N., Xue, S.: Impact of energy intensity, renewable energy, and economic growth on CO2 emissions: evidence from Africa across regions and income levels. Renew. Sustain. Energy Rev. **147**, 111233 (2021). https://doi.org/10.1016/j.rser.2021.111233
7. Guang, F., Wen, L., Sharp, B.: Energy efficiency improvements and industry transition: an analysis of China's electricity consumption. Energy **244**, 122625 (2022). https://doi.org/10.1016/j.energy.2021.122625
8. Jafari, M., Botterud, A., Sakti, A.: Decarbonizing power systems: a critical review of the role of energy storage. Renew. Sustain. Energy Rev. **158**, 112077 (2022). https://doi.org/10.1016/j.rser.2022.112077
9. Azam, M., Hunjra, A.I., Bouri, E., Tan, Y., Al-Faryan, M.A.S.: Impact of institutional quality on sustainable development: evidence from developing countries. J. Environ. Manage. **298**, 113465 (2021). https://doi.org/10.1016/j.jenvman.2021.113465
10. Dau Jarama, G., Vilema-Escudero, S.F., Samaniego López, J.M., Orellana, M.M.: Impact of foreign direct investment on energy intensity: evidence from latin american countries. In: Inga Ortega, E.M., García Herranz, N., Robles-Bykbaev, V.E., Gallego Diaz, E. (eds.) Systems, Smart Technologies, and Innovation for Society, pp. 336–348. Springer Nature Switzerland, Cham (2025). https://doi.org/10.1007/978-3-031-87065-1_31

11. Vilema-Escudero, S.F., Ruiz-Martínez, M., Manya-Orellana, M., Arguello-Núñez, L.: Energy Intensity and Employment: evidence from Latin American countries. IOP Conf. Ser.: Earth Environ. Sci. **1434**, 012013 (2024). https://doi.org/10.1088/1755-1315/1434/1/012013
12. Brito Muniz, G.F., Santos, K.M.G., Santos, A.M.: Trends in renewable and decentralized energy grids: a comparative study between Brazil and Central America. In: 2024 IEEE 42nd Central America and Panama Convention (CONCAPAN XLII), pp. 1–6 (2024). https://doi.org/10.1109/CONCAPAN63470.2024.10933872
13. Hampl, N.: Equitable energy transition in Latin America and the Caribbean: reducing inequity by building capacity. Renew. Sustain. Energy Transit. **2**, 100035 (2022). https://doi.org/10.1016/j.rset.2022.100035
14. Cui, S., Wang, Y., Zhu, Z., Zhu, Z., Yu, C.: The impact of heterogeneous environmental regulation on the energy eco-efficiency of China's energy-mineral cities. J. Clean. Prod. **350**, 131553 (2022). https://doi.org/10.1016/j.jclepro.2022.131553
15. García-Quevedo, J., Jové-Llopis, E.: Environmental policies and energy efficiency investments. an industry-level analysis. . Energy Policy **156**, 112461 (2021). https://doi.org/10.1016/j.enpol.2021.112461
16. Panait, M., Janjua, L.R., Apostu, S.A., Mihăescu, C.: Impact factors to reduce carbon emissions evidences from Latin America. Kybernetes. **52**, 5669–5686 (2022). https://doi.org/10.1108/K-05-2022-0712
17. Arocena, P., Orcos, R., Zouaghi, F.: The impact of ISO 14001 on firm environmental and economic performance: the moderating role of size and environmental awareness. Bus. Strateg. Environ. **30**, 955–967 (2021). https://doi.org/10.1002/bse.2663
18. Cárdenas, J.A.T.: El Ecuador y su cumplimiento internacional en materia de derecho al medio ambiente sano a través de la incorporación de los principios internacionales del Derecho Ambiental en la normativa nacional. USFQ Law Rev. **8**, 43–75 (2021). https://doi.org/10.18272/ulr.v8i2.2267
19. Liu, X., Hang, Y., Wang, Q., Chiu, C.-R., Zhou, D.: The role of energy consumption in global carbon intensity change: a meta-frontier-based production-theoretical decomposition analysis. Energy Econ. **109**, 105968 (2022). https://doi.org/10.1016/j.eneco.2022.105968
20. East, J., Montealegre, J.S., Pachon, J.E., Garcia-Menendez, F.: Air quality modeling to inform pollution mitigation strategies in a Latin American megacity. Sci. Total Environ. **776**, 145894 (2021). https://doi.org/10.1016/j.scitotenv.2021.145894
21. Bhattacharya, R., Bose, D., Mitra, G., Sarkar, A.: Prospects of sustainability for carbon footprint reduction. In: Clean and Renewable Energy Production, pp. 335–353. John Wiley & Sons, Ltd (2024). https://doi.org/10.1002/9781394174805.ch14
22. Mehmood Mirza, F., Sinha, A., Rehman Khan, J., Kalugina, O.A., Wasif Zafar, M.: Impact of energy efficiency on CO2 emissions: empirical evidence from developing countries. Gondwana Res. **106**, 64–77 (2022). https://doi.org/10.1016/j.gr.2021.11.017
23. Schleussner, C.-F., Ganti, G., Rogelj, J., Gidden, M.J.: An emission pathway classification reflecting the Paris Agreement climate objectives. Commun. Earth Environ. **3**, 1–11 (2022). https://doi.org/10.1038/s43247-022-00467-w
24. Khan, A., Min, J., Hassan Shah, W.U., Li, Q., Sun, C.: Efficacy of CO2 emission reduction strategies by countries pursuing energy efficiency, nuclear power, and renewable electricity. Energy **300**, 131418 (2024). https://doi.org/10.1016/j.energy.2024.131418
25. Georgescu, I.A., Bâra, A., Oprea, S.-V.: Challenges for low-carbon economies in Latin America. testing pollution haven hypothesis in developing countries. Energy Rep. **12**, 5280–5299 (2024). https://doi.org/10.1016/j.egyr.2024.11.009
26. Rezaei Sadr, N., Bahrdo, T., Taghizadeh, R.: Impacts of Paris agreement, fossil fuel consumption, and net energy imports on CO2 emissions: a panel data approach for three West European countries. Clean Tech. Environ. Policy. **24**, 1521–1534 (2022). https://doi.org/10.1007/s10098-021-02264-z

27. Asongu, S.A., Odhiambo, N.M.: Enhancing governance for environmental sustainability in sub-Saharan Africa. Energy Explor. Exploit. **39**, 444–463 (2021). https://doi.org/10.1177/0144598719900657
28. Yuan, H., Zou, L., Feng, Y.: How to achieve emission reduction without hindering economic growth? The role of judicial quality. Ecol. Econ. **209**, 107839 (2023). https://doi.org/10.1016/j.ecolecon.2023.107839
29. Asongu, S.A., Odhiambo, N.M.: Governance, CO2 emissions and inclusive human development in sub-Saharan Africa. Energy Explor. Exploit. **38**, 18–36 (2020). https://doi.org/10.1177/0144598719835594
30. CEPAL: CEPALSTAT Bases de Datos y Publicaciones Estadísticas. https://statistics.cepal.org/portal/cepalstat/index.html?lang=es
31. Angrist, J.D., Pischke, J.-S.: Mostly Harmless Econometrics: An Empiricist's Companion. Princeton University Press (2009)
32. Baltagi, B.H.: Econometric Analysis of Panel Data. Springer Texts in Business and Economics (2021)

The Emerging Relationship Between Supply Chain Resilience Management and Circular Economy: A Fuzzy Cognitive Map Analysis

Lia Cedeño-Zambrano, Antonio Zavala-Alcívar(✉), Lindsay Rangel-Anchundia, and Karen Zambrano

IPGCS Group, Faculty of Engineering, Industry, and Architecture, Universidad Laica Eloy Alfaro de Manabí, Manta 130214, Ecuador
{e1313710541,antonio.zavala,lindsay.rangel, karen.zambrano}@uleam.edu.ec

Abstract. Global disruptive events are becoming more common, on hand with importance given to sustainable efforts toward reducing pollution and waste. Supply chains must grapple with both areas in trying to enhance their resilience while reducing waste born from their operations. The purpose of this study was to analyze the relationship between the principles of circular economy and supply chain resilience management with the use of fuzzy cognitive maps (FCM) in a canned fish preserve supply chain. Seven principles from both subjects were identified from the extant literature and then given a causal relationship numerical value through the opinions of five experts on the field, which were used to develop the causal model through the Mental Modeler software and to create the FCM with the use of the "*NetworkX*" Python package. Results include the influence of SCREM principles over CE ones, with importance given to leadership and trust as key agents in transitioning efforts towards a CE that improves resilience in supply chains despite potential structural obstacles.

Keywords: Circular Economy · Fuzzy cognitive maps · Supply chain resilience management

1 Introduction

Profitability is the vital component that dictates whether a business's operations will continue or not, but this approach opposes the concepts of a circular economy (CE). Geisendorf (2018) defines a circular economy as one where "the value of products and materials is maintained, waste is avoided, and resources are kept within the economy when a product has reached the end of its life" [1].

By contrast, the current production model plans for products to have an artificial lifespan, forcing consumers to buy and replace items more frequently. This business strategy bolsters private profit by designing products with linear life cycles and is at odds with sustainable development models, and with circular economy as a whole [1, 2]. In other words, the de-facto business model that is currently used isn't compatible with the

J. L. Ballesteros et al. (Eds.): CITIS 2025, LNNS 1780, pp. 501–514, 2027.
https://doi.org/10.1007/978-3-032-14287-0_49

concepts of circular economy. However, that does not mean a CE cannot be employed; businesses can even benefit from it. Circular economy is more easily adapted if Industry 4.0's Digital Technologies (DTs) are applied, in what is called a Smart Circular Economy (SCE) By using said DTs, resource productivity [3] is increased as well as supply chain agility, which is one of the principles of supply chain resilience management (SCREM) [4]. Resilience allows profits to be steady in unexpected and disruptive situations by letting production continue.

Current research has studied the link between circular economy and I4.0 [5]. Conversely, there is a lack of studies relating circular economy and SCREM. Table 1 shows a comparative of the top ten most relevant research conducted on 2023 that relate both circular economy and SCREM. Most of the research conducted was qualitative in nature, and none of the papers utilized fuzzy cognitive maps for data analysis, nor did they classify CE and SCREM principles to determine and prioritize strategies. To address this gap, the main objective of this research is to analyze the relationship between the principles of circular economy and supply chain resilience management using fuzzy cognitive maps.

Table 1. Comparison of the top ten most relevant research conducted on 2023 that relate both circular economy and SCREM.

Paper	RT	Research methodology	Experts involved	Analyzed principles	TM	DAT		
						Other techniques	FCN	PC
[6]	O	Systematic literature review (SLR), Fuzzy-Delphi, Fuzzy SECA, Data clustering	15	16	Q1, Q2	Several Fuzzy MCDM methods (TOPSIS, SAW, COPRAS, MULTIMOORA, SECA)		
[7]	O	Literature review, Grey DEMATEL	10	18	Q2			
[8]	O	Framework development through systematic literature review (SLR)			Q1			
[9]	O	Survey-based	280		Q2	Structural Equation Modelling (SEM)		

(*continued*)

Table 1. *(continued)*

Paper	RT	Research methodology	Experts involved	Analyzed principles	TM	DAT		
						Other techniques	FCN	PC
[10]	O	Framework development based on non-market-based valuation methods (NMVM)			Q1			
[11]	O	Document co-citation analysis			Q2			
[12]	O	Semi-structured interviews	40		Q1	In Vivo coding and Pattern coding		
[13]	R	Systematic literature review (SLR)			Q1			
[14]	R	Systematic literature review (SLR)			Q1			
[15]	R	Literature review			Q1			
This paper	O	Fuzzy-Delphi, Fuzzy Cognitive Maps (FCM), Data clustering	5	7	Q2		X	X

RT: Research type; O: Original article; R: Review article, TM: Type of methodology; Q1: Qualitative; Q2: Quantitative; DAT: Data analysis techniques; FCN: Fuzzy Cognitive Map; PC: Principles classification.
The following search string was used: TITLE-ABS-KEY ("circular economy" AND ("supply chain") AND ("resilience")) AND (EXCLUDE (DOCTYPE, "ch") OR EXCLUDE (DOCTYPE, "bk")) AND (LIMIT-TO (PUBYEAR, 2023)),

The structure of the paper is as follows: Sect. 2 reviews the extant literature as a preamble to the topic, describing the core principles and concepts involved in CE and SCREM. Section 3 describes the methodology to be used. Section 4 points out the obtained results with the use of a FCM connecting the principles of both CE and SCREM. Finally, Sect. 5 presents the conclusions of the study, limitations, and proposed future lines of research.

2 Theoretical Background

2.1 Circular Economy (EC)

Circular economy began as an idea for circular material flows proposed decades ago by Boulding (1966) in his book *The Economics of the Coming Spaceship Earth.* This was expanded by other pioneers who, referring to the second law of thermodynamics, stated that "the traditional linear economy without recycling elements cannot be sustainable and consequently must be replaced by a circular system" [1]. At present, circular economy is based on three circular design principles [16]: (1) eliminate waste and pollution; (2) circulate products and materials at their highest value; and (3) regenerate nature.

The current literature [1] considers as core elements of a CE the presence of circular design in both the product development and transportation stages, including: material sourcing; production consumption; Customer Relationship Management (CRM); and its end of life. Four conditions that enable the transition or feasibility of a CE in a real context frame these core elements: (1) measurability tools to determine the degree of circularity; (2) business model perspective to determine profitability; (3) policies to support not yet profitable circular economic business models (CEBM); and (4) sustainability implications, particularly on social fields.

CE has been critiqued as an idealized concept that can't be fully translated to reality, even stating circularity as a "theoretically, practically and ideologically questionable notion" given its unclear theoretical grounds and structural obstacles faced with its implementation [17]. Such critiques also include its opposing nature with economic growth [1], which requires more expenditure in resources to continue ever-expanding. This pose doubts as to whether it is useful as a means to a sustainable end [18].

CE literature is diverse and carries opposite points of view depending on the focus of each study. For this research, the CE principles proposed by [16] will be considered for the development and creation of the fuzzy cognitive map, and the definitions and conceptual framework proposed by [1] will be emphasized for the results and discussion section.

2.2 Supply Chain Resilience Management (SCREM)

The concept of resilience, which draws from fields like psychology and ecology [19], has been incorporated into business management systems and supply chain management systems, where the concept of SCREM was created. SCREM scope was initially limited to the management of risk and vulnerability in isolated and individual companies [4] and later included the prevention of disruptions by managing the supply chain. Numerous studies have demonstrated how the term "*resilience*" has evolved within the context of supply chain management [19, 20].

According to the literature [21], five capabilities are necessary for supply chains to develop to manage resilience. These capabilities include: (1) prevention; (2) resistance; (3) response; (4) recovery and continuity; and (5) learning and continuous improvement. These capabilities are linked to the four principles and elements of resilience: supply chain reengineering, collaboration, agility, and SCREM culture [4, 20]. For this research, the SCREM principles proposed by [4] will be considered for the development and creation of the fuzzy cognitive map and emphasized for the results and discussion section.

2.3 Principles of Supply Chain Resilience Management and Circular Economy

The literature on both SCREM and CE is broad and encompasses a range of perspectives, which can vary depending on the specific focus of each study. The main principles pertaining to both subjects that were utilized in the current study are presented in Table 2. Various articles that discuss the current state of the art refer to said principles.

Table 2. Principles of circular economy and supply chain resilience management.

Subject	Principle	Description	Related strategies	References
Circular Economy	P1 – Elimination of waste and pollution	To design the materials to re-enter the economy at the end of their use, thereby making regenerative processes with circular design	Circular design, reverse logistics, eco-friendly packaging	[1, 16]
	P2 – Circulate products and materials	To circulate products and materials at their highest value, with the use of a restorative cycle for technical materials (reuse, repair, remanufacture, recycle) and a regenerative cycle for biological materials (composting, anaerobic digestion, etc.)	9Rs' (refuse, rethink, reduce, reuse, repair, refurbish, remanufacture, repurpose, recycle, recover), Closed supply chains (CSC)	[1, 16]

(*continued*)

Table 2. (*continued*)

Subject	Principle	Description	Related strategies	References
	P3 – Regenerate nature	To emulate natural systems with a regenerative, non-extractive economy that employs circular strategies that return biological materials to the earth	Eco-friendly packaging, regenerative food production practices (agroecology, conservation agriculture, agroforestry), governmental policies	[16]
Supply chain resilience management	P4 – Supply chain reengineering	Refers to redundancy and flexibility. Redundancy is the ability to resist the disturbing event by designing and managing security units or replacing critical nodes when a disruptive event occurs. Flexibility is the ability of the supply chain to adapt and deal with the consequences of the disruptive event in the minimum time and effort possible	Local suppliers, redundant inventories	[4, 19, 20]

(*continued*)

Table 2. (*continued*)

Subject	Principle	Description	Related strategies	References
	P5 – Collaboration	Refers to information sharing and trust. Information sharing is the ability of the supply chain to have relevant, efficient, and timely information from all its members for joint decision making. Trust is the belief of the supply chain members that their partners in the chain are willing and able to fulfill their responsibilities	Supply chain connectivity and integration, technological innovation, collaborative communication	[4, 19, 20]
	P6 – Agility	Refers to visibility and velocity. Visibility is the ability of the supply chain to know the identity, location, and status of its members in the face of any disruption to the supply chain. Velocity is the ability to respond rapidly to disruptive events by efficiently distributing your critical resources	Digital supply chain twin, interoperability in operations and decision making, lean production practices, demand planning and control, collaborative operations	[4, 19, 20]

(*continued*)

Table 2. (*continued*)

Subject	Principle	Description	Related strategies	References
	P7 – SCREM culture	Refers to leadership and innovation. Leadership is the ability to guide the establishment of risk mitigation strategies. Innovation is the ability to create joint strategies to manage risk more efficiently. Openness to learning and joint decision making	Leadership substitute theory, leader empowerment, improvement of capacities in new technologies, implementation of continuous improvement of processes	[4, 19, 20]

3 Materials and Methods

The structured research approach proposed in this work consists of four stages. First, the scope and focus of the study were specified. Second, the principles of CE and SCREM were identified in the extant literature. To this end, a general search was conducted in the Scopus database using two separate search strings: *("circular economy"); ("supply chain") AND ("resilience")*. Inclusion criteria were limited to English language results, with recent literature reviews and conceptual frameworks being prioritized as information sources, as well as their corresponding primary references.

Third, a fuzzy cognitive map was developed, which provides a graphical model of the behavior of a dynamic system through a graph representation, where the links between the nodes indicate the degree of influence in a causal direction from a range of [-1, 1]. To this end, the Fuzzy-Delphi approach was employed: a panel of five (5) experts identified the links existing among the principles of CE and SCREM through a survey, classifying their causal relationships from a range of $[-5, 5]$, corresponding to a nominal response of *Very Low* and *Very High*, respectively. The expert's given link values were then processed and introduced into the web-based tool *Mental Modeler*, where the Fuzzy Weight Matrix and indegree, outdegree, and centrality of each node were obtained. The Fuzzy Weight Matrix was subsequently converted into a Fuzzy Cognitive Map (FCM) utilizing the "NetworkX" Python package [22].

Finally, the principles' indegree and outdegree values were normalized on a range of [0, 1] to obtain their indegree and outdegree percentages, and segment the principles accordingly. This facilitated the classification of the principles, which in turn allowed the determination and prioritization of strategies.

3.1 Case Study

This research analyzed a supply chain pertaining to the canned fish preserves industry in Ecuador, which distributes its products internationally across both Latin American and European markets. This supply chain must adhere to strict international regulations in order to maintain its exportations, and is influenced by ongoing disruptions typical of the industry, such as: 1) national environmental regulations restricting raw material extraction; 2) rising production costs; and 3) social disruptions and insecurity that impact the supply chain both directly and indirectly, particularly in light of Ecuador's current social, political, and economic instability [23].

4 Results and Discussion

After processing the expert's link values on the Mental Modeler software, the fuzzy weight matrix and FCM were obtained. The degree of influence from one principle to another in numerical and graphical form is shown in Table 3 and Fig. 1 respectively. Table 4 shows the indegree, outdegree, and centrality of each node (principle). A node's indegree measures how strongly it is influenced by other nodes while its outdegree measures how influencing it is. The centrality of a node is the sum of its outdegree and indegree values and indicates how central it is to the model, given its higher number of node connections and link values [24].

Table 3. Fuzzy Weight Matrix.

Principles	*P1*	*P2*	*P3*	*P4*	*P5*	*P6*	*P7*
P1	0.00	0.40	0.20	−0.20	0.00	0.20	0.00
P2	0.80	0.00	0.40	-0.20	0.00	−0.20	0.00
P3	0.20	0.40	0.00	0.00	0.00	0.00	0.00
P4	−0.40	−0.20	0.00	0.00	0.60	0.80	0.40
P5	0.20	0.40	0.00	0.40	0.00	0.80	0.40
P6	−0.20	−0.20	0.00	0.40	0.20	0.00	0.20
P7	0.60	0.60	0.20	0.80	0.80	0.80	0.00

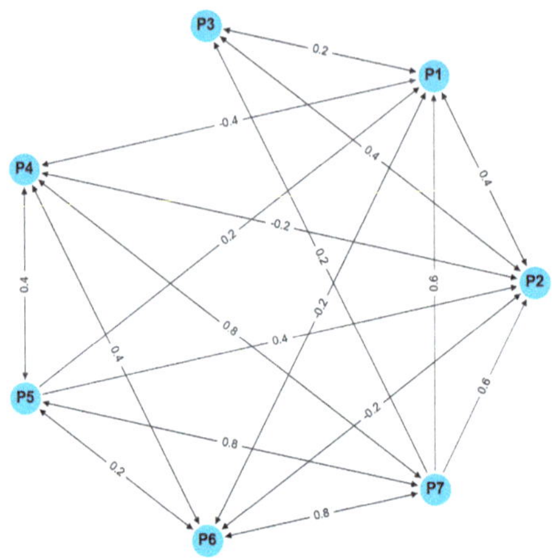

Fig. 1. Fuzzy cognitive map of CE and SCREM principles.

Table 4. Indegree, outdegree and centrality of each node.

Principles	Indegree	Outdegree	Centrality	Indegree %	Outdegree %
P1	2.4	1	3.4	0.8	0.125
P2	2.2	1.6	3.8	0.7	0.3125
P3	0.8	0.6	1.4	0	0
P4	2	2.4	4.4	0.6	0.5625
P5	1.6	2.2	3.8	0.4	0.5
P6	2.8	1.2	4	1	0.1875
P7	1	3.8	4.8	0.1	1

The results are illustrated in four regions in Fig. 2, with all indegrees and outdegrees scaled normalized to the interval [0,1]. Node P4 has high indegree and outdegree, and needs to be carefully managed; nodes P5 and P7 have low indegree and high outdegree, and have high potential to trigger a change in the model; nodes P1, P2, and P6 have high indegree and low outdegree, and are easily influenced by other nodes, making them difficult to manage individually; node P3 has low indegree and outdegree, being a highly isolated principle with limited management implications. The principles "SCRM culture" (P7) and "regenerate nature" (P3) have the highest and lowest degree of centrality, respectively. Most principles of SCREM have higher centrality compared to CE principles, apart from "circulate products and materials" (P3), which is more closely related to supply chain risk management strategies (e.g., use of recycled materials as raw material input, supplier diversity).

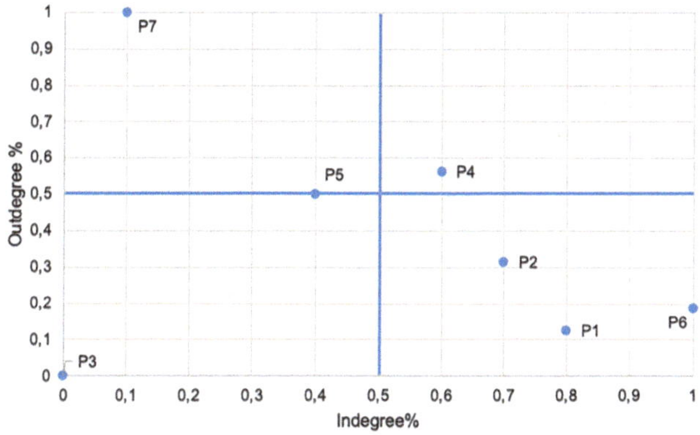

Fig. 2. Segmented principles by indegree% and outdegree%.

4.1 Managerial Implications

As shown in Table 3, Collaboration (P5) and SCREM culture (P7) have a positive influence on CE principles. Trust and leadership are key elements of said principles, and stakeholders are vital agents of both parts, serving an important role in the incorporation of CE into SCM and thus SCREM. Two notable concepts currently exist as methods of transition to a CE in an operational sense: Circular Supply Chain Management (CSCM) which is the incorporation of the CE into supply chain management [25], and the Smart Circular Economy (SCE) which is defined as "an industrial system that uses DTs to provide intelligent functions for implementing value-added circular strategies" [26].

However, stakeholders and companies face several obstacles when transitioning toward a CE model. In the early steps of change, financial constraints are present, which represents a considerable barrier for small and medium-sized businesses. Technological hurdles, such as lack of DTs integration, arise in later steps [26].

Social issues also appear when considering the CE model from a management and technocentric perspective [17]. Jobs involving salvaging, saving, repairing, and reusing products and materials are created to generate circularity. These jobs are generally undertaken by socially marginalized groups [27], resulting in overlooked labor practices, working conditions, power asymmetries, interdependencies, political and economic constraints [28], and equity and inclusion [27]. Despite these potential barriers, integrating CE and lean practices into risk management allows companies to create a more resilient supply chain, which benefits businesses. Diversification and dematerialization have significant potential for improving SCREM and reducing energy consumption, waste, pollution, and production costs. Diversification of the supply chain opens additional circularity pathways that alleviate damage during critical material supply issues. Related strategies include redundant inventory, recycling, nearby suppliers, and innovative sourcing options such as reverse channels and forward contracts. Dematerialization in industrial economies can be achieved by reducing the intensity of certain critical materials or components at the product level [17].

Considering all this, a multi-level implementation approach, beginning with the stakeholders, is relevant in the hopes of generating a CE that is: (1) modest, which solves actual problems; (2) concrete, on its professed circularity and the conflicts that arise from it, (3) inclusive, accounting not only for materials and waste but also people and energy on a global scale; and (4) transparent, on both its achievements and shortcomings on all domains of change: economic, social and environmental ones [17].

5 Conclusions

This research analyzed the relationship between the principles of circular economy and supply chain resilience management with the use of fuzzy cognitive maps. Key findings include the influence of SCREM principles over CE ones; particularly SCREM culture, collaboration, and supply chain reengineering. Leadership and trust are key components of these principles for transitioning towards a CE that, despite potential obstacles, improves resilience in supply chains and reduces energy consumption and costs.

This study has limitations regarding the modest number of experts and principles used for developing the FCM. Future research can utilize this study as a template to develop causal models incorporating additional principles and expert criteria, as well as analyze specific scenarios within a simulated dynamic system, given the availability of more data for simulation purposes.

Furthermore, this study can be replicated in another supply chain case study. This opens the possibility to analyze key differences between supply chain resilience and circular economy principles across various industries.

Acknowledgements. Authors of this publication acknowledge the contribution of the Projects P-INV-1011961-2024 and P-INV-1014084-2025, funded by Universidad Laica Eloy Alfaro de Manabí.

References

1. Geisendorf, S., Pietrulla, F.: The circular economy and circular economic concepts—a literature analysis and redefinition. Thunderbird Int. Bus. Rev. **60**, 771–782 (2018). https://doi.org/10.1002/tie.21924
2. Bisschop, L., Hendlin, Y., Jaspers, J.: Designed to break: planned obsolescence as corporate environmental crime. Crime Law Soc Change. **78**, 271–293 (2022). https://doi.org/10.1007/s10611-022-10023-4
3. Kristoffersen, E., Blomsma, F., Mikalef, P., Li, J.: The smart circular economy: a digital-enabled circular strategies framework for manufacturing companies. J. Bus. Res. **120**, 241–261 (2020). https://doi.org/10.1016/j.jbusres.2020.07.044
4. Zavala-Alcívar, A., Verdecho, M.-J., Alfaro-Saíz, J.-J.: A conceptual framework to manage resilience and increase sustainability in the supply chain. Sustainability. **12**, 6300 (2020). https://doi.org/10.3390/su12166300
5. Godinho Filho, M., Monteiro, L., de Oliveira Mota, R., dos Santos Leite Gonella, J., de Souza Campos, L.M.: The relationship between circular economy, industry 4.0 and supply chain performance: a combined ISM/Fuzzy MICMAC approach. Sustainability **14**, 2772 (2022). https://doi.org/10.3390/su14052772

6. Amoozad Mahdiraji, H., Govindan, K., Yaftiyan, F., Garza-Reyes, J.A., Razavi Hajiagha, S.H.: Unveiling coordination contracts' roles considering circular economy and eco-innovation toward pharmaceutical supply chain resiliency: evidence of an emerging economy. J. Clean. Prod. **382**, 135135 (2023). https://doi.org/10.1016/j.jclepro.2022.135135
7. Dwivedi, A., Chowdhury, P., Paul, S.K., Agrawal, D.: Sustaining circular economy practices in supply chains during a global disruption. Int. J. Logistics Manag. **34**, 644–673 (2023). https://doi.org/10.1108/IJLM-04-2022-0154
8. Senna, P., Marujo, L.G., Santos, A.C. de S.G. dos, Ferreira, A.C., da Silva, L.A.A.: E-waste supply chain risk management: a framework considering omnichannel and circular economy. Benchmarking: Int. J. **31**, 3429–3458 (2024). https://doi.org/10.1108/BIJ-05-2023-0341
9. Dey, P.K., Chowdhury, S., Abadie, A., Vann Yaroson, E., Sarkar, S.: Artificial intelligence-driven supply chain resilience in Vietnamese manufacturing small- and medium-sized enterprises. Int. J. Prod. Res. **62**, 5417–5456 (2024). https://doi.org/10.1080/00207543.2023.2179859
10. Nandi, S., Hervani, A.A., Helms, M.M., Sarkis, J.: Conceptualising circular economy performance with non-traditional valuation methods: lessons for a post-pandemic recovery. Int. J. Log. Res. Appl. **26**, 662–682 (2023). https://doi.org/10.1080/13675567.2021.1974365
11. Kirbac, A., Burmaoglu, S., Ozdemir Gungor, D.: Technological intelligence for circular supply chain: a co-citation analysis approach. Foresight **25**, 320–333 (2023). https://doi.org/10.1108/FS-11-2021-0226
12. Zerbino, P., Stefanini, A., Aloini, D., Dulmin, R., Mininno, V.: Does supply chain sustainability benefit from formal scavenging? A case study in circular settings. J. Clean. Prod. **385**, 135669 (2023). https://doi.org/10.1016/j.jclepro.2022.135669
13. Asante, M., Epiphaniou, G., Maple, C., Al-Khateeb, H., Bottarelli, M., Ghafoor, K.Z.: Distributed ledger technologies in supply chain security management: a comprehensive survey. IEEE Trans. Eng. Manag. **70**, 713–739 (2023). https://doi.org/10.1109/TEM.2021.3053655
14. Ada, E., Kazancoglu, Y., Lafcı, Ç., Ekren, B.Y., Çimitay Çelik, C.: Identifying the drivers of circular food packaging: a comprehensive review for the current state of the food supply chain to be sustainable and circular. Sustainability. **15**, 11703 (2023). https://doi.org/10.3390/su151511703
15. Abedsoltan, H.: <scp>COVID</scp> -19 and the chemical industry: impacts, challenges, and opportunities. J. Chem. Technol. Biotechnol. **98**, 2789–2797 (2023). https://doi.org/10.1002/jctb.7531
16. Ellen MacArthur Foundation: Circular economy introduction. What Is a Circular Economy?
17. Corvellec, H., Stowell, A.F., Johansson, N.: Critiques of the circular economy. J. Ind. Ecol. **26**, 421–432 (2022). https://doi.org/10.1111/jiec.13187
18. Skene, K.R.: Circles, spirals, pyramids and cubes: why the circular economy cannot work. Sustain. Sci. **13**, 479–492 (2018). https://doi.org/10.1007/s11625-017-0443-3
19. Ali, A., Mahfouz, A., Arisha, A.: Analysing supply chain resilience: integrating the constructs in a concept mapping framework via a systematic literature review. Supply Chain Manag.: Int. J. **22**, 16–39 (2017). https://doi.org/10.1108/SCM-06-2016-0197
20. Christopher, M., Peck, H.: Building the resilient supply chain. Int. J. Logistics Manag. **15**, 1–14 (2004). https://doi.org/10.1108/09574090410700275
21. Kamalahmadi, M., Parast, M.: A review of the literature on the principles of enterprise and supply chain resilience: major findings and directions for future research. Int. J. Prod. Econ. **171**, 116–133 (2016). https://doi.org/10.1016/j.ijpe.2015.10.023
22. Hagberg, A.A., Schult, D.A., Swart, P.J.: Exploring Network structure, dynamics, and function using NetworkX (2008). https://doi.org/10.25080/TCWV9851
23. Pérez, F.R., Vina, M.: From below and from within: fishing communities under the COVID-19 pandemic and other globalizations in southern Manabí. Ecuador. Territ Politic Gov. **10**, 917–936 (2022). https://doi.org/10.1080/21622671.2021.1960595

24. Mirghafoori, S.H., Sharifabadi, A.M., Takalo, S.K.: Development of causal model of sustainable hospital supply chain management using the intuitionistic fuzzy cognitive map (IFCM) method. J. Ind. Eng. Manag. **11**, 588–605 (2018). https://doi.org/10.3926/jiem.2517
25. Lahane, S., Kant, R., Shankar, R.: Circular supply chain management: a state-of-art review and future opportunities. J. Clean. Prod. **258**, 120859 (2020). https://doi.org/10.1016/j.jclepro.2020.120859
26. Lobo, A., Trevisan, A.H., Liu, Q., Yang, M., Mascarenhas, J.: Barriers to transitioning towards smart circular economy: a systematic literature review (2022). https://doi.org/10.1007/978-981-16-6128-0_24
27. Isenhour, C., Reno, J.: On materiality and meaning: ethnographic engagements with reuse repair & care. Worldwide Waste **2**, 1 (2019). https://doi.org/10.5334/wwwj.27
28. Schulz, C., Hjaltadóttir, R.E., Hild, P.: Practising circles: studying institutional change and circular economy practices. J. Clean. Prod. **237**, 117749 (2019). https://doi.org/10.1016/j.jclepro.2019.117749

Author Index

J. L. Ballesteros et al. (Eds.): CITIS 2025, LNNS 1780, pp. 515–517, 2027.
https://doi.org/10.1007/978-3-032-14287-0

Kolibri 360 GmbH
Ferdinand-Jühlke-Straße 7
99095 Erfurt, Deutschland
produktsicherheit@kolibri360.de